Interdisciplinary Applied Mathematics

Volumes published are listed at the end of the book

Marco Pettini

Geometry and Topology in Hamiltonian Dynamics and Statistical Mechanics

 Springer

Marco Pettini
Osservatorio
  Astrofisico di Arcetri
50125 Firenze
ITALY
pettini@arcetri.astro.it

Series Editors

S.S. Antman
Department of Mathematics and
  Institute for Physical Science and
  Technology
University of Maryland
College Park, MD 20742
USA
ssa@math.umd.edu

J.E. Marsden
Control and Dynamical Systems
Mail Code 107-81
California Institute of Technology
Pasadena, CA 91125
USA
marsden@cds.caltech.edu

L. Sirovich
Laboratory of Applied Mathematics
Department of Biomathematics
Mt. Sinai School of Medicine
Box 1012
NYC 10029
USA
chico@camelot.mssm.edu

ISBN-13: 978-1-4419-2164-2   e-ISBN-13: 978-0-387-49957-4

Mathematics Subject Classification (2000): 53B20, 53B21, 53C21, 37K, 70H

Printed on acid-free paper.

9 8 7 6 5 4 3 2 1

springer.com

To the memory of my father

Foreword

It is a special pleasure for me to write this foreword for a remarkable book by a remarkable author. Marco Pettini is a deep thinker, who has spent many years probing the foundations of Hamiltonian chaos and statistical mechanics, in particular phase transitions, from the point of view of geometry and topology.

It is in particular the quality of mind of the author and his deep physical, as well as mathematical insights which make this book so special and inspiring. It is a "must" for those who want to venture into a new approach to old problems or want to use new tools for new problems.

Although topology has penetrated a number of fields of physics, a broad participation of topology in the clarification and progress of fundamental problems in the above-mentioned fields has been lacking. The new perspectives topology gives to the above-mentioned problems are bound to help in their clarification and to spread to other fields of science.

The sparsity of geometric thinking and of its use to solve fundamental problems, when compared with purely analytical methods in physics, could be relieved and made highly productive using the material discussed in this book.

It is unavoidable that the physicist reader may have then to learn some new mathematics and be challenged to a new way of thinking, but with the author as a guide, he is assured of the best help in achieving this that is presently available.

The major mathematical tool used by the author to tackle the problems mentioned in the title is Riemannian differential geometry, the same as is used in general relativity. This way a geometric based theory of Hamiltonian chaos and thermodynamic phase transitions is pursued. Moreover, a connection is made between the origin of Hamiltonian chaos and phase transitions. In this approach the origin of both is related to curvature fluctuations of the phase space of the system.

I note that for the mathematically inclined reader the use of a coordinate-dependent formulation based on Riemannian geometry may be less satisfactory than for the physicist reader and might be considered a lack of mathematical elegance. After all, geometry's and topology's virtue is a global approach to the structure of manifolds and their properties. However, from a physicist's point of view one might invoke Boltzmann's dictum that "elegance is for tailors."

The above-mentioned curvature variations in the Riemannian description of phase space lead then, on the one hand, to Hamiltonian chaos through a parametric instability mechanism. On the other hand, when they are also due to the additional cause of a strongly and suddenly changing complex topology, they are also closely related to phase transitions. In fact, numerical studies show that a phase transition is invariably marked by a peak in the curvature fluctuations and by "cuspy" energy dependencies of Lyapunov exponents.

Thus these catastrophic events, due to the highly irregular, "bumpy" landscape of phase space, trigger on the deeper level of the topology of phase space itself the singularities occurring in the usual description of phase transitions on a higher level.

A remarkable achievement is the proof of two theorems, giving, for a large class of Hamiltonian systems, a necessary topological condition for a first- or second-order phase transition to take place. Roughly speaking, these theorems say, "no topology change in phase space, no phase transition." However, there is at present no theorem that gives a sufficient topological condition for the occurrence of a phase transition.

This may be related to the fact that not every topological transition in phase space leads to a phase transition, so that the question arises, what kinds of topological transitions are related to phase transitions and what is, from a topological point of view, the difference between various types of phase transitions?

Clearly the elucidation of these questions would deepen our basic understanding of two of the most striking phenomena in nature: that of chaos and that of phase transitions.

It is my conviction that this book makes a courageous attempt to clarify these fundamental phenomena in a new way.

Therefore, I highly recommend this refreshing and very original book not only for its factual content but also for the privilege one has in sharing the author's deep insights and new approaches and results to some unsolved problems in physics. I have no doubt that the reader will find this book highly stimulating and rewarding.

E. G. D. Cohen, professor
The Rockefeller University
New York, July 2006

Preface

Phase transitions are among the most impressive phenomena occurring in nature. They are an example of *emergent* behavior, i.e., of collective properties having no direct counterpart in the dynamics or structure of individual atoms or molecules: to give a familiar example, the molecules of ice and liquid water are identical and interact with the same laws of force, despite their remarkably different macroscopic properties.

That these macroscopic properties must have some relation to microscopic dynamics seems obvious, for example the molecules in a drop of water are free to move everywhere in the drop, in contrast to what happens in a crystal of ice.

However, according to a widespread point of view, when a large number of particles is involved, since we are unable to follow all their individual histories, we are compelled to get rid of dynamics and to replace it by a statistical description. For a long time only a marginal role has been thus attributed to microscopic dynamics: the large number of particles and our ignorance of their initial conditions have been considered enough to provide a solid ground to statistical mechanics.

More recently, much attention has been paid to another source of unpredictability, which is intrinsic to the dynamics itself: deterministic chaos, and, in particular, Hamiltonian chaos.

The present book is a monograph committed to a synthesis of two basic topics in physics: Hamiltonian dynamics, with all its richness unveiled since the famous numerical experiment of Fermi and coworkers at Los Alamos, and statistical mechanics, mainly for what concerns phase transition phenomena in systems described by realistic interatomic or intermolecular forces.

The novelty of the theoretical proposal put forward in this monograph stems from a well-known fact: the natural motions of a Hamiltonian system are geodesics of appropriately defined Riemannian manifolds. Whence the possibility of deepening our understanding of the microscopic dynamical foundations of macroscopic physics of many-particle systems. In fact, the geometrization of dynamics allows questions like, can we "read" in the

geometry of these mechanical manifolds something relevant to the understanding of basic properties of the dynamics? A major issue is undoubtely
to understand the origin of the chaotic instability of dynamics. The first part
of the book contains what we can call the beginning of a Riemannian theory of
Hamiltonian chaos, which works strikingly well when applied to models (like
the Fermi–Pasta–Ulam model) fulfilling the simplifying hypotheses introduced
to analytically compute the largest Lyapunov exponent. In the spirit of the
Springer Series in Interdisciplinary Applied Mathematics, I have made explicit
in what direction further developments of the theory should go. The second
part of the book stems from another question, again rooted in the Riemannian
theory of chaos: what happens to these mechanical manifolds when a Hamiltonian system undergoes a phase transition? and how can we "geometrically
read" the occurrence of a phase transition? It is at this point that topology comes into play, and, roughly speaking, considering certain submanifolds
of configuration space, the answer is that necessarily a phase transition can
occur only at a point where the topology of these submanifolds undergoes a
transition, and this is true at least for a large class of systems.

The presentation of the book follows the logic of the historical development of a successful ten-year research program that I carried out with
the help of several collaborators. The many open points are at the same time
highlighted, giving the material presented more the form of an intermediate
stage of publication than the form of a monograph on a mature and already
concluded research program. And it is just this characteristic that, I hope, will
make this book attractive for those, mathematicians or physicists, who might
be interested in contributing to the general theoretical framework, its physical
applications, or the mathematics necessary in the context of applications.

The mathematics involved is not used to clean up or rephrase already
existing results, rather it is constructively used to gain insight. The language of
differential geometry and differential topology is not familiar to the majority of
physicists and has almost never entered statistical mechanics, a circumstance
that might induce skepticism and/or could be discouraging.

Thus, in order to make this book accessible to as wide a readership as
possible, including both mathematicians and physicists, and since it makes
use of concepts that might be not known to everyone, the following format
has been chosen.

The first part of the book is aimed at a reader who is familiar with the
basics of Riemannian geometry, for example at the level of a course in general relativity. As to the second part, a knowledge of Morse theory and de
Rham's cohomology theory at an elementary level is assumed. However, for
those physicists who are not familiar with these branches of mathematics, I
have provided in appendices the main points that are needed to follow the
exposition. Similarly, I assume that the reader is familiar with the basics of
Hamiltonian dynamical systems (theory and phenomenology) and statistical
mechanics, but I have summarized in Chapter 2 the main concepts needed
throughout the book. In all cases references to the literature for the details

are made. I hope that a reader familiar with the basic mathematical tools and with the basic physical meaning of the topics treated will be able to read the book straightforwardly.

I have made a special effort to emphasize logical coherence and the excellent consistency already attained by the ensemble of results presented in the book. Nevertheless, as mentioned above, these results constitute the starting of a new theory rather than its completion. This is the reason why this monograph has no pretence at mathematical rigor (with the exception of Chapter 9), nor at mathematical elegance (the geometrization of Hamiltonian flows, their integrability and instability, in Chapters 3, 4, and 5, respectively, is written in a coordinate-dependent style in view of applications and thus of explicit computations). Nevertheless, I hope that this will not prevent mathematicians from understanding the meaning of what has been achieved in applying geometrical and topological methods to the study of the relationship between dynamical systems and statistical mechanics, with special emphasis on phase transitions. In fact, I would like this monograph to allow the reader, mathematician or physicist, to familiarize herself or himself with this new field and to stimulate new developments and contributions to the many points that are still open and explicitly evidenced throughout the text.

The theoretical scenario depicted in this book is based on the outcomes of a research program inspired and coordinated by the author. However, this research program has been successfully developed only thanks to the collective effort of several collaborators and friends. Therefore, among the most senior of them, my warmest thanks go to Monica Cerruti-Sola, whose continual and precious collaboration during fifteen years has been of invaluable help. My warmest acknowledgments also go to Giulio Pettini for having contributed during a crucial period. I have been honored by the active interest in this research program demonstrated by E.G.D. Cohen and Raoul Gatto, with whom stimulating and fruitful collaborations were carried on during several years.

At the very beginning of my interest in the connection between Hamiltonian dynamics and statistical mechanics, there was a collaboration, a long time ago, with Roberto Livi, Antonio Politi, Stefano Ruffo, and Angelo Vulpiani, friends and colleagues with whom useful discussions and scientific interchanges have never ceased.

I had the chance to work with several gifted and very brilliant PhD students. Among them, my warmest acknowledgments go to Lapo Casetti, who has creatively, brilliantly, and courageously contributed to most of the fundamental steps of this research program since its very beginning; as well, my warmest acknowledgments go to Roberto Franzosi, whose brilliant, creative, and continual collaboration during the last ten years has been of invaluable help in making crucial leaps forward in the topological theory of phase transitions.

It is with a feeling of deepest sorrow that my memory goes to another student and dear friend of mine, Lando Caiani, who died while he was at

SISSA-ISAS in Trieste for his PhD. Lando was an outstanding, very promising, and cultivated young physicist.

A precious contribution to the Riemannian approach to the study of Hamiltonian chaos and to the early developments of the topological approach to phase transitions was given by Cecilia Clementi, whose intelligence and productivity were nothing but absolutely impressive.

It is a pleasure to acknowledge the precious help of Lionel Spinelli in working out rigorous results on the topological theory of phase transitions, of Guglielmo Iacomelli, with whom we worked on extensions of these methods to quantum systems, and of Guido Ciraolo, with whom we worked on the Riemannian theory of Hamiltonian chaos of low-dimensional systems.

A timely and very fruitful collaboration with Luca Angelani, Giancarlo Ruocco, and Francesco Zamponi is warmly acknowledged.

I have profited from many helpful discussions about mathematics with Gabriele Vezzosi, whose friendly and continuous interest for this work has been an effective encouragement.

I warmly thank another friend, A.M. Vinogradov, for many illuminating discussions on several topics in mathematics.

During many years, I have profited from useful suggestions, remarks, comments by, and discussions with V.I. Arnold, E. van Bejieren, G. Benettin, S. Caracciolo, P. Cipriani, P. Collet, E. Del Giudice, R. Dorfman, J.P. Eckmann, Y. Elskens, D. Escande, L. Galgani, P. Giaquinta, C. Giardinà, A. Giorgilli, T. Kambe, M. Kastner, J. Lebowitz, A. Lichtenberg, R. Lima, C. Liverani, H. Posch, M. Rasetti, D. Ruelle, S. Schreiber, R. Schilling, Ya. Sinai, A. Tenenbaum, S. Vaienti, M. Vittot; my thanks to all of them.

My scientific activity has been supported by the Osservatorio Astrofisico di Arcetri, Firenze, Italy (now part of the Istituto Nazionale di Astrofisica, I.N.A.F.). Its former director, Franco Pacini, is warmly acknowledged for his collaboration and support. For many years, this scientific activity has been financially supported by the Istituto Nazionale di Fisica Nucleare (I.N.F.N.), which is here warmly acknowledged.

While writing this book I have been supported in many ways by my beloved children Eleonora and Leonardo.

Last, but not least, this book would have not seen the light of day without the invaluable help of Massimo Fagioli, psychiatrist and eminent scientist, who, having unveiled fundamental dynamical processes of the unconscious mind, in Rome has been conducting, since 32 years, the so-called *Analisi Collettiva*, a very large group in which an emergent phenomenon (as in the case of phase transitions!), due to the unconscious interactions among people, has a strong healing power. In this way Massimo Fagioli drew me out of what T.S. Eliot would have called a "waste land," where I was wandering after my father's passing away.

Florence, September 2006   *Marco Pettini*

Contents

Chapter 1

Introduction

This book reports on an unconventional explanation of the origin of chaos in Hamiltonian dynamics and on a new theory of the origin of thermodynamic phase transitions. The mathematical concepts and methods used are borrowed from Riemannian geometry and from elementary differential topology, respectively. The new approach proposed also unveils deep connections between the two mentioned topics.

Written as a monograph on a new theoretical framework, this book is aimed at stimulating the active interest of both mathematicians and physicists in the many still open problems and potential applications of the theory discussed here.

Thus we shall focus only on those particular aspects of the subjects treated that are necessary to follow the main conceptual construction of this volume. Many topics that would naturally find their place in a textbook, despite their general relevance will not be touched on if they are not necessary for following the *leitmotif* of the book.

In order to ease the reading of the volume and to allow the reader to choose where to concentrate his attention, this introduction is written as a recapitulation of the content of the book, giving emphasis to the logical and conceptual development of the subjects tackled, and drawing attention to the main results (equations and formulas) worked out throughout the text.

In this book, we shall consider physical systems described by N degrees of freedom (particles, classical spins, quasi-particles such as phonons, and so on), confined in a finite volume (therein free to move, or defined on a lattice), whose Hamiltonian is of the form

$$H = \frac{1}{2} \sum_{i=1}^{N} p_i^2 + V(q_1, \ldots, q_N) , \qquad (1.1)$$

which we call "standard," where the q's and the p's are, respectively, the coordinates and the conjugate momenta of the system. Our emphasis is on systems with a large number of degrees of freedom. The dynamics of the

1

system (1.1) is defined in the $2N$-dimensional phase space spanned by the q's and the p's.

Historically, long before the atomistic nature of matter was ascertained, Hamiltonian dynamics described the motion of celestial bodies in the solar system, which apparently move regularly, at least on human time scales of observation. This regularity, however, eluded analytic integrability of the equations of motion even when only three interacting bodies were considered, as Poincaré admirably proved. As a consequence, an ensemble of approximate methods, known as classical perturbation theory, was developed.

Since the formulation of the kinetic theory of gases and then with the birth of statistical mechanics, Hamiltonian dynamics has had to cope with dynamical behaviors of a qualitatively opposite kind with respect to those of celestial mechanics. In fact, Boltzmann's *Stosszahlansatz* (the hypothesis of molecular chaos), is not simply the consequence of our ignorance of the positions and momenta of the atoms or molecules of a system at some conventionally initial time, but, as was later understood by Krylov, there is an intrinsic instability of the dynamics.

This dynamical instability, which in the present context is called *Hamiltonian chaos*, is a phenomenon that makes finite the predictability time scale of the dynamics. Cauchy's theorem of existence and uniqueness of the solutions of the differential equations of motion formalizes the *deterministic* nature of classical mechanics; however, *predictability* stems from the combination of determinism and *stability* of the solutions of the equations of motion. Roughly speaking, stability means that in phase space the trajectories group into bundles without any significant spread as time passes, or with an at most linearly growing spread with time. In other words, small variations of the initial conditions have limited consequences on the future evolution of the trajectories, which remain close to one another or at most separate in a nonexplosive fashion.

Conversely, Hamiltonian chaos is synonymous with *unpredictability* of a *deterministic* but *unstable* Hamiltonian dynamics. A locally exponential magnification with time of the distance between initially close phase space trajectories is the hallmark of deterministic chaos.

A concise introduction to Hamiltonian mechanics is contained in Chapter 2, where the basic definitions and concepts are given, the framework of classical perturbation theory is outlined together with some of its most important results, and the classical explanation of the origin of Hamiltonian chaos based on homoclinic intersections, as well as the definition of Lyapunov characteristic exponents, is outlined.

The natural differential geometric language for Hamiltonian dynamics is that of *symplectic geometry*. However, in the present book, we resort to a geometrization of Hamiltonian dynamics by means of Riemannian geometry, whose basic elements are given in Appendix B, and we sketch the possibility of using Finsler geometry. The Riemannian geometrization of Hamiltonian dynamics, outlined in Chapter 3, is actually possible because for standard

Hamiltonians (1.1) the Legendre transform to a Lagrangian formulation always exists, and from the Lagrangian

$$L = \frac{1}{2} \sum_{i=1}^{N} \dot{q}_i^2 - V(q_1, \ldots, q_N) \, , \tag{1.2}$$

the equations of motion are derived in the Newtonian form

$$\ddot{q}_i = -\frac{\partial V}{\partial q^i} \, , \quad i = 1, \ldots, N \, . \tag{1.3}$$

The use of symplectic geometry in Hamiltonian mechanics is very elegant and powerful, for example, to investigate Hamiltonian systems with symmetries. Then why do we neglect it in favor of Riemannian geometry? the reason is that on Riemannian manifolds we know how to measure the distance between two points of the manifold, which we cannot do with symplectic manifolds. Moreover, the equations of motion (1.3) stem from the stationarity condition for the action functional (for isoenergetic paths)

$$\delta \int_{q(t_0)}^{q(t_1)} dt \, W(q, \dot{q}) = 0 \, , \tag{1.4}$$

where $W(q, \dot{q}) = \{[E - V(q)] \, \dot{q}_i \dot{q}^i\}^{1/2}$, which is equivalent to the variational definition of a *geodesic* line on a Riemannian manifold, which is a line of stationary or minimum length joining the points A and B:

$$\delta \int_{A}^{B} ds = 0 \, . \tag{1.5}$$

If configuration space is given the non-Euclidean metric of components

$$g_{ij} = 2[E - V(q)]\delta_{ij} \, , \tag{1.6}$$

whence the infinitesimal arc element $ds^2 = 4[E - V(q)]^2 dq_i \, dq^i$, then Newton's equations (1.3) are retrieved from the geodesic equations

$$\frac{d^2 q^i}{ds^2} + \Gamma^i_{jk} \frac{dq^j}{ds} \frac{dq^k}{ds} = 0 \, . \tag{1.7}$$

This is a nice well-known fact since the time of Levi-Civita. However, this would not be so useful without the equation stemming from the *second variation* of the length functional (1.5), the Jacobi–Levi-Civita (JLC) equation for the geodesic deviation vector field J (J locally measures the distance between nearby geodesics), which in a parallel-transported frame reads

$$\frac{d^2 J^k}{ds^2} + R^k_{\ ijr} \frac{dq^i}{ds} J^j \frac{dq^r}{ds} = 0 \, . \tag{1.8}$$

Applied to the configuration space of a physical system, this is a powerful tool to investigate the (in)stability of the phase space trajectories by relating (in)stability to the curvature features of the configuration space manifold; $R^i{}_{jkl}$ are the components of the Riemann curvature tensor.

For the sake of completeness, before discussing the Riemannian geometric approach to chaos, in Chapter 4 we briefly discuss how the problem of *integrability* fits into the Riemannian framework. Integrability is a vast field in Hamiltonian mechanics, and reviewing it here, even in a sketchy fashion, would be out of place. We just show that with the aid of Riemannian geometry some constructive work can also be done about integrability, and in doing this we also understand the reason why integrability is so exceptional.

We remark that in Chapter 4, as well as throughout the book, we mainly adopt a geometric language that is coordinate-dependent. Though less elegant than an intrinsic formulation, it has the advantage of a direct link with the constructive analytic expressions to be used in practical computations.

In Chapter 5 the core of the Riemannian theory of Hamiltonian chaos is discussed. No matter in which metric equation (1.8) is explicitly computed,[1] it requires the simultaneous numerical integration of both the equations of motion and the (in)stability equation. Using the Eisenhart metric in an extended-configuration space-time, (1.8) yields the standard tangent dynamics equation, which is currently used to compute Lyapunov exponents, whereas using the so-called Jacobi metric (1.6) one obtains equation (3.84) of Chapter 3, which is definitely more complex. At first sight the Riemannian geometrization of the dynamics could seem a not very helpful rephrasing of things. However, an equation relating (in)stability with geometry makes one hope that some global information about the average degree of instability (chaos) of the dynamics is encoded in global geometric properties of the mechanical manifolds.[2] That this might happen is proved by the special case of constant-curvature (isotropic) manifolds, for which the JLC equation simplifies to

$$\frac{d^2 J^i}{ds^2} + K J^i = 0 \;, \tag{1.9}$$

where K is any of the constant sectional curvatures of the manifold. On a positively curved manifold, the norm of the separation vector J does not grow, whereas on a negatively curved (hyperbolic) manifold, that is, with $K < 0$, the norm of J grows exponentially in time, and if the manifold is compact, so that its geodesics are sooner or later obliged to fold, this provides an example of chaotic geodesic motion.

The remarkable properties of geodesic flows on hyperbolic manifolds have been known to mathematicians since the first decades of last century [1];

---

[1] As explained in Chapter 3, there are different ambient spaces and different metrics to rephrase Newtonian dynamics in the Riemannian geometric language.

[2] Since the JLC equation involves the Riemann curvature tensor, the relevant geometric properties must be related to curvature.

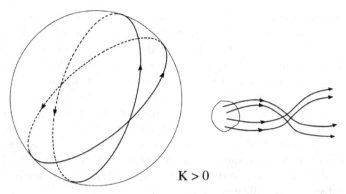

Fig. 1.1. On a manifold of constant positive curvature, the distance between any pair of nearby geodesics—issuing from a neighborhood—is oscillating and bounded from above (right). This is illustrated in the case of a 2D sphere (left), whose geodesics are great circles.

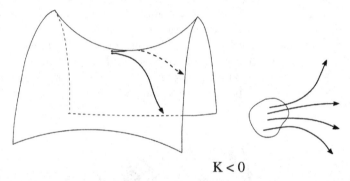

Fig. 1.2. On a manifold of constant negative curvature, the geodesics issuing from any neighborhood exponentially separate from each other (right). This is pictorially illustrated in the case of a 2D saddle (left).

it was Krylov who thought of using these results to account for the fast phase space mixing of gases and thus for a dynamical justification of the ergodic hypothesis in *finite times*, which is necessary to make statistical mechanics useful to physics [2]. Krylov's work has been very influential on the development of the so-called abstract ergodic theory [3], where Anosov flows [4] (e.g., geodesic flows on compact manifolds with negative curvature) play a prominent role. Ergodicity and mixing of these flows have been thoroughly investigated. From time to time, Krylov's intuitions have been worked out further. An incomplete excerpt of the outcomes of these developments can be found in [6–14]. What has been invariably discovered, is that, surprisingly, geodesic flows associated with chaotic physical Hamiltonians do not live on everywhere negatively curved manifolds. Few exceptions are known, in particular two low dimensional models [15,16], where chaos is actually associated with negative curvature.

The somewhat biased search for negative curvatures has been the main obstacle to an effective use of the geometric framework originated by Krylov to explain the source of chaos in Hamiltonian systems. On the other hand, it is true that the Jacobi equation, which describes the (in)stability of a geodesic flow, is in practice only tractable on negatively curved manifolds, formidable mathematical difficulties are encountered in treating the (in)stability of geodesic flows on manifolds of non-constant and not everywhere negative curvature. Nevertheless, a successful theory of Hamiltonian chaos can be started by giving up the idea that chaos must stem from negative curvature, and by initially accepting to work under some restrictive assumptions.

In fact, in Chapter 5, we discuss a successful strategy to work out from (1.8) the effective instability equation (5.27),

$$\frac{d^2\psi}{ds^2} + \langle k_R \rangle_\mu \, \psi + \langle \delta^2 k_R \rangle_\mu^{1/2} \, \eta(s) \, \psi = 0 \, , \qquad (1.10)$$

where ψ is such that $\|\psi^2(t)\| \sim \|J^2(t)\|$, k_R is the Ricci curvature of the mechanical manifold, $\langle \cdot \rangle_\mu$ stands for averaging on it, and $\eta(s)$ is a Gaussian-distributed Markov process. This equation is *independent of the dynamics*, it holds only if some suitable geometric conditions, which we call *quasi-isotropy*, are fulfilled by the given system of interest, and it puts in evidence the existence of another mechanism, besides hyperbolicity, to make chaos: the variability of the curvature probed by a geodesic activates *parametric instability*.

Fig. 1.3. Pictorial representation of how two geodesics—γ_1 and γ_2 issuing respectively from the close points A and B—separate on a 2D "bumpy" manifold where the variations of curvature activate parametric instability.

This seems to be an ubiquitous mechanism responsible for chaos in physical Hamiltonians. A sort of "statistical-mechanical" treatment of the dynamics itself is the nice outcome of the geometric theory of Hamiltonian chaos, hence the possibility of an analytic computation of the largest Lyapunov exponent through the general formula (5.40) for the rate of the exponential growth of $\|\psi^2(t)\| + \|\dot{\psi}^2(t)\|$, which is

$$\lambda(k_0, \sigma_k, \tau) = \frac{1}{2}\left(\Lambda - \frac{4k_0}{3\Lambda}\right) , \tag{1.11}$$

$$\Lambda = \left(\sigma_k^2\tau + \sqrt{\left(\frac{4k_0}{3}\right)^3 + \sigma_k^4\tau^2}\right)^{1/3} ,$$

where $k_0 = \langle k_R \rangle_\mu$, $\sigma_k = \langle \delta^2 k_R \rangle_\mu$ and τ is a characteristic time defined through a geometric argument. Three applications are considered: to a chain of harmonic oscillators also coupled through a quartic anharmonic potential (the FPU model), to a chain of coupled rotators, and to the so-called mean-field XY model. In the first two cases, an impressively excellent fitting of the numerical values of the largest Lyapunov exponents, as a function of the energy per degree of freedom, has been obtained, whereas for the mean-field XY model the quantitative agreement is less good.

Chapter 5 contains the *beginning* of a Riemannian-geometric theory of Hamiltonian chaos, the excellent results therein reported suggest that it is worthwhile to pursue research in this framework. The successful analytic computation of Lyapunov exponents proves that our understanding of the origin of Hamiltonian chaos is correct, whereas it would be reductive and wrong to consider this geometric approach as a mere recipe for analytically estimating Lyapunov exponents or the formula (1.11) as always valid.

The numerical test of the hypotheses that lead to (1.10), as well as the somewhat tricky correction to the bare result obtained for the chain of coupled rotators, hints at future developments beyond the assumption of quasi-isotropy, involving also configuration-space topology. In fact, in the FPU case, for which the hypothesis of quasi-isotropy seems well confirmed by numerical tests and for which the straightforward application of (1.11) leads to the correct result, the configuration space manifolds are topologically trivial at any energy value. In contrast, in the case of coupled rotators, the straightforward application of (1.11) leads to a mismatch between numerical and analytical results for the Lyapunov exponents in the strongly chaotic phase. This seems reasonable in the light of the numerical tests of the quasi-isotropy assumption for this model. At the energy density where the mentioned mismatch starts, critical points of the potential appear, that is, points $q_c = [\bar{q}_1, \ldots, \bar{q}_N]$ such that $\nabla V(q)|_{q=q_c} = 0$. From Morse theory we know that the occurrence of critical points (of a suitable function defined on a manifold) is generically associated with a non-trivial topology. In order to grasp why topology affects the degree of instability of the dynamics, let us consider the tangent dynamics

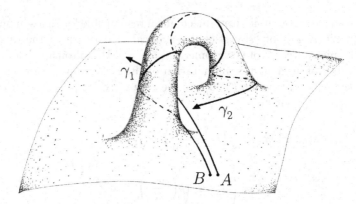

Fig. 1.4. Pictorial representation of how topology can affect geodesic separation. Two initially close geodesics, γ_1 and γ_2, respectively issuing from the points A and B, have very different evolutions because γ_2 is temporarily "trapped" by the handle, and, after some windings, it can be released in any direction.

equation (JLC equation in the Eisenhart metric) commonly used to numerically compute Lyapunov exponents,

$$\frac{d^2 J^i}{dt^2} + \left(\frac{\partial^2 V}{\partial q_i \partial q^l} \right)_{q(t)} J^l = 0 \,, \tag{1.12}$$

and remembering that in the neighborhood of any critical point q_c, by the *Morse lemma*, there always exists a coordinate system for which

$$V(\tilde{q}) = V(q_c) - \tilde{q}_1^2 - \cdots - \tilde{q}_k^2 + \tilde{q}_{k+1}^2 + \cdots + \tilde{q}_N^2 \,, \tag{1.13}$$

where k is the index of the critical point, i.e., the number of negative eigenvalues of the Hessian of V, let us note that in the neighborhood of a critical point, (1.13) yields $\partial^2 V / \partial_i \partial q_l = \pm \delta_{il}$, which, substituted into (1.12), gives k unstable directions that contribute to the exponential growth of the norm of the tangent vector J. In other words, the neighborhoods of critical points are "*scatterers*" of the trajectories, which enhance chaos by adding to parametric instability another instability mechanism, reminiscent of local hyperbolicity. However, if in the case of the chain of coupled rotators a nontrivial topology is responsible for the enhancement of chaos, with respect to the prediction based only on the quasi-isotropy assumption, things seem to go in the opposite direction for the mean-field XY model, though also in this case configuration-space topology is highly nontrivial (as discussed in Chapter 10). This is to say that a lot of interesting work remains to be done.

We can surmise that a first step forward, beyond the restrictive assumption of quasi-isotropy and encompassing the role of nontrivial topology, should lead to a generalization of the instability equation (1.10) that could be of the form

$$\frac{d^2}{ds^2} \begin{pmatrix} \psi \\ \phi \end{pmatrix} + \begin{pmatrix} \kappa(s) & \alpha \\ \beta & \gamma \end{pmatrix} \begin{pmatrix} \psi \\ \phi \end{pmatrix} = 0 \,, \tag{1.14}$$

where $\kappa(s) = \langle k_R \rangle_\mu \, \psi + \langle \delta^2 k_R \rangle_\mu^{1/2} \, \eta(s)$ and where α, β, γ are functions, to be specified by the future developments of the theory, accounting for the relative frequency of encounters of neighborhoods of critical points, for the average number of unstable directions, and for the interplay between the two instability mechanisms (parametric modulations can have also stabilizing effects, as is the case of the reversed pendulum stabilized by a fast oscillation of its pivotal point). Then the instability exponent from (1.14) would be the average growth rate of $\|\psi^2\| + \|\dot{\psi}^2\| + \|\phi^2\| + \|\dot{\phi}^2\|$.

To summarize, let us compare the advantages of the Riemannian theory over the conventional one based on homoclinic intersections of perturbed separatrices. The traditional explanation of the origin of Hamiltonian chaos requires the use of action-angle coordinates; it is of a perturbative nature and thus applies only to quasi-integrable models; it works constructively only for 1.5 or 2 degrees of freedom; and even the basic result on which it relies—the Poincaré–Birkhoff theorem—has no known extension at $N > 2$. Last but not least, no computational relationship exists between homoclinic intersections and Lyapunov exponents. In contrast, the Riemannian theory works with the natural coordinates of a system, it is valid at any energy, it explains the cause of chaotic instability in a clear and intuitive way, and it makes a natural link between the explanation of the origin of chaos and the quantitative way of detecting it through Lyapunov exponents.

Chapter 6 bridges the first part to the second part of the book. Therein the attention begins to focus on the geometry of the dynamics of systems with phase transitions. The logical connections proceed as follows. As we have recalled in Chapter 2, the crossover in the energy dependence of the largest Lyapunov exponent λ, first observed in the Fermi–Pasta–Ulam model, has been phenomenologically attributed to a transition from weak to strong chaos, or slow and fast phase space mixing, respectively. This is called the strong stochasticity threshold (SST). We have surmised in the past that this transition has to be the consequence of some "structural" change occurring in configuration space, and in phase space as well. This dynamical transition has been observed in every nonintegrable many-degrees-of-freedom Hamiltonian system for which $\lambda(\varepsilon)$, $\varepsilon = E/N$, has been computed. Then, a natural question arises: could some kind of dynamical transition between weak and strong chaos (possibly a very sharp one) be the microscopic dynamical counterpart of a thermodynamic phase transition? And if this were the case, what kind of difference in the $\lambda(\varepsilon)$ pattern would discriminate between the presence or absence of a phase transition? And could we make a more precise statement about the kind of "structural" change to occur in configuration space when the SST corresponds to a phase transition and when it does not?

Actually, the $\lambda(\varepsilon)$ patterns, numerically found for the φ^4 model with symmetry groups $O(1), O(2), O(4)$ in two- and three-dimensional lattices, for the classical Heisenberg XY model in two and three dimensions, and analytically computed in the so-called mean-field XY model, show abrupt transitions between different regimes of chaoticity. Typically, a "cuspy" point in $\lambda(\varepsilon)$ appears in correspondence with the phase transition point. This is observed also for other models studied in the literature. Then, since Lyapunov exponents are tightly related to the geometry of the mechanical manifolds in configuration space, it is natural to try to characterize the above-mentioned "structural" changes through the geometric ingredients that enter the analytic formula (1.11) for λ, namely the average Ricci curvature and its variance. The intriguing surprise is that the phase transition point is invariably marked by a peak in the curvature fluctuations. What do we learn from this?

The answer is given in Chapter 7. Here we start with the observation that the topology change driven by a continuously varying parameter in a family of two dimensional–surfaces is accompanied by a sharp peak in the variance of the Gaussian curvature. This is confirmed by computing the variance of the curvature of the level sets of a generic function in the neighborhood of one of its critical points. This is an example at large dimension.

In other words, the tempting idea was that of attributing to the deeper level of configuration space *topology* the responsibility for the appearance of the strong and sudden "structural" change necessary to entail a phase transition. An important step forward in this direction was obtained by studying the Ricci curvature fluctuations of the configuration-space manifolds (M_u, g) of one and two dimensional lattice φ^4 models equipped with different Riemannian metrics $g^{(k)}$, having nothing to do with the "dynamical" metric (1.6). In the manifolds $(M_u, g^{(k)})$, M_u is defined by the potential function $V(q)$ of the model, i.e., $M_u = \{q = (q_1, \ldots, q_N) \in \mathbb{R}^N | V(q) \le u\}$, and the metrics $g^{(k)}$ used are arbitrary and independent of $V(q)$. The results strongly support the idea that at the phase transition point in the two-dimensional model, something happens that is to some extent independent of the metric structure imposed on the configuration–space submanifolds M_u.

This is resumed in the formulation of a *topological hypothesis*. Concisely, consider the microcanonical volume

$$\Omega(E) = \int_0^E d\eta \; \frac{(2\pi\eta)^{N/2}}{\eta\Gamma(\frac{N}{2})} \int_0^{E-\eta} du \int_{\Sigma_u} \frac{d\sigma}{\|\nabla V\|} \; ; \qquad (1.15)$$

the larger N the closer to some Σ_u are the microscopic configurations that significantly contribute to the statistical averages, and therefore the idea is that in order to observe the development of singular behaviors of thermodynamic observables computed through $\Omega(E)$ in (1.15), it is *necessary* that a value u_c exist such that $\Sigma_{u<u_c}$ are not diffeomorphic to (have a different topology from) the $\Sigma_{u>u_c}$.

Chapter 7 ends with a *direct* and remarkable confirmation of this working hypothesis. Confirmation is achieved by means of the numerical computation, again for the one- and two-dimensional lattice φ^4 models, of a *topologic invariant* of the equipotential hypersurfaces of configuration space, i.e., $\Sigma_u = V^{-1}(u) \equiv \{q = (q_1, \ldots, q_N) \in \mathbb{R}^N | V(q) = u\}$. The topologic invariant, a diffeomorphism invariant, is the Euler–Poincaré characteristic $\chi(\Sigma_u)$

of equipotential hypersurfaces computed through the Gauss–Bonnet–Hopf formula

$$\chi(\Sigma_u) = \gamma \int_{\Sigma_u} K_G \, d\sigma \ , \qquad (1.16)$$

where $\gamma = 2/\mathrm{vol}(\mathbb{S}_1^n)$ is twice the inverse of the volume of an even n-dimensional sphere of unit radius, $n = N - 1$; K_G is the Gauss–Kronecker curvature of the manifold; $d\sigma = \sqrt{\det(g)} dx^1 dx^2 \cdots dx^n$ is the invariant volume measure of Σ_u; and g is the Riemannian metric induced from $\mathbb{R}^N$.

Two things are evident from the numerical computations: the first is that the u-pattern of $\chi(\Sigma_u)$ clearly makes a big difference between presence and absence of a phase transition; moreover, it unambiguously locates the transition point. The second fact is that topology changes considerably even in the absence of a phase transition; it is its way of changing with u that is suddenly modified at the transition point.

What we are after is the possible deepening of our mathematical understanding of the origin of phase transitions. In fact, the topological properties of configuration space submanifolds, both of equipotential hypersurfaces Σ_u and of the regions M_u bounded by them, are already determined when the microscopic potential V is assigned and are completely *independent* of the statistical measures. The appearance of singularities in the thermodynamic observables could then be the *effect* of a deeper cause: a suitable topological transition in configuration space.

With Chapters 8 and 9 we put forward the fundamental elements for a topological theory of phase transition phenomena. In Chapter 10 we go back to models, but this time working out exact analytic results.

In Chapter 8 we unveil the existence of a quantitative connection between geometry and topology of the energy landscape in phase space, or in configuration space, and thermodynamic entropy defined as $S_N(E) = (k_B/2N) \log[\int_{\Sigma_E} d\sigma \, / \|\nabla H\|]$:

$$S(E) \approx \frac{k_B}{2N} \log \left[\mathrm{vol}(\mathbb{S}_1^{2N-1}) \sum_{i=0}^{2N-1} b_i(\Sigma_E) + \int_{\Sigma_E} d\sigma \frac{\widetilde{\mathcal{R}}(E)}{N!} \right] + r(E) \ , \quad (1.17)$$

where $b_i(\Sigma_E)$ are the Betti numbers of the constant energy hypersurfaces in phase space. Betti numbers are fundamental topological invariants of a manifold. Another version of this formula reads

$$S(v) \approx \frac{k_B}{2N} \log \left[\mathrm{vol}(\mathbb{S}_1^{2N-1}) \left(\mu_0 + \sum_{i=1}^{N-1} 2\mu_i(M_v) + \mu_N \right) + \overline{\mathcal{R}}(E(v)) \right] + r(E(v)),$$

$$(1.18)$$

which now holds in configuration space and where the $\mu_i(M_v)$ are the Morse indexes (in one-to-one correspondence with topology changes) of the submanifolds M_v of configuration space. These formulas are approximate, but following a different reasoning, and using the definition $S_N^{(-)}(v) = k_B/N \log[\int_{M_v} d^N q]$, also an exact formula can be derived, which reads

$$
S_N^{(-)}(v) = \frac{k_B}{N} \log \left[\mathrm{vol}[M_v \setminus \bigcup_{i=1}^{\mathcal{N}(v)} \Gamma(x_c^{(i)})] + \sum_{i=0}^{N} w_i\, \mu_i(M_v) + R(N, v) \right],
$$
(1.19)

where the first term in the square brackets is the configuration-space volume minus the sum of volumes of certain neighborhoods of the critical points of the interaction potential, the second term is a weighed sum of the Morse indexes, and the third term is a smooth function. This formula is proved in Chapter 9 as Theorem 9.39.

Since the above formula provides an exact relation between a thermodynamic function and some quantities peculiar to the mathematics that we are using, it is of special interest. In fact, it is thanks to this formula that we can convince ourselves, with the aid of Theorem 9.39 of Chapter 9, that topology is relevant to phase transitions. So we come to Chapter 9, which contains a major leap forward: the proof of two theorems that establish a *necessary* topological condition for the occurrence of first- or second-order phase transitions. A thermodynamic phase transition point necessarily stems from a corresponding topological transition point in configuration space. The theorems apply to a wide class of smooth, finite-range, and confining potentials V_N bounded below, describing systems confined in finite regions of space with continuously varying coordinates. The relevant configuration-space submanifolds are both the level sets $\{\Sigma_v := V_N^{-1}(v)\}_{v \in \mathbb{R}}$ of the potential function V_N and the configuration space submanifolds enclosed by the Σ_v defined by $\{M_v := V_N^{-1}((-\infty, v])\}_{v \in \mathbb{R}}$, where N is the number of degrees of freedom and v is the potential energy. The proof of Theorem 9.14 proceeds by showing that under the assumption of diffeomorphicity of the equipotential hypersurfaces $\{\Sigma_v\}_{v \in \mathbb{R}}$, as well as of the $\{M_v\}_{v \in \mathbb{R}}$, in an arbitrary interval of values for $\bar{v} = v/N$, the Helmholtz free energy is uniformly convergent in N to its thermodynamic limit, at least within the class of twice differentiable functions, in the corresponding interval of temperature. This theorem is used to prove that in (1.19) the origin of the possible unbound growth with N of a derivative of the entropy, that is, of the development of an analytic singularity in the limit $N \to \infty$ and thus of a phase transition, can be due only to the topological term $\sum w_i\, \mu_i(M_v)$. Thus the *topological hypothesis* turns into a necessity theorem.

As already seen at the end of Chapter 7, where we have reported the results of the computation of $\chi(\Sigma_v)$ for one- and two-dimensional lattice φ^4 models, there is not a one-to-one correspondence between topology variations and phase transitions. This means that the converse of the just-mentioned theorems is not true, and this can be easily understood by inspection of (1.19). In fact, if we keep in mind that it is only the topological term that can induce nontrivial behaviors of $S(v)$, we see that "soft" variations with

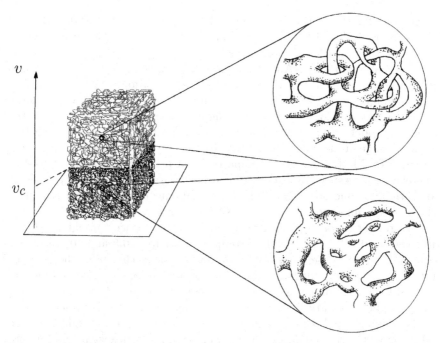

Fig. 1.5. Pictorial representation of a transition beteween complex topologies. Two cobordant surfaces of high genus are represented. From the ground level up to the crossover level v_c, the manifolds M_v have a genus which increases with v. Above the crossover level v_c, the manifolds M_v have also a nonvanishing linking number which increases with v. This is a low-dimensional metaphor of what could be at the origin of a phase transition.

v of the topology of the M_v cannot be transformed into something much different by $S(v)$. As a consequence, the problem of the mathematical definition of *sufficiency* conditions is open. In Chapter 10 we report the results obtained for some exactly solvable models. Fortunately, there is a number of models (all of mean-field kind) for which we can exactly compute the canonical partition function, hence their thermodynamic behavior, *and* for which we can exactly compute also a topological invariant, the Euler characteristic of the submanifolds M_v of configuration space. The models considered in Chapter 10 are the mean-field XY model, the k-trigonometric model, the mean-field spherical model, the mean-field lattice φ^4 model. All these systems are described by long-range forces (all the degrees of freedom interact with all the others) and thus are out of the validity domain of the present formulation of the theorems of Chapter 9; by the way, the limitation to short-range interactions has been introduced to ease the proof of Theorem 9.14, but we strongly suspect that the truly important assumption should be that of additivity, relaxing the restrictive assumption of short-range forces. Nevertheless, the exact results reported in Chapter 10 together with the numerical results worked out for the lattice φ^4 model (which belongs to the validity domain of Theorems 9.14 and 9.39), provide precious hints to orient future investigations of these challenging questions: What kind of topological transitions entail a phase transition? And what, from the topological standpoint, makes the difference between different kinds of phase transitions?

A preliminary concise summary of the hints hitherto available is as follows.

By probing topology changes in configuration space through $\chi(\Sigma_v)$ for the lattice φ^4 model, we have seen that the phase transition point corresponds to a sudden change in the rate of change of $\chi(v)$. Remarkably, already with a small system of 7×7 lattice sites the $\chi(v)$ pattern sharply marks a major qualitative difference with respect to the one-dimensional case (see Chapter 7), whereas no thermodynamic observable (like the specific heat) is able to discriminate which system undergoes a phase transition with only $N = 49$.

By probing topological changes in configuration space through $\chi(M_v)$ in the mean-field XY model, and by comparing Figures 10.5 and 10.3, we have seen that at the phase transition point the topology change of the M_v corresponds to the simultaneous attachment of handles of $\frac{N}{2}$ different types, in contrast to the one-dimensional nearest-neighbor XY model, in which no such abrupt topological variation occurs and no phase transition is present.

By probing topological changes in configuration space through $\chi(M_v)$ in the k-trigonometric model, we have found also in this case sharp and unambiguous differences among the $k = 1$ case (no phase transition), the $k = 2$ case (second-order transition), and the $k = 3$ case (first-order transition). By computing $\mu(v) = \sum_i \mu_i(M_v)$, we observe that also $\mu(v)$ clearly discriminates among these three different possibilities; moreover, very similar patterns of $\mu(v)$ are found for both the XY mean-field and k-trigonometric models, as is evident by comparing Figures 10.7, 10.8, and 10.20. Remarkably, both $\chi(v)$ and $\mu(v)$ make a clear distinction between first- and second-order phase transitions.

Chapter 10 ends with some comments on recently appearing articles on topology and phase transitions. Though providing interesting results that can contribute to the advancement of the subject, some of these papers are misleading for what concerns the proposed interpretations.

Let us conclude with a few general comments. Earlier attempts at introducing topological concepts in statistical mechanics concentrated on *macroscopic* low-dimensional parameter spaces. Actually, this happened after Thom's remark that the critical point shown by the van der Waals equation corresponds to the Riemann–Hugoniot catastrophe [17]. Hence some applications of the theory of singularities of differentiable maps to the study of phase transitions followed [18]. Other approaches resorted to geometric concepts [19] or topological concepts [20] in macroscopic low-dimensional parameter spaces. An elegant formulation of phase transitions as due to a topological change of some abstract manifold of macroscopic variables was obtained using the Atiyah–Singer index theorem [21, 22], and this has been applied to the 2D Ising model.

Throughout the present book we establish a completely new kind of link between the study of phase transitions and elementary differential topology. In fact, here we deal with the high-dimensional *microscopic* configuration space of a physical system. The level sets of the microscopic interaction potential among the particles, or the manifolds bounded by them, are the configuration space submanifolds that necessarily have to change their topology in correspondence with a phase transition point. The topology changes implied here are those described within the framework of Morse theory through *attachment of handles* [23].

Notice that in all the cases considered so far the role of the potential V is twofold: it determines the relevant submanifolds of configuration space and it is a good Morse function on the same space. However, for example, in the case of entropy-driven phase transitions occurring in hard-sphere gases, the fact that the (singular) interaction potential can no longer play the role of Morse function does not mean that the connection between topology and phase transitions is lost; it rather means that other Morse functions are to be used.[3]

This topological approach also makes a subtle link between dynamics and thermodynamics because it affects both of them, the former because it can be seen as the geodesic flow of a suitable Riemannian metric endowing configuration space [24], the latter because we have worked out an analytic relation between thermodynamic entropy and Morse indexes of the configuration-space submanifolds.

Though at present in the framework that we have developed, including the theorems of Chapter 9, we have considered only first- and second-order phase transitions, the topological approach seems to have the potential of unifying the mathematical description of very different kinds of phase transitions. For example, there are "exotic" kinds of transitional phenomena in statistical physics, such as the glassy transition of amorphous systems to a supercooled liquid regime, or the folding transitions in polymers and proteins, that are *qualitatively* unified through the so-called *landscape paradigm* [25, 26], which is based on the idea that the relevant physics of these systems can be understood through the study of the properties of the potential energy hypersurfaces and, in particular, of their stationary points, usually called "saddles." That this landscape paradigm naturally goes toward a link with Morse theory and topology has been hitherto overlooked.

Last but not least. Sometimes it happens that a physical phenomenon is to some extent identified with its explanation. This seems to be the case of phase transitions, which are often identified with nonanalytic behaviors of

---

[3] Perhaps for hard spheres a good Morse function could be the sum of all the pairwise Euclidean distances between the hard spheres of a system: it is real-valued; and it has a minimum when the density is maximum, that is, for close packing, meaning that this function is bounded below. It is unclear whether it is nondegenerate, but Morse functions are dense and degeneracy is easily removed if necessary.

thermodynamic observables. Historically, this is due to the fact that these phenomena occur in thermal equilibrium and have been observed at a macroscopic level, so that thermodynamic observables have been the natural candidates to experimentally characterize a phase transition. In particular, on the basis of the experimental phenomenology, the translation in mathematical terms of the discontinuity of the physical properties (condensation, melting, and so on) has been that of the development of nonanalytic singular behaviors at the transition point. But as is well known, the statistical measures are analytic at any N, and the only way out is to work in the $N \to \infty$ limit. However, let us think of a small snowflake that melts into a droplet of water, or of a filamentary polymer chain that acquires a globular configuration, or of a protein that folds to its native structure, not to mention Bose–Einstein condensation or other transitional phenomena in small quantum systems at nanoscopic or mesoscopic levels; all these systems (and many others) display major qualitative physical changes also at very small N, much smaller than the Avogadro number, and perhaps thermodynamic observables are no longer so relevant. Once we have understood that the above-mentioned mathematical singularities are due to a deeper phenomenon, at least for a wide class of systems, the basic mathematical cause of the appearance of a phase transition is already there at finite, even small, N, and one can find it by looking at the microscopic-configuration-space topology.

Chapter 2

Background in Physics

In this first chapter we will give an outline of some fundamental elements of statistical mechanics, of Hamiltonian dynamics, and of the relationship between them.

The general problem of statistical physics is the following. Given a collection–in general a large collection–of atoms or molecules, given the interaction laws among the constituents of this collection of particles, and given the dynamical evolution laws, how can we predict the macroscopic physical properties of the matter composed of these atoms or molecules?

At the epoch of the founding fathers of statistical physics, Boltzmann and Gibbs, there were no means to attack the problem from the side of microscopic dynamics, and thus all the efforts aimed at getting rid of it. We will briefly recall what kind of arguments have been used to achieve this goal.

The Gibbs ensemble formulation of statistical mechanics accounts for the laws of thermodynamics, and, in principle, should allow one to derive all the macroscopic equilibrium properties of a system on the basis of the knowledge of the interatomic or intermolecular forces. Moreover, there are experimental facts (even belonging to our common everyday experience) such as the phenomenon of condensation of a gas, or solidification of a liquid, in general a *change of state* of matter, that also call for an explanation in the same framework of statistical mechanics. This last topic is part of a huge and fascinating field encompassing many other collective phenomena: *phase transitions*. A few basic notions on this phenomenon will be given, with some emphasis given to the Yang–Lee theory.

At the dawning of the computer era, as admirably realized by E. Fermi, J. von Neumann, and S. Ulam, a new insight into the foundations of statistical mechanics became possible through the "long-time" numerical solution of the differential equations of motion of a collection of up to a few hundreds of interacting particles, nothing with respect to the Avogadro number, but surprisingly enough to discover an unsuspected richness of the dynamics and, later on, enough to give birth to so-called molecular dynamics, a numerical *ab initio* computational method to estimate macroscopic properties (such as viscosity, specific heat, magnetic susceptibility, and elastic constants) of real materials.

The dynamical phenomenology of many-particle Hamiltonian systems has been widely investigated during the last decades. Among other phenomena, phase transitions have been investigated from the point of view of microscopic dynamics. An outline of some of these results will be given in the final part of this chapter, where our choice is biased by the need to provide the reader with the basic "experimental facts" about dynamics that have stimulated the development of a geometric statistical theory of chaotic dynamics and the topological theory of phase transitions. Chapters from 3 to 6, and from 7 to 10, are devoted to a detailed discussion of these two topics.

2.1 Statistical Mechanics

In the following we will simply speak of particles or of degrees of freedom,[1] instead of atoms or molecules. Consider N point particles, of mass m, contained in a volume $\mathcal{V}$, described by a Hamiltonian function

$$H = \sum_{i=1}^{N} \left[\frac{\mathbf{p}_i^2}{2m} + U(\mathbf{r}_i) \right] + \sum_{i<j} \Phi(|\mathbf{r}_i - \mathbf{r}_j|) , \qquad (2.1)$$

where the interaction potential $\Phi(r)$ is in general assumed to consist of an attractive region with a *finite range* and a repulsive region at short distances that, if the potential is very steeply increasing, represents a *hard core*. The forces are assumed to have the additivity property and are not allowed, at short distances, to become infinitely attractive (as is the case of pure Coulomb or gravitational forces). The potential $U(\mathbf{r}_i)$ accounts for the external forces acting on the ith particle as well as for the forces exerted by the walls of the container constraining the particles to move inside the volume $\mathcal{V}$.

Now, if the thermal de Broglie wavelength of the particles is much smaller than the average interparticle distance, i.e.,

$$\frac{\hbar}{(2mk_\mathrm{B}T)^{1/2}} \left(\frac{N}{\mathcal{V}} \right)^{1/3} \ll 1 ,$$

where $\frac{3}{2}k_\mathrm{B}T$ is the average kinetic energy per particle, then instead of describing the microscopic motions through the Schrödinger equation, we can use classical Newtonian mechanics. Nevertheless, a nontrivial trace of the fact that the microscopic world is surely governed by the laws of quantum mechanics is concealed in the existence of an attractive part in typical interparticle interaction potentials, an essential feature for the stability of matter that cannot be explained by classical physics.

---

[1] This terminology is more general and encompasses, for example, spin variables on a lattice, and discretized versions of field models.

The basic assumptions about the Hamiltonian in (2.1) can be insufficient to describe the great richness of the physical properties of matter at macroscopic and mesoscopic levels, properties that probably can be explained only if more-complicated internal structures of molecules, as well as other kinds of a forces of quantum nature or the interaction with a self-consistent electromagnetic field, are considered. However, the domain of application of classical statistical mechanics can be extended very far from that of its original formulation. To give an important example, let us mention that a mathematical relationship can be established between classical statistical mechanics and quantum field theory, a seminal relationship that has been widely exploited in recent years and that witnesses that the theoretical relevance of classical statistical mechanics, in particular for what concerns phase transitions, is not limited to classical physics.

In what follows, to ease the notation and to underline that we refer to a larger class of systems with respect to a collection of atoms or molecules, we assume that the standard Hamiltonian function for a system with N degrees of freedom reads as

$$H(p,q) = \sum_{i=1}^{N} \frac{1}{2} p_i^2 + V(q_1, \dots, q_N) \,, \tag{2.2}$$

where $V(q)$ is the potential energy and the q_i's and the p_i's are, respectively, the canonically conjugated coordinates and momenta. From now on, $p = (p_1, \dots, p_N)$ and $q = (q_1, \dots, q_N)$.

2.1.1 Invariant Measure for the Dynamics

The Γ-space, or *phase space* of a Hamiltonian system with N degrees of freedom, is the $2N$-dimensional space whose coordinates are $q_1, \dots, q_N, p_1, \dots, p_N$. A given point $x^0 = (q_1^0, \dots, q_N^0, p_1^0, \dots, p_N^0) \in \Gamma$ represents a microscopic state of a system, and in the course of time, it moves in Γ-space according to the Hamilton equations of motion

$$\dot{q}_i = \frac{\partial H}{\partial p_i} \,, \qquad \dot{p}_i = -\frac{\partial H}{\partial q_i} \,, \qquad i = 1, 2, \dots, N. \tag{2.3}$$

A fundamental theorem is Liouville's theorem. Consider any measurable set (in the sense of Lebesgue) A_0 of points of the phase space Γ of a given system. This set is transformed by the natural motion into another set A_t at time t; the theorem asserts that for any t the measure of the set A_t coincides with the measure of A_0.

To prove this assertion, let us use the notation $x_i = q_i$ and $x_{N+i} = p_i$, and $X_i = \frac{\partial H}{\partial p_i}$, $X_{N+i} = -\frac{\partial H}{\partial q_i}$, for $i = 1, \dots, N$. With this notation the (2.3) simply reads as

$$\frac{dx_i}{dt} = X_i(x_1, \dots, x_{2N}) \,. \tag{2.4}$$

Let us observe that the vector field X is divergence free, i.e.,

$$\sum_{i=1}^{2N} \frac{\partial X_i}{\partial x_i} = \sum_{i=1}^{N} \left(\frac{\partial^2 H}{\partial p_i \partial q_i} - \frac{\partial^2 H}{\partial q_i \partial p_i} \right) = 0 . \tag{2.5}$$

Since the system of differential equations (2.4) is of the first order, if at some initial time t_0 we assign the initial conditions $x_i^{(0)}$, $i = 1, \ldots, 2N$, the solutions of the equations of motion will be given as a set of functions

$$x_i = f_i(t; x_1^{(0)}, \ldots, x_{2N}^{(0)}) . \tag{2.6}$$

Let us now denote by $\mu(A_t)$ the measure of the set A_t given by

$$\mu(A_t) = \int_{A_t} dx_1 \cdots dx_{2N} , \tag{2.7}$$

and introduce the coordinate change

$$x_i = f_i(t; y_1, \ldots, x_{2N}) , \tag{2.8}$$

where $(y_1, \ldots, y_{2N})$ is any point of A_0 and $(x_1, \ldots, x_{2N}) \in A_t$. Then $\mu(A_t)$ is also given by

$$\mu(A_t) = \int_{A_0} \det(J) \, dy_1 \cdots dy_{2N} , \tag{2.9}$$

where $\det(J)$ is the determinant of the Jacobian matrix of the coordinate change due to the natural motion, that is,

$$\det(J) = \left| \frac{\partial(x_1, \ldots, x_{2N})}{\partial(y_1, \ldots, y_{2N})} \right| . \tag{2.10}$$

The time variation of $\mu(A_t)$ is given by

$$\frac{d\mu(A_t)}{dt} = \int_{A_0} \frac{\partial \det(J)}{\partial t} \, dy_1 \cdots dy_{2N} . \tag{2.11}$$

The derivative of the determinant of a matrix B is computed by means of Jacobi's formula $d \det(B) = \text{tr}[\text{Adj}(B) \, dB]$, where $B = \{\beta_{ij}\}$ and $\text{Adj}(B) = \{\alpha_{ij}\}$ with $\alpha_{ij} = (-1)^{i-j} \det(B_{[ji]})$, having set $B_{[ji]} = B$ (without the jth row and ith column), so that $\det(B) = \sum_k \beta_{ik} \alpha_{ki}$. Jacobi's formula then yields $d \det(B) = \sum_j \sum_i \alpha_{ji} \, d\beta_{ij}$ whence

$$\frac{\partial \det(J)}{\partial t} = \sum_{i=1}^{2N} \left| \frac{\partial(x_1, \ldots, \partial x_i/\partial t, \ldots, x_{2N})}{\partial(y_1, \ldots, y_{2N})} \right| = \sum_{i=1}^{2N} \left| \frac{\partial(x_1, \ldots, X_i, \ldots, x_{2N})}{\partial(y_1, \ldots, y_{2N})} \right| ,$$

where we have used $\frac{\partial x_i}{\partial t} = \frac{dx_i}{dt} = X_i$. If now $\{\alpha_{ij}\} = \text{Adj}(J)$, we have

$$\left| \frac{\partial(x_1, \ldots, X_i, \ldots, x_{2N})}{\partial(y_1, \ldots, y_{2N})} \right| = \sum_{k=1}^{2N} \frac{\partial X_i}{\partial y_k} \alpha_{ki}$$

$$= \sum_{k=1}^{2N} \left(\sum_{r=1}^{2N} \frac{\partial X_i}{\partial x_r} \frac{\partial x_r}{\partial y_k} \right) \alpha_{ki} = \sum_{r=1}^{2N} \frac{\partial X_i}{\partial x_r} \sum_{k=1}^{2N} \frac{\partial x_r}{\partial y_k} \alpha_{ki}$$

$$= \sum_{r=1}^{2N} \frac{\partial X_i}{\partial x_r} \left| \frac{\partial(x_1, \ldots, x_{i-1}, x_i, x_{i+1}, \ldots, x_{2N})}{\partial(y_1, \ldots, y_{2N})} \right| ,$$

$$(2.12)$$

but if $r \neq i$, then the Jacobian matrix has two rows that are equal and thus its determinant vanishes. In conclusion,

$$\frac{\partial \det(J)}{\partial t} = \sum_{i=1}^{2N} \sum_{r=1}^{2N} \frac{\partial X_i}{\partial x_r} \left| \frac{\partial(x_1, \ldots, x_r, \ldots, x_{2N})}{\partial(y_1, \ldots, y_{2N})} \right| \delta_{ri}$$

$$= \sum_{i=1}^{2N} \frac{\partial X_i}{\partial x_i} \det(J) = \det(J) \sum_{i=1}^{2N} \frac{\partial X_i}{\partial x_i} = 0 \qquad (2.13)$$

which inserted into (2.11) yields

$$\frac{d\mu(A_t)}{dt} = 0 , \qquad (2.14)$$

which proves that the Liouville measure $\mu(A_t) = \int_{A_t} dx_1 \cdots dx_{2N}$ in Γ-space is invariant under the natural Hamiltonian motion. Any given measurable set of initial conditions in phase space is deformed by the dynamics, but its measure is kept constant.

In other words, a measurable set of initial conditions evolves in phase space as if it were an incompressible fluid whose "particles" are the representative points of different microscopic realizations of a same macroscopic state. With Gibbs, this will be called an *ensemble* of mechanical systems in Γ-space. The distribution of this ensemble, which later will be given the meaning of a probability distribution, is defined for an arbitrary number $\mathcal{N}$ of representative points whose relative distribution, that is, the fraction $\Delta\mathcal{N}/\Delta\Gamma$, where $\Delta\mathcal{N}$ is the number of them contained in the phase volume $\Delta\Gamma = \Delta^N p \Delta^N q$ located at the point (p, q) at time t, has a density function $\rho(p, q, t)$ as its continuum limit. A consequence of the invariance of the Liouville measure is that $d\mathcal{N}/dt = 0$, whence $d\rho(p, q, t)/dt = 0$, or equivalently,

$$\frac{\partial \rho}{\partial t} + \{\rho, H\} = 0 , \qquad (2.15)$$

which is the Liouville equation, where $\{\rho, H\}$ are the Poisson brackets

$$\{\rho, H\} = -\sum_{i=1}^{N} \left(\frac{\partial H}{\partial q_i} \frac{\partial \rho}{\partial p_i} - \frac{\partial H}{\partial p_i} \frac{\partial \rho}{\partial q_i} \right) .$$

Note that in the language of fluid dynamics, Liouville's theorem expressed by (2.14) and (2.15) corresponds to the "Lagrangian" and "Eulerian" descriptions respectively.

A simple consequence of Liouville's theorem is that for any function $f(\rho)$ of the density, the integral

$$\int_{\Gamma} f(\rho) \, dp_1 \cdots dp_N \, dq_1 \cdots dq_N$$

is independent of time.

Moreover, it can be readily verified that any density distribution that depends on the coordinates p, q only through the Hamiltonian $H(p, q)$,

$$\rho_{eq}(p, q) = f(H(p, q)) , \tag{2.16}$$

is *stationary*, and thus we will call it an *equilibrium* density distribution and the ensemble described by ρ_{eq} an *equilibrium ensemble*.

2.1.2 Invariant Measure Induced on Σ_E

The Hamiltonian dynamics of an isolated system corresponds to the constrained motion of its representative point on a constant-energy hypersurface $\Sigma_E \subset \Gamma$ defined by $H(x) \equiv H(p, q) = E$. Therefore we wonder how to build an invariant measure on Σ_E out of the Liouville invariant measure (2.7).

To this end, consider any measurable set $A \subset \Sigma_E$. Then, with the outward normal at each point, define the volume element $\gamma \subset \Gamma$ bounded by A and $A' \subset \Sigma_{E+\Delta E}$ and filled by the outward normals to A joining Σ_E and $\Sigma_{E+\Delta E}$. By Liouville's theorem, the volume

$$\int_{\gamma} dx_1 \cdots dx_{2N} = \int_{E < H(x) < E+\Delta E} \chi(x) \, dx_1 \cdots dx_{2N}$$

is left invariant by the natural motions; the function $\chi(x)$ is defined by $\chi(x) = 1$ if $x \in \gamma$, and $\chi(x) = 0$ otherwise. Also, the ratio $(1/\Delta E) \int_{\gamma} dx_1 \cdots dx_{2N}$ is a Liouville invariant in Γ for any ΔE, however small. Hence, in the limit $\Delta E \to 0$, by applying the derivation formula (8.13) given in Chapter 8, we get

$$\lim_{\Delta E \to 0} \frac{1}{\Delta E} \int_{E < H(x) < E+\Delta E} \chi(x) \, dx_1 \cdots dx_{2N} = \int_{\Sigma_E} \chi(x) \frac{d\sigma}{\|\nabla H\|} = \int_A \frac{d\sigma}{\|\nabla H\|} ,$$

and the invariant measure on Σ_E is readily obtained by defining

$$\mu_E(A) = \int_A \frac{d\sigma}{\|\nabla H\|} \;.$$

The volume of the whole energy surface is

$$\Omega(E) = \int_{\Sigma_E} \frac{d\sigma}{\|\nabla H\|} \;,$$

and, using again the derivation formula (8.13), we see that

$$\Omega(E) = \frac{d}{dE} M_E \;,$$

where $M_E = \int_{H(x)\leq E} dx_1 \cdots dx_{2N}$.

For any measurable function $f(y)$, with $y \in \Sigma_E$, the expression

$$\langle f \rangle = \frac{1}{\Omega(E)} \int_{\Sigma_E} f(y) \frac{d\sigma}{\|\nabla H\|} \tag{2.17}$$

is given the meaning of the average of f on Σ_E.

2.1.3 The Irreversible Approach to Equilibrium. The Zeroth Law of Thermodynamics

A long debate animated the initial development of statistical mechanics about an apparent conflict between the time-reversibility of microscopic dynamics and the irreversible approach to equilibrium at the macroscopic level. The problem is that of explaining why an isolated (conservative) mechanical system composed of a large number of atoms or molecules, independently of its initial state, always approaches thermal equilibrium, that is, its microscopic dynamics is such that all macroscopic observables relax toward steady values. This is referred to as the *zeroth law of thermodynamics*, and expresses the commonly observed typical behavior of macroscopic systems that irreversibly evolve to thermal equilibrium.

The mentioned conflict is drastically shown by an important consequence of the existence of an invariant measure for Hamiltonian flows φ_t^H, expressed by the *Poincaré recurrence theorem*. This states that for any measurable set $A \subset \Sigma_E$, with $\Sigma_E \subset \Gamma$, $\mu(\Sigma_E) < \infty$, and $\mu(A) > 0$, almost all the points $x \in A$ return to A infinitely many times.

We can see how this happens by considering an arbitrary measurable set A and choosing an arbitrary time unit $\tau \in \mathbb{R}^+$ so that the set $\varphi_{-n\tau}(A)$, that is, the preimage of A at the time $t = -n\tau$, represents the set of points that in n time "steps" will enter the set A. Then the set $T_n = \bigcup_{i=n}^{\infty} \varphi_{-i\tau}(A)$, $i, n \in \mathbb{Z}^+$, is the set of points of Σ_E that will enter A after n or more time steps; T_0 is the set of points that already belong to A or will enter it after an arbitrary number of steps. Thus $A \subset T_0$. Since by increasing the index n of T_n we have fewer sets in the union defining T_n, the following ordering holds: $T_n \subset T_{n-1} \subset \cdots \subset T_1 \subset T_0$. Now, the set of points starting from A that return to A after an arbitrarily long time $m\tau$ or more is $A \cap (\bigcap_{n=0}^{m} T_n)$. We observe that $\mu(T_n) = \mu(\varphi_\tau(T_n))$ because μ is an invariant measure for the flow. Then we observe that $\varphi_\tau(T_n) = T_{n-1}$, so that

$$\mu(T_n) = \mu(\varphi_\tau(T_n)) = \mu(T_{n-1}) < \infty \ .$$

Then we use the ordering by inclusion of the T_n's to write[2]

$$\mu(A \cap (\bigcap_{n=0}^{m} T_n)) = \mu(A \cap T_0) - \sum_{n=1}^{m} \mu(A \cap (T_{n-1} \backslash T_n))$$

and compute its limit for $m \to \infty$. Since $A \cap T_0 = A$, $T_n \subset T_{n-1}$ and $\mu(T_n) = \mu(T_{n-1})$, we have $\mu(T_{n-1} \backslash T_n) = 0$. In conclusion, the measure of the arbitrary set $A \subset \Sigma_E$ equals the measure of the set of points that will return to A, while the measure of the points of A that will never return to A is zero.

A direct consequence of the Poincaré recurrence theorem is known as the *Zermelo paradox*, which states that for almost all initial states, an arbitrary function of phase space will infinitely often assume its initial value within an arbitrarily small error, provided that the system remains in a finite region of phase space. Zermelo's paradox makes more precise the reason why microscopic reversibility of the dynamics seems incompatible with macroscopic irreversibility of thermalization processes. To give an example of its physical meaning, consider the situation in which, at some initial time, a piece of ice is put in a pot of hot water. The ice melts and the reversibility of Hamilton's equations of motion together with the Poincaré recurrence theorem tell us that sooner or later we should observe a piece of ice that pops up in the pot of hot water, an event which–we can safely bet–we will never observe and nobody has ever observed. A way out of this paradox is as follows. Suppose that $A \subset \Sigma_E$ is the set of representative points in phase space that correspond to the initial conditions of the water molecules, some belonging to the piece of ice and the others to the hot liquid water, the probability of observing the recurrence of the piece of ice is bounded from above by the estimate for less "extreme" recurrences by $\exp(-CN)$, where C depends on the total energy, density, and so on. Thus if $N \approx 10^{23}$, that is, on the order of Avogadro's number, the waiting time $\propto [\mu(A)]^{-1}$ can be huge, even with respect to the estimated age of the universe. In contrast, if N is very small, say a few tens, irreversibility becomes a somewhat fuzzy concept.

The symmetry of Hamilton's equations of motion with respect to time inversion, $t \to -t$, entails another apparently paradoxical situation known as *Loschmidt's paradox*. This states that for each process there exists a corresponding time-reversed process; thus for each thermalization process there

---

[2] Note that if $C \subset B \subset A$, then $\mu(A \cap B \cap C) = \mu(A \cap B) - \mu(A \cap (B \backslash C))$.

exists a corresponding process of spontaneous departure from equilibrium, in contradiction to the existence of irreversible processes.

The solution of Loschmidt's paradox is to invoke the chaotic instability of microscopic dynamics. In fact, a chaotic system loses memory of the initial condition in a finite time, on the order of the inverse of its largest Lyapunov exponent,[3] and this breaks the time inversion symmetry: given two configurations with the same positions and opposite velocities, the two corresponding phase space trajectories can have a reasonably good superposition only of a finite length; that is, after a finite time they will separate from each other. Earlier evidence of this fact was given by an old numerical experiment [27] on Loschmidt's paradox. Following the dynamics of a collection of 100 hard disks colliding in a box, which is a strongly chaotic system, it was observed that the Boltzmann entropy quickly relaxed to its equilibrium value. By inverting the velocities, after a certain number of collisions, the entropy was observed to retrace back its decay pattern. However, the larger the number of collisions before the velocity inversion, the smaller the maximum value attained by the entropy with respect to its initial value.

More generally, the Boltzmann–Gibbs picture of the approach to equilibrium is based on the *assumption* that the motion of the representative Γ-point on an energy hypersurface Σ_E has no preference for any of its regions and that sooner or later any accessible part of Σ_E will be reached. A reasonable consequence of this assumption of *a priori equiprobability* of microstates is that the Γ-point will spend in a region $A \subset \Sigma_E$ a time $\tau(A)$ proportional to the measure $\mu(A)$ of A. Then one assumes that the macroscopic state[4] is specified by giving the values of some phase functions $f_i(y)$, $y \in \Sigma_E$, such that (i) each macroscopic state corresponds to all the microscopic states belonging to a *region* of Σ_E; (ii) at large N, there is a set of values of the f_i that corresponds to a region of Σ_E that is overwhelmingly the largest in measure. This latter condition characterizes the thermal equilibrium of the system. In fact, if a system is *not* in thermal equilibrium, i.e., its initial condition belongs to a region of Σ_E of small measure, then it almost surely will go into the state of very large measure. If the system is *in* the equilibrium state, i.e., its initial condition belongs to the region of overwhelmingly large measure, then it will almost surely remain there, though fluctuations will possibly drive the system a bit out of equilibrium for short time intervals.

As a final remark, consider the phase space density function $\rho(p, q, t = 0)$– normalized to unity to give it the meaning of a probability–representing at some initial time $t = 0$ a system prepared in a nonequilibrium macroscopic state. Therefore, $\rho(p, q, t = 0)$ will be nonvanishing only in some restricted

---

[3] This time scale is shorter than and distinct from the relaxation time to equilibrium.

[4] The definition of macroscopic observables is somewhat arbitrary, so we assume that macroscopic observables are those quantities that are measured in the experiments that one wants to explain.

region R of Σ_E. The time evolution of the normalized $\rho(p, q, t)$ will be given by the Liouville equation (2.15), and thus the volume of R will be conserved but not its shape. We have to think of an extremely complicated filamentation of R, diffusing everywhere on Σ_E and finely sampling it, so that in the course of time the distribution $\rho(p, q, t)$ will become more and more uniform on Σ_E, at least in a suitably coarse-grained sense.

A more precise formulation of the idea that in the course of time the Γ-point fills the whole energy hypersurface, so that if $A \subset \Sigma_E$

$$\lim_{T \to \infty} \frac{\tau(A)}{T} = \frac{\mu(A)}{\mu(\Sigma_E)} \ ,$$

can be given by means of the ergodic theorem discussed in the following section.

2.1.4 Ergodicity

In order to quantitatively derive the macroscopic properties–that is, to compute the expected values of macroscopic observables–of a large collection of atoms or molecules, one should compute some suitable time average along the microscopic phase space trajectory. Let f be a macroscopic observable, mathematically representable as a function of the microscopic states $x = (p_1, \ldots, p_N, q_1, \ldots, q_N) \in \Gamma$. Ideally, for a given initial condition $x = (p_1^{(0)}, \ldots, p_N^{(0)}, q_1^{(0)}, \ldots, q_N^{(0)})$ and for a time interval T shorter than some characteristic observational time scale, a prediction of the outcome of a laboratory measurement would be obtained by computing

$$\overline{f}^T = \frac{1}{T} \int_0^T f[x(t)] \, dt \ , \tag{2.18}$$

where $x(t)$ is the solution of the equations of motion (2.4). Needless to say, until the advent of electronic computers there was no hope of working out such computations, not even for a few particles.[5] Therefore, statistical mechanics has been formulated just to get rid of microscopic dynamics. This is made possible by the *ergodic hypothesis*, the fundamental assumption. Roughly speaking, if the time needed to measure a macroscopic physical quantity is sufficiently long (in a sense to be made more precise), instead of computing averages as in (2.18), the prescription is to compute "static" phase space averages as in (2.17). Of course, the ergodic hypothesis has raised an *ergodic problem*, and thus it has stimulated an enormous effort to find rigorous

---

[5] As we shall see at the end of the present chapter, digital computers have made possible a major leap forward in the study of the microscopic-dynamical origin of macroscopic properties of physical systems. Besides a deepening of our theoretical understanding of the subject, this has also led to the development of molecular dynamics, a precious practical tool for ab initio computations.

arguments in its favor. It is conceptually intriguing and mathematically formidable problem that has led to the formulation of several ergodic theorems. However, though interesting from the mathematical viewpoint, the existing results do not allow one to decide whether a given Hamiltonian corresponds to an ergodic flow. Despite this status of the art, statistical mechanics has grown into a successful and powerful theory; in other words, the ergodic problem has never represented a practical problem for physically relevant Hamiltonians that for the purposes of statistical mechanics can be generically considered truly ergodic.[6] On the basis of some of the existing results in ergodic theory and in KAM theory (see Section 2.2.1), we can understand the reason why the ergodic hypothesis works so well.

First of all there is Birkhoff's ergodic theorem, a major achievement. Let us give, without proof, its two main steps.

Theorem 2.1. *Let* $\Sigma_E \subset \Gamma$ *be a finite-volume subset of phase space, invariant for the natural motions, that is,* $\varphi_t^H \Sigma_E = \Sigma_E$, *where* φ_t^H *represents the Hamiltonian flow,* $x(t) = \varphi_t^H x(0)$, *and let* $f(x)$ *be a measurable phase function defined on* Σ_E. *The limit*

$$\overline{f} = \lim_{T \to \infty} \frac{1}{T} \int_0^T f[x(t)]\, dt$$

exists for almost all $x \in \Sigma_E$, *that is, with the exception of at most a set of vanishing measure. Moreover,* $\overline{f}$ *is independent of the choice of the initial point of the given trajectory.*

This limit is the *time average* of f. Now consider the situation in which Σ_E cannot be decomposed into two invariant subsets $A \subset \Sigma_E$ and $B \subset \Sigma_E$, with $\varphi_t^H A = A$ and $\varphi_t^H B = B$. In this case Σ_E is called *metrically transitive*. Setting $\mu_E(\Sigma_E) = 1$, the measure of any invariant subset of a metrically transitive set is either 1 or 0. Having defined the phase average $\langle f \rangle$ as in (2.17), the following ergodic theorem holds.

Theorem 2.2. *If a measurable invariant space* $\Sigma_E \subset \Gamma$ *is metrically transitive, then for any measurable function* f, *we have*

$$\overline{f} = \langle f \rangle \,,$$

and also the converse is true; that is, metrical transitivity is equivalent to ergodicity.

A widespread opinion about Birkhoff's theorem was that it did not really advance our understanding of the ergodic problem, because it converts the ergodic problem into another equally difficult problem, that of proving the metric indecomposability of Σ_E. However, this is not entirely true,

---

[6] Obviously, integrable systems are nonergodic, but these are the exception, not the rule.

at least from the physicist's viewpoint. Let us see why. After the Poincaré–Fermi theorem (see Section 2.2.1), the generic situation for nonintegrable Hamiltonian systems is that there are no invariants of motion besides the energy, which makes the whole Σ_E topologically accessible. In principle one could argue that after the KAM theorem (see Section 2.2.1), and despite the Poincaré–Fermi theorem, invariant regions of positive measure can exist on Σ_E. However, as is mentioned in Section 2.2.1, this occurs for exceedingly tiny deviations from integrability that are unphysical.

Moreover, metrical transitivity is entailed by the *mixing* property of the dynamics. Mixing is defined as follows. For any pair of measurable sets $A \subset \Sigma_E$ and $B \subset \Sigma_E$, with the normalization $\mu_E(\Sigma_E) = 1$, if

$$\lim_{t \to \infty} \mu_E[(\varphi_t^H A) \cap B] = \mu_E(A)\mu_E(B) \qquad (2.19)$$

then the dynamics is mixing. This means that after a lapse of time equal to t, the fraction of A in B is $\mu_E[(\varphi_t^H A) \cap B]/\mu_E(B)$, and it converges to the measure of A as a whole. With a classical example, mix together two different liquids, say "red" and "blue," in a given proportion, and then stir the mixture. After a sufficiently long time, any portion of the mixture will contain the "red" and "blue" liquids in the same initially assigned proportion.

If we take $A \subset \Sigma_E$ and $B = A$, with A an invariant measurable set, we have

$$(\varphi_t^H A) \cap A = A \ ,$$

so that from (2.19), asymptotically $\mu_E(A) = [\mu_E(A)]^2$, that is, $\mu_E(A) = 0, 1$, the condition for metrical transitivity. We conclude that a mixing dynamics is also ergodic. Apart from special systems, such as geodesic flows on compact hyperbolic manifolds, for which a rigorous proof of mixing can be given, we know from early investigations of N.S. Krylov [2] that a requisite for the dynamics to be mixing is the exponential sensitivity to the variation of initial conditions. In modern terms, a positive largest Lyapunov exponent, that is, a chaotic dynamics (see Section 2.2), is required to have a mixing dynamics. Notice that a mixing dynamics, rather than a simply ergodic dynamics, is necessary to ensure a finite-time convergence of time averages to ensemble averages, as was noticed already by Krylov, and this is a physically fundamental condition, since any laboratory measurement is performed during a finite time. In conclusion, since generic nonintegrable Hamiltonian systems are chaotic, and since at large N, that is, $N \gg 3$, their phase space trajectories fill the whole constant-energy hypersurface on which the system has been prepared, from the *physicist's standpoint* this is sufficient to ensure the *bona fide* mixing property–and thus after Birkhoff's theorem also ergodicity–of the dynamics of generic many-particle systems. Nothing seems to seriously threaten the general validity of statistical mechanics. Nevertheless, as we shall see at the end of the present chapter, the mixing characteristic time can be a nontrivial function of the energy of the system, so that also very slow relaxations (even apparent freezing) of time to ensemble averages can be observed.

2.1.5 From Micro to Macro: The Link with Thermodynamics

We have seen that the time averages of a physical observable can be replaced by static ensemble averages under rather generic physical conditions. As far as thermodynamic macroscopic observables are concerned, making the link with a thermodynamic potential is enough, because all the other observables can be worked out using the standard macroscopic relations (such as Maxwell's relations). The usual approach is to define entropy in the microcanonical ensemble as the logarithm of phase space volume. Then one shows that the proposed definition has all the properties that entropy is expected to have. Then one proceeds by considering a large-N subsystem of a larger system (to be called thermostat), so that the total energy of the subsystem is no longer a constant, and by working out the ensemble density in the phase space of the subsystem, one is led to define a canonical ensemble where the basic mathematical object, the partition function, is directly related to the Helmholtz free energy. Then, letting also N fluctuate, one defines the grand-canonical ensemble whose grand-partition function directly gives the pressure. Of course all this is fine because a consistent theory is built. Loosely speaking, this is a "bottom up" approach to the problem of making a link between microscopic and macroscopic descriptions of a physical system.

Instead of sketching here this standard presentation of the bases of ensemble theory of statistical mechanics, we propose an equivalent conceptual construction, which we could define as a "top-down" approach, based on an old and almost forgotten work by L. Szilard [28], which, in the author's words, allows one "to construct statistical thermodynamics on the basis of the second law from purely thermodynamic considerations." In other words, the second law of thermodynamics is the founding physical principle of ensemble statistical mechanics. We propose this unusual approach because it has the advantage of making clearer the physics that is incorporated in the theory and, consequently, the limits of its validity.[7]

Let us now recall Szilard's work, which surprised many, including Einstein and von Laue, in which the author showed that the second law of thermodynamics provides information not only about the mean values of macroscopic observables but also about their fluctuation properties.

---

[7] To give an example, if we were to use statistical mechanics to cope with some basic *energy conversion* mechanism in living matter, we would come up against a paradox concerning the high efficiency of energy production in mammals and humans: according to the estimates provided by electrochemistry (see, for example, [29]), the efficency of energy-conversion processes at microscopic level is about 50%. On the other hand, the human body has a temperature T sligthly above 300 K with an excursion ΔT of a few degrees, whence—according to the second law of thermodynamics—the thermodynamic efficiency $\Delta T/T$ should be about 1%, much lower indeed. This is already enough to affirm that ensemble statistical mechanics is of no or only little use to describe these kinds of fundamental processes in living matter at the microscopic level, which therefore must stem from a strongly correlated and coherent dynamics.

In his work, Szilard, disregarding the kinetic substratum of macroscopic systems, assumes that any system, after sufficiently long contact with an infinite thermal reservoir (thermostat), is at equilibrium.

Depending on the moment at which the contact with the thermostat is interrupted, the energy content E of the system will be a random function of the temperature T of the reservoir, distributed with a probability $P(E,T)$. The same distribution describes the energy statistics of a series of an infinite number of replicas of the same system after sufficiently long contact with a reservoir at the same temperature.

Then it is shown that the distribution $P(E,T)$ must be a *stable* distribution, that is, the energy statistics cannot be changed, keeping the mean energy fixed, without some compensation; this energy statistics is also called *normal* statistics. The property of $P(E,T)$ of being stable is a major consequence of the second law of thermodynamics. In fact, if this were not the case, one could use infinitesimal thermal cycles exploiting the fluctuation phenomena to violate the second law by constructing a perpetuum mobile of the second kind.

Let us consider two ensembles $\mathcal{E}_1$ and $\mathcal{E}_2$ of many replicas of systems $\mathcal{S}_1$ and $\mathcal{S}_2$, with energy distributions $P_1(E)$ and $P_2(E)$ respectively. To each sample of $\mathcal{E}_1$ we associate a sample of $\mathcal{E}_2$ and bring them into thermal contact. The probability distribution $P(E)$ for the composite system is given by

$$P(E) = \int_0^E d\eta \; P_1(\eta) \, P_2(E - \eta) \; .$$

The stability requirement implies that the original distributions are retained after contact, provided that the two sets were initially given in their normal statistics. This is enough to work out the equilibrium distribution. Moreover, if the equilibrium of each system in $\mathcal{E}_1$ has been reached independently of each system in $\mathcal{E}_2$, no information about the energy content of a system in $\mathcal{E}_2$ can be obtained by knowledge of the energy content of its companion in $\mathcal{E}_1$. This initial statistical independence is assumed to remain true after the contact between $\mathcal{S}_1$ and $\mathcal{S}_2$ has occurred.

Now let us establish a thermal contact between some $\mathcal{S}_1$ and its associated $\mathcal{S}_2$. The energy of both systems will fluctuate. After a sufficiently long lapse of time the connection is broken. The energy of $\mathcal{S}_1$ belongs to the interval $(E_1, E_1 + dE_1)$ with a probability

$$W_{12}(E_1, E) \, dE_1 \; ,$$

and similarly, the energy of $\mathcal{S}_2$ belongs to the interval $(E_2, E_2 + dE_2)$ with a probability $W_{21}(E_2, E) \, dE_2$. After sufficiently long contact, the distribution W will depend only on the total energy E, independently of its initial repartition between the two systems.

The function W_{12}, independent of T, can be computed by means of $P_1(E_1, T)$ and $P_2(E_2, T)$ as follows. Under the hypothesis of stability of P, the distributions P_1 and P_2 do not change before and after the contact between $\mathcal{S}_1$ and $\mathcal{S}_2$. Moreover, under the hypothesis of statistical independence, after removing the contact between the members of a pair, the probability that the energy of $\mathcal{S}_1$ is within E_1 and $E_1 + dE_1$ and that the energy of $\mathcal{S}_2$ is at the same time within E_2 and $E_2 + dE_2$ is

$$P_1(E_1, T)P_2(E_2, T) \, dE_1 \, dE_2 \; .$$

The state of a pair of systems can be also characterized by E_1, the total energy $E = E_1 + E_2$, and the probability

$$p = P_1(E_1, T)P_2(E - E_1, T) \, dE \, dE_1 \; . \tag{2.20}$$

This probability can be also computed by choosing all the pairs of systems whose energy is between E and $E + dE$; and then by selecting those for which the energy of $\mathcal{S}_1$ belongs to the interval $(E_1, E_1 + dE_1)$, we will find these latter pairs with a probability $W_{12}(E_1, E)$. On the other hand, the probability of finding a pair of systems in the above-mentioned subset is given by $P(E, T)dE$. In order to satisfy both conditions, we will find suitable pairs with probability

$$p = W_{12}(E_1, E)P(E, T) \, dE \, dE_1 \; , \tag{2.21}$$

so that by equating (2.20) and (2.21), one obtains

$$W_{12}(E_1, E) = \frac{P_1(E_1, T)P_2(E - E_1, T)}{\int_0^E d\eta \, P_1(\eta) \, P_2(E - \eta)} \; . \tag{2.22}$$

Hence we can conclude that[8]

$$P_2(E_2, T) = C(T)g(E) \exp[\varphi(T)E] \; . \tag{2.23}$$

Similarly, using W_{21}, we can find the same expression for P_1. Since $g(E)$ remains indeterminate, the normalization condition $\int_0^\infty dE \, P(E, T) = 1$ yields the following expression for the function $C(T)$:

$$C(T) = \left[\int_0^\infty dE \, g(E) \, e^{\varphi(T)E} \right]^{-1} \; . \tag{2.24}$$

---

[8] By taking the logarithms of both members of (2.22) we get $\ln W_{12}(E_1, E) + \ln P(E, T) = \ln P_1(E_1, T) + \ln P_2(E - E_1, T)$; then we take the partial derivatives of both sides with respect to E_1 at constant T and evaluate the result at $E_1 = 0$, obtaining

$$\frac{\partial}{\partial E} \ln P_2 = - \left. \frac{\partial}{\partial E_1} \ln W_{12} \right|_{E_1=0} + \left. \frac{\partial}{\partial E_1} \ln P_1 \right|_{E_1=0} = \psi(E) + \varphi(T) \; .$$

Integrating on E we obtain $\ln P_2(E, T) = \Psi(E) + \varphi(T)E + c(T)$, whence, setting $C(T) = e^{c(T)}$ and $g(E) = e^{\Psi(E)}$, the result of (2.23) follows.

From the distribution function $P(E,T)$ we can compute

$$\langle E \rangle = \int_0^\infty dE\ E\ P(E,T) \tag{2.25}$$

and

$$\sigma_E^2 = \int_0^\infty dE\ (E - \langle E \rangle)^2\ P(E,T)\ . \tag{2.26}$$

If the fluctuations are small, from the relation

$$\int_0^\infty dE\ f(E)\ P(E,T) \approx f(\langle E \rangle) + \left(\frac{d^2 f}{dE^2}\right)_{E=\langle E \rangle} \frac{\sigma_E^2}{2}$$

one finds that

$$\frac{d\langle E \rangle}{dT} = \frac{d\varphi}{dT} \int_0^\infty dE\ E(E - \langle E \rangle)C(T)g(E)e^{\varphi(T)E}\ ,$$

whence the identity

$$\frac{d\varphi}{dT} = \frac{1}{\sigma_E^2}\frac{d\langle E \rangle}{dT}\ \left(\equiv \frac{1}{\chi(T)}\right)\ . \tag{2.27}$$

It is at this point that the second law of thermodynamics enters the game. In fact, it can be shown that $\chi(T)$ must be the same for all the systems, and thus $d\varphi/dT$ is a universal function of the temperature. By showing that during a "free" adiabatic expansion the statistics must remain normal if the second law is to hold, a closed differential equation for $\chi(T)$ is obtained, which gives

$$\varphi(T) = -\frac{1}{kT}\ , \tag{2.28}$$

where k is an integration constant; thermodynamics does not say anything about its numerical value; it demands only that it be the same for all systems. Thus, for obvious reasons, it has to be identified with the Boltzmann constant.

The final expression for the distribution $P(E,T)$ is thus

$$P(E,T) = \frac{g(E)\exp[-E/k_B T]}{\int_0^\infty dE\ g(E)\exp[-E/k_B T]}\ . \tag{2.29}$$

Nothing can be said about $g(E)$. It is here, however, that the link can be made between the macroscopic level of phenomenological thermodynamics and the microscopic level of atomic and molecular description of matter. In fact, $g(E)$ is a weight function that we can naturally interpret as a density of states of energy E. To make an ansatz for $g(E)$, we can proceed as follows: in order to link the macroscopic and microscopic levels, $g(E)$ must be built using the positions and momenta of all the particles, and at any given E these variables need to be constrained on a constant-energy surface. Thus we have to consider an integral on Σ_E, and the natural candidate for this purpose satisfying Liouville's theorem whose measure is bona fide ergodic for the microscopic dynamics and naturally accounts for the energy dependence of the relative variation of the microscopic states compatible with the macroscopic value E of the energy is

$$g(E) \equiv \Omega_N(E) = \int_{\Sigma_E} \frac{d\sigma}{\|\nabla H\|} \ . \tag{2.30}$$

The normalization factor of $P(E, T)$, that is, $Z_N(T) = C^{-1}(T)$, now reads

$$Z_N(T) = \int_0^\infty dE \ e^{-E/k_B T} \int_{\Sigma_E} \frac{d\sigma}{\|\nabla H\|} \equiv \int d^N p \ d^N q \ e^{-H(p,q)/k_B T} \ , \tag{2.31}$$

where we have assumed that $E = 0$ is the energy minimum and where a standard coarea formula[9] has been used to write the right-hand side. Thus we obtain the equilibrium density function $\rho_{eq}(p, q)$ in Γ-space

$$\rho_{eq}(p, q) = \frac{1}{Z_N(T)} e^{-H(p,q)/k_B T} \tag{2.32}$$

which is "imposed" by the second law of thermodynamics. Yet the link with thermodynamic observables is lacking. To this end, let us note that $Z_N(T)$ in (2.31) is formally the Laplace transform of $\Omega_N(E)$ in (2.30), so that we can invert the relation between $Z_N(T)$ and $\Omega_N(E)$ by Laplace antitransforming

$$
\begin{aligned}
\Omega_N(E) &= \frac{1}{2\pi i} \int_{\beta' - i\infty}^{\beta' + i\infty} Z_N(\beta) e^{\beta E} d\beta \\
&= \frac{1}{2\pi} \int_{-\infty}^{+\infty} Z_N(\beta' + i\beta'') \ e^{(\beta' + i\beta'')E} d\beta'' \\
&= \frac{1}{2\pi} \int_{-\infty}^{+\infty} \exp\{\log[Z_N(\beta' + i\beta'')] + (\beta' + i\beta'')E\} d\beta'' \ , \tag{2.33}
\end{aligned}
$$

where $\beta' > \lambda_0$, with λ_0 the convergence abscissa of the Laplace transform. Since E and $Z_N(\beta)$ are proportional to N, in the limit $N \to \infty$ the only contribution to the integral above comes from the neighborhood of the maximum of the integrand. This occurs for $\beta = \beta^\star$, with $\beta^\star$ determined by the equation

$$\frac{\partial \log Z_N(\beta^\star)}{\partial \beta^\star} = -E \ , \tag{2.34}$$

---

[9] This formula is used repeatedly throughout this book, in Chapters 7, 8, 9 the reader will find an appropriate reference.

and so $\beta^\star = \beta^\star(E)$. The exponent in the last integral above is then expanded in the neighborhood of $\beta^\star$. Thus

$$\Omega_N(E) = \frac{1}{2\pi} \int_{-\infty}^{\infty} \exp\left[\log Z_N(\beta^\star) + \beta^\star E - \frac{1}{2}\frac{\partial^2 \log Z_N(\beta^\star)}{\partial \beta^{\star 2}}\beta''^2 + \cdots\right] d\beta''$$

$$= \exp\left[\log Z_N(\beta^\star) + \beta^\star E\right]\left[2\pi\frac{\partial^2 \log Z_N(\beta^\star)}{\partial \beta^{\star 2}}\right]^{-\frac{1}{2}} \{1 + \cdots\} . \quad (2.35)$$

Using (2.31) we find that $\partial^2 \log Z_N(\beta^\star)/\partial \beta^{\star 2} = \langle H^2 \rangle - \langle H \rangle^2 > 0$, and since at very large N the exponent has an extremely steep maximum at $\beta'' = 0$, the most important contribution to the integral (2.35) comes from the neighborhood of $\beta'' \approx 0$. Then, performing the integration over β'', we obtain the second equation in (2.35). Taking the logarithm of both sides gives

$$\log \Omega_N(E) = \left[\log Z_N(\beta^\star) + \beta^\star E\right] + \left[\log\left(2\pi\frac{\partial^2 \log Z_N(\beta^\star)}{\partial \beta^{\star 2}}\right)^{-\frac{1}{2}} + \cdots\right].$$
$$(2.36)$$

The first term in the right-hand side of this equation is $O(N)$, whereas the second term in square brackets is $O(\log N)$, so that at large N it can be ignored in comparison with the first. Hence

$$\log \Omega_N(E) \approx \log Z_N(\beta^\star) + \beta^\star E . \quad (2.37)$$

In the thermodynamic limit $N \to \infty$, this approximate relation becomes exact in the form

$$\lim_{N\to\infty} \frac{1}{N} \log \Omega_N(E) = \lim_{N\to\infty} \left[\frac{1}{N} \log Z_N(\beta^\star) + \beta^\star \frac{E}{N}\right] . \quad (2.38)$$

Now, by putting $\beta^\star = 1/k_B T$, we immediately recognize that (2.34) and (2.37) coincide with the Legendre transformation

$$S(E) = \frac{1}{T}[E - F(T)] , \qquad \frac{\partial(F/T)}{\partial(1/T)} = E , \quad (2.39)$$

which relates the entropy $S(E)$ with the Helmholtz free energy $F(T)$, provided that we make the identifications[10]

$$S(E) = k_B \log \Omega(E) \quad (2.40)$$

and

$$F(T) = -k_B T \log Z(T) . \quad (2.41)$$

This makes the desired link between the microscopic description of a many-body system and thermodynamics.

---

[10] These are extensive definitions; the corresponding intensive ones, $S_N(E) = \frac{k_B}{N} \log \Omega(E)$ and $F_N(T) = -\frac{1}{N}k_B T \log Z(T)$, are in general more practical.

The two quantities $\Omega_N(E)$ and $Z_N(T)$ are the fundamental mathematical objects in the Gibbsian *statistical ensemble theory*.

More precisely, when both the total energy E and the number N of degrees of freedom of a system are given, the representative statistical ensemble is the so-called *microcanonical ensemble*, and $\Omega_N(E)$ is the state density of the system. By solving $E = E(S, V)$ from (2.40), the thermodynamic internal energy is $U(S, V) = E(S, V)$, and thus

$$T = \left(\frac{\partial U}{\partial S}\right)_V , \qquad P = -\left(\frac{\partial U}{\partial V}\right)_S ,$$

are the absolute temperature and pressure, respectively. Moreover, $F = U - TS$ gives the Helmholtz free energy, $G = U + PV - TS$ gives the Gibbs potential, and

$$C_V = \left(\frac{\partial U}{\partial T}\right)_V$$

is the heat capacity at constant volume. Equivalently, we have also

$$\frac{1}{T} = \frac{\partial S(E, V)}{\partial E} , \qquad C_V^{-1} = \frac{\partial T(E)}{\partial E} .$$

It is worth mentioning that in order to avoid Gibbs paradox,[11] the correct expression for the microcanonical volume is

$$\Omega_N(E) = \frac{1}{h^N N!} \int_{\Sigma_E} \frac{d\sigma}{\|\nabla H\|} ,$$

where h is a dimensional constant (an action) to be numerically set equal to Planck's constant, and $N!$ overcomes the mentioned paradox.

An equivalent definition [up to additive constant terms $\mathcal{O}(\log N)$] of entropy in the microcanonical ensemble is

$$S(E) = k_B \log \omega(E) , \tag{2.42}$$

where

$$\omega(E) = \int_{H(p,q) \le E} d^N p \, d^N q \equiv \int_0^E d\eta \int_{\Sigma_\eta} \frac{d\sigma}{\|\nabla H\|} . \tag{2.43}$$

When N is assigned but energy fluctuates because the system under consideration is put in contact with a heat bath at temperature T, the representative ensemble is the *canonical ensemble* and $Z_N(T)$ is the canonical partition function. Now the internal energy $U(V, T)$ is given by

$$U(V, T) \equiv \langle H \rangle = \frac{1}{Z_N(T)} \int d^N p \, d^N q \, H(p,q) \, e^{-H(p,q)/k_B T} , \tag{2.44}$$

---

[11] Gibbs paradox consists of the unreasonable prediction of an entropy increase when two identical systems made of the same ideal gas are allowed to mix.

and using Maxwell's relations, entropy and pressure read

$$S = -\left(\frac{\partial F}{\partial T}\right)_{\mathcal{V}} , \qquad P = -\left(\frac{\partial F}{\partial \mathcal{V}}\right)_T ;$$

the Gibbs potential is $G = F + P\mathcal{V}$; and the heat capacity at constant volume is found to be

$$C_{\mathcal{V}} = \frac{\langle H^2 \rangle - \langle H \rangle^2}{k_{\mathrm{B}} T^2} .$$

Also the canonical partition function has to be divided by $h^N N!$ to avoid paradoxes. This so-called correct Boltzmann counting factor has no practical consequence in the computation of averages, such as, for example, in (2.44).

In general, and at large N, canonical and microcanonical ensembles are *equivalent*. In fact, the distribution of energy in the canonical ensemble is a Gaussian distribution centered at $\langle H \rangle$ whose variance is

$$\sigma_E = \sqrt{k_{\mathrm{B}} T^2 C_{\mathcal{V}}} ,$$

and since $\langle H \rangle \propto N$ and $C_{\mathcal{V}} \propto N$, the ratio $\sigma_E / \langle H \rangle$ goes to 0 as $N \to \infty$. In other words, the larger N, the closer to a δ-function is the canonical energy distribution, so that the microstates that really contribute to the canonical statistical averages are those lying on the energy hypersurface $\Sigma_{\langle H \rangle}$.

However, equivalence no longer holds true in the case of long-range forces and when additivity of energy, entropy, and other thermodynamic functions is lost [30, 31]. In the case of long-range forces, for example, negative heat capacity is allowed in the microcanonical ensemble, whereas it is strictly forbidden in the canonical ensemble. Said differently, entropy is always a concave function of energy in the canonical ensemble, whereas it can be both convex and concave as a function of the energy in the microcanonical ensemble.

If we assume that only the *average number* of particles of the system under consideration is known, while the actual number N of degrees of freedom fluctuates, and the energy E fluctuates because of the contact with a heat bath, the representative statistical ensemble is the *grand-canonical ensemble*. The Γ-space of this ensemble is spanned by all the coordinates and momenta of all the realizations of a physical system obtained with $0, 1, 2, 3, \dots, N$ degrees of freedom. The distribution function $\rho(p, q, N)$ of the representative points in Γ-space, giving the density of points representing systems with N degrees of freedom with momenta and coordinates (p, q), can be obtained in the form[12]

$$\rho(p, q, N) = \frac{z^N}{h^N N!} e^{-\beta P\mathcal{V} - \beta H(p, q)} , \tag{2.45}$$

where $z = e^{\beta \mu}$ is called the *fugacity* and μ the chemical potential. Then one allows the volume of the system to become infinite, and hence N varies in the

---

[12] One has to consider a canonical ensemble of $\mathcal{N}$ particles, volume $\tilde{\mathcal{V}}$, and temperature T. Then one focuses on a small subsystem with N degrees of freedom in a volume $\mathcal{V}$ to compute $\rho(p, q, N)$ for the subsystem.

range $0 \leq N < \infty$. Again the internal energy of the system will be given by the ensemble average of the Hamiltonian $H(p, q)$, while all the other thermodynamic functions are derived from the grand-canonical partition function $\mathcal{Q}(z)$ defined as the power series

$$\mathcal{Q}(z, \mathcal{V}, T) = \sum_{N=0}^{\infty} z^N Z_N(\mathcal{V}, T) . \qquad (2.46)$$

After integration of both sides of (2.45) over (p, q) for a given N, and after summation on N from 0 to ∞, one obtains

$$\frac{P\mathcal{V}}{k_{\mathrm{B}}T} = \log \mathcal{Q}(z, \mathcal{V}, T) , \qquad (2.47)$$

which directly gives the pressure as a function of $z, \mathcal{V}, T$. Given a volume $\mathcal{V}$, the average number of particles contained in it is

$$N_{\mathrm{av}} = \frac{\sum_{N=0}^{\infty} N \, z^N Z_N(\mathcal{V}, T)}{\sum_{N=0}^{\infty} z^N Z_N(\mathcal{V}, T)} = z \frac{\partial}{\partial z} \log \mathcal{Q}(z, \mathcal{V}, T) . \qquad (2.48)$$

The probability that a system in a grand-canonical ensemble has N degrees of freedom in a volume $\mathcal{V}$ is proportional to

$$W(N) = z^N Z_N(\mathcal{V}, T) = e^{\beta \mu N - \beta F(N, \mathcal{V}, T)} , \qquad (2.49)$$

where $F(N, \mathcal{V}, T)$ is the Helmholtz free energy computed in the canonical ensemble.

Using

$$\left[\frac{\partial F(N, \mathcal{V}, T)}{\partial N} \right]_{N=N_{\mathrm{av}}} = \mu , \qquad \gamma := \left[\frac{\partial^2 F(N, \mathcal{V}, T)}{\partial N^2} \right]_{N=N_{\mathrm{av}}} ,$$

we can Taylor expand about N_{av} the argument of the exponential in the right-hand side of (2.49) to get

$$W(N) \approx W(N_{\mathrm{av}}) e^{\frac{1}{2} \beta \gamma (N - N_{\mathrm{av}})^2} ,$$

which is a Gaussian distribution centered at N_{av} with variance

$$\sigma_N = \sqrt{\frac{1}{\beta \gamma}} = \sqrt{\frac{k_{\mathrm{B}} T N}{v^2 (-\partial P / \partial v)}} , \qquad (2.50)$$

where $v = \mathcal{V}/N$ is the specific volume (inverse density), and where we have used $P(v) = -(1/N)(\partial F / \partial v)$ for the pressure of the system. It is an experimental fact that $(\partial P / \partial v) \leq 0$ holds always true. From (2.50) we have $\sigma_N / N \to \infty$ as $N \to \infty$. This implies that at large N almost all the systems in a grand-canonical ensemble have the same number N_{av} of particles; hence the grand-canonical ensemble is trivially equivalent[13] to the canonical ensemble and the following relation holds,

$$Q(z, \mathcal{V}, T) \approx z^{N_{av}} Z_{N_{av}}(\mathcal{V}, T),$$

so that

$$F(N, \mathcal{V}, T) = -k_{\mathrm{B}} T \log Q(z, \mathcal{V}, T) + k_{\mathrm{B}} T N \log z$$

and eliminating z with the aid of (2.48), we finally obtain the Helmholtz free energy, from which all the other thermodynamic functions can be derived.

2.1.6 Phase Transitions

Phase transitions involve abrupt major changes of the physical properties of macroscopic objects when a thermodynamic parameter is even slightly varied across a critical value. These are both fascinating and challenging physical phenomena. They are fascinating because they are rather mysterious; for example, in the familiar condensation phenomenon, how do the molecules of a vapor "know" that at a precise value of the specific volume they must condense and form two phases? and what makes the transition so sharp? Sometimes these phenomena are definitely spectacular, as are the cases of superfluidity and superconductivity. And they are challenging because to explain the observed phenomenology, they have raised, and to some extent still raise, hard theoretical problems.

Historically, the first known phase-transition phenomena were the transitions between the solid, liquid, and gaseous phases due to temperature and/or pressure variations. Melting, freezing, boiling, condensation, were very well known phenomena long before the development of modern physics. But it was only in 1873, with the formulation of the Van der Waals equation of state for real gases, that a successful phenomenological description of the liquid–gas transition was given. This equation of state reads

$$\left(P + \frac{a}{V^2}\right)(V - b) = RT,$$

and some isotherms described by this equation are shown in Figure 2.1. The isotherms display a kink, which disappears when the inflection point of $P = P(V)$ becomes horizontal. This defines the *critical* isotherm and correspondingly a critical temperature T_c, a critical pressure P_c, and a critical volume V_c. A simple computation gives $RT_c = 8a/27b$, $P_c = a/27b^2$ and

---

[13] Actually, the discussion about the equivalence between canonical and grand-canonical ensembles requires a thorough investigation of the cases for which $(\partial P/\partial v) = 0$, typically encountered in the coexistence region of a first-order phase transition.

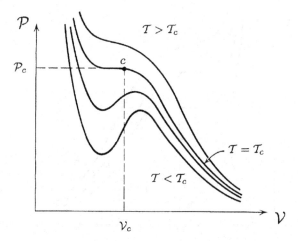

Fig. 2.1. Van der Waals isotherms obtained from 2.51. Below T_c a kink appears signaling the condensation transition (gas–liquid).

$V_c = 3b$. Then we define $\mathcal{P} = P/P_c$, $\mathcal{V} = V/V_c$, $\mathcal{T} = T/T_c$, so that the Van der Waals equation becomes

$$\left(\mathcal{P} + \frac{3}{\mathcal{V}^2}\right)\left(\mathcal{V} - \frac{1}{3}\right) = \frac{8}{3}\mathcal{T} , \tag{2.51}$$

which is an equation valid for any substance: one says that this equation expresses the *law of corresponding states*. We find here a property of *universality* that is well verified experimentally [32].

Experimentally, the transition between the gas phase and the liquid phase occurs at constant temperature and pressure. If we start from a state in which only liquid is present, the addition of heat results in the conversion of some liquid into gas until all the liquid is converted into gas. During the phase transition both P and T remain constant. Below T_c, the isotherms in the $P-V$ plane join two branches, describing the liquid and the gas phases respectively, through a horizontal line representing the gas–liquid mixture: when a certain amount of liquid is converted into gas, the volume expands.

This phenomenology is typical of a *first-order phase transition*. The kinks displayed by the isotherms of the Van der Waals equation are thus unphysical, and in fact they imply a negative compressibility. This can be attributed to the implicit assumption of homogeneity, excluding phase coexistence. Through the Maxwell construction this problem can be cured (see [33]). On the $P-T$ plane there are transition lines separating different phases. Let ℓ be the transition line that separates the stability domain of phase I from that of phase II. The Clapeyron equation

$$\left(\frac{dP}{dT}\right)_\ell = \frac{\Delta S}{\Delta v} \tag{2.52}$$

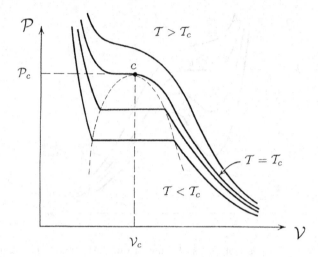

Fig. 2.2. Van der Waals isotherms obtained from 2.51 plus Maxwell's construction. The horizontal lines, below T_c, correspond to the gas–liquid coexistence.

then holds. It describes the vapor pressure in any first-order transition, where $\Delta S = S_{II} - S_I$ and $\Delta v = v_{II} - v_I$ are the discontinuities of entropy and specific volume, respectively, passing from phase I to phase II. The quantity $T\Delta S$ is the *latent heat of transition*, which is directly proportional to the discontinuity of the first-order derivative of the Gibbs thermodynamic potential $G = F + PV$, where F is the free energy, since $(\partial G/\partial T)_P = -S$.

Other transition phenomena, experimentally observed and theoretically studied since the beginning of the twentieth century, were not considered to have any relation to phase transitions because of the absence of a latent heat. For example, to mention a few of them, the phenomenon known today, of polymorphism of crystals, induced by suitable variations of temperature and pressure, was first observed in sulfur in the eighteenth century. Also, the ferromagnetic–paramagnetic transition, observed as a loss of spontaneous magnetization of ferromagnetic materials heated at high temperature, was already known to Faraday, who observed this phenomenon before the systematic investigations by P. Curie. Langevin, who first attempted a theoretical explanation of the phenomenon, considered ferromagnetism and paramagnetism as two distinct states somehow analogous to the gaseous and liquid states. However, the absence of a discontinuity associated with the transition between ferromagnetism and paramagnetism prevented him from establishing a deeper connection. The same occurred with the investigations of ordered structures in metallic alloys. The description of how the degree of order changes with temperature was proposed by Bragg and Williams and first applied to β-brass. This is an equiatomic alloy of copper and zinc whose atoms–at high temperature–are randomly distributed at the sites of a cubic-centered lattice, whereas at low temperature copper and zinc tend to preferably occupy the sites of two simple cubic sublattices. At $T = 0$, the ordering is perfect. At increasing temperature the degree of order lowers until it vanishes for a critical value of the temperature. This critical temperature is the analogue of the Curie temperature for which the spontaneous magnetization of a ferromagnet vanishes. Also in this case, the transition between order and disorder is continuous, in contrast to an ordinary change of state. Thus Bragg and Williams defined this as a "continuous transition."

The study of the physical properties of liquid helium, which was produced by H. Kammerling Onnes in 1908, brought about an advancement of the problem of classifying those transitional phenomena that are not ordinary changes of state.

Several experiments concurred to indicate that liquid helium could exist in two different forms: a sort of polymorphism, occurring at the transition temperature of 2.19 K. In particular, at this temperature discontinuities were observed of the thermal dilatation coefficient, of the dielectric constant, and of the constant-volume specific heat. The two forms of liquid helium were denoted by He I and He II. Above 2.19 K, He I was stable, whereas He II was stable below this temperature. This was called the λ-transition because of the shape of the graph of specific heat as a function of T.

The discontinuities of the above-mentioned physical parameters suggested that the transition between He I and He II perhaps should have been considered a change of state. However, this was in contrast to the experimental datum of vanishing latent heat, which made the Clapeyron equation inapplicable. A tiny thermodynamic cycle, experimentally implemented by W.H. Keesom in 1933 across the transition line in the $P - T$ plane (made of two isotherms and two isobars), led to the following replacement of the Clapeyron equation:

$$\left(\frac{dP}{dT}\right)_\lambda = \frac{\Delta C_P}{Tv\Delta\alpha_P} , \qquad (2.53)$$

where the subscript λ denotes the transition line, ΔC_P is the jump of constant-pressure specific heat passing from He I to He II, $\Delta\alpha_P$ is the jump of the thermal dilatation coefficient at constant pressure, and v is the specific volume.

The experimental study of the transition between normal (ohmic) conduction and the superconductive state in metals, cooled at very low temperatures, also revealed a discontinuity in the specific heat at the transition point. Furthermore, the Weiss "molecular field" theory of the paramagnetic–ferromagnetic transition yielded a similar discontinuous pattern of the specific heat at the transition point.

These results, and the comparison between (2.52) and (2.53), led P. Ehrenfest to attempt a generalization of the concept of change of state. He proposed to define *first-order phase transitions* as those transition phenomena that are characterized by discontinuities of physical observables related to the *first derivative* of the Gibbs potential, as is the case of the entropy, and to define *second-order phase transitions* as those transition phenomena for which the discontinuities of physical observables are related to the *second derivative* of the Gibbs potential, as is the case of specific heat.

Ehrenfest's classification of phase transitions had the great merit of putting together different kinds of transition phenomena within a common framework. However, it soon turned out to be not sufficiently accurate. In fact, take, for example, the ferromagnetic transition associated with the Ising model.[14] While the Weiss mean-field theory predicts for the two-dimensional model a finite discontinuity of the specific heat at the transition point, Onsager's exact solution for the same model predicts a logarithmic divergence to infinity of specific heat at the transition point. A way out of this difficulty is provided by an important contribution given to the theory of phase transitions by L.D. Landau in 1937.

Spontaneous Symmetry-breaking and Phase Transitions

Landau noted that phase transitions with vanishing latent heat were accompanied by a symmetry change of the physical states of a system. For example, the disordered phase of β-brass has a cubic-centered structure, while in the ordered phase it is simply cubic, that is, less symmetric. In fact, the two simply cubic sublattices, equivalent in the disordered phase, are no longer equivalent in the ordered phase where the permutation invariance of the two sublattices is violated. A symmetry has been lost in the transition. Actually, many (though not all) phase transitions are such that one phase is more symmetric than the other, and the transition from the more-symmetric phase to the less-symmetric one is called a spontaneous *symmetry-breaking* phenomenon. The maximal set of the possible symmetries that a physical system can have is represented by all the symmetries of the Hamiltonian describing it. In general, at low temperatures the accessible states of a system can lack some of the symmetries of the Hamiltonian, so that the corresponding phase is the less symmetric one, whereas at higher temperatures the thermal fluctuations allow the access to a wider range of energy states having more, and eventually all, the symmetries of the Hamiltonian.[15]

In the broken-symmetry phase, an extra variable is required to characterize the physical states belonging to it. Such a variable, of extensive nature, is called an *order parameter*. The order parameter vanishes in the more-symmetric phase and is different from zero in the less-symmetric phase.

---

[14] The Ising model is described by the Hamiltonian $H = -J \sum_{\langle i,j \rangle} s_i s_j$, where $\langle i,j \rangle$ stands for nearest-neighbors on a d-dimensional lattice and $s_i = \pm 1$.

[15] In view of the topological theory of phase transitions that will be presented in the second part of the book, let us make the following remark. Any state of a system undergoing a symmetry-breaking phase transition either possesses or does not possess the relevant symmetry. Thus, it is not possible to analytically deform

In view of another classification (beyond Ehrenfest's) of phase transitions, which will be summarized in the next section, it is important to remark that there are also systems for which one can define an order parameter in the presence of a latent heat, as is the case of the para-ferroelectric transition in some materials (such as $BaTiO_3$) for which the electric polarization is the order parameter.

Let us briefly sketch some basic facts of the Landau theory. Consider a thermodynamic system whose free energy F is a function of temperature T, pressure P, and some other extensive macroscopic parameters m_i, that is, $F = F(P, T, m_i)$. The parameters m_i are defined to be all vanishing in the most symmetric phase. Thus, as a function of the m_i, $F(P, T, m_i)$ is invariant with respect to the transformations of the symmetry group G_0 of the most-symmetric phase of the system. A state of the system is represented by a vector $|m\rangle = |m_1, \ldots, m_n\rangle$ belonging to a vector space $\mathbb{E}$ of all the states of the system. In $\mathbb{E}$ we can construct a linear representation of G_0 that associates with any $g \in G_0$ a matrix $M(g)$ of rank n. In general, the representation $M(g)$ is reducible. By decomposing $\mathbb{E}$ into invariant irreducible subspaces $\mathbb{E}_1, \mathbb{E}_2, \ldots, \mathbb{E}_k$, of basis vectors $|e_i^{(n)}\rangle$, with $n = 1, 2, \ldots, n_i$ and $n_i = \dim \mathbb{E}_i$, the state variables m_i are transformed into new variables $\eta_i^{(n)} = \langle e_i^{(n)}|m\rangle$. In terms of irreducible representations $D_i(g)$ induced by $M(g)$ in $\mathbb{E}_i$ one has

$$M(g) = D_1(g) \oplus D_2(g) \oplus \cdots \oplus D_k(g) .$$

If at least one of the $\eta_i^{(n)}$ is non zero, then the system no longer has the symmetry G_0. This symmetry has been broken, and the new symmetry group is G_i, associated with the representation $D_i(g)$ in $\mathbb{E}_i$. The variables $\eta_i^{(n)}$ are order parameters, and now the free energy is $F = F(P, T, \eta_i^{(n)})$. The actual values of the $\eta_i^{(n)}$ as functions of P and T are variationally determined by imposing the condition of thermodynamic equilibrium, that is, by minimizing F.

In order to make a continuous transition from the symmetry G_0 to another symmetry G_i, the $\eta_i^{(n)}$ must vanish at the transition point, assuming in its vicinity infinitesimal values. Thus $F(P, T, \eta_i^{(n)})$ can be expanded in power series of the $\eta_i^{(n)}$ in the neighborhood of the transition point. Moreover, since the free energy must be invariant with respect to any coordinate transformation, and in particular with respect to the transformations of the group G_0, the power expansion of F must contain only invariant combinations of the $\eta_i^{(n)}$. Now, no irreducible representation of a group has a linear invariant,

---

a state belonging to a phase into a state belonging to another phase possessing a different symmetry. This is suggestive of a similar impossibility of analytically deforming one into the other the level sets of the Hamiltonian of a system [say $\Sigma_E = H^{-1}(E)$ and $\Sigma'_E = H^{-1}(E')$] that correspond to different phases. In other words, it seems natural to surmise that these level sets associated with different phases are nondiffeomorphic, that is, topologically inequivalent.

while it has a quadratic invariant that is a positive quadratic form of the $\eta_i^{(n)}$. As a consequence, the lowest-order expansion of F reads

$$F(P,T,\eta_i^{(n)}) = F_0(P,T,0) + \sum_n A^{(n)}(P,T) \sum_i [\eta_i^{(n)}]^2 + O([\eta_i^{(n)}]^3) \,, \quad (2.54)$$

where the $A^{(n)}(P,T)$ are positive functions of P and T for all n above T_c. A broken symmetry phase, that is, the occurrence of nonzero $\eta_i^{(n)}$ below T_c, requires that at least one of the coefficients $A^{(n)}(P,T)$ change its sign at $T = T_c$. Since only one of the representations of $G_i \subset G_0$ corresponds to thermodynamic instability of a disordered phase, in the expansion (2.54) we can keep only the term that changes its sign at T_c. The quantities $\eta_i^{(n)}$ that are nonzero below T_c, that is, in the ordered state, are called the *order parameters*. Henceforth we drop the suffix n with the assumption that the relevant representation is the one that corresponds to the thermodynamic instability at T_c.[16] Moreover, we generically denote by $\mathcal{G}$ the subgroup G_i of G_0. With the notation $\eta = \sum_i \eta_i^2$, and considering higher orders in the expansion (2.54), we have

$$F(P,T,\eta) = F_0(P,T,0) + A(P,T)\eta^2 + C(P,T)\eta^3 + B(P,T)\eta^4 + O(\eta^5) \,. \quad (2.55)$$

Though the transition point is a singular point for F, it is, however, assumed that this does not conflict with the regular expansion given above. In the symmetric phase the minimum of F, given by $(\partial F/\partial \eta) = 0$, has to correspond to $\eta = 0$, while in the broken-symmetry phase the stable state minimizing F has to correspond to $\eta \neq 0$. This entails that at the critical temperature T_c of the transition point we must have

$$A_c(P,T_c) = 0 \,, \quad C_c(P,T_c) = 0 \,, \quad B_c(P,T_c) > 0 \,.$$

When the index[17] of $\mathcal{G} \subset G_0$ is 2, no cubic invariant of $\mathcal{G}$ is allowed; thus $C(P,T) = 0$ identically, and we are left with the expansion

$$F(P,T,\eta) = F_0(P,T,0) + A(P,T)\eta^2 + B(P,T)\eta^4$$

with $B > 0$.

---

[16] Note that a simultaneous change of sign of more than one single coefficient $A^{(n)}$ can occur only at isolated points in the (P,T) plane. Moreover, an order parameter is not only a quantity that vanishes in a phase and is different from zero in the broken-symmetry phase; it is also necessary that its fluctuations grow with N at the transition point. This leads to the concept of *dominant representation* within the set of irreducible representations of the symmetry group. Fluctuations are not accounted for in the Landau theory.

[17] The number of cosets of a finite group G_0, generated by a subgroup $\mathcal{G}$, is called the index of $\mathcal{G}$ in G_0: if $G_0 = \mathcal{G} \cup a\mathcal{G}$, with $a \notin \mathcal{G}$, then $\mathcal{G}$ is of index 2 in G_0; if $G_0 = \mathcal{G} \cup a\mathcal{G} \cup b\mathcal{G}$, with $a \notin \mathcal{G}$ and $b \notin \mathcal{G} \cup a\mathcal{G}$, then $\mathcal{G}$ is of index 3 in G_0.

We have assumed that, near T_c, A is regular, so that it can be expanded as $A(P,T) = A_c(P,T_c) + a(P)(T-T_c) = a(P)(T-T_c)$, where T_c is the transition temperature, whence, setting $B(P) = B(P,T_c)$, it follows that

$$F(P,T,\eta) = F_0(P,T,0) + a(P)(T - T_c)\eta^2 + B(P)\eta^4 . \qquad (2.56)$$

In order to find the temperature-dependence $\overline{\eta}(T)$ of the order parameter, we have to minimize the free energy, i.e., to compute $(\partial F/\partial \eta) = 0$, which gives

$$\overline{\eta}^2 = -\frac{A}{2B} = \frac{a}{2B}(T_c - T) \qquad (2.57)$$

and the root $\eta = 0$, which corresponds to the symmetric phase (in this case, with $A < 0$, F has a maximum and not a minimum at $\eta = 0$). Neglecting higher orders of η, we compute

$$S = -\frac{\partial F}{\partial T} = S_0 - \frac{\partial A}{\partial T} = S_0 + \frac{a^2}{B}(T - T_c) \qquad (2.58)$$

while in the symmetric phase, being $\eta = 0$, it is $S = S_0$. Thus it is evident that the entropy is continuous at $T = T_c$. Then we can compute the behavior of the specific heat $C_P = T(\partial S/\partial T)_P$ at the transition point. In the broken-symmetry phase it is

$$C_P = (C_P)_0 + \frac{a^2}{B}T \bigg|_{T=T_c} = (C_P)_0 + \frac{a^2}{B}T_c , \qquad (2.59)$$

whereas in the symmetric phase, $S = S_0$, so that $C_P = (C_P)_0$. In other words, at the critical point of a second-order phase transition, the specific heat makes a finite jump, since $B > 0$, $C_P > (C_P)_0$, that is, the specific heat is larger in the broken-symmetry phase than in the symmetric phase. This jump of C_P brings about discontinuous jumps of many thermodynamic quantities (such as the compressibility and the thermal dilatation coefficient).

A graphical presentation of the families of curves $F(\eta)$, plotted at different temperatures, whose functional forms are those given by (2.56) and by (2.55), helps in understanding the difference between a first- and a second-order phase transition. In Figure 2.3 the patterns $F(\eta)$ are reported for $T > T_c$, $T = T_c$ and $T < T_c$. The stability conditions for the thermodynamic equilibrium require that near T_c one have $(\partial^2 F/\partial \eta^2)|_{\eta=0} > 0$ for $T > T_c$, and $(\partial^2 F/\partial \eta^2)|_{\eta=0} < 0$ for $T < T_c$. Hence at the transition point $(\partial^2 F/\partial \eta^2)|_{\eta=0,T=T_c} = 0$. The function $\overline{\eta}(T)$ is continuous at T_c, as prescribed by (2.57).

In Figure 2.4, the patterns of $F(\eta)$, containing also a term proportional to η^3, are reported at different temperatures. In this case, the function $\overline{\eta}(T)$ is discontinuous at T_c, as can be easily grasped by observing that below T_c, $F(\eta)$ has an absolute minimum for $\eta \neq 0$ that at T_c rises to the same level of the relative minimum of $F(\eta)$ at $\eta = 0$, so there is coexistence of the two phases. For $T > T_c$ there is a reversal between these minima, so that the absolute minimum of $F(\eta)$ occurs at $\eta = 0$, hence the jump of the order parameter.

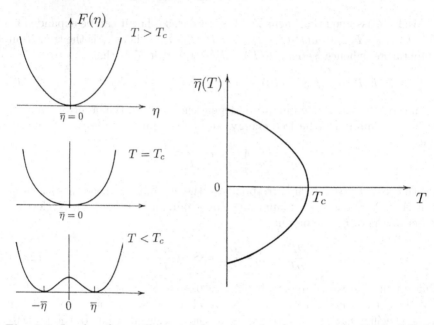

Fig. 2.3. Landau theory illustrated for a second-order transition. To the left the shapes are shown of the Helmholtz free energy $F(\eta)$ as a function of the order parameter η, above, at, and below the critical temperature T_c. To the right the bifurcation diagram of the order parameter values $\overline{\eta}$ corresponding to the minima of $F(\eta)$ is reported as a function of the temperature.

Critical Exponents and Universality

Landau's theory of phase transitions was an important achievement. However, being a mean-field theory that neglects thermodynamic fluctuations,[18] it is inaccurate in the vicinity of phase transition points. As a matter of fact, fluctuations play a major role in second-order phase transitions. To clarify this, let us return to the Van der Waals isotherms. We have already seen that a critical temperature exists (which depends on the fluid), for which the critical isotherm has a horizontal inflection point where the V-interval of phase coexistence shrinks to zero, this means that here the latent heat vanishes and the transition from vapor to liquid is continuous. The vanishing of the derivative $(\partial P/\partial V)$ at the critical point entails the divergence of the isothermal compressibility

$$\kappa_T = \frac{1}{V(-\partial P/\partial V)_T} \, ,$$

---

[18] The applicability of the Landau theory is subject to the satisfaction of the Ginzburg criterion. This criterion is satisfied when the fluctuations are sufficiently weak [34].

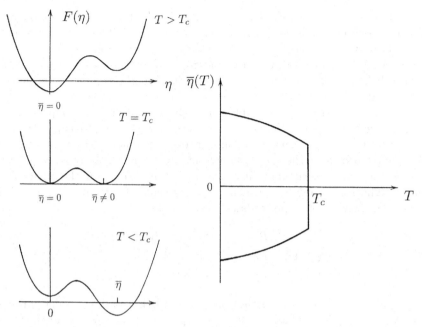

Fig. 2.4. Landau theory illustrated for a first-order transition. To the left the shapes are shown of the Helmholtz free energy $F(\eta)$ as a function of the order parameter η, above, at, and below the critical temperature T_c. To the right the bifurcation diagram of the order parameter values $\overline{\eta}$ corresponding to the minima of $F(\eta)$ is reported as a function of the temperature. The asymmetric shapes of $F(\eta)$ are such that at T_c a jump of $\overline{\eta}(T)$ is produced.

and since κ_T is related to the density fluctuations through the so-called fluctuation-response theorem, that is, $\langle N^2 \rangle - \langle N \rangle^2 = \langle N \rangle^2 kT\kappa_T/V$, the divergence of κ_T implies the divergence of the amplitude of the density fluctuations. By the same theorem, the compressibility is also related to the correlation function $G(\mathbf{r})$ by

$$\int d\mathbf{r}\ G(\mathbf{r}) = kT\kappa_T\ ,$$

and for a fluid in d dimensions, $G(\mathbf{r})$ is of the form

$$G(\mathbf{r}) \propto \frac{e^{-|\mathbf{r}|/\xi}}{|\mathbf{r}|^{(d-1)/2}\xi^{(d-3)/2}}\ .$$

The quantity ξ is the *correlation length* and is a measure of the effective range of interaction of the particles in the system. Now, the important consequence of the relation between the integral of the correlation function and the isothermal compressibility is that the divergence of κ_T implies also the divergence of the correlation length ξ at the critical point. Therefore, since the number of degrees of freedom effectively interacting with each other is

$\propto \xi^d$, this number diverges at the critical point, so that there is no longer a characteristic length scale in the system except for ξ. The assumption that ξ is the only characteristic scale left when T is very close to T_c—so that the critical behavior of all the thermodynamic functions is derived from the critical behavior of ξ—is known as the *scaling hypothesis* and has important theoretical consequences.

When the microscopic variables of a system are linearly coupled to an external field B, we can define the susceptibility $\chi = (\partial \eta / \partial B)|_{B=0}$, which, again through the fluctuation-response theorem, is functionally related to the Fourier transform of the correlation function. Thus, also the susceptibility diverges at the critical point because of the divergence of the correlation length. The remarkable experimental fact is that all the diverging physical quantities, near T_c obey power laws, for example,

$$C \sim |T - T_c|^{-\alpha} \qquad \text{specific heat,}$$
$$\chi \sim |T - T_c|^{-\gamma} \qquad \text{susceptibility,}$$
$$\xi \sim |T - T_c|^{-\nu} \qquad \text{correlation length,}$$

and a power law also describes the nondivergent behavior of the order parameter near T_c: $\eta \sim |T_c - T|^{\beta}$ for $T < T_c$. The exponents $\alpha, \beta, \gamma, \nu$ are called *critical exponents*. The great relevance of critical exponents is due to the observation that phase transitions occurring in different systems happen to have the same critical exponents, a phenomenon that is known as *universality* and that allows a unified treatment of critical phenomena in a large number of physical systems. Actually, experiments show that the values of these exponents do not depend on the details of microscopic interactions, nor on the kind of lattice (for on-lattice models), nor on the value of T_c, nor on other physical details of the system. The only parameters that seem to affect the critical exponents are the spacial dimension of the system, the dimensionality of the order parameter, and the kind of symmetry (discrete or continuous) of the Hamiltonian. All the systems possessing the same set of values of the critical exponents are said to belong to the same *universality class*. Moreover, experience has largely confirmed that the critical exponents, even though they belong to different universality classes, and therefore they take different numerical values, satisfy universal relations called *scaling laws*. An example that involves the above-mentioned critical exponents is $\alpha + 2\beta + \gamma = 2$.

Finally, we just mention that a powerful theoretical method to systematically reduce the number of degrees of freedom, by integrating over short-wavelength fluctuations, was devised by Kadanoff, Wilson, and others, who have developed the *renormalization group* theory of critical phenomena.

Classification of Phase Transitions

To summarize, beyond the old Ehrenfest classification scheme of phase transitions, a more modern one is the following:

- *First-order phase transitions.* These transitions involve a latent heat and are associated with phase-coexistence.[19] Belonging to this class we find (i) Phase transitions without an order parameter for which the symmetry groups of the two phases are such that none of them is strictly included in the other; (ii) phase transitions for which the symmetry group of the less-symmetric phase is a subgroup of index 3 of the symmetry group of the more-symmetric phase, entailing that the order parameter is *discontinuous* at the transition point.

- *Second-order phase transitions.* These are continuous transitions with no associated latent heat. In this case the symmetry group of the less-symmetric phase is a subgroup of index 2 of the symmetry group of the more-symmetric phase, whence the order parameter is *continuous* at the transition point. Continuous transitions have many interesting properties in the vicinity of the transition point. These properties and the associated phenomena are called *critical phenomena* (see the preceding section).

- *Infinite-order phase transitions.* These are continuous transitions that do not break any symmetry. The Kosterlitz–Thouless transition occurring in the two-dimensional classical Heisenberg XY model provides a paradigmatic example.

There is a great variety of phase transition phenomena observed in physical systems and predicted by theoretical models. They extend from quantum field theory of elementary particles to cosmology, with the breaking of symmetries in the laws of physics during the early life of the universe. Many, or even most, of these transition phenomena satisfactorily fit in the above given classification scheme and are well understood with the aid of the existing theories. Nevertheless, there are several problems in traditional condensed-matter and soft-matter systems that still challenge our understanding of those transitional phenomena that occur, for example, in glasses, spin glasses, and polymers, not to speak of living-matter systems, where a wide variety of equilibrium and nonequilibrium transitional phenomena take place at a mesoscopic or nanoscopic level; for example, just to mention a few of them, the hard problem of protein folding; the structural changes in biomembranes; the formation of cytoskeletal microtubules; the formation of electrets of the so-called vicinal water around proteins, around microtubules, and other filamentary structures; the formation of arrays of pseudo-Josephson junctions of these electrets; the emergence of qualitatively new properties in networks (of neurons, of biochemical reactions, of genetic regulation, and so on) modeled as phase transitions, for instance, of random graphs.

Many of the above-mentioned systems undergoing phase transitions that are not yet well understood are often put in the family of *complex systems,*

---

[19] For example, in a pot of boiling water, vapor bubbles coexist with liquid water; in a pot of freezing water, pieces of ice can coexist with liquid water. In both cases, adding or draining heat does not result in a change of temperature but in a modification of the relative amounts of water in the coexisting phases.

which seem naturally to call for the topological approach to phase transitions that will be discussed in the second part of this book.

The Yang–Lee Theory of Phase Transitions

We have seen in the preceding sections that a common property of all phase transitions is that the thermodynamic variables, as functions of the state variables, seem to lose their analyticity properties at the transition points. These nonanalytic points are the boundaries between different phases of the system.

However, the microcanonical volume, the canonical partition function, and the grand-canonical partition function are analytic, and thus the thermodynamic functions derived from them are necessarily analytic as well (with the possible exception of the case of vanishing temperature, i.e., the point $T = 0$).

As first suggested by Kramers, the *thermodynamic limit*, where the number N of degrees of freedom and the volume of the sample both tend to infinity while the density is kept fixed, must be invoked to explain the existence of true singularities of the thermodynamic variables.[20] A first confirmation of this hypothesis was given about ten years later by the Onsager's exact solution of the 2D Ising model. From a mathematical point of view, then, no singularities or no phase transitions can exist as long as N is finite. Real systems are always finite,[21] so that the phase transition point is not a singular point in a *strict* mathematical sense. The unusual behavior of thermodynamic variables at phase transitions in real systems can, however, be understood as a consequence of the true singularity that would be present in the thermodynamic limit [37].

Kramers's suggestion has been put on a rigorous basis by Yang and Lee [36], whose theory relies on the well-known mathematical fact that a sequence of smooth functions can have a nonsmooth limit, provided that the convergence is *not uniform*.

The Yang–Lee theory applies to generic systems of N particles in a volume $\mathcal{V}$ interacting through a Van Hove potential $\phi(r)$:

$$\phi(r) = \begin{cases} +\infty & \text{if } r \leq r_0 \,, \\ < 0 \text{ and } > -\varepsilon & \text{if } r_0 < r < r_1 \,, \\ 0 & r \geq r_1 \,, \end{cases} \qquad \begin{array}{l} r = \|\mathbf{r}_i - \mathbf{r}_j\| \,, \\ \forall \, i, j \in \{1, 2, \ldots, N\} \,. \end{array} \qquad (2.60)$$

This is a generic short-range attractive potential, bounded below, with a repulsive hard core. This kind of potential prevents the particles from getting closer

---

[20] According to G.E. Uhlenbeck [35], the use of the thermodynamic limit as an explanation of the singularities of the partition function was suggested for the first time by Kramers in the 1938 Leiden conference on statistical mechanics.

[21] Common wisdom says that the Avogadro number is so large that it is "practically" close to infinity. However, apart from phase transitions occurring in small systems with N of a few hundreds or a few thousands, there are cases in which the relevant quantities approach their thermodynamic limit with a logarithmic N-dependence, and the logarithm of the Avogadro number is a small number.

to one another than a minimum distance r_0. Therefore, in a given volume $\mathcal{V}$, the number of particles that can be closely packed into it has a maximum $M(\mathcal{V})$.

For $N > M(\mathcal{V})$, the partition function $Z_N(\mathcal{V})$ vanishes because of the divergence of the interaction energy. Thus, the grand-partition function $\mathcal{Q}(z, \mathcal{V})$ is a polynomial of degree $M(\mathcal{V})$ of the fugacity z:

$$\mathcal{Q}(z,\mathcal{V},T) = 1 + Z_1(\mathcal{V},T)z + Z_2(\mathcal{V},T)z^2 + \cdots + Z_M(\mathcal{V},T)z^M$$

$$= \prod_{k=1}^{M(\mathcal{V})} \left(1 - \frac{z}{\zeta_k}\right) , \tag{2.61}$$

where $Z_N = \int \exp(-\beta\phi(r))d^N r$ is the canonical partition function with N degrees of freedom, and the ζ_k are the $M(\mathcal{V})$ roots of the algebraic equation

$$\sum_{N=0}^{M(\mathcal{V})} Z_N(\mathcal{V},T)\, z^N = 0 . \tag{2.62}$$

Since all the Z_N are positive, the roots of this equation are complex, and the equation of state, expressed in parametric form as

$$\frac{P}{kT} = \frac{1}{\mathcal{V}} \log \mathcal{Q}(z,\mathcal{V},T) ,$$

$$\frac{1}{v} = \frac{1}{\mathcal{V}} z \frac{\partial}{\partial z} \log \mathcal{Q}(z,\mathcal{V},T) , \tag{2.63}$$

entails that for any finite volume $\mathcal{V}$, the pressure $P = P(z)$ and the specific volume $v = v(z)$ are analytic in a region of the complex fugacity plane that contains the positive real axis. As a consequence, also the function $P = P(v)$ is analytic, and all the other thermodynamic functions are free of singularities. Moreover, the following properties hold: (i) $P(v) \geq 0$, (ii) $\mathcal{V}/M(\mathcal{V}) \leq v < \infty$, (iii) $(\partial P/\partial v) \leq 0$.

Now, in the limit $\mathcal{V} \to \infty$, the equation of state is given by

$$\frac{P}{kT} = \lim_{\mathcal{V}\to\infty} \frac{1}{\mathcal{V}} \log \mathcal{Q}(z,\mathcal{V},T) ,$$

$$\frac{1}{v} = \lim_{\mathcal{V}\to\infty} \frac{1}{\mathcal{V}} z \frac{\partial}{\partial z} \log \mathcal{Q}(z,\mathcal{V},T) , \tag{2.64}$$

where the operations $\lim_{\mathcal{V}\to\infty}$ and $(z\,\partial/\partial z)$ can be interchanged only if the limit is uniform. We see here that a nonuniform limit could offer a possibility of finding singularities that would correspond to phase transitions. More precisely, when $\mathcal{V}$ increases, the number of roots of (2.62) also increases, and if some of these zeros of $\mathcal{Q}(z,\mathcal{V},T)$ converge to a point z_0 on the real axis of z, then, by Vitali's theorem in complex analysis,[22] for $\mathcal{V} \to \infty$ the radius

---

[22] Consider a sequence of regular functions $f_n(z)$ in a domain D such that for every n and z in D, $|f_n(z)| \leq A$, and such that as $n \to \infty$, $f_n(z)$ converges to a limit point in D. Then $f_n(z)$ converges uniformly to an analytic limit inside D.

of the neighborhood of a point z_0 on the real axis, a neighborhood where $F_V(z) = V^{-1} \log \mathcal{Q}(z, V)$ converges uniformly, would shrink to zero, and the limit function could be nonanalytic. This is the fundamental idea of the Yang–Lee theory, which is rigorously expressed by the following two theorems:

Theorem 2.3. *The limit*

$$F_\infty \equiv \lim_{V \to \infty} \frac{1}{V} \log \mathcal{Q}(z, V)$$

exists for any $z > 0$, and is a continuous nondecreasing function of z. This limit is also independent of the shape of the volume. It is assumed that as $V \to \infty$, the area of the surface of V does not increase with the volume faster than $V^{2/3}$.

Theorem 2.4. *Let us consider a region R of the complex plane of the variable z that contains a segment of the real axis. If in R there are no zeros of the grand-partition function, then for any $z \in R$ and $V \to \infty$, the quantity $V^{-1} \log \mathcal{Q}(z, V)$ converges uniformly to a limit function of z that is analytic for any $z \in R$.*

On the basis of the second theorem, a single thermodynamic phase will be identified by all the values of z belonging to the same region R. In fact, in such a region the two operations of limit and of derivation can be interchanged, so that by the first theorem, the equation of state in parametric form

$$\frac{P(z)}{kT} = F_\infty(z) , \tag{2.65}$$

$$\frac{1}{v} = z \frac{\partial}{\partial z} F_\infty(z) ,$$

is expressed in terms of analytic functions. If the region R contains the whole real axis, then no phase transition can exist; this situation is represented in Figure 2.5. In contrast, if in the limit $V \to \infty$ the zeros of $\mathcal{Q}(z, V)$ converge to a point z_0 on the real axis, then two distinct regions R_1 and R_2 are formed where Theorem 2.4 holds separately; this situation is represented in Figure 2.6. In this case, by Theorem 2.3, the pressure function $P(z)$ is still continuous at $z = z_0$, while the specific volume $v(z)$ can be discontinuous at the same point. In such a case, the system would have two distinct thermodynamic phases, corresponding to the regions with $z < z_0$ and $z > z_0$. With the aid of (2.48) and (2.63), one finds that for any finite V,

$$z \frac{\partial}{\partial z} \left[\frac{1}{v(z)} \right] = \langle (N/V)^2 \rangle - \langle N/V \rangle^2 \geq 0 .$$

That is, $1/v(z)$ is a nondecreasing function of z, and if $z_1 \in R_1$ and $z_2 \in R_2$ with $z_2 > z_1$, then $1/v(z_2) \geq 1/v(z_1)$. A discontinuity that $1/v(z)$ can develop

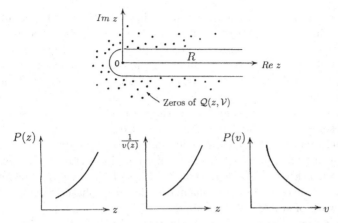

Fig. 2.5. Absence of phase transitions. The region R of the complex fugacity plane is free of zeros of the grand-partition function, correspondingly, the $P - v$ diagram is regular and describes a single thermodynamic phase.

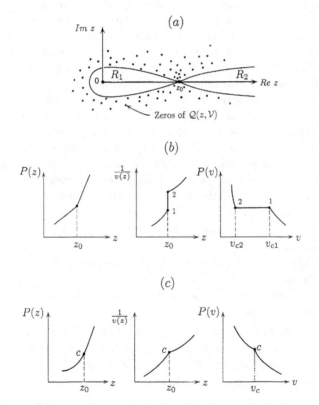

Fig. 2.6. Figure (a) shows the complex fugacity plane separated into two regions, R_1 and R_2, by a zero of the grand-partition function located at z_0. Each region corresponds to regular behaviors of thermodynamic functions, that, however, match at z_0 in a nonanalytic way. Figure (b) describes a two-phase system undergoing a first-order phase transition; the $P - v$ diagram displays a coexistence line. Figure (c) describes a two-phase system undergoing a second-order phase transition.

in the thermodynamic limit, together with this monotonic character of $1/v(z)$, describes a first-order phase transition when the horizontal coexistence line in the $P - v$ plane is generated. This is illustrated by Figure 2.6b, where we have to keep in mind that the jump in the pattern of $1/v(z)$ is a limiting feature; at finite volume this is a sharply increasing but continuous function. In Figure 2.6c the case is shown in which both $P(z)$ and $dP(z)/dz$ are continuous at $z = z_0$, but it is the second derivative, $d^2P(z)/dz^2$, that is discontinuous, thus representing a second-order phase transition.

In conclusion, the equation of state (2.64) can account for the existence of phase transition phenomena, provided that phase transitions are mathematically identified with a loss of analyticity of thermodynamic functions, hence the *thermodynamic limit dogma* to describe phase transitions within the framework of statistical mechanics.

The fact that the distribution of the zeros of the grand-partition function completely determines the equation of state is a remarkable one. However, this could remain too abstract, because of the a priori great difficulty of computing the distribution of the roots in the complex plane. Fortunately, there is a class of physically interesting systems for which the roots are not dispersed everywhere in the complex plane but distribute themselves on a fixed circle. This is the case of the Ising model in an external magnetic field and of a lattice gas. The gas on a lattice is such that on each lattice site there can be one atom or a vacancy. To prevent the occupation of a lattice site by more than one atom, the interaction energy u between two atoms is $u = +\infty$ if two atoms occupy the same site; if two atoms occupy nearest-neighboring sites their interaction energy is assumed to be $u = -2\varepsilon$, otherwise $u = 0$. This lattice gas and the Ising model with an external magnetic field are completely equivalent, in the sense that the thermodynamic properties of one of these systems can be derived from those of the other, provided that a table of correspondence among the different quantities is introduced.

Lee and Yang [38] proved the following remarkable result:

Theorem 2.5 (circle theorem). *If the interaction u between two atoms of a lattice gas is such that*

$u = +\infty$ *if the atoms occupy the same lattice site,*

$u \leq 0$ *otherwise,*

then all the zeros of the grand-partition function lie on the unit circle in the complex fugacity plane.

In the $N \to \infty$ limit, the roots of $\mathcal{Q}(z, \mathcal{V}) = 0$ can be described by means of a density function $g(\theta)$ such that $Ng(\theta)d\theta$ is the number of roots with z belonging to the interval $[e^{i\theta}, e^{i(\theta+d\theta)}]$. One obviously has $\int_0^{2\pi} g(\theta)d\theta = 1$.

For the lattice gas it turns out that

$$\frac{P}{kT} = \int_0^\pi g(\theta) \log(z^2 - 2z \cos\theta + 1)d\theta \;,$$

$$\frac{1}{v} = 2z \int_0^\pi g(\theta) \frac{z - \cos\theta}{z^2 - 2z \cos\theta + 1} d\theta \;. \tag{2.66}$$

Thus knowledge of the distribution function $g(\theta)$ allows one to compute the isotherms in the P–v plane for the lattice gas to describe the phenomenon of condensation. After the replacements $P = -F - B$ (F is the Helmholtz free energy and B the external field) and $v = 2/(1 - M)$ (M is the magnetization), the same equations give the isotherms in the $M - B$ plane for the Ising model.

For a chain of N spins with periodic boundary conditions, Lee and Yang computed the angular distribution $g(\theta)$ and found that the roots do not close in on the real axis: a finite-arc segment, crossing the real axis, remains void of roots for any N. This is in agreement with the absence of a ferromagnetic transition in the one-dimensional Ising model. In contrast, in the case of the two-dimensional Ising model with vanishing external field, for which a phase transition exists, the roots on the unit circle are distributed so as to get arbitrarily close to the real axis in the limit of arbitrarily large N.

2.2 Hamiltonian Dynamics

As we have seen above, statistical mechanics is a successful theory of the average behavior of a large collection of atoms or molecules; it gets rid of microscopic dynamics, which is the very origin of macroscopic physics, under the apparently reasonable hypotheses of ergodicity and mixing. These hypotheses spawned an "ergodic problem," which, according to different lines of thought, has been given different degrees of emphasis.

As is witnessed by the content of this book, the efforts to deepen our understanding of the validity limits of the basic assumptions of statistical mechanics are far reaching, and go far beyond a mere confirmation of the solidity of its grounds. Some scholars, among whom we mention L. Landau and A.I. Khinchin, claimed that the ergodic problem had nothing to do with microscopic dynamics and that its solution would rather stem from probability theory, due to the huge number of particles in macroscopic systems.

Others, among whom E. Fermi played a key role, were rather inclined to look for a justification of the ergodic hypothesis through a development of our understanding of generic properties of Hamiltonian dynamics. There are some fundamental analytical results, which will be sketchily reviewed in this section, and some fundamental "experimental" results, worked out by numerically solving the equations of motion of paradigmatic models and concisely reviewed in Section 2.3, that convincingly support the physical relevance and soundness of the assumption of generic validity of ergodicity and mixing.

2.2.1 Perturbative Results for Quasi-integrable Systems

The motions of a Hamiltonian system are the solutions of the equations (2.3).

In general, the integration by quadrature of $2N$ first-order ordinary differential equations needs the knowledge of $2N$ first integrals of motion. However, since the Hamiltonian flow has to preserve the symplectic structure of phase space, the canonical equations (2.3) require the knowledge of only N first integrals of motion, $\mathcal{F}_1, \ldots, \mathcal{F}_N$, provided that these are in involution, that is, their Poisson brackets $\{\mathcal{F}_i, \mathcal{F}_j\} = 0$ with each other vanish. In this case, the phase space motions belong to an invariant subspace (fixed by the initial conditions) that is isomorphic to an N-torus $\mathbb{T}^N$, so that by means of a canonical transformation to action-angle variables, it is always possible to represent the phase space motion as a translation on the torus. More precisely, we have the following:

Theorem 2.6 (Liouville-Arnold). *Let $H(p, q)$ be the Hamiltonian of an N-degrees-of-freedom system.[23] Assume that this system has N independent first integrals of motion in involution, $\mathcal{F}_1, \ldots, \mathcal{F}_N$, with $\mathcal{F}_1 = H$. If we consider the space M_f that is implicitly defined by the level sets of the functions $\mathcal{F}$ by $\mathcal{F}_1 = f_1, \ldots, \mathcal{F}_N = f_N$, then this space is an invariant set for the motion of the system. Moreover, if the space M_f is connected and compact, then in a neighborhood of M_f a completely canonical transformation exists from the variables (p, q) to action-angle variables (J, θ) such that the transformed Hamiltonian depends only on $J = (J_1, \ldots, J_N)$, and such that the motion of the system with respect to the new variables is a translation on the torus $\mathbb{T}^N$ with frequency $\omega = (\omega_1, \ldots, \omega_N) = (\partial H(J)/\partial J_1, \ldots, \partial H(J)/\partial J_N)$. Thus Hamilton's equations can be integrated by quadrature.*

Integrability will be discussed in Chapter 4. Let us now make a quick survey of some fundamental results concerning the dynamics of those systems (the overwhelming majority) for which the exact solution by quadrature of the equations of motion is ruled out.

Generically, adding to an integrable Hamiltonian, thus representable as a function of the actions only, $H = H(J)$, an arbitrarily small term $H_1(J, \theta)$ that depends also upon the angle variables θ canonically conjugate to J integrability is lost.

Definition 2.7. *A Hamiltonian system is said to be quasi-integrable if its Hamiltonian function has the form*

$$H(J, \theta, \epsilon) = H_0(J) + \sum_{r=1}^{\infty} \epsilon^r H_r(J, \theta), \qquad (2.67)$$

where $(J, \theta) \in \mathbb{R}^N \times \mathbb{T}^N$, ϵ is a small parameter and H_0 is the Hamiltonian function of a completely integrable system.

The perturbation parameter ϵ has to be *small* in the sense that, a norm $\| \cdot \|$ in the space of Hamiltonians having been defined, the condition

$$\|H - H_0\| \ll \|H_0\| \qquad (2.68)$$

---

[23] Here again, we use the notation $p = (p_1, \ldots, p_N)$ and $q = (q_1, \ldots, q_N)$, and with N degrees of freedom we mean N pairs of canonically conjugate variables.

has to be satisfied. The above given definition of quasi-integrability makes it possible to develop a perturbative treatment of nonintegrable systems. However, from the very beginning, the domain of application of the method that we are going to sketch is severely limited by the above given condition on the strength of the perturbation amplitude and by the need of an unperturbed reference system expressed in action-angle coordinates. This is due to the fact that Poincaré [39] originally developed the perturbative method to tackle problems in celestial mechanics, where only few bodies are involved, which move quite regularly on human observational time scales.

The perturbative method consists in searching a sequence of canonical coordinate transformations, which we denote by[24] $W_\epsilon^{(1)} : (J, \theta) \to (P^{(1)}, Q^{(1)})$ for the first-order step, and $W_\epsilon^{(n)} : (P^{(n-1)}, Q^{(n-1)}) \to (P^{(n)}, Q^{(n)})$ at generic order n, which, by means of successive iterations, eliminate the functional dependence on the angular variables in the transformed Hamiltonians at successive orders in ϵ. The first-order step will be achieved by a function $W_\epsilon^{(1)} : (J, \theta) \to (P^{(1)}, Q^{(1)})$ that is ϵ-near to the identity because in the limit $\epsilon \to 0$ the Hamiltonian H is already independent of the variables θ. Thus we can assume that the generating function has the form

$$W_\epsilon^{(1)}(P^{(1)}, \theta, \epsilon) = \sum_{i=1}^N P_i^{(1)} \theta_i + \epsilon W^{(1)}(P^{(1)}, \theta) \tag{2.69}$$

whence the following condition for the vanishing of the second-order terms is obtained:

$$\sum_{i=1}^N \left[\frac{\partial H_0}{\partial J_i} \frac{\partial W^{(1)}}{\partial \theta_i} \right]_{P^{(1)}, Q^{(1)}} + H_1(P^{(1)}, Q^{(1)}) = 0. \tag{2.70}$$

This is the *fundamental equation of perturbation theory*. In view of the periodic dependence on the angular variables of the functions we are dealing with, we can use Fourier series developments like

$$H_1(P^{(1)}, \theta) = \sum_{\kappa \neq 0} H_{1,\kappa}(P^{(1)}) e^{i\kappa \cdot \theta} , \tag{2.71}$$

$$W_1^{(1)}(P^{(1)}, \theta) = \sum_{\kappa \neq 0} W_\kappa(P^{(1)}) e^{i\kappa \cdot \theta} , \tag{2.72}$$

where $\kappa = (k_1, \ldots, k_N)$ and $\mathbf{0}$ stands for the null vector, and substituting these Fourier developments into (2.70) we obtain

$$i\kappa \cdot \omega \, W_\kappa^{(1)}(P^{(1)}) + H_{1,\kappa}(P^{(1)}) = 0 \tag{2.73}$$

$$\forall \, \kappa \in \mathbb{Z}^N \setminus \{\mathbf{0}\},$$

---

[24] We use $P^{(n)}$ and $Q^{(n)}$ to denote N-components vectors. To denote the single components we use subscripts, e.g., $P_i^{(n)}$.

where $\omega = (\omega_1, \ldots, \omega_N) = (\partial H_0(J)/\partial J_1, \ldots, \partial H_0(J)/\partial J_N)$. It immediately follows that if $\kappa \cdot \omega \neq 0$, we get the following *formal* solution[25] for (2.70):

$$W_\kappa^{(1)}(P^{(1)}) = i\frac{H_{1,\kappa}}{\kappa \cdot \omega} . \tag{2.74}$$

The possibility of performing the transformation to the variables $(P^{(1)}, Q^{(1)})$ crucially depends on the unperturbed frequency that we are considering.

Definition 2.8. *Let* $\omega \in \mathbb{R}^N$. *The subset of* $\mathbb{Z}^N$ *given by*

$$\mathcal{M}_\omega = \{\kappa \in \mathbb{Z}^N \mid \kappa \cdot \omega = 0\} \tag{2.75}$$

is called the resonance module $\mathcal{M}_\omega$ *of the frequency vector* ω.

According to the dimension $0 \leq \dim \mathcal{M}_\omega \leq N-1$ of the resonance module, the orbit on the unperturbed torus $\mathbb{T}^N$ can be *periodic*, and in this case $\dim \mathcal{M}_\omega = N-1$ corresponds to *complete resonance*; or *dense* on the whole $\mathbb{T}^N$, in which case $\dim \mathcal{M}_\omega = 0$, corresponding to *nonresonance*; or again, *dense* on a torus of dimension $N - d$, immersed in $\mathbb{T}^N$, and now $\dim \mathcal{M}_\omega = d, 0 < d < N - 1$.[26] Thus, if the unperturbed motion is periodic, then there is no way to work out an analytic form for $W_\epsilon^{(1)}$. Yet, if the unperturbed motion is quasi-periodic, there will always be values of $\kappa \in \mathbb{Z}^N$ that can make the denominator in the right-hand side of (2.74) arbitrarily small, thus preventing the series from converging. This is the famous problem of *small denominators*, which has plagued perturbation theory since its beginning. Actually, Poincaré proved a theorem, later generalized by Fermi, that under generic conditions denies the possibility of convergence for the perturbative series and is enunciated as follows:

Theorem 2.9 (Poincaré–Fermi). *Let an* N-degrees-of-freedom *Hamiltonian system, with* $N \geq 3$, *be described by*

$$H(J, \theta, \epsilon) = H_0(J) + \epsilon H_1(J, \theta). \tag{2.76}$$

If H_0 *satisfies the nondegeneracy condition* $\det\left|\frac{\partial^2 H_0}{\partial J_i \partial J_k}\right| \neq 0$, *and if all the coefficients in the Fourier development of* H_1 *do not vanish (hypothesis of generic perturbation),[27] then there are no analytic* (C^ω) *first integrals of* H,

---

[25] The fundamental equation of perturbation theory is formally solved by the treatment that we are sketching in the sense that we ignore the problem of convergence of the series involved.

[26] When $\dim \mathcal{M}_\omega \neq N - 1$ the motions are said to be *quasi-periodic*.

[27] Actually, both hypotheses–of generic perturbation and of nondegeneracy of the unperturbed part–can be considerably weakened by exploiting Birkhoff's perturbation theory [41] so that the theorem applies to a very broad class of systems of physical interest.

defined on an open set of phase space, that are functionally independent of H.

This asserts that neither analytic (Poincaré) [39] nor smooth (Fermi) [40] integrals of motion besides the energy can survive a generic perturbation of an integrable system with three or more degrees of freedom. Thus, in the absence of other isolating integrals of motion, any constant energy surface of these generic systems is expected to be everywhere accessible to the phase space trajectory. In fact, were the perturbative series convergent, it would be possible to recover a functional dependence of the Hamiltonian on the action variables only, which is possible only for integrable systems. The nonexistence of analytic prime integrals of motion for a generic quasi-integrable system hinders complete integrability, that is, it is no longer possible that *all* the motions are bounded and quasi-periodic. On the other hand, there is no hindrance to the survival of *some* of the unperturbed quasi-periodic motions also in the presence of the perturbation. In other words, phase space is no longer regularly foliated into invariant tori, but it could happen that some of them survive for $\epsilon \neq 0$, provided that we give up their regular dependence on J, for J varying in an open subset $A \subset \mathbb{R}^N$. Such a result can be worked out under suitable hypotheses about the frequencies and amplitudes of the perturbation: this is the celebrated KAM theorem, enunciated as follows:

Theorem 2.10 (Kolmogorov–Arnold–Moser). *Consider a quasi-integrable Hamiltonian system described by*

$$H(\theta, J) = H_0(J) + \epsilon H_1(\theta, J) \tag{2.77}$$

if H_0 satisfies the nondegeneracy condition,[28] *and if the perturbation is sufficiently small ($\epsilon < \epsilon_c$). Then some of the invariant tori associated with H_0 are deformed under the action of the perturbation but are still invariant for the flow of the complete Hamiltonian H. These tori are those corresponding to the unperturbed frequencies that satisfy the Diophantine condition*[29]

$$|\kappa \cdot \omega| \geq \gamma(\epsilon) \|\kappa\|^{-N} \quad \forall \, \kappa \in \mathbb{Z}^N \setminus \{\mathbf{0}\} \,. \tag{2.78}$$

The ensemble Ω_γ of the unperturbed frequencies that satisfy this condition is a Cantor set. Moreover, the measure of the complement of Ω_γ is small with the perturbation

$$\lim_{\epsilon \to 0} \mu(\Omega \setminus \Omega_\gamma) = 0 \,. \tag{2.79}$$

---

[28] The original formulations of this theorem are found in [42]. An extension to the case of Hamiltonians of the form $H(\theta, J) = \omega \cdot J + \epsilon H_1(\theta, J)$, representing a collection of perturbed harmonic oscillators, can be found in [41].

[29] This condition means that we are considering the effect of a nonintegrable perturbation acting on *non-resonant* and sufficiently irrational tori of H_0. The degree of irrationality of a real number is higher the slower the convergence of its continued fraction development.

Summarizing: The consequence of the Poincaré–Fermi theorem is that the whole constant-energy hypersurface Σ_E of the phase space of a generic non-integrable Hamiltonian system is topologically accessible; the KAM theorem specifies that for quasi-integrable Hamiltonian systems, below a threshold value for the perturbation parameter ϵ and for particular choice of the initial condition the phase space trajectory can be confined to a restricted region of Σ_E. Apparently, this theorem could seriously threaten the foundations of statistical mechanics because it goes against ergodicity. In order to assess the actual relevance of this menace, for a casual choice of the initial conditions we have to estimate the probability of being in the domain of validity of the KAM theorem. If the number N of degrees of freedom is large, the theorem has in practice no physical relevance because of the strong N-dependence of the threshold value ϵ_c on the perturbation amplitude [43], typically[30]

$$\epsilon_c(N) \sim e^{-N \log N}, \tag{2.80}$$

so that for large-N systems–which are dealt with by statistical mechanics–the admissible perturbation amplitudes for the KAM theorem to apply drop down to exceedingly tiny values of no physical meaning.

A remarkable improvement of KAM theory started with the Nekhoroshev theorem, which deals with *finite-time* stability of regular orbits in phase space instead of *infinite-time* stability as in the KAM theorem. Still considering quasi-integrable systems, we can wonder how long a nonintegrable trajectory–issuing from an initial condition close to an invariant torus of $H_0(J)$–will remain almost "stuck" to it. An estimate for this confinement time is given by the following theorem:

Theorem 2.11 (Nekhoroshev). *Let the analytic quasi-integrable N degrees of freedom Hamiltonian*

$$H(\theta, J) = H_0(J) + \epsilon H_1(\theta, J) \tag{2.81}$$

satisfy the nondegeneracy condition[31] $\det \left| \frac{\partial^2 H_0}{\partial J_i \partial J_k} \right| \neq 0$. *There exist a critical value ϵ_c and suitable constants C, α, τ, β such that for $\epsilon < \epsilon_c$, every solution $(\theta(t), J(t))$ of the equations of motion of H with initial conditions at least at a distance $C\epsilon^\alpha$ from the boundary of the domain where the variables J are defined, one has*

$$|J(t) - J(0)| \leq C\epsilon^\alpha \quad \text{for } t \leq \tau e^{(1/\epsilon)^{\beta(N)}}. \tag{2.82}$$

---

[30] The N-dependence of ϵ_c depends, to some extent, on general properties of the interaction potential; for example, in [44] it is found that $\epsilon_c \sim N^{-\delta}$, but the value worked out for δ is so huge that the physical consequences are the same.

[31] The original and refined formulations of this theorem can be found in [45]; the theorem can be extended also to the case of Hamiltonians of the form [46] $H(\theta, J) = \omega \cdot J + \epsilon H_1(\theta, J)$.

A comment about the relevance of this result for statistical mechanics is in order. The absence of a truly invariant set for the dynamics now prevents any conflict with the ergodic hypothesis. However, since any experiment or measure lasts only a finite time, also quasi-invariance of the actions could lead to the inadequacy of the statistical-mechanical predictions. Observation time scales and "ergodization" time scales now enter the game, and their comparison could in principle lead to a practical violation of ergodicity. Also in this case it is necessary to focus attention on the dependence on the number of degrees of freedom of the constants entering the theorem. Again the perturbation amplitude ϵ must be smaller than some critical value ϵ_c, and even the optimal estimates give ϵ_c very small and

$$\beta(N) \sim \frac{1}{N} \, ,$$

so that the lower bound for the stability time drops down to $\mathcal{O}(1)$ already at rather small N.

There are exceptions: for a special Hamiltonian system, a Nekhoroshev-like result [47] gives just 1 for the relevant constant independently of N, but this requires a very large gap between the acoustic and optical branches of the Brillouin frequency spectrum of the unperturbed system.

However, the Nekhoroshev theorem has in general only little relevance, if any, for statistical mechanics.

2.2.2 Hamiltonian Chaos

For a long time, the equations of Newtonian mechanics have been the paradigm of classical determinism, conceptually identified with the notion of predictability at any time. But during the last decades, it has been realized that "determinism" and "predictability" are far from being the same concept, and that predictability crucially depends upon the *stability* of the solutions of the dynamical differential equations. Determinism implies that once an initial condition is given, the trajectory is uniquely determined for all later times; stability means that two initially close trajectories will remain close in the future. If this is not true, it becomes impossible to predict the evolution of a system even for very small times.

As long as nonlinear dynamical systems are considered, stability is the exception rather than the rule. Even if this relies—at least from a conceptual point of view—upon mathematical results that have been known since the end of the nineteenth century, its importance has been completely realized only with the aid of a new and powerful approach: numerical simulation. The very complicated structure of some trajectories that can arise in nonlinear dynamical systems was discovered by Poincaré [39] in the late nineteenth century, but the physics community became fully aware of the existence and the meaning of these structures only when they were visualized by computer simulation in the work of Hénon and Heiles [48].

The instability we are referring to is known as intrinsic stochasticity of the dynamics, or *deterministic chaos*. These terms mean that the dynamics, although completely deterministic, yet exhibits some features that make it indistinguishable from a random process. The characteristic feature of a chaotic system, which is at the basis of the unpredictability of its dynamics, is the sensitive (exponential) dependence on initial conditions: locally, the distance between two trajectories that originate in very close-by points in phase space grows exponentially in time, so that the system loses the memory of its initial conditions. Regular dynamics, i.e., quasiperiodic motion, is—as far as conservative systems are considered—a "weak" property, because it is destroyed by very small perturbations of the system. In contrast, chaos is a strong property, because given a dynamical system in which chaos is present, in many cases it will be present even after the system has been subjected to significant perturbations [49].

As we have seen in the preceding section, the most relevant results—at least in view of their potential impact on statistical mechanics—worked out in the framework of classical perturbation theory apply to those regions of phase space where the motions are sufficiently far from resonance of the involved frequencies. While the N-tori of an integrable Hamiltonian can survive to sufficiently weak nonintegrable perturbations, provided that these tori are sufficiently irrational, completely different is the fate of *resonant* tori. In fact, no matter how small the perturbation ϵH_1 is, the resonant tori of H_0 are destroyed. According to a theorem due to Poincaré and Birkhoff, which holds for $N = 2$, under the action of a generic perturbation, resonant tori break into an even number of fixed points (on the Poincaré surface of section), half of them elliptic and half hyperbolic. Near these hyperbolic points, the nonintegrable perturbation originates the phenomenon of *homoclinic intersections*, which entail a chaotic dynamics. Let us now take a glance at what happens in the resonant regions of phase space.

Consider the simplest situation of a one-degree-of-freedom system subject to a time-dependent perturbation

$$H(\theta, J) = H_0(J) + \epsilon V(\theta, J, t) , \quad \epsilon \ll 1, \quad N = 1. \tag{2.83}$$

Let $V(t) = V(t + T)$ be a periodic time-dependent perturbation of period $T = 2\pi/\nu$ such that we can write

$$V(\theta, J, t) = \frac{1}{2} \sum_{k,l} V_{k,l}(J) e^{i(k\theta - l\nu t)} + c.c. , \tag{2.84}$$

with $V_{k,l} = -V^\star_{-k,-l}$. Hence the equations of motion

$$\dot{J} = -\epsilon \frac{\partial V}{\partial \theta} = -\frac{i}{2} \epsilon \sum_{k,l} k V_{k,l}(J) e^{i(k\theta - l\nu t)} + c.c. ,$$

$$\dot{\theta} = \frac{\partial H}{\partial J} = \frac{dH_0}{dJ} + \epsilon \frac{\partial V}{\partial J} = \omega(J) + \frac{1}{2} \epsilon \sum_{k,l} \frac{dV_{k,l}(I)}{dJ} e^{i(k\theta - l\nu t)} + c.c. .$$

$$\tag{2.85}$$

For any pair of integers such that

$$k\omega - l\nu \approx 0 ,$$

equations (2.85) have a resonant solution. The approach to tackle resonances consists in "extracting" out of (2.85) the "dangerous" resonance and then studying the dynamics as if it were essentially determined by this "dangerous" resonance. Then one tries to evaluate the effect of the initially neglected nonresonant terms. Of course, more than a single "dangerous" resonance can exist. However, for the sake of simplicity, we assume that one of them has a dominant role. Let k_0, l_0, J_0 be the parameters of the dominant resonance, that is, such that $k_0\omega(J_0) - l_0\nu = 0$. Thus we can drastically simplify the Fourier developments in (2.85) by retaining only the resonant harmonic

$$\dot{J} = \epsilon V_0 k_0 \sin(k_0\theta - l_0\nu t + \varphi) ,$$
$$\dot{\theta} = \omega(J) + \epsilon \frac{dV_0}{dJ} \cos(k_0\theta - l_0\nu t + \varphi) , \qquad (2.86)$$

having put $V_{k_0,l_0} = |V_{k_0,l_0}|e^{i\varphi} = V_0 e^{i\varphi}$. Now we introduce a new phase variable $\psi = k_0\theta - l_0\nu t + \varphi$ and rewrite the equations above as

$$\dot{J} = \epsilon k_0 V_0(J) \sin\psi ,$$
$$\dot{\psi} = k_0\omega(J) - l_0\nu + \epsilon \frac{dV_0(J)}{dJ} \cos\psi . \qquad (2.87)$$

Though these equations considerably simplify the initial problem, they are still rather difficult to study. A further simplification can be introduced by noting that if ϵ is sufficiently small, then $\dot{J}$ is small as well, and the deviation of J from its resonant value J_0 is also small. Suppose that J is very close to J_0. The effect of the resonance will be that of making the action J grow with time, and since the frequency ω is a function of J, the growth of J switches off the resonance. The nonlinearity of the system saturates the growth of the action variable near a resonance, and for small ϵ we have to expect a small-amplitude oscillation of J around the resonant value J_0. Now put $\Delta J = J - J_0$, and assuming that it is small note that equations (2.87) practically correspond to a Hamiltonian $H = H_0(J_0) + \epsilon V_0(J) \cos\psi$. Then we make a series expansion of $H_0(J)$ in powers of ΔJ up to the second order, i.e.,

$$H(\psi, J) = H_0(J_0) + \omega(J_0)\Delta J + \frac{1}{2}\left[\frac{d\omega(J)}{dJ}\right]_{J=J_0}(\Delta J)^2 + \epsilon V_0(J_0)\cos\psi . \quad (2.88)$$

By means of the canonical transformation

$$(\theta, J) \rightarrow (k_0\theta - l_0\nu t, \Delta J) , \qquad \nu = \omega(J_0) ,$$

equation (2.88) becomes

$$H_R = k_0 H - l_0\nu\Delta J , \qquad (2.89)$$

which, apart from an irrelevant constant, is

$$H_R = \frac{1}{2}a(\Delta J)^2 + b\sin\psi \tag{2.90}$$

with

$$a = k_0 \left(\frac{d\omega}{dJ}\right)_{J=J_0} \quad , \quad b = \epsilon V_0 k_0 \ .$$

The Hamiltonian H_R in (2.90) is the *universal* Hamiltonian for a *nonlinear resonance*. The Euler–Lagrange equation of motion associated with H_R is

$$\frac{d^2 q}{dt^2} + \Omega_0^2 \sin q = 0 \tag{2.91}$$

with $q = \psi + \pi$ and $\Omega_0^2 = [\epsilon V_0(d\omega/dJ)_{J_0}]^{1/2}$. Equation (2.91) is the equation of motion of a nonlinear pendulum whose phase space is sketched in Figure 2.7.

A simple estimate from (2.90) of the largest oscillation width, the so-called resonance width, is

$$\max_q \{\Delta J\} = \left[\epsilon V_0 \left(\frac{d\omega}{dJ}\right)_{J_0}^{-1}\right]^{1/2} \ ,$$

where we can observe that, in contrast to what usually happens with perturbation theory, the variation of ΔJ is proportional to $\epsilon^{1/2}$ instead of being proportional to ϵ, and since $\epsilon \ll 1$ we have $\epsilon^{1/2} \gg \epsilon$, and thus it is much larger.

Let us now consider the case of internal resonances, that is, resonances that are due not to an external forcing term but to the existence of two or

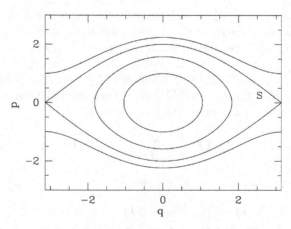

Fig. 2.7. Phase-space trajectories of a simple pendulum. The closed curves are the oscillations; the curves above and below the separatrix S are clockwise and counterclockwise rotations, respectively.

more degrees of freedom. Consider the case of $N = 2$ and a system described by the Hamiltonian

$$H = \sum_{i=1}^{2} H_{0i}(J_i) + \epsilon V(J_1, J_2, \theta_1, \theta_2).$$

In this case the resonance can take place between the two degrees of freedom. For this to occur it is necessary that

$$n\omega_1(J_1^0) - m\omega_2(J_2^0) = 0 \ , \quad n, m \in \mathcal{N}.$$

Also in this case we start by considering the Fourier series development of the perturbation, out of which we shall extract only the resonant term; thus we write

$$V = \frac{1}{2} \sum_{n_1 n_2} V_{n_1 n_2}(J_1, J_2) e^{i(n_1\theta_1 - n_2\theta_2)} + c.c. \ ,$$

and we also develop H_{0i}, and the associated frequencies ω_i, about (J_1^0, J_2^0). As in the preceding case, the Hamiltonian (2.2.2) reads

$$H_R = \frac{1}{2} \sum_{i=1}^{2} \omega_i'(\Delta J_i)^2 + \epsilon V_0 \cos \psi \ ,$$

where

$$\omega_i' = \left(\frac{d\omega(J_i)}{dJ_i} \right)_{J_i^0} \ , \quad \Delta J_i = J_i - J_i^0 \ ,$$

the phase is given by $\psi = n\theta_1 - m\theta_2 + \varphi$, and

$$V_{nm}(J_1^0, J_2^0) = |V_{nm}(J_1^0, J_2^0)| \, e^{i\varphi} \equiv V_0 e^{i\varphi} \ .$$

The equations of motion derived from (2.2.2) are

$$\dot{J}_1 = \frac{d(\Delta J_1)}{dt} = \epsilon n V_0 \cos \psi \ ,$$

$$\dot{J}_2 = \frac{d(\Delta J_2)}{dt} = -\epsilon m V_0 \cos \psi \ ,$$

$$\dot{\psi} = n\omega_1' \Delta J_1 - m\omega_2' \Delta J_2 \ . \tag{2.92}$$

Multiplying the first equation by m, the second one by n, and summing them, one immediately sees that the quantity

$$m J_1 + n J_2 = \text{cost} \tag{2.93}$$

is a constant of motion. Moreover, by introducing again $q = \psi + \pi$ and differentiating the third of equations (2.92) we get

$$\frac{d^2 q}{dt^2} + \Omega_0^2 \sin q = 0 \ , \tag{2.94}$$

where $\Omega_0^2 = \epsilon V_0(n^2\omega_1' + m^2\omega_2')$, which is again the pendulum equation of motion. In the $N > 2$ case, more than two degrees of freedom can resonate, that is,

$$\sum_{i=1}^{M \leq N} m_i\omega_i(J_1 \cdots J_N) = 0 \ .$$

This entails the existence of several equations like (2.93), while the resonance process is still described by a single equation for the oscillating phase ψ again in the form of (2.94). There is a major limitation to this method due to the generic possibility of multiple resonances, so that considering the motion of a system near a nonlinear resonance constitutes an oversimplification of the problem, acceptable, at most, in very special regions of phase space.

Homoclinic Intersections near a Perturbed Separatrix

We have seen that in the presence of a nonlinear resonance the local representation of the phase space orbits coincides, apart from deformations, with the phase space orbits of a pendulum.

The limiting phase space trajectory, separating librations from rotations, is called a *separatrix*. The cuspy points of the separatrix are unstable equilibrium points, called hyperbolic points because of the shape of linearized motions close to them. Each hyperbolic point is associated with eigendirections corresponding to real positive and negative eigenvalues of the linearized motion. These directions are the asymptotes of the hyperbolic trajectories of the linearized motion, and are also the sets of points attracted by the fixed points for $t \to \infty$ or $t \to -\infty$. In contrast, the sets of points that, in the exact motion converge toward the fixed point for $t \to \infty$ or $t \to -\infty$ are called the *stable manifold* W^s and *unstable manifold* W^u of the fixed point, respectively. These manifolds are tangent to the eigendirections of the fixed point. For a pendulum, or in the case of Hamiltonians describing nonlinear resonances, the two manifolds, W^s and W^u are superposed and jointly form the separatrix. However, there is no reason for this to happen in general. To the contrary, we learn from Melnikov's theory [50] that the addition of an arbitrarily small perturbation to an integrable system entails transverse intersections of W^u and W^s.

Suppose that W^u and W^s intersect at h_0 (see Figure 2.8b); this point is called *homoclinic*. The image h_1 of h_0 (transformed by the action of the Hamiltonian flow or of the Poincaré application on the surface of section) by definition of W^u and W^s must belong to both W^u and W^s, thus also h_1 is a homoclinic point.

The oriented tangents T_0^+ and T_0^- at h_0 are transformed into oriented tangents T_1^+ and T_1^- at h_1. The conservation of the orientation (which is related to general properties of canonical transformations) implies the existence of a structure as shown in Figure 2.8c, that is, the existence of an intermediate homoclinic point h_0'. By iterating forward and backward in time, an infinity

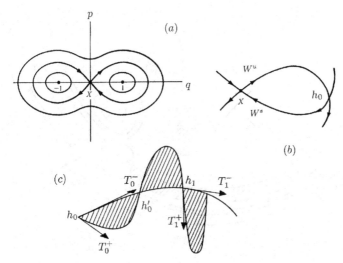

Fig. 2.8. Homoclinic intersections. In Figure (a) the phase portrait of a quartic oscillator—described by $H(p,q) = \frac{1}{2}p^2 - \frac{1}{2}q^2 + \frac{1}{4}q^4$—is shown. The stable and unstable manifolds issuing from the hyperbolic point X are superposed to form an eight-shaped separatrix. After the addition of a perturbation, the stable and unstable manifolds no longer superpose but intersect transversally; Figure (b) shows one intersection point, h_0, out of a countable infinity of them. In (c) the image h_1—under the Hamiltonian flow— of h_0 is shown.

of images and preimages of h_0 and h_0' are generated. At large n, the images h_n of h_0 along W^s converge to the fixed point and their separation exponentially decreases.[32] Therefore, in order to conserve the areas of the lobes of W^u to satisfy Liouville's theorem, the lobes must be exponentially stretched. Moreover, self-intersections of W^s and W^u are forbidden because otherwise the conservation of areas would prevent the points belonging to a lobe produced by such an intersection from converging to the fixed point for $t \to \pm\infty$. As a consequence the lobes are constrained to take on increasingly contorted shapes, getting thinner and thinner (see Figure 2.9) and generating new sets of homoclinic points having an infinity of images and preimages of their own. A formidably complex structure is thus generated, of which Poincaré wrote:

> These intersections form a sort of texture, or of a net whose meshes are infinitely tight; each of these two curves can never intersect itself, but has to fold in a complicated way as to intersect all the meshes of the net an infinite number of times. One is amazed by the complexity of this picture, which I do not even attempt to draw.

The intersections between W^s and W^u are called *homoclinic intersections*. Their existence, together with the just-mentioned stretching and folding of the

---

[32] In fact, the velocity of the motion along an unperturbed separatrix is, for a pendulum, $p^0(t) = 2\,\mathrm{sech}\,t$.

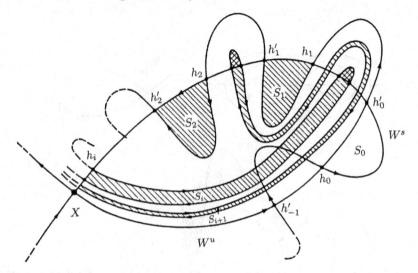

Fig. 2.9. Homoclinic intersections of W^u and W^s for a quartic oscillator plus a generic perturbation that breaks integrability.

lobes formed by W^s and W^u, makes us understand the origin of the erratic appearance of the orbits belonging to a *stochastic layer* that replaces a separatrix destroyed by a generic perturbation. Any trajectory in the stochastic layer tends to stick to W^s or W^s and thus to assume all its complexity. An abstract version of the effect of the combined action of stretching and folding is provided by the *Smale horseshoe* [51], and in fact, homoclinic intersections–by the Smale–Birkhoff homoclinic theorem–generate a hyperbolic invariant set [52–54].

To better understand the effect of a generic perturbation on a nonlinear resonance, we consider a forced pendulum

$$\ddot{q} + \sin q = \epsilon \cos \omega t \ , \tag{2.95}$$

where ω is a constant angular frequency and ϵ is a small parameter. For any small but finite value of ϵ, equation (2.95) is not integrable, i.e., it has no analytic integral of motion. There is a method, due to Melnikov, to prove that (2.95) is not integrable. Rewrite (2.95) in the form

$$\dot{x} = X_0(x) + \epsilon X_1(x, t) \ , \tag{2.96}$$

where $x \in \mathbb{R}^2$, X_0 is a Hamiltonian vector field of Hamiltonian H_0, and X_1 is a periodic Hamiltonian field with period T and Hamiltonian H_1. Assume that X_0 has a homoclinic orbit $\overline{x}(t)$ such that for $t \to \infty$, $x(t) \to x_0$, with x_0 a hyperbolic fixed point, and consider the *Melnikov function*

$$M(t_0) = \int_{-\infty}^{\infty} \{H_0, H_1\} \left(\overline{x}(t - t_0), t \right) dt \ ,$$

where $\{,\}$ stands for the Poisson bracket. There is a theorem [54, 55] stating that if $M(t_0)$, considered as a periodic function of t_0, has simple zeros, then (2.96) has transversally intersecting manifolds W^s and W^u, that is, homoclinic intersections. To apply the method to the forced pendulum, set $x = (q, \dot{q})$ so that (2.95) becomes

$$\frac{d}{dt}\begin{pmatrix} q \\ \dot{q} \end{pmatrix} = \begin{pmatrix} \dot{q} \\ -\sin q \end{pmatrix} + \epsilon \begin{pmatrix} 0 \\ \cos \omega t \end{pmatrix} .$$

The homoclinic orbit is given by

$$\bar{x}(t) = \begin{pmatrix} q(t) \\ \dot{q}(t) \end{pmatrix} = \begin{pmatrix} \pm 2 \tan^{-1}(\sin ht) \\ \pm 2 \operatorname{sech} t \end{pmatrix} ,$$

and with

$$H_0(q, \dot{q}) = \frac{1}{2}\dot{q}^2 - \cos q ,$$
$$H_1(q, \dot{q}, t) = \epsilon q \cos \omega t ,$$

the Melnikov function is

$$M(t_0) = \pm \int_{-\infty}^{\infty} \left(\frac{\partial H_0}{\partial q}\frac{\partial H_1}{\partial \dot{q}} - \frac{\partial H_0}{\partial \dot{q}}\frac{\partial H_1}{\partial q} \right) dt$$
$$= \pm \epsilon \int_{-\infty}^{\infty} \dot{q} \cos \omega t\, dt = \pm \epsilon \int_{-\infty}^{\infty} [2 \operatorname{sech}(t - t_0) \cos \omega t]dt .$$

After a change of variables, since the hyperbolic secant is even and the cosine is odd, one gets

$$M(t_0) = \pm \epsilon \left[\int_{-\infty}^{\infty} \operatorname{sech} t \cos \omega t\, dt \right] \cos (\omega t_0) ,$$

whence finally

$$M(t_0) = \pm 2\pi\epsilon \operatorname{sech} \left(\frac{\pi\omega}{2} \right) \cos(\omega t_0) ,$$

which clearly has simple zeros.

As a consequence of the presence of these intersections, in a neighborhood of the region in phase space that was occupied by the separatrix in the integrable case, a chaotic layer suddenly appears. The chaotic layer is the region irregularly filled by dots in Figure 2.10, where a two-dimensional section[33] of the 3D phase space of the system is shown. If one follows the evolution of two initially close points in the chaotic layer, one finds that their separation locally grows *exponentially* in time, so that the dynamics in the chaotic sea is unpredictable.

---

[33] This section has been obtained as a stroboscopic Poincaré section [56], so that each point on the plot corresponds to an intersection of a trajectory of the system with the planes $t = 2n\pi/\omega$.

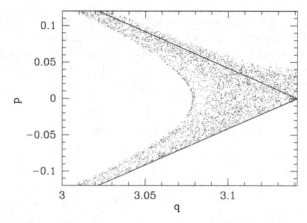

Fig. 2.10. Section of the phase space of a perturbed pendulum, showing the appearance of chaotic seas close to the separatrix of the unperturbed system (the solid line in the figure). The dots are obtained from a single trajectory issuing from a point very close to the unperturbed separatrix. The amplitude of the perturbation is $\epsilon = 10^{-4}$.

The appearance in phase space of irregular regions such as the chaotic layer could justify by itself the use of the term "chaotic dynamics." However, there are also other properties of the dynamics described by (2.95) that justify the use of such a term. For example, if we introduce a symbolic coding of the dynamics in which the symbol 0 is associated with each passage through the point $q = 0$ with $\dot{q} > 0$ and the symbol 1 to each passage through the same point with $\dot{q} < 0$, then given any bi-infinite sequence of zeros and ones, for example generated by coin tosses, this sequence corresponds to a real trajectory of the system (2.95). Aspects of the motion of the system, though deterministic, are thus indistinguishable from a random process.

This example is extremely simple but contains the essential features of the problem. However, even in the simple low-dimensional cases, by means of concepts like homoclinic intersections, it is possible only to give a qualitative description of the onset of chaos, but a quantitative description of the stochastic regions is not at hand, i.e., there is no recipe to compute *how fast* two initially close-by points separate. For N-dimensional systems the situation is obviously even worse.

To obtain quantitative information on chaotic dynamics we must introduce Lyapunov exponents.

2.2.3 Lyapunov Exponents

The standard *operational* way to detect chaos and to quantify its strength is through Lyapunov characteristic exponents (LCE). A trajectory is said to be *chaotic* if its (largest) Lyapunov exponent is *positive*. The mathematical definition of LCE relies on Oseledeč's multiplicative theorem [57].

Let us recall that for a generic flow $\varphi^t : M \to M$ on a manifold M, given an invariant measure, μ, and denoting by $d\varphi_x^t : T_x M \to T_{(\varphi^t x)} M$ its tangent dynamics, Oseledeč's theorem ensures that $\forall x \in M_1$, $\forall e \in T_x M$ ($e \neq 0$), with $M_1 \subset M$ and such that $\mu(M_1) = 1$, the quantity

$$\lambda(x, e) = \lim_{t \to \infty} \frac{1}{t} \ln \| d\varphi_x^t(e) \|,$$

which is independent of the metric of M, exists and is finite. The definition relies of the largest Lyapunov exponent relies on this theorem. More explicitly, let $J(x)$ be the Jacobian matrix of the Hamiltonian flow φ^t at the point $x = (p_1, \ldots, p_N, q_1, \ldots, q_N)$. Denoting by ξ the variation vector, the tangent dynamics $d\varphi^t$ is described by

$$\frac{d\xi_i}{dt} = J_{ik}(x(t))\xi_k.$$

Thus $J : T_x M \to T_{(\varphi^t x)} M$.

Take a vector $\xi_x \in T_x M$ and its transformed $\xi_{(\varphi^t x)} \in T_{(\varphi^t x)} M$; notice that

$$\langle \xi_{(\varphi^t x)}, J \, \xi_x \rangle = \langle J^* \xi_{(\varphi^t x)}, \xi_x \rangle$$

and that the product appearing at the left-hand side belongs to $T_{(\varphi^t x)} M$, while the product of the right-hand side necessarily belongs to $T_x M$, i.e., $J^* \xi_{(\varphi^t x)} \in T_x M$, and in general $\| \xi_{(\varphi^t x)} \| \neq \| \xi_x \|$, whence $J^* \neq J^{-1}$. Therefore J maps the tangent vectors forward in time, while J^* maps these vectors backward in time but not retracing the forward evolution of $\xi(t)$. Having defined $G = \prod_{i=1}^n J_i(\varphi^{i-1} x)$, Oseledeč's theorem states that the limiting matrix

$$\Lambda_x = \lim_{n \to \infty} (G^* G)^{1/2n}$$

exists and is finite, and thus for any n the product $(\prod_{i=1}^n J_i^*)(\prod_{k=1}^n J_k)$ maps an arbitrary vector $\xi_x^0 \in T_x M$ into a vector $\xi_x^n \in T_x M$, i.e., into the *same* tangent space $T_x M$. Thus an eigenvalue problem for Λ_x is well defined in the vector space $T_x M$. Since the matrix Λ_x is positive semidefinite by construction, its eigenvalues $\{\gamma_i\}$ are real and non-negative. Moreover, since Λ_x is symplectic (since the Jacobians J_i and their products are symplectic), its $2N$ eigenvalues are paired [34] as $\{\gamma_i, \gamma_i^{-1}\}$. The $2N$ Lyapunov characteristic exponents $\lambda_1, \ldots, \lambda_{2N}$ are then defined as

$$\lambda_i = \log \gamma_i \, ,$$
$$\lambda_{2N-i+1} = -\lambda_i \, , \quad i = 1, \ldots, N \, . \tag{2.97}$$

---

[34] This symmetry in the Lyapunov spectrum is necessary to ensure the satisfaction of Liouville's theorem and thus holds for Hamiltonian flows, though it has been recently generalized also to a class of non-Hamiltonian systems [58].

The set of $2N$ values

$$\lambda_1 \geq \lambda_2 \geq \cdots \geq \lambda_{2N} \tag{2.98}$$

is called the *Lyapunov spectrum*. The tangent space $T_{x(0)}M$, where $x(0)$ is the initial condition in phase space, admits a decomposition into linear subspaces,

$$T_{x(0)}M = E_1 \oplus E_2 \oplus \cdots \oplus E_{2N} , \tag{2.99}$$

and each λ_i is associated with the corresponding subspace E_i in that a vector $\xi(0) \in E_i$ will grow exponentially with the exponent λ_i. If there exists on the phase space a probability measure μ that is ergodic and invariant for the dynamics, then the numbers λ_i do *not* depend on the initial condition $x(0)$, apart from a possible subset of initial conditions of measure zero with respect to μ.

In practice, the evolution of the norm of a tangent vector is sensitive only to the first—the largest—exponent, because a generic initial vector $\xi(0)$ will have a nonvanishing component in the E_1 subspace, so that the largest exponent λ_1 will always dominate in the long-time limit. In fact,

$$\|\xi(t)\|^2 = \langle \xi(t), \xi(t) \rangle = \langle G(t)\xi(0), G(t)\xi(0) \rangle = \langle G^*(t)G(t)\xi(0), \xi(0) \rangle ,$$

and due to the fact that

$$G^*(t)G(t)\xi(0) = (e^{2\lambda_1 t}\xi_1(0), e^{2\lambda_2 t}\xi_2(0), \ldots, e^{2\lambda_{2N} t}\xi_{2N}(0))$$

one asymptotically has

$$\|\xi(t)\| = e^{\lambda_1 t}\|\xi_1(0)\| \left(1 + \frac{\xi_2^2}{\xi_1^2}e^{2(\lambda_2 - \lambda_1)t} + \cdots + \frac{\xi_{2N}^2}{\xi_1^2}e^{2(\lambda_{2N} - \lambda_1)t} \right)^{1/2}$$
$$\sim e^{\lambda_1 t}\|\xi_1(0)\| .$$

Hence, the asymptotic definition of the largest Lyapunov exponent immediately follows as

$$\lambda_1 = \lim_{t \to \infty} \frac{1}{t} \log \frac{\|\xi(t)\|}{\|\xi(0)\|} , \tag{2.100}$$

which measures the degree of instability of a trajectory: if λ_1 is positive, the trajectory is unstable with a characteristic instability time λ_1^{-1}.

According to this definition, Lyapunov exponents are not given a *local* meaning because they are defined as *asymptotic* quantities. In other words, this definition does not allow us to properly think of Lyapunov exponents as averages of local divergence rates of phase space trajectories. This can be illustrated by means of an example borrowed from random matrices. An infinite product of matrices representing elliptic rotations with a random change of the rotation parameters (semiaxes of the ellipse), has a limiting matrix with positive Lyapunov exponent, i.e., it represents a hyperbolic rotation (Fürstenberg's theorem) [59]. Indeed, it is evident that the average of the largest eigenvalue of the single matrices in the product (average of the local exponents) necessarily gives a vanishing Lyapunov exponent.

This is to warn about the apriori nonnegligible difference between the abstract definition of LCE and their practical evaluation in numerical simulations. In fact, as we shall see below, the standard numerical method used to estimate Lyapunov exponents is never asymptotic; it is more like an average of the local divergence rates of nearby trajectories, and one never computes LCE through the eigenvalues of a product matrix Λ_x, which could be computed for a finite, possibly long, number of time steps.

As we shall see in the following chapters, at least for Hamiltonian flows, pseudo-Lyapunov exponents, as we might call numerical Lyapunov exponents, can be given a proper geometrical definition and description independently of Oseledeč's theorem.

Let us now give a more intuitive definition of Lyapunov exponents. Our discussion will be aimed at showing how to define and compute the (pseudo)[35] Lyapunov exponents for a flow. For a more general discussion of Lyapunov exponents, see Eckmann and Ruelle [60].

Let us consider a dynamical system whose trajectories in an n-dimensional phase space M are the solutions of the following system of ordinary differential equations:

$$
\begin{aligned}
\dot{x}_1 &= X_1(x_1, \ldots, x_n) , \\
\vdots \qquad & \qquad \vdots \\
\dot{x}_n &= X_n(x_1, \ldots, x_n) .
\end{aligned}
\tag{2.101}
$$

If we denote by $x(t) = (x_1(t), \ldots, x_n(t))$ a given *reference* trajectory whose initial condition is $x(0)$, and by $y(t)$ another trajectory that is initially close to $x(t)$, and we denote by $\xi(t)$ the vector

$$
\xi(t) = y(t) - x(t) ,
\tag{2.102}
$$

then the evolution of ξ describes the *local* separation of the two trajectories in phase space. The vector ξ is assumed to obey the linearized equations of motion, because it is assumed to be initially small. These equations are, as can be shown by inserting (2.102) into the equations of motion (2.101) and expanding in a power series up to the linear terms,

$$
\begin{aligned}
\dot{\xi}_1 &= \sum_{j=1}^{n} \left(\frac{\partial X_1}{\partial x_j} \right)_{x(t)} \xi_j , \\
\vdots \qquad & \qquad \vdots \\
\dot{\xi}_n &= \sum_{j=1}^{n} \left(\frac{\partial X_n}{\partial x_j} \right)_{x(t)} \xi_j ,
\end{aligned}
\tag{2.103}
$$

---

[35] Henceforth, no distinction will be made between "true" and "pseudo" Lyapunov exponents.

and are referred to as the *tangent dynamics equations*.[36] Note that (2.103) is a system of linear differential equations, whose coefficients, however, depend on time. According to the definition (2.102), the norm $\|\xi\|$ of the vector ξ, i.e.,

$$\|\xi(t)\| = \left[\sum_{i=1}^{n} \xi_i^2(t)\right]^{1/2} , \qquad (2.104)$$

locally measures the distance of the two trajectories as a function of t. If the trajectory $x(t)$ is unstable, all its perturbations grow exponentially, so that $\|\xi(t)\| \propto \exp(\lambda t)$. If the elements of the Jacobian matrix $\partial X_i / \partial x_j$, which are the coefficients of the linear equations (2.103), were either constant or periodic, it would be possible to solve the system, but since the Jacobian matrix depends on the trajectory $x(t)$, its entries are in general neither constant nor periodic, so that the rate of exponential divergence varies with time.

Using the definition (2.100) of λ_1, and the compact notation

$$\dot{\xi}_i = J_{ik}[x(t)]\xi_k ,$$

observing that

$$\frac{1}{2}\frac{d}{dt}\log(\xi^T \xi) = \frac{\xi^T \dot{\xi} + \dot{\xi}^T \xi}{2\xi^T \xi} = \frac{\xi^T J[x(t)]\xi + \xi^T J^T[x(t)]\xi}{2\xi^T \xi} ,$$

and setting $\mathcal{J}[x(t), \xi(t)] = \{\xi^T J[x(t)]\xi + \xi^T J^T[x(t)]\xi\}/(2\xi^T \xi)$, we have

$$\lambda_1 = \lim_{t\to\infty} \frac{1}{t} \log \frac{\|\xi(t)\|}{\|\xi(0)\|} = \lim_{t\to\infty} \frac{1}{t} \int_0^t d\tau \, \mathcal{J}[x(\tau), \xi(\tau)] , \qquad (2.105)$$

which *formally* gives λ_1 as a time average.

Let us now apply the above to a standard Hamiltonian system whose Hamiltonian is of the form (1.1); the dimension of the phase space is $n = 2N$, and the equations of motion (2.101) are Hamilton equations

$$\dot{q}_k = p_k ,$$
$$\dot{p}_k = -\frac{\partial V}{\partial q_k} , \qquad k = 1, \ldots, N , \qquad (2.106)$$

and also the linearized dynamics (2.103) can be cast in the canonical form

$$\dot{\xi}_1 = \xi_{N+1} ,$$
$$\vdots \quad \vdots$$
$$\dot{\xi}_N = \xi_{2N} ,$$

---

[36] This notation follows from the fact that the dynamics of the vector ξ takes place in the tangent space $T_{x(t)}M$ of the phase space M.

$$\dot{\xi}_{N+1} = -\sum_{j=1}^{N} \left(\frac{\partial^2 V}{\partial q_1 \partial q_j}\right)_{q(t)} \xi_j , \qquad (2.107)$$

$$\vdots \quad \vdots$$

$$\dot{\xi}_{2N} = -\sum_{j=1}^{N} \left(\frac{\partial^2 V}{\partial q_N \partial q_j}\right)_{q(t)} \xi_j .$$

These equations describe the tangent dynamics for a Hamiltonian flow. To measure the largest Lyapunov exponent λ in a numerical simulation, one integrates numerically both (2.106) and (2.108), and then makes use of the definition (2.100), which can be rewritten in this case as

$$\lambda_1 = \lim_{t \to \infty} \frac{1}{t} \log \frac{\left[\xi_1^2(t) + \cdots + \xi_N^2(t) + \dot{\xi}_1^2(t) + \cdots + \dot{\xi}_N^2(t)\right]^{1/2}}{\left[\xi_1^2(0) + \cdots + \xi_N^2(0) + \dot{\xi}_1^2(0) + \cdots + \dot{\xi}_N^2(0)\right]^{1/2}} , \qquad (2.108)$$

where we have used that $\dot{\xi}_i = \xi_{i+N}$ (see (2.108)). More precisely, in a numerical simulation one uses the discretized version of (2.108), i.e.,

$$\lambda_1 = \lim_{m \to \infty} \frac{1}{m} \sum_{i=1}^{m} \frac{1}{\Delta t} \log \frac{\|\xi(i\Delta t + \Delta t)\|}{\|\xi(i\Delta t)\|} , \qquad (2.109)$$

where, after a given number of time steps Δt, for practical numerical reasons [61] it is convenient to renormalize the value of $\|\xi\|$ to a fixed one.

The definition (2.100) does not allow one to measure the other exponents of the Lyapunov spectrum. To measure them, one has to observe that they can be related to the growth of *volumes* in the tangent space. A two-dimensional area V_2 in the tangent space spanned by two linearly independent tangent vectors $\xi^{(1)}$ and $\xi^{(2)}$ will expand according to

$$V_2(t) \propto \exp[(\lambda_1 + \lambda_2)t] , \qquad (2.110)$$

a three-dimensional volume, as

$$V_3(t) \propto \exp[(\lambda_1 + \lambda_2 + \lambda_3)t] , \qquad (2.111)$$

and so on, so that, choosing $k \leq n$ linearly independent and normalized vectors $\xi^{(1)}, \xi^{(2)}, \ldots, \xi^{(k)} \in T_x M$, we obtain

$$\lim_{t \to \infty} \frac{1}{t} \log \|\xi^{(1)}(t) \wedge \xi^{(2)}(t) \wedge \cdots \wedge \xi^{(k)}(t)\| = \sum_{i=1}^{k} \lambda_i . \qquad (2.112)$$

Therefore the algorithm (2.109) can be generalized to obtain an algorithm to compute numerically the whole Lyapunov spectrum [62]. However, such a computation is rather hard when the number N is large.

The sum of *all* the n Lyapunov exponents in the Lyapunov spectrum, $\sum_{i=1}^{n} \lambda_i$, measures the expansion rate of n-volumes in phase space. Therefore, for a Hamiltonian system, this sum has to vanish.

The numerical integration of (2.103) and the consequent measure of λ—or of the spectrum $\{\lambda_i\}$ when it is possible in practice—is the standard technique to characterize Hamiltonian chaotic dynamics. An operative definition of a chaotic dynamical system can be stated as follows: a system is chaotic if it has at least one positive and one negative Lyapunov exponent. In fact this ensures that the system shows (almost everywhere with respect to the ergodic measure μ used to define the Lyapunov exponents) the distinctive features of chaos as described in the example of the forced pendulum. In fact, the presence of a positive exponent ensures the presence of an exponential divergence of nearby orbits, and the presence of a negative one ensures that they also fold and mix in a very complicated way, so that they can produce those structures we referred to as "chaotic seas." However, as long as autonomous Hamiltonian systems are considered, the antisymmetry of the spectrum (2.97) ensures that the presence of a positive exponent implies the presence of a negative one with the same absolute value, so that a single (the largest) positive exponent is sufficient to have chaos; indeed, if the largest exponent vanishes, the dynamics will be regular. These facts, together with that the largest Lyapunov exponent λ measures the smallest instability time scale, show how natural the use of the value of λ is to measure chaos in such systems.

It is important to specify with respect to what invariant ergodic measure μ the Lyapunov exponents are defined: this may be also a δ-measure concentrated on a single trajectory, in which case we could speak of a chaotic trajectory rather than of a chaotic system. In Hamiltonian systems with a large number of degrees of freedom we expect the microcanonical measure of the chaotic regions to be overwhelmingly larger than the measure of the regular regions; the existence of these regular regions is ensured—as we have seen in Section 2.2.1—by the KAM theorem. However, as we have already commented, these regular regions can exist only for exceedingly tiny, thus unphysical, deviations from integrability. Therefore, already for N of a few tens, the relevant measure in the definition of Lyapunov exponents is indeed the microcanonical measure. Numerical experiments are in agreement with this expectation, since no relevant dependence of the Lyapunov exponent on the initial conditions has ever been detected in large systems, and this is the reason why throughout this book we never refer to any possible dependence of λ on μ, treating the Lyapunov exponent as any other "thermodynamic" observable. It is only for small systems (especially $N = 2$, which is the best-known case) that the simulations show that the measure of the chaotic regions may be very small in a very large energy range; these systems are often referred to as *mixed* systems, since they are in between completely chaotic and regular ones.

Since we are interested in large systems, up to the thermodynamic limit, a number of questions naturally arises: What is the behavior of the Lyapunov exponents as n increases? Does a thermodynamic limit exist for the Lyapunov spectrum? And so on. Numerical results [63] have shown that as $n \to \infty$ the Lyapunov spectrum $\{\lambda_i\}$ appears indeed to converge to a well-behaved function

$$\lambda(x) = \lim_{n \to \infty} \lambda_{xn} . \tag{2.113}$$

The function $\lambda(x)$ is a nonincreasing function of $x \in [0,1]$. Some rigorous work in this respect has been done by Sinai [64]. The existence of a limiting Lyapunov spectrum in the thermodynamic limit has many important consequences that we will not review here; a good discussion can be found in [65]. We only want to remark here that the existence of a thermodynamic limit for the Lyapunov spectrum implies that the largest Lyapunov exponent is expected to behave as an intensive quantity as N increases.

2.3 Dynamics and Statistical Mechanics

After World War II, the advent of electronic digital computers opened the possibility of numerically tackling problems whose solutions were not only unknown but also far from intuition and hard to grasp even heuristically.

Already in the 1950s, these highly nontrivial and far-reaching potentialities of electronic computers were clearly present to the minds of scientists such as J. von Neumann, S. Ulam, and E. Fermi.

In a foreword to their coauthored work reprinted in the Fermi collected papers [66], S. Ulam wrote:

> After the war, during one of his frequent summer visits to Los Alamos, Fermi became interested in the development and potentialities of the electronic computing machines. He held many discussions with me on the kind of future problems which could be studied through the use of such machines. We decided to try a selection of problems for heuristic work where in the absence of closed analytic solutions experimental work on a computing machine would perhaps contribute to the understanding of properties of solutions. This could be particularly fruitful for problems involving the asymptotic—long time or "in the large"—behaviour of nonlinear physical systems. In addition, such experiments on computing machines would have at least the virtue of having the postulates clearly stated. This is not always the case in an actual physical object or model where all the assumptions are not perhaps explicitly recognized. Fermi expressed often the belief that future fundamental theories in physics may involve nonlinear operators and equations, and that it would be useful to attempt practice in the mathematics needed for the understanding of nonlinear systems. The plan was then to start with the possibly simplest such physical model and to study the results of the calculation of its long-time

behaviour. Then one would gradually increase the generality and the complexity of the problem calculated on the machine...perhaps problems of pure kinematics, e.g. the motion of a chain of points subject only to constraints but no external forces, moving on a smooth plane convoluting and knotting itself indefinitely. These were to be studied preliminary to setting up ultimate models for motion of systems where "mixing" and "turbulence" would be observed. The motivation then was to observe the rates of mixing and "thermalization" with the hope that the computational results would provide hints for a future theory. One could venture a guess that one motive in the selection of problems could be traced to Fermi's early interest in the ergodic theory....

These words were a lucid anticipation of the since then ever increasing importance of numerical experiments in filling theoretical or mathematical gaps. For example, statistical mechanics is based on the ergodic hypothesis, and we have already explained why the stronger assumption of phase mixing is necessary. But what about the actual properties of Hamiltonian dynamics? Numerical simulations have made possible a substantial advance of our understanding of generic properties of Hamiltonian dynamics. For instance, Poincaré grasped the complexity of the motions near perturbed separatrices, but it was only after the numerical experiment of Hénon and Heiles that the consequences of homoclinic intersections became clear, and thereafter a lot of work on Hamiltonian chaos was begun.

By means of numerical experiments we can compute time averages of physical observables along nontrivial phase space trajectories, thus making possible a direct comparison with statistical ensemble averages. Of course, numerical simulations can be performed with N values much smaller than the Avogadro number, but it soon became evident that already with N of a few hundreds many observables attain their thermodynamic limit values. This circumstance, already several decades ago, stimulated the development of a practical application of numerical simulations of Hamiltonian dynamics, called *molecular dynamics*, which was devised to perform ab initio computations in condensed-matter physics and in chemical physics.

As far as the foundations of statistical mechanics are concerned, numerical simulations went—from the very beginning—far beyond a simple confirmation of the validity of the basic assumptions of statistical mechanics, revealing a rich and unexpected phenomenology. Three different kinds of problems have been and can be studied: (i) transient nonequilibrium phenomena (thermalization processes); (ii) stationary nonequilibrium phenomena (gradient-driven transport phenomena); (iii) equilibrium phenomena.

The numerical study of transient nonequilibrium phenomena aims at following the zeroth law of thermodynamics at work; its early developments are sketchily given in the next section.

The numerical study of stationary nonequilibrium phenomena has its paradigmatic problem in the Fourier law of heat conduction in the presence of a temperature gradient. The experimental datum of a linear variation of the spatial temperature profile between the two extrema cannot be reproduced by considering harmonic systems (this is a classical result due to E. Schrödinger). Nonlinearities and strong chaos play a crucial role in explaining this kind of nonequilibrium phenomena [67] and are at the grounds of recent developments of a general formulation of nonequilibrium statistical mechanics [68].

The numerical study of equilibrium phenomena is based on phase space trajectories issuing from "equilibrium initial conditions," that is, from microstates compatible with a macroscopic equilibrium property (for example energy equipartition). Dynamical equilibrium computations can convey more information than statistical computations because there are observables, like Lyapunov exponents, that have no statistical ensemble counterpart. In Section 2.3.1 this fact is briefly illustrated through the discovery of what has been called strong stochasticity threshold, a dynamical property of large-N flows with very interesting physical consequences. Finally, a major topic among equilibrium phenomena is represented by phase transitions, and in Section 2.3.2 we briefly illustrate how their dynamical investigation proceeds.

2.3.1 Numerical Hamiltonian Dynamics at Large N

The dawning of the "numerical simulation epoch" dates back to the problem devised originally by E. Fermi, J. Pasta, and S. Ulam (FPU) in 1954 [66]. Their purpose was to check numerically that a generic but very simple nonlinear many-particle dynamical system would indeed behave for large times as a statistical mechanical system, that is, it would approach equilibrium if initially not in equilibrium. In particular, their purpose was to obtain the usual equipartition of energy over all the degrees of freedom of a system for generic initial conditions. To their surprise, for the system FPU considered—a one-dimensional anharmonic chain of 32 or 64 particles with fixed ends and in addition to harmonic, cubic (α-model), or quartic (β-model) anharmonic forces between nearest neighbors—this was not observed. The Hamiltonians of the two models read

$$H(p,q) = \sum_{k=1}^{N} \left[\frac{1}{2}p_k^2 + \frac{1}{2}(q_{k+1} - q_k)^2 + \frac{\alpha}{3}(q_{k+1} - q_k)^3 \right] \tag{2.114}$$

for the α-model, and

$$H(p,q) = \sum_{k=1}^{N} \left[\frac{1}{2}p_k^2 + \frac{1}{2}(q_{k+1} - q_k)^2 + \frac{\beta}{4}(q_{k+1} - q_k)^4 \right] \tag{2.115}$$

for the β-model. If we switch off the anharmonic terms, the orthogonal coordinate transformation (for fixed endpoints, i.e., $q_0 = q_N = 0$)

$$Q_k(t) = \sqrt{\frac{2}{N}} \sum_{n=1}^{N} q_n(t) \sin \frac{\pi k n}{N} \qquad (2.116)$$

diagonalizes both Hamiltonians to

$$H(P, Q) = \frac{1}{2} \sum_{k=1}^{N} (P_k^2 + \omega_k^2 Q_k^2) ,$$

where $\omega_k = 2 \sin(\pi k/2N)$ are the frequencies of the so called normal modes. At equilibrium, statistical mechanics predicts equipartition of energy among these normal modes. However, if the system is prepared in an arbitrary state, it will remain there forever. In contrast, in the presence of the anharmonic terms, all the normal modes interact with each other and—according to the value of the coupling constant—they will exchange energy more or less efficiently. Eventually thermal equilibrium should set in.

After the coordinate change in (2.116), the α-model Hamiltonian becomes

$$H(P, Q) = \frac{1}{2} \sum_{k=1}^{N} (P_k^2 + \omega_k^2 Q_k^2) + \alpha \sum_{i,j,k=1}^{N} C_{ijk} Q_i Q_j Q_k \qquad (2.117)$$

and that of the β-model

$$H(P, Q) = \frac{1}{2} \sum_{k=1}^{N} (P_k^2 + \omega_k^2 Q_k^2) + \beta \sum_{i,j,k,l=1}^{N} D_{ijkl} Q_i Q_j Q_k Q_l , \qquad (2.118)$$

where the coefficients have a complicated dependence on the indexes.

Contrary to expectation, a variety of manifestly nonequilibrium and non-equipartition behaviors was seen, including quasi-periodic recurrences to the initial state. In Figure 2.11 the oscillation in time is displayed of the energy content of a few normal modes of the α-model. At $t = 0$ all the energy was concentrated in the longest wave-length mode. In fact, a behavior reminiscent of that of a dynamical system with few degrees of freedom was found, rather than the expected statistical-mechanical behavior. These results raised the fundamental question about the validity—or at least the generally assumed applicability—of statistical mechanics to nonlinear systems, of which the system considered by FPU seemed to be a typical example.

Fermi's early interest in ergodic theory is witnessed by his contribution to the theorem due to Poincaré and thenceforth known as the Poincaré–Fermi theorem that we have already met in Section 2.2.1. After this theorem, no hindrance to ergodicity seems to be possible, whence the surprise of Fermi, Pasta, and Ulam (FPU), when no apparent tendency to equipartition was observed

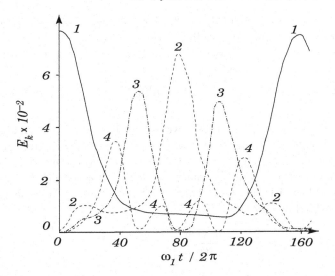

Fig. 2.11. FPU α-model. Oscillation of the energy content of the normal modes. All the energy is initially concentrated in the lowest mode.

in their numerical experiment. Fermi himself considered what they found a "little discovery." The effort to resolve the so-called FPU problem has led to enormous advances in our understanding of nonlinear dynamical systems— for a review we refer to [69]—because most of the attempts to clarify this difficulty have approached the problem as one in dynamical systems theory. These analyses have revealed many very interesting properties of the FPU system and uncovered a number of possible explanations for the resolution of the observed conflict with statistical mechanics. For example, a seminal idea was to explain the FPU recurrences as echoes due to the free stream- ing of coherent nonlinear excitations, since then called *solitons* [70], which are stable solutions of the Korteweg–de Vries (KdV) field equation with cu- bic nonlinearity (which is a continuum limit of the lattice FPU α-model), or of the modified Korteweg–de Vries (mKdV) field equation with quartic nonlinearity (which is a continuum limit of the lattice FPU β-model). The stability of the solutions of these nonlinear field equations would correspond to a regular dynamics of the lattice models from which they are derived and, consequently, to a lack of equipartition. On the other hand, the stability loss of these solutions would correspond to the onset of a chaotic dynamics of the corresponding lattice models and of their good statistical behavior [71]. Though very interesting, this approach does not provide a conclusive answer to the FPU problem because it relies on a somewhat arbitrary, and certainly not unique, continuum limit of the lattice models.

Another approach to the problem, after pioneering work by Ford and Waters [72], ascribed the lack of equipartition in the FPU β-model to the absence of stochasticity, an idea that was made more precise and explicit in a paper by Izrailev and Chirikov [73]. After an approximation based on an averaging technique due to Bogoliubov and Krylov, the equations of motion for the normal mode amplitudes Q_k in (2.116) were cast in the form

$$\ddot{Q}_k + \omega_k^2 Q_k \left[1 - \frac{3\beta}{4N} \omega_k^2 (2 - \omega_k^2) Q_k^3 \right] = \frac{\beta}{8N} \sum_m F_{km} \cos \theta_{km} ,$$

where the coordinates Q_k are those given in (2.116), ω_k is the normal mode linear frequency, $\dot{\theta}_{km}$ are nonlinear mode frequencies, and F_{km} are complicated combinations of the mode amplitudes. By applying Chirikov's resonance overlap criterion for the onset of stochasticity, for weak anharmonicity and initial excitation of low-order modes, Izrailev and Chirikov worked out the prediction of the threshold energy value E_c,

$$E_c \geq \frac{N}{\beta k} , \qquad k \ll N ,$$

for the transition to stochasticity. Though pioneering, this prediction was somewhat too local in normal-mode space to draw convincing conclusions on the global properties of phase space.

The almost contemporary announcement by Kolmogorov of the starting of what would later become the celebrated KAM theorem seemed to provide an explanation to the unexpected FPU results. But later developments of KAM theory, including optimal estimates of the N-dependence of the perturbation thresholds and the Nekhoroshev theorem, revealed that this is not really an adequate framework to explain the FPU problem (see Section 2.2.1).

The rich variety of the numerical phenomenology accumulated over time seemed to keep off "the hope that the computational results would provide hints for a future theory." In fact, "rates of mixing and thermalization" have a startling and complicated dependence on energy, number of degrees of freedom, and initial conditions. Actually, any dynamical evolution of the system depends on the starting point in phase space and on the "landscape" of its surroundings. Thus, there can be a huge variety of dynamical behavior entailed by the preparation of the system in an initial condition out of equilibrium. As a consequence, in order to get some global information on the phase space structure, independently of the initial conditions, one has to look at the chaotic component of phase space. This way of tackling the FPU problem is very illuminating and leads to the conclusion that the FPU problem does not threaten the validity of statistical mechanics. Indeed, this has stimulated the starting of the theory of Hamiltonian chaos that is discussed in the present book.

Stochasticity Threshold at Large N

By focusing on chaos rather than on energy equipartition, a modern revisitation [74] of Fermi's original numerical experiment on the α-model has given strong support to the physicist' viewpoint that generic Hamiltonian flows are, at large N, bona fide ergodic and mixing, a point of view on which we have already commented in the preceding sections.

Notice that the FPU α-model can be derived as a third-order truncation of the power series expansion of the Toda lattice potential, a nonlinear integrable system described by the Hamiltonian

$$H(p,q) = \sum_{k=1}^{N} \frac{1}{2} p_k^2 + \frac{a}{b} \sum_{k=1}^{N} \left[e^{-b(q_{k+1}-q_k)} + b(q_{k+1} - q_k) - 1 \right] . \quad (2.119)$$

Thus the deviation from integrability of the FPU α-model is $\mathcal{O}[(q_{k+1} - q_k)^4]$, a weak deviation indeed, whose consequence is the smallness of the largest Lyapunov exponent, and hence the need for heavy numerical computations.[37]

The outcome $\lambda(\varepsilon, N)$ of the numerical computation of the largest Lyapunov exponent at different values of the energy per degree of freedom ε, and at different N values, is shown in Figure 2.12. These results strongly suggest the existence of a threshold value $\varepsilon_c(N)$ of the energy density, such that

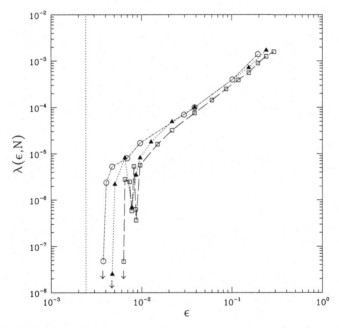

Fig. 2.12. FPU α-model. The largest Lyapunov exponents $\lambda(\varepsilon, N)$ are shown for different values of the energy density ε and a sine wave initially excited. Open squares refer to $N = 32$, solid triangles refer to $N = 64$, open circles refer to $N = 128$, respectively; here the arrows are upper bounds for λ. From [74].

---

[37] The reliability of long-time numerical computations is ensured by the use of symplectic integration schemes; see Section 2.3.2.

above this threshold the motion is chaotic, and below it the trajectories belong to an apparently regular region of phase space. Thus it seems natural to call this threshold a stochasticity threshold (ST).

Fermi and coworkers chose an initial condition well below this ST (the energy density corresponding to their initial condition is marked by the vertical dotted line in Figure 2.12); had they taken a ten times larger amplitude of the initial excitation, they would have observed equipartition during the integration time they used. This is the simple but nontrivial explanation of the lack of statistical-mechanical behavior observed in the original FPU numerical experiment.

The main question, from the point of view of statistical mechanics, concerns the stability of the stochasticity threshold with respect to N. Unambiguous information about this point is provided by computing the Lyapunov exponents with random (i.e., generic) initial conditions, and then by comparing the patterns of $\lambda(\varepsilon)$ obtained for different values of N (see Figure 2.13).

At large ε, there is a tendency of all the sets of points to join, while they tend to separate at small ε: the larger N, the smaller the energy density

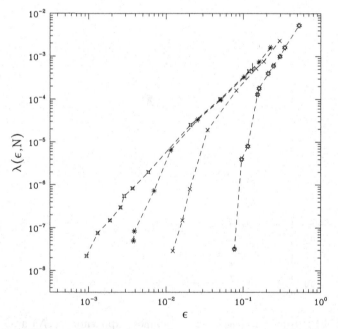

Fig. 2.13. FPU α-model. The largest Lyapunov exponents $\lambda(\varepsilon, N)$ are plotted vs. the energy density ε, for different values of N. Random initial conditions are chosen. Starlike polygons refer to $N = 8$, crosses to $N = 16$, asterisks to $N = 32$, and starlike squares to $N = 64$, respectively. The arrows have the same meaning as in Figure 2.12. From [74].

at which the separation occurs. The "critical" energy density ε_c at which the separation occurs shows the N-dependence $\varepsilon_c(N) \propto 1/N^2$. Independent evidence of the vanishing with N of the ST of the FPU α-model is reported in [75]. Similarly, also the equipartition threshold for a class of initial excitations is found to vanish as $1/N^2$ in the FPU β-model in [76]. In [76] it has been shown that the equipartition threshold corresponds to the transition between two different N-dependences of the relaxation times: from a power-law dependence to an exponential dependence. We could thus think that the ST is a transition between a weak chaos and an even weaker chaotic regime, for example, from an Arnold-like diffusion along bands of overlapping resonances (originating the so-called modulational diffusion; see [49]) to a true and bare Arnold diffusion that is extremely slow.

The lack of equipartition in the original FPU experiment is not representative of a global property of phase space: apparently regular, soliton-like structures, similar to those of Zabusky and Kruskal, have a very long, possibly infinite, lifetime below the stochasticity threshold, whereas, above the same threshold, they have only a finite lifetime [74].

Since the threshold energy density for the onset of chaos shows a clear tendency to vanish at an increasing number of degrees of freedom ($\sim 1/N^2$), the so-called FPU problem does not invalidate the (generic) approach to equilibrium and the validity of equilibrium statistical mechanics.

Notice that in contrast to the case of the FPU α-model, no ST has been numerically detected for the FPU β-model. The explanation is as follows. The FPU α-model can be seen as a fouth-order perturbation (truncation) of the Toda model (integrable), while the FPU β-model can be seen as a third-order perturbation (removal of the third order term by a counterterm) of the same Toda model; in other words, the FPU α is "more" integrable than the FPU β. Thus an ST in the β-model is expected to exist at much lower energy density than in the α-model, with the consequence that, in order to detect a ST in the β-model, one should measure very small Lyapunov exponents, and this would make the computing time exceedingly long. However, through a different kind of numerical analysis, based on the approach to equipartition of initial excitations of a few low-frequency modes, it has been found in [76] that an equipartition threshold does exist at $R = (N+1)\frac{6\beta}{\pi^2}E \approx 1$, where R is the ratio of nonlinear to linear energy in the system. Since the equipartition threshold—which is identified with a stochastic transition—occurs at $R \approx 1$, we get that the transition energy per degree of freedom goes as $\varepsilon_c \approx \pi^2/(6\beta N^2)$, whence $\varepsilon_c \approx 0.001$ at $N = 128$, $\varepsilon_c \approx 0.00025$ at $N = 256$, for which the upper bounds of the largest Lyapunov exponents are $\lambda \approx 4 \times 10^{-8}$ and $\lambda \approx 2.5 \times 10^{-9}$ respectively (by extrapolating the low-energy part of the curve in Figure 2.14), actually exceedingly small values. [76] contains also an interesting theoretical explanation for the numerically found N-dependence of the equipartition/stochasticity threshold.

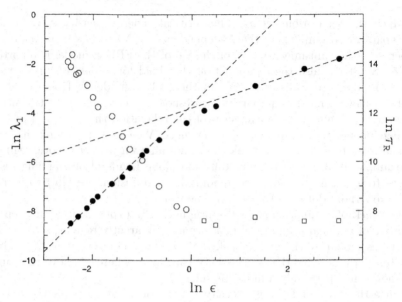

Fig. 2.14. FPU β-model. The equipartition time τ_R (open circles and open squares), in units of the fastest linear frequency, versus the energy density $\epsilon = E/N$, for $N = 128$ and four initially excited modes of low frequency. Full circles represent the largest Lyapunov exponents λ_1, and the dashed lines are power laws ϵ^2 and $\epsilon^{2/3}$. From [78].

Strong Stochasticity Threshold

Several years ago, in [77] was introduced a time dependent spectral entropy $S(t) = -\sum_i w_i(t) \log w_i(t)$, where $w_i(t) = E_i(t)/\sum_k E_k(t)$ is the normalized energy content of the ith harmonic normal mode, defined so as to detect energy equipartition (when $S(t)$ attains its maximum value) and to measure the time needed to reach it. By means of this spectral entropy, in [78,79] the relationship was investigated between equipartition times, measured through the time relaxation patterns of this spectral entropy, and the chaotic properties of the dynamics in nonlinear large Hamiltonian systems. For the FPU β-model, it turns out that at different initial conditions and at long times, the spectral entropy always relaxes toward its maximum value signaling equipartition. However, depending on the value of the total energy density, the relaxation occurs with quite different rapidity. The relaxation time is approximately constant for energy densities greater than some threshold value ε_c, but it steeply grows by decreasing the energy density below this threshold. Moreover, the largest Lyapunov exponent shows a crossover in its ε-dependence corresponding to this threshold value. This phenomenological result can be interpreted as the (smooth) transition, at ε_c, between two different regimes of chaoticity, weak chaos and strong chaos, whence the definition of this transition as a strong stochasticity threshold (SST) [79]. Weak and strong chaos are qualitative terms to denote slow and fast phase space mixing respectively. In [78, 79] a random matrix model for the tangent dynamics was introduced to try to make more precise and quantitative the definitions of weak and strong chaos.

Applied to the FPU β-model, for the largest Lyapunov λ, this random-matrix approximation predicts the scaling $\lambda(\varepsilon) \approx \varepsilon^{2/3}$ in the range $\varepsilon = 0$ up to $\varepsilon \approx 50$. However, at low energy density, the ε-scaling of λ is numerically found to be much steeper, $\lambda(\varepsilon) \approx \varepsilon^2$, so that λ is much smaller, and vanishes faster with ε, than the random-matrix prediction. For this reason we say that here *chaos is weak*. In the energy density range where also the numerical results give $\lambda(\varepsilon) \approx \varepsilon^{2/3}$, we say that *chaos is strong* because the random-matrix model assumes that the dynamics looks like a random uncorrelated process (if suitably sampled in time). At very high energy density, the numerically observed pattern of $\lambda(\varepsilon)$ gradually changes to $\lambda(\varepsilon) \sim \varepsilon^{1/4}$, which is still associated with a strongly stochastic regime though not explained by the random-matrix model; the reason is that a free parameter, a time scale of unknown ε dependence, entering the random matrix model, is arbitrarily assumed constant. Figure 2.15 shows the behavior of $\lambda(\varepsilon)$ for $N = 128$ and different initial conditions. The pattern of $\lambda(\varepsilon)$ is observed to be independent of the initial conditions.

Since the SST is independent of the initial conditions, it has to be ascribed to some change in the global properties of the phase space. For this reason it has to have major consequences on the dynamics. An interesting explanation based on a model for phase space diffusion is given in [80].

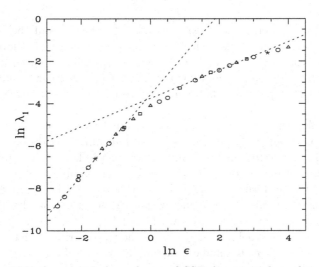

Fig. 2.15. FPU β-model. Independence of SST from initial conditions. Largest Lyapunov exponents λ vs. energy density ϵ at $N = 128$ and at different initial conditions: random at equipartition (circles), wave packets at different average wave numbers (squares, triangles, and asterisks). From [79].

A remarkable feature of the SST is also that it is found to be *independent of the number N of degrees of freedom*, which makes it relevant for *equilibrium statistical mechanics*. The stability with N of the SST is an outcome of numerical simulations confirmed by an analytical prediction for $\lambda(\varepsilon)$ worked out in the limit $N \to \infty$, as is evident in Figure 5.4, where a perfect agreement is shown among the numerical values of $\lambda(\varepsilon)$ obtained for $N = 256$ and $N = 2000$ and the analytic prediction (see Chapter 5). Let us remark again that in numerical simulations equipartition is *always* attained, both in the weakly and strongly chaotic regimes, provided that ε exceeds the ST. On the other hand, we have seen that the ST vanishes at increasing N.

The SST has been found to be correlated with changes in the *transient nonequilibrium* behavior (e.g., relaxation to equipartition) [78,79,81], and has been found also to be correlated with *stationary nonequilibrium* phenomena such as heat conduction [67]. Among the model-dependent consequences of the existence of the SST, it is worth mentioning that in the FPU β model, at $\varepsilon < \varepsilon_c^{SST}$ high-frequency excitations yield longer relaxation times with respect to low frequencies. This is in agreement with the common belief that high frequencies have the tendency to freeze; at $\varepsilon > \varepsilon_c^{SST}$ the situation is reversed. High-frequency excitations yield a quicker relaxation than low frequencies [79].

By combining the FPU α and β models into an FPU $(\alpha + \beta)$ model, described by the Hamiltonian

$$H(p,q) = \sum_{k=1}^{N} \left[\frac{1}{2}p_k^2 + \frac{1}{2}(q_{k+1} - q_k)^2 + \frac{\alpha}{3}(q_{k+1} - q_k)^3 + \frac{\beta}{4}(q_{k+1} - q_k)^4 \right] ,$$

it is possible to observe the coexistence of both the ST and the SST. This model Hamiltonian, with the choice of $\alpha = 0.25$ and $\beta = \frac{2}{3}\alpha^2$, is a fourth-order expansion of the Toda model (2.119). Consequently, its potential function is very close to interatomic potentials of the Morse or Lennard-Jones type in condensed matter, provided that a suitably restricted energy density range is considered.

This model, with particles of unit mass, unit harmonic coupling constant, fixed end-points ($q_1 = q_{N+1} = 0$), and random initial conditions, has been studied [82]. The results of the computation of the largest Lyapunov exponents at different energy densities and for different values of N are shown in Figure 2.16. The patterns of $\lambda(\varepsilon, N)$ therein reported display some remarkable features. For small values of the energy density, there is a sudden drop of λ, which, in close analogy with [74], allows us to define an ST below which we can assume that the overwhelming majority of the trajectories in phase space are regular. This ST moves to smaller and smaller values of ε as N is increased.

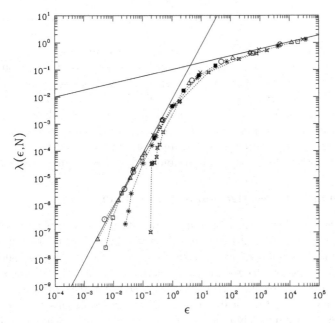

Fig. 2.16. FPU $(\alpha + \beta)$ model. The largest Lyapunov exponents $\lambda(\epsilon, N)$ are shown for different values of the energy density ϵ for various values of N. Starlike squares refer to $N = 8$, asterisks to $N = 16$, open squares to $N = 32$, open triangles to $N = 64$, open circles to $N = 128$, starlike polygons to $N = 512$ and crosses to $N = 1024$, respectively. Full squares refer to $N = 32$ and excitation amplitudes A ranging from 5 to 50. Solid lines are the theoretically expected asymptotic scalings ϵ^2 and $\epsilon^{1/4}$ at low and high energy density, respectively. From Ref. [82].

Slightly below $\varepsilon \approx 1.0$, the pattern of $\lambda(\varepsilon, N)$ is observed to enter the $\sim\varepsilon^2$ scaling (Figure 2.16), due to a crossover between two asymptotic power-law behaviors, $\sim\varepsilon^2$ at small ε and $\sim\varepsilon^{1/4}$ at large ε, where the latter has been attributed to the existence of an SST [78,79]. This crossover is the signature of the transition from weak to strong chaos, as already discussed in [78,79].

It is remarkable that the existence of the SST is not only a characteristic of the FPU β model. In fact, it has been detected in the following one-dimensional lattices: with diatomic Toda interactions (i.e., with alternating masses that break integrability) [83]; with single-well φ^4 interactions [78]; with smoothed Coulomb interactions [83]; with Lennard-Jones interactions [83]; in an isotropic Heisenberg spin chain [84]; in a coupled-rotators chain that displays two thresholds separating two regions of weak chaos (occurring at low and high energies) from an intermediate region of strong chaos [85,86]; in a "mean-field" XY chain [87]; and in homopolymeric chains [88]. It has also been detected in two- and three-dimensional lattices, with two-well φ^4 interactions [89,90], and with XY Heisenberg interactions [91,92]. Therefore the SST seems to be a generic property of Hamiltonian systems with many degrees of freedom.

Despite the existence of a vast literature on Hamiltonian dynamics, we have focused our attention only on stochastic transitions. The reason for this is that the attempt at explaining the origin of the SST has stimulated the development of the Riemannian theory of Hamiltonian chaos, which is discussed the next chapters. Moreover, as we shall also see in the following chapters, the systematic observation that in the presence of a phase transition the SST becomes very sharp, with the appearance of a "cuspy" pattern of $\lambda(\varepsilon)$, has stimulated the development of the topological theory of phase transitions.

2.3.2 Numerical Investigation of Phase Transitions

The theoretical description of phase transitions is a topic of statistical mechanics (see Section 2.1.6), which makes no reference to microscopic dynamics, though in real physical systems dynamics is always there. Even when it is necessary to resort to numerical computations, the standard numerical approach to phase transitions is based on the Monte Carlo method, which, to compute statistical averages of macroscopic observables, resorts to a special sampling technique of the canonical Gibbs measure.[38] In other words, in numerical statistical mechanics, the ergodic invariant measure of the commonly used Monte Carlo method is by construction the canonical ensemble one.

But we can also approach the study of phase transitions by means of direct numerical simulation of the microscopic (Hamiltonian) dynamics. As we have already seen at the beginning of the present chapter, in this case the ergodic invariant measure is the microcanonical one. Therefore, investigating phase transitions by means of Hamiltonian dynamics is equivalent to a microcanonical approach to the subject. For a wide class of systems, the well-known equivalence—in the thermodynamic limit—among all the statistical ensembles warrants an a priori equivalence of the approaches of canonical Monte Carlo and Hamiltonian dynamics.

However, sometimes the ensemble equivalence fails to be true, as is the case of long-range interactions [31], as was shown analytically for a particular model by Hertel and Thirring [30]. This inequivalence is mainly revealed by the appearance of negative values of the specific heat [20, 94], and in some cases, such as a self-gravitating N-body system, the inequivalence is so strong that a phase transition, observed in the microcanonical ensemble, disappears with canonical Monte Carlo computations [95].

The microcanonical description of phase transitions, and thus their numerical study through Hamiltonian dynamics, seems to offer some advantages in

---

[38] The ensemble averages are computed along suitable Markov chains generated by means of the so-called *Metropolis importance sampling* of the Gibbs canonical measure in phase space (in Chapter 9 this method of sampling a given measure is constructively defined).

tackling first-order phase transitions [96], and seems considerably less affected by finite-size scaling effects with respect to the canonical ensemble computation [97]. This notwithstanding, a natural question arises: what do we learn through Hamiltonian dynamics that we didn't already know through Monte Carlo? The answer is given in Chapter 6, where it is shown how Lyapunov exponents and other related geometric quantities bring about a nontrivial seminal novelty.

Models

There is a great variety of physical systems and of theoretical models displaying phase transitions. Among them, only those described by continuous configurational variables admit a Hamiltonian dynamical description. These Hamiltonian models displaying phase transitions can be divided into two main families: lattice models (describing either discretized versions of field theories or condensed-matter systems), and off-lattice models (describing fluids, amorphous systems, polymers). Below, we give a few model Hamiltonians that have been used to study the microscopic dynamical counterpart of phase transitions and, as we shall see in Chapter 6, stimulated the early development of the topological theory of phase transitions.

The first model is a lattice version of a classical φ^4 field model. This can be studied in one, two, and three spatial dimensions, in scalar and vector versions. All these different versions of the model are defined by the Hamiltonian

$$H[\varphi, \pi] = a^d \sum_\alpha \sum_{\mathbf{i}} \left[\frac{1}{2}(\pi_{\mathbf{i}}^\alpha)^2 + \frac{J}{2a^2} \sum_{\mu=1}^d (\varphi_{\mathbf{i}+\mathbf{e}_\mu}^\alpha - \varphi_{\mathbf{i}}^\alpha)^2 - \frac{1}{2}m^2(\varphi_{\mathbf{i}}^\alpha)^2 \right]$$
$$+ \frac{\lambda}{4} \sum_{\mathbf{i}} \left[\sum_\alpha (\varphi_{\mathbf{i}}^\alpha)^2 \right]^2 , \qquad (2.120)$$

where the index α runs from 1 to n for an $O(n)$ symmetry group, the index $\mathbf{i}$ labels the spatial lattice sites, $\mathbf{e}_\mu$ is the unit vector in the direction μ, $(\pi_{\mathbf{i}}, \varphi_{\mathbf{i}})$ are the canonically conjugated variables, a is the lattice spacing, and d is the number of spatial dimensions of the lattice.

Another interesting model that has been considered describes a collection of planar, classical "spins" (in fact rotators) with the ferromagnetic coupling $V = -\sum_{\langle i,j \rangle} J\mathbf{S}_i \cdot \mathbf{S}_j$ (where $|\mathbf{S}_i| = 1$). In two dimensions, for example on a square lattice of $N = n \times n$ sites, with the addition of a standard kinetic energy term, the Hamiltonian is

$$H = \sum_{i,j=1}^n \left\{ \frac{1}{2}p_{i,j}^2 + J\left[2 - \cos(q_{i+1,j} - q_{i,j}) - \cos(q_{i,j+1} - q_{i,j}) \right] \right\} , \qquad (2.121)$$

where $q_{i,j}$ are the angles with respect to a fixed direction on the reference plane of the system. This system, known as the classical Heisenberg XY model, by the Mermin–Wagner theorem [98] cannot have a symmetry-breaking phase transition because of the combined conditions of short-range interactions, continuous symmetry, and two spatial dimensions. Actually, it undergoes a Kosterlitz–Thouless phase transition. In order to elude the "no go" conditions of the Mermin–Wagner theorem, one has to consider the same system on a cubic lattice of $N = n \times n \times n$ sites and described by the Hamiltonian

$$
H = \sum_{i,j,k=1}^{n} \left\{ \frac{1}{2} p_{i,j,k}^2 + J \left[3 - \cos(q_{i+1,j,k} - q_{i,j,k}) - \cos(q_{i,j+1,k} - q_{i,j,k}) \right. \right.
$$
$$
\left. \left. - \cos(q_{i,j,k+1} - q_{i,j,k}) \right] \right\} . \tag{2.122}
$$

Since the "spins" are constrained on planes, this is called the anisotropic Heisenberg XY model.

An example of an off-lattice model is provided by a system of N gravitationally interacting point masses described by the Hamiltonian

$$
H = \sum_{i=1}^{N} \frac{1}{2m_i} \left(p_{xi}^2 + p_{yi}^2 + p_{zi}^2 \right) - \frac{G}{2} \sum_{i,j=1}^{N} (1 - \delta_{ij}) \frac{m_i m_j}{|\mathbf{r}_i - \mathbf{r}_j|} \tag{2.123}
$$

where $\mathbf{r}_i \equiv (x_i, y_i, z_i)$, which undergoes a clustering phase transition when a well-known rescaling invariance (of time, lengths, and energy) is broken [95].

Numerics

There is a wide literature concerning numerical integration of systems of ordinary differential equations. However, the numerical integration of Hamilton's equations of motion (2.3) associated with a standard Hamiltonian (2.2) is correctly performed only by means of symplectic integration schemes. These algorithms satisfy energy conservation (with zero mean fluctuations around a reference value of the energy) for arbitrarily long times as well as the conservation of all the other Poincaré invariants, among which there is the phase space volume, so that also Liouville's theorem is satisfied by a symplectic integration. The simplest symplectic integrator is the *leap-frog* scheme

$$
q_i(t + \Delta t) = q_i(t) + p_i(t) \, \Delta t ,
$$
$$
p_i(t + \Delta t) = p_i(t) + F_i(t + \Delta t) \, \Delta t , \tag{2.124}
$$

where the $F_i(t + \Delta t) = -[\partial V(q)/\partial q_i]_{t+\Delta t}$ are the forces. This discrete-time mapping is a canonical (thus symplectic) coordinate change from the variables $q_i(t)$, $p_i(t)$ to the variables $Q_i = q_i(t + \Delta t)$, $P_i = p_i(t + \Delta t)$, as is immediately recognized with the aid of the generating function

$$
F(Q, p) = - \sum_i Q_i p_i + \Delta t H(Q, p) .
$$

Higher-order very precise and efficient (that is, very accurate even with "large" time steps Δt) symplectic algorithms are available [99–101]. From a theoretical viewpoint, in spite of the ubiquitous chaotic instability of Hamiltonian dynamics, an *interpolation theorem* due to J. Moser [102] ensures that the numerical phase space trajectories are always close to a true phase space trajectory of the Hamiltonian flow with a slightly modified initial condition [103].

Given any observable $A = A(p, q)$, one computes its time average as

$$\langle A \rangle_t = \frac{1}{t} \int_0^t d\tau \, A[p(\tau), q(\tau)]$$

along the numerically computed phase space trajectories. For sufficiently long integration times, and for generic nonlinear (chaotic) systems, these time averages are used as estimates of microcanonical ensemble averages in all the expressions given below.

Temperature

The basic macroscopic thermodynamic observable is temperature. In the microcanonical ensemble, as already mentioned in Section 2.1.5, temperature is derived from entropy—the basic thermodynamic potential in microcanonical ensemble—as

$$\frac{1}{T} = \left(\frac{\partial S}{\partial E} \right)_V , \tag{2.125}$$

where the entropy is given by either

$$S(N, E, V) = k_B \log \int dp_1 \cdots dp_N dq_1 \cdots dq_N \, \delta[E - H(p, q)] \tag{2.126}$$

or

$$S^{(-)}(N, E, V) = k_B \log \int dp_1 \cdots dp_N dq_1 \cdots dq_N \, \Theta[E - H(p, q)] . \tag{2.127}$$

By means of a Laplace transform technique [104], from (2.125) and (2.127) one obtains

$$T = \frac{2}{k_B N} \langle K \rangle , \tag{2.128}$$

where $\langle K \rangle$ is the microcanonical ensemble average of $K = E - V(q)$, that is, of the kinetic energy. In practical numerical simulations, one has to compute (setting $k_B = 1$)

$$T = \frac{2}{N} \frac{1}{t} \int_0^t d\tau \sum_{i=1}^N \frac{1}{2} p_i^2(\tau) \tag{2.129}$$

for t sufficiently large, so that T has attained a stable value (in general this is a rapidly converging quantity).

The equivalent use of (2.125) and (2.126), which in principle is more accurate for Hamiltonian dynamical computations because of the constant energy constraint, gives [104]

$$T = 2\left[(N-2)\langle K^{-1}\rangle\right]^{-1} . \tag{2.130}$$

At large N the two definitions are perfectly equivalent.

Order Parameter

We have already discussed in Section 2.1.6 how the definition of an order parameter is related to the relevant symmetry of a system that is broken below the phase transition point.

For what concerns the lattice φ^4 models, because of the Mermin–Wagner theorem, the interactions being of short range, in two dimensions with the symmetry group $O(1)$, which is the same as $\mathbb{Z}_2$, a symmetry-breaking phase transition is allowed, whereas, for $n > 1$ the $O(n)$ symmetry is a continuous one and thus a second-order phase transition can exist only on three-dimensional lattices. The order parameter for these models is

$$\langle\varphi\rangle = \left\langle \left(\sum_\alpha \langle\varphi\rangle_\alpha^2\right)^{1/2}\right\rangle ,$$

$$\langle\varphi\rangle_\alpha = \sum_{\mathbf{i}} \varphi_{\mathbf{i}}^\alpha .$$

The order parameter for the Heisenberg XY model in two dimensions, since its Hamiltonian is invariant under the action of the group $O(2)$, is the bidimensional vector

$$\langle\mathbf{M}\rangle = \left(\sum_{i,j=1}^n \mathbf{S}_{i,j}^x, \sum_{i,j=1}^n \mathbf{S}_{i,j}^y\right) \equiv \left(\sum_{i,j=1}^n \cos q_{i,j}, \sum_{i,j=1}^n \sin q_{i,j}\right), \tag{2.131}$$

which describes the mean spin orientation field.

The order parameter for the anisotropic Heisenberg XY model in three dimensions, whose Hamiltonian is still invariant under the action of the group $O(2)$, is the bidimensional vector

$$\langle\mathbf{M}\rangle = \left(\sum_{i,j,k=1}^n \mathbf{S}_{i,j,k}^x, \sum_{i,j,k=1}^n \mathbf{S}_{i,j,k}^y\right) \equiv \left(\sum_{i,j,k=1}^n \cos q_{i,j,k}, \sum_{i,j,k=1}^n \sin q_{i,j,k}\right) . \tag{2.132}$$

After the numerical computation of $\langle\varphi\rangle(\varepsilon)$, or of $\langle|\mathbf{M}|\rangle(\varepsilon)$, the temperature dependence of the order parameter $\langle\varphi\rangle(T)$, or $\langle|\mathbf{M}|\rangle(T)$, is parametrically given by $\langle\varphi\rangle(\varepsilon)$, or $\langle|\mathbf{M}|\rangle(T)$, with $T = T(\varepsilon)$.

By following the time evolution of the order parameter one can obtain very interesting information, for example, in [105] a Goldstone mode at work has been visualized. Bifurcations of the order parameter are in general very clear and allow a good numerical estimate of the corresponding critical exponent [89].

However, the best way to detect a phase transition and to locate the transition point is described in the following section.

Locating the Transition by Binder Cumulants

The critical properties of the infinite system can be inferred from the values of the thermodynamic observables in finite samples of different sizes using finite-size scaling [106, 107]. In particular, the transition point can be located by means of the so-called *Binder cumulants* [106]. The Binder cumulant g is defined as

$$g = 1 - \frac{\langle \phi^4 \rangle}{3 \langle \phi^2 \rangle^2} , \tag{2.133}$$

where $\langle \phi \rangle$ is a canonical average of the order parameter ϕ. For the lattice φ^4 models it is $\langle \phi \rangle \equiv \langle \varphi \rangle$, while for the XY models it is $\langle \phi \rangle \equiv \langle |\mathbf{M}| \rangle$. By computing $g = g(T, N)$ for at least three different values of N, the intersection point g^* of the corresponding temperature patterns $g = g(T)$ is a universal quantity at the critical point [107]. As long as canonical and microcanonical ensembles are equivalent, one computes the averages $\langle \phi^{2n} \rangle$ as time averages along the numerical solutions of the Hamilton equations of motion of the model under investigation; hence $g = g(\varepsilon, N)$. The critical energy density of a phase transition point, ε_c^∞, is then defined from the intersection of the curves $g(\varepsilon, N)$ worked out at different N values. Energy is the fundamental control parameter in Hamiltonian numerical simulations, and therefore the critical temperature T_c of a phase transition is obtained from $g^* \rightarrow \varepsilon_c^\infty$ and the caloric curve $T(\varepsilon)$.

Specific Heat

Another important macroscopic observable to characterize phase transitions is the specific heat. The numerical computation, by means of Hamiltonian dynamics, of the constant-volume specific heat C_V can proceed in different ways. Let us begin with a *microcanonical estimate* of the *canonical specific heat*. To this end, one considers a well-known formula that relates the average fluctuations of a generic observable computed in canonical and microcanonical ensembles [108], which specialized to the kinetic energy fluctuations reads

$$\langle \delta K^2 \rangle_{\text{micro}} = \langle \delta K^2 \rangle_{\text{can}} - \frac{\beta^2}{C_V} \left(\frac{\partial \langle K \rangle_{\text{can}}}{\partial \beta} \right)^2 , \tag{2.134}$$

where $C_V = (\partial E/\partial T)$, and where $\langle \cdot \rangle_{\text{micro}}$ and $\langle \cdot \rangle_{\text{can}}$ stand for microcanonical and canonical averages respectively. The quantity $\langle \delta K^2 \rangle_{\text{micro}}$ can be computed as a time average along the numerical trajectories. Considering that $\langle K \rangle_{\text{micro}} = \langle K \rangle_{\text{can}} = N/2\beta$, $\langle \delta K^2 \rangle_{\text{can}} = N/(2\beta^2)$, by inverting (2.134), one has

$$c_V(T) = \frac{C_V}{N} \rightarrow \begin{cases} c_V(\epsilon) = \dfrac{k_B d}{2}\left[1 - \dfrac{Nd}{2}\dfrac{\langle K^2 \rangle - \langle K \rangle^2}{\langle K \rangle^2}\right]^{-1}, \\ T = T(\epsilon), \end{cases} \qquad (2.135)$$

where d is the number of degrees of freedom for each particle. Time averages of the kinetic-energy fluctuations computed at any given value of the energy density ε yield $C_V(T)$, according to its parametric definition in (2.135).

In order to compute a *microcanonical* constant-volume specific heat, we have to start from its microcanonical definition $1/C_V = \partial T(E)/\partial E$, and then, according to the definition of the entropy that is adopted, in (2.126) and (2.127), two different formulas can be worked out [104], which are both exact at *any* value of N [in contrast to the expression (2.135)], and which coincide in the limit of arbitrarily large N. The first formula, derived from the entropy in (2.127), reads

$$c_V(\varepsilon) = \frac{C_V}{N} = [N - (N-2)\langle K \rangle \langle K^{-1} \rangle]^{-1}, \qquad (2.136)$$

whereas that derived from the entropy in (2.126) is

$$c_V(\varepsilon) = \frac{C_V}{N} = \frac{N(N-2)}{4}\left[(N-2) - (N-4)\frac{\langle K^{-2} \rangle}{\langle K^{-1} \rangle^2}\right]^{-1}, \qquad (2.137)$$

and these are the natural expressions to work out the microcanonical specific heat, the second one being more appropriate to Hamiltonian dynamical simulations.

Let us remark that, as is well known, all the thermodynamic quantities remain regular functions of the temperature, or of the energy, as long as N is finite, and no breaking of ergodicity and symmetry appears. Nevertheless, some marks of the transition show up neatly also in a finite system. The specific heat does not diverge, but exhibits a peak at some temperature $T_c^{C_V}(N)$. The height of the peak grows with the size of the system. Rigorously, the order parameter ϕ is expected to vanish on the whole temperature range for any finite value of N. In practice, e.g., in a canonical Monte Carlo simulation as well as in Hamiltonian dynamics (where the length of the sampling of ϕ is necessarily finite), the system is trapped in one phase for a time that grows exponentially with N [109]. A fictitious symmetry breaking is thus observed at a temperature $T_c^\phi(N)$, different in general from $T_c^{C_V}(N)$, even if

$$\lim_{N \to \infty} T_c^{C_V}(N) = \lim_{N \to \infty} T_c^\phi(N) = T_c^\infty. \qquad (2.138)$$

The Equation of State

The equation of state, worked out using the definition of entropy given in (2.126), is (again having chosen $k_B = 1$)

$$P = \frac{N}{V}T - \left\langle \left(\frac{\partial V}{\partial V}\right)_E K^{-1} \right\rangle \langle K^{-1}\rangle^{-1} , \qquad (2.139)$$

where P is the pressure, T is the microcanonical temperature computed according to (2.130), V is the potential function of the microscopic interactions entering the Hamiltonian, and K is again the kinetic energy. The brackets $\langle \cdot \rangle$ again stand for microcanonical averages to be numerically computed by means of time averages along the dynamical trajectories. In order to compute the volume derivative of the potential V, a standard method is to make the coordinate transformation $q_i = V^{1/d}q_i'$, $i = 1, \ldots, N$, where d is the space dimension (see [104]).

This is very useful for fluids or, in general, off-lattice systems. For example it has been successfully used to study the clustering transition in a self-gravitating N-body system [95] putting in evidence the absence of latent heat.

An Excerpt of Numerical Studies of Phase Transitions

To exemplify how clean and unequivocal the outcomes of the above-described dynamical study of phase transitions are, a few pictures are reported here (more details can be found in the cited papers). These refer to the models defined in (2.120) and studied in [89, 90], in (2.121) and (2.122) and studied in [105], in (2.123) and studied in [95]. Other interesting studies are reported in [110–116].

Let us begin with the 2D lattice φ^4 model with $O(1)$ symmetry group obtained by specializing (2.120) accordingly. In Figures 2.17, 2.18, and 2.19 it is shown what the order parameter, Binder cumulants, and the specific heat look like. The tendency of both the order parameter and specific heat to sharpen at increasing number of lattice sites is evident. These results look very similar to those commonly obtained by means of canonical Monte Carlo simulations.

On three-dimensional lattices one finds similar but sharper results in the case of $O(1)$ symmetry (see Figure 2.23, where the order parameter versus temperature is reported for a cubic lattice of $N = 8^3$ sites, results to be compared with those of Figure 2.17; the same figure also shows a comparison with canonical Monte Carlo simulations). For $O(2)$ and $O(4)$ symmetries, the φ^4 model has no symmetry-breaking phase transition on two-dimensional lattices (by the Mermin–Wagner theorem) and one must necessarily consider the 3D case in order to observe a second-order phase transition.

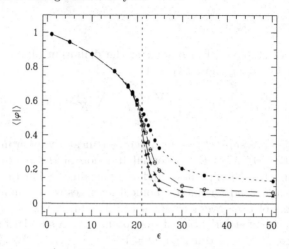

Fig. 2.17. 2D lattice φ^4 model with $O(1)$ symmetry. Absolute magnetization $\langle|\varphi|\rangle$ versus energy density ϵ. Symbols: $N = 10^2$ (solid circles), $N = 20^2$ (open circles), $N = 30^2$ (solid triangles), $N = 50^2$ (open triangles). From [89].

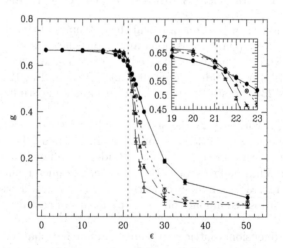

Fig. 2.18. Binder cumulants g versus energy density ϵ. The vertical dotted line marks the estimated transition value $\epsilon_c \approx 21.1$. The transition region is magnified in the inset. Symbols are the same as in Figure 2.17. From [89].

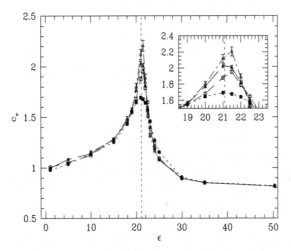

Fig. 2.19. Constant-volume specific heat c_V versus energy density ϵ. Symbols are the same as in Figure 2.17. From [89].

A similar situation holds in the case of the 2D classical Heisenberg XY model: since the model has a continuous symmetry, $O(2)$, one has to consider the three-dimensional case in order to observe a second-order phase transition.

Notice that both the 2D φ^4 and 2D XY models undergo a Kosterlitz–Thouless phase transition, that is, an infinite-order transition (see the classification given in Section 2.1.6). In Figures 2.20 and 2.21 the magnetization and the specific heat are respectively reported for the 3D XY model. Again a phenomenology is found that is familiar when the same models are tackled using Monte Carlo canonical computations. Finally, an example of the outcome of the computation of an isoenergetic (instead of isothermal) curve in the $P-V$ plane is given. This has been obtained for a self-gravitating N-body system in three spatial dimensions and off-lattice using the equation of state (2.139). This system, when a spatial scale is fixed, for example through some external confinement potential (the simplest way to realize it is through a "box"), undergoes a clustering phase transition at some critical energy value scaled with $N^{5/3}$ (because of the nonadditivity of the system [95]). Apart from the other ways of detecting it, this transition is also signaled by a sudden change from a polytropic curve ($P \propto V^{-4/3}$) corresponding to the clustered phase to a standard perfect gas curve ($P \propto V^{-1}$) proper to the nonclustered phase.

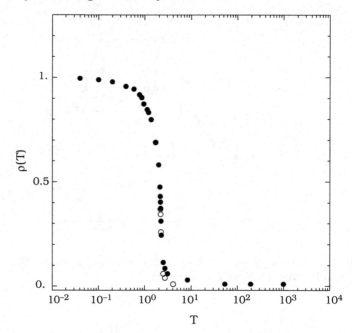

Fig. 2.20. Order parameter $\rho(T) = \langle |\mathbf{M}| \rangle (T)$ versus temperature T for the 3D XY model. $N = 10 \times 10 \times 10$ (full circles) and $N = 15 \times 15 \times 15$ (open circles). From [105].

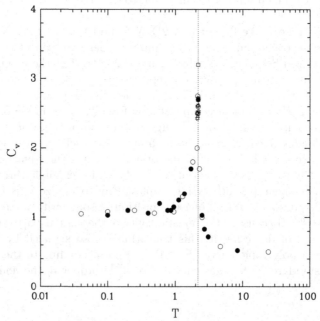

Fig. 2.21. Constant-volume specific heat versus temperature for the 3D XY model. $N = 8 \times 8 \times 8$ (open triangles); $N = 10 \times 10 \times 10$ (open circles); $N = 12 \times 12 \times 12$ (open stars); $N = 15 \times 15 \times 15$ (open squares). The vertical dotted line points out the critical temperature $T_c = 2.17$ at which the phase transition occurs. From [105].

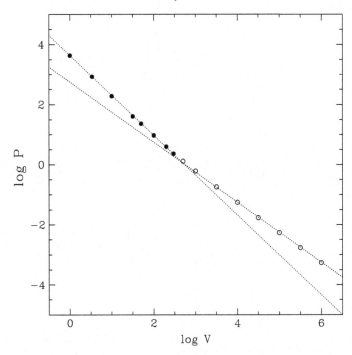

Fig. 2.22. Equation of state, pressure versus volume, for a self-gravitating N-body system with $N = 100$. Full circles correspond to the negative specific heat clustered phase and follow a polytropic law $P \propto \mathcal{V}^{-4/3}$. Open circles refer to the nonclustered (gas) phase and the data fit on a $P \propto \mathcal{V}^{-1}$ curve. From [95].

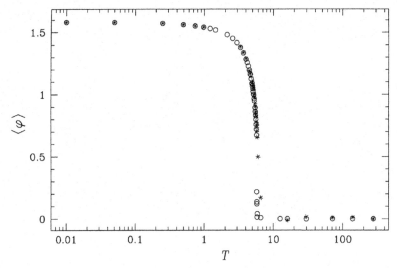

Fig. 2.23. Order parameter $\langle|\varphi|\rangle$ versus temperature for the $O(1)$ φ^4 model on a three-dimensional lattice of $N = 8 \times 8 \times 8$ sites. Open circles refer to the outcomes of Hamiltonian dynamics, stars refer to canonical Monte Carlo simulations. From [117].

Chapter 3

Geometrization of Hamiltonian Dynamics

A Hamiltonian system whose kinetic energy is a quadratic form in the velocities is referred to as a *standard*, or natural, Hamiltonian system. Every Newtonian system, that is, a system of particles interacting through forces derived from a potential, i.e., of the form (1.1), belongs to this class. The trajectories of a standard system can be seen as geodesics of a suitable Riemannian manifold. This classical result is based on the variational formulation of dynamics. In fact, Hamilton's principle states that the natural motions of a Hamiltonian system are the extrema of the functional (Hamiltonian action $\mathcal{S}$)

$$\mathcal{S} = \int L \, dt \, , \qquad (3.1)$$

where L is the Lagrangian function of the system, and the geodesics of a Riemannian manifold are the extrema of the length functional

$$\ell = \int ds \, , \qquad (3.2)$$

where s is the arc-length parameter. Once a connection between length and action is established, by means of a suitable choice of the metric, it will be possible to identify the geodesics with the physical trajectories.

3.1 Geometric Formulation of the Dynamics

The Riemannian formulation of classical dynamics is not unique, even if we restrict oureselves to the case of natural systems. There are several possible choices for the ambient space and its metric. The most commonly known choice—dating back to the nineteenth century—is the so-called Jacobi metric on the configuration space of the system. Actually this was the geometric framework of Krylov's work. Among other possibilities, we will also consider a metric originally introduced by Eisenhart on an enlarged configuration space-time. The choice of the metric to be used will be dictated mainly by practical convenience in performing computations.

These choices do not exhaust all the possibilities of geometrizing conservative dynamics. For instance, with regard to systems whose kinetic energy is not quadratic in the velocities—the classical example is a particle subject to conservative as well as velocity-dependent forces, such as the Lorentz force—it is impossible to give a Riemannian geometrization, but this becomes possible in the more general framework of a Finsler geometry [118]. In this chapter, a quick survey is given of some basic ways of putting into correspondence standard Hamiltonian flows with geodesic flows.

For a summary of the notation and the concepts of differential geometry that will be used in the following we refer the reader to Appendix B. The summation convention over repeated indices will be always used, if not explicity stated otherwise.

3.1.1 Jacobi Metric on Configuration Space M

Let us consider an autonomous dynamical system, i.e., a system with interactions that do not explicitly depend on time, whose Lagrangian can be written as

$$L = T - V = \frac{1}{2} a_{ik} \dot{q}^i \dot{q}^k - V(q) , \qquad (3.3)$$

where the dot stands for a derivative with respect to the parameter on which the q's depend,[1] and q is a shorthand notation for all the coordinates $q_1, \ldots, q_N$. Both these conventions will be used throughout the following chapters when there is no possibility of confusion.

As is well known, according to Hamilton's least action principle, the natural motions of the system are the phase space paths satisfying the integral condition

$$\delta \int_{t_0}^{t_1} L(q, \dot{q}) dt = 0 , \qquad (3.4)$$

or, equivalently, the Lagrange equations of motion, which are derived from (3.4).

The autonomous Hamiltonian $H(p, q) = T(p) + V(q)$, with $p_i = a_{ik}\dot{q}^i$, is a constant of motion equal to the energy value E of the initial conditions. The fact that the motions must be isoenergetic with energy E implies that the accessible part of the configuration space M is not the whole space, but only the subspace $M_E \subset M$ defined by

$$M_E = \{q \in M : V(q) \le E\} . \qquad (3.5)$$

---

[1] Such a parameter is the time t here, but could also be the arc length s in the following.

In fact, a curve γ' that lies outside M_E will never be parametrizable in such a way that the energy is E, because γ' will then pass through points where $V > E$ and the kinetic energy is positive.[2]

Now, using $L = p_i \dot{q}^i - E$ we can rewrite (3.4) in terms of the simplified action as

$$\delta \int_{t_0}^{t_1} p_i \dot{q}^i dt = 0 \tag{3.6}$$

together with the condition $\frac{1}{2} a_{ik} \dot{q}^i \dot{q}^k + V = E = \text{const.}$ Hence (3.6) reads also as

$$\delta \int_{t_0}^{t_1} a_{ik} \dot{q}^i \dot{q}^k dt = 0 , \tag{3.7}$$

and from the condition $E = \text{const}$,

$$a_{ik} \dot{q}^i \dot{q}^k = 2(E - V) = 2T . \tag{3.8}$$

Now, with a time reparametrization $t = t(\tau)$ such that

$$a_{ik}[q(\tau)] \frac{dq^i}{d\tau} \frac{dq^k}{d\tau} = 1 , \tag{3.9}$$

from

$$a_{ik}[q(\tau)] \frac{dq^i}{d\tau} \frac{dq^k}{d\tau} \left(\frac{d\tau}{dt} \right)^2 = 2T \tag{3.10}$$

one gets

$$dt = \frac{d\tau}{\sqrt{2T}} . \tag{3.11}$$

In such a way the trajectories are constrained on constant-energy surfaces, and from (3.7) and (3.11),

$$\delta \int_{t_0}^{t_1} a_{ik}[q(t)] \dot{q}^i(t) \dot{q}^k(t) dt = \delta \int_{t_0}^{t_1} 2T[q(t)] dt$$

$$= \delta \int_{\tau_0}^{\tau_1} 2T[q(t)] \frac{dt}{d\tau} d\tau = \delta \int_{\tau_0}^{\tau_1} \{2T[q(\tau)]\}^{1/2} d\tau = 0 . \tag{3.12}$$

Then, substituting the implicit definition of τ given in (3.9), the last integral in (3.12) is rewritten in a form that is independent of the time parameterization of the trajectories, i.e.,

$$\delta \int_{\gamma} \{2[E - V(q)] a_{ik} dq^i dq^k\}^{1/2} = 0 , \tag{3.13}$$

---

[2] The accessible configuration space M_E can then be seen as the union of all the "configuration subspaces" $\{q \in M : V(q) = E - T\}$ that one gets for all the possible values of T, $0 \leq T \leq E$.

where γ labels any isoenergetic curve joining two fixed endpoints $q_{(0)}$ and $q_{(1)}$; the varied paths are now asynchronous (with respect to physical time).[3] The meaning of (3.13) is that the mechanical motions derived from Hamilton's principle satisfy the condition

$$0 = \delta \int 2T \, dt = \delta \int \left(g_{ij} \dot{q}^i \dot{q}^j \right)^{1/2} dt = \delta \int ds \; . \tag{3.14}$$

That is, the natural motions are the geodesics of M_E, provided that its arc-length element ds is such that $ds^2 = g_{ik}(q) dq^i dq^k = 2[E - V(q)] a_{ik} dq^i dq^k$. In other words, the region M_E of the configuration space M of a dynamical system with N degrees of freedom has a differentiable manifold structure, and the Lagrangian coordinates $(q_1, \ldots, q_N)$ can be regarded as local coordinates on M. The latter, endowed with a proper metric stemming from the considerations given above, is a Riemannian manifold. For the sake of simplicity, considering systems of the form (1.1), for which the kinetic energy matrix is diagonal and the masses are all equal to one, i.e., $a_{ik} = \delta_{ik}$, the metric tensor is

$$g_{ik}(q) = 2[E - V(q)] a_{ik} \equiv 2[E - V(q)] \, \delta_{ik} \; . \tag{3.15}$$

This metric is referred to as the *Jacobi metric*, and its arc-length element is

$$ds^2 \equiv g_{ij} dq^i dq^j = 2[E - V(q)] \frac{dq^i}{dt} \frac{dq_i}{dt} dt^2 = 4[E - V(q)]^2 \, dt^2 \; . \tag{3.16}$$

We denote by (M_E, g_J) the mechanical manifolds endowed with the Jacobi metric.

In slightly more formal terms, let us consider the configuration space as a smooth N-dimensional manifold M. Let $TM = \bigcup_{q \in M} T_q M$ be its tangent bundle. The function $L : TM \to \mathbb{R}$ defined by $L = \frac{1}{2} \langle v, v \rangle$ is a free Lagrangian describing a free motion on M. In this case $L \equiv T$, i.e., the Lagrangian has only a kinetic part. The scalar product $\langle v, v \rangle$ for all the tangent vectors v provides a Riemannian metric on M. In local coordinates one has $L = g_{ij} \dot{q}^i \dot{q}^j$, where g_{ij} is the metric tensor on M.

Let ω_L be the Lagrangian closed 2-form on TM; ω_L is associated with the canonical symplectic form $\omega_0 = \sum_{i=1}^{N} dq^i \wedge dp_i$ defined on T^*M (phase space) by means of the Legendre transform $\mathbf{F}L : TM \to T^*M$, so: $\omega_L = (\mathbf{F}L)^* \omega_0$. In local coordinates, putting $L_{\dot{q}^i \dot{q}^j} = (\partial^2 L / \partial \dot{q}^i \partial \dot{q}^j)$, one has $\omega_L = \sum_{i,j=1}^{N} (L_{\dot{q}^i q^j} dq^i \wedge dq^j + L_{\dot{q}^i \dot{q}^j} dq^i \wedge d\dot{q}^j)$. We denote by X_E the unique Lagrangian vector field on TM such that

$$\omega_L(X_E, Y) = dE(Y)$$

for each arbitrary vector field Y on TM; E is the "energy" given by $E = S - L$ with $S : TM \to \mathbb{R}$ defined by $S(v_x) = \mathbf{F}L(v_x)v_x$.

---

[3] This version of the least action principle is known as Maupertuis's least action principle.

The solutions of the Euler–Lagrange equations are the natural motions of the system described by the Lagrangian L. Moreover, the natural motions are the integral curves of the vector field X_E. Notice that second-order equations are possible on TM but not on T^*M.

The so-called base integral curves of X_E are given by the canonical projection of the integral curves of X_E from TM to M. Conversely, for each curve $\gamma_0 : \mathbb{R} \to M$ a natural lift exists from M to TM, that is, $\gamma_0 \to (\gamma_0, \dot{\gamma}_0)$.

It can be proved that $\gamma_0 : \mathbb{R} \to M$ is a base integral curve of X_E if and only if γ_0 is a geodesic for M, that is, $\nabla_{\dot{\gamma}_0} \dot{\gamma}_0 = 0$ with $\nabla_{\dot{\gamma}_0}$ the covariant derivative of the canonical Levi-Civita connection associated to g_{ij}.

In local coordinates, having set $\gamma_0(s) = (q^1(s), \dots, q^N(s))$, one has

$$\frac{d^2 q^i}{ds^2} + \Gamma^i_{jk} \frac{dq^j}{ds} \frac{dq^k}{ds} = 0 \ ,$$

where s is the proper time and, as usual, summation over repeated indices is implicitly assumed; Γ^i_{jk} are the Christoffel coefficients of the Levi-Civita connection associated with g_{ij} and are given by

$$\Gamma^i_{jk} = \frac{1}{2} g^{im} \left(\frac{\partial g_{mk}}{\partial q^j} + \frac{\partial g_{mj}}{\partial q^k} - \frac{\partial g_{jk}}{\partial q^m} \right) \ . \tag{3.17}$$

Now, if $V : M \to \mathbb{R}$ is a potential energy function on M, we can incorporate it into the Lagrangian by defining

$$L_V = \frac{1}{2} \langle v, v \rangle - V(\pi_M v) \ , \tag{3.18}$$

where $\pi_M : TM \to M$ is the canonical projection of the tangent bundle, and then define energy as $E(v) = \frac{1}{2} \langle v, v \rangle + V(\pi_M v)$. Then $\gamma_0(s)$ is a base integral curve of the corresponding Lagrangian vector field X_E if and only if $\nabla_{\dot{\gamma}_0} \dot{\gamma}_0(s) = -\mathrm{grad} V(\gamma_0(s))$ that is, in local coordinates,

$$\frac{d^2 q^i}{ds^2} + \Gamma^i_{jk} \frac{dq^j}{ds} \frac{dq^k}{ds} = -g^{ij} \frac{\partial V}{\partial q^j} \ ,$$

which are the Euler–Lagrange equations.

Assuming M to be a compact manifold, there exists a number E such that $E > V(q)$ for $q \in M$. Then with such a number E one can associate the kinetic energy metric, or Jacobi metric, on M, by putting $\tilde{g} = (E - V(q)) g$; evidently $\tilde{g}$ is conformally equivalent to g, and in coordinates

$$\tilde{g}_{ij} = (E - V(q)) g_{ij} \ .$$

It can be shown that the base integral curves of the Lagrangian (3.3) coincide with geodesics of the Jacobi metric (3.15) up to a reparametrization with energy 1.

If we denote by $\tilde{\Gamma}^i_{jk}$ the connection coefficients derived from the metric (3.15), the corresponding geodesics are given by

$$\frac{d^2q^i}{ds^2} + \tilde{\Gamma}^i_{jk}(q)\frac{dq^j}{ds}\frac{dq^k}{ds} = 0 \ . \tag{3.19}$$

Let us restrict to those systems whose kinetic-energy term, with a suitable choice of local coordinates, can be diagonalized, and so let us assume that $\tilde{g}_{ij} = (E - V(q))\delta_{ij}$; hence, with $T = E - V$ and $T_{,i} = \partial T/\partial q^i = -\partial V/\partial q^i$, (3.19) gives

$$\frac{d^2q^i}{ds^2} + \frac{1}{2T}\left(2T_{,j}\frac{dq^j}{ds}\frac{dq^i}{ds} - \tilde{g}^{ij}T_{,j}\tilde{g}_{kl}\frac{dq^k}{ds}\frac{dq^l}{ds}\right) = 0 \ ;$$

finally, using $ds^2 = 2T^2dt^2$, this yields

$$\frac{d^2q^i}{dt^2} = -\frac{\partial V}{\partial q^i} \ ,$$

i.e., Newton's equations associated with L_V of (3.18).

To summarize, a general result for the Riemannian geometrization of natural Hamiltonian dynamics is the following:

Theorem 3.1. *Given a dynamical system on a Riemannian manifold (M, a), i.e., a dynamical system whose Lagrangian is*

$$L = \frac{1}{2}a_{ij}\dot{q}^i\dot{q}^j - V(q) \ ,$$

then it is always possible to find a conformal transformation of the metric

$$g_{ij} = e^{\varphi(q)}a_{ij}$$

such that the geodesics of (M, g) are the trajectories of the original dynamical system; this transformation is defined by

$$\varphi(q) = \log[E - V(q)] \ .$$

The proof proceeds as above, simply replacing all the δ_{ij} matrices with the kinetic-energy matrix a_{ij}; for details, see, e.g., [7].

3.1.2 Eisenhart Metric on Enlarged Configuration Space $M \times \mathbb{R}$

A first alternative choice of the ambient space and Riemannian metric, to reformulate Newtonian dynamics in a geometric language, was proposed in [119]. This makes use of an enlarged configuration space $M \times \mathbb{R}$, with local coordinates $(q^0, q^1, \ldots, q^N)$, where a proper Riemannian metric G_e is defined to give

$$ds^2 = (G_e)_{\mu\nu}\,dq^\mu dq^\nu = a_{ij}\,dq^i dq^j + A(q)\,(dq^0)^2 \ , \tag{3.20}$$

where μ and ν run from 0 to N and i and j run from 1 to N, and the function $A(q)$ does not explicitly depend on time. With the choice $1/[2A(q)] = V(q)+\eta$ and under the condition

$$q^0 = 2 \int_0^t V(q) \, d\tau + 2\eta t \, , \qquad (3.21)$$

for the extra variable it can easily be seen that the geodesics of the manifold $(M \times \mathbb{R}, G_e)$ are the natural motions of standard autonomous Hamiltonian systems. Since $\frac{1}{2}g_{ij}\dot{q}^i\dot{q}^j + V(q) = E$, where E is the energy constant along a geodesic, we can see that the following relation exists between q^0 and the action:

$$q^0 = -2 \int_0^t T \, d\tau + 2(E + \eta)t \, . \qquad (3.22)$$

Explicitly, the metric G_e is

$$G_e = \begin{pmatrix} [2V(q) + 2\eta]^{-1} & 0 & \cdots & 0 \\ 0 & a_{11} & \cdots & a_{1N} \\ \vdots & \vdots & \ddots & \vdots \\ 0 & a_{N1} & \cdots & a_{NN} \end{pmatrix}, \qquad (3.23)$$

and together with the condition (3.22), this gives an affine parametrization of the arc length with the physical time, i.e., $ds^2 = 2(E + \eta)dt^2$, along the geodesics that coincide with natural motions. The constant η can be set equal to an arbitrary value greater than the largest value of $|E|$ so that the metric G_e is nonsingular. Although this metric is a priori very interesting, because it seems to have some better property than the Jacobi metric and than another metric also due to Eisenhart and defined in the next section, we have not yet investigated how it works in practical applications.[4] In the case of a diagonal kinetic-energy metric, i.e. $a_{ij} \equiv \delta_{ij}$, the only non vanishing Christoffel symbols are

$$\Gamma^i_{00} = \frac{(\partial V/\partial q^i)}{[2V(q) + 2\eta]^2}, \qquad \Gamma^0_{i0} = -\frac{(\partial V/\partial q^i)}{[2V(q) + 2\eta]} \, , \qquad (3.24)$$

---

[4] Let us consider a few examples. In contrast to the Jacobi metric g_J, on the boundary $V(q) = E$ the metric G_e is nonsingular; moreover, by varying E we get a family of different metrics g_J, whereas by choosing a convenient value of η, at different values of the energy the metric G_e remains the same. The consequence is that a comparison among the geometries of the submanifolds of $(M \times \mathbb{R}, G_e)$—where the geodesic flows of different energies "live"—is sensible. In contrast, this is not true with (M_E, g_J). In some cases, the possibility of making this kind of comparison can be important (see Chapter 7). With respect to the Eisenhart metric g_e on $M \times \mathbb{R}^2$ in the next section, the metric G_e on $M \times \mathbb{R}$ defines a somewhat richer geometry, for example the scalar curvature of g_e is identically vanishing, which is not the case of G_e.

whence the geodesic equations

$$\frac{d^2q^0}{ds^2} + \Gamma_{i0}^0 \frac{dq^i}{ds}\frac{dq^0}{ds} + \Gamma_{0i}^0 \frac{dq^0}{ds}\frac{dq^i}{ds} = 0 \ , \tag{3.25}$$

$$\frac{d^2q^i}{ds^2} + \Gamma_{00}^i \frac{dq^0}{ds}\frac{dq^0}{ds} = 0 \ , \tag{3.26}$$

which, using the affine parametrization of the arc length with time, i.e., $ds^2 = 2(E + \eta)dt^2$, with $(dq^0/dt) = 2[V(q) + \eta]$ from (3.21), give

$$\frac{d^2q^0}{dt^2} = 2\frac{dV}{dt} \ ,$$
$$\frac{d^2q^i}{dt^2} = -\frac{\partial V}{\partial q_i}, \qquad i = 1, \ldots, N \ , \tag{3.27}$$

respectively. The first equation is the differential version of (3.21), and equations (3.27) are Newton's equations of motion.

3.1.3 Eisenhart Metric on Enlarged Configuration Space-Time $M \times \mathbb{R}^2$

Eisenhart also proposed a geometric formulation of Newtonian dynamics that makes use, as ambient space, of an enlarged configuration space-time $M \times \mathbb{R}^2$ of local coordinates $(q^0, q^1, \ldots, q^i, \ldots, q^N, q^{N+1})$. This space can be endowed with a nondegenerate pseudo-Riemannian metric [119] whose arc length is

$$ds^2 = (g_e)_{\mu\nu} \, dq^\mu dq^\nu = a_{ij} \, dq^i dq^j - 2V(q)(dq^0)^2 + 2 \, dq^0 dq^{N+1} \ , \tag{3.28}$$

where μ and ν run from 0 to $N + 1$ and i and j run from 1 to N, and which, from now on, will be referred to as the *Eisenhart metric*, and whose metric tensor will be denoted by g_e. The relation between the geodesics of this manifold and the natural motions of the dynamical system is contained in the following theorem [120]:

Theorem 3.2 (Eisenhart). *The natural motions of a Hamiltonian dynamical system are obtained as the canonical projection of the geodesics of $(M \times \mathbb{R}^2, g_e)$ on the configuration space-time, $\pi : M \times \mathbb{R}^2 \mapsto M \times \mathbb{R}$. Among the totality of geodesics, only those whose arc lengths are positive definite and are given by*

$$ds^2 = c_1^2 dt^2 \tag{3.29}$$

correspond to natural motions; the condition (3.29) can be equivalently cast in the following integral form as a condition on the extra coordinate q^{N+1}:

$$q^{N+1} = \frac{c_1^2}{2}t + c_2^2 - \int_0^t L \, d\tau \ , \tag{3.30}$$

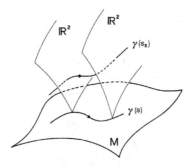

Fig. 3.1. Enlarged configuration space-time.

where c_1 and c_2 are given real constants. Conversely, given a point $P \in M \times \mathbb{R}$ belonging to a trajectory of the system, and given two constants c_1 and c_2, the point $P' = \pi^{-1}(P) \in M \times \mathbb{R}^2$, with q^{N+1} given by (3.30), describes a geodesic curve in $(M \times \mathbb{R}^2, g_e)$ such that $ds^2 = c_1^2 dt^2$.

For the full proof, see [120]. Since the constant c_1 is arbitrary, we will always set $c_1^2 = 1$ in order that $ds^2 = dt^2$ on the physical geodesics.

From (3.28) it follows that the explicit table of the components of the Eisenhart metric is given by

$$
g_e = \begin{pmatrix}
-2V(q) & 0 & \cdots & 0 & 1 \\
0 & a_{11} & \cdots & a_{1N} & 0 \\
\vdots & \vdots & \ddots & \vdots & \vdots \\
0 & a_{N1} & \cdots & a_{NN} & 0 \\
1 & 0 & \cdots & 0 & 0
\end{pmatrix} ,
\tag{3.31}
$$

where a_{ij} is the kinetic energy metric. The non vanishing Christoffel symbols, in the case $a_{ij} = \delta_{ij}$, are only

$$
\Gamma^i_{00} = -\Gamma^{N+1}_{0i} = \partial_i V ,
\tag{3.32}
$$

so that the geodesic equations read

$$
\frac{d^2 q^0}{ds^2} = 0 ,
\tag{3.33}
$$

$$
\frac{d^2 q^i}{ds^2} + \Gamma^i_{00} \frac{dq^0}{ds} \frac{dq^0}{ds} = 0 ,
\tag{3.34}
$$

$$
\frac{d^2 q^{N+1}}{ds^2} + \Gamma^{N+1}_{0i} \frac{dq^0}{ds} \frac{dq^i}{ds} = 0 ;
\tag{3.35}
$$

using $ds = dt$ one obtains

$$\frac{d^2 q^0}{dt^2} = 0 \ , \tag{3.36}$$

$$\frac{d^2 q^i}{dt^2} = -\frac{\partial V}{\partial q_i} \ , \tag{3.37}$$

$$\frac{d^2 q^{N+1}}{dt^2} = -\frac{dL}{dt} \ . \tag{3.38}$$

Equation (3.36) states only that $q^0 = t$. The N equations (3.37) are Newton's equations, and (3.38) is the differential version of (3.30).

The fact that in the framework of the Eisenhart metric the dynamics can be geometrized with an affine parametrization of the arc length, i.e., $ds = dt$, will be extremely useful in the following, together with the remarkably simple curvature properties of the Eisenhart metric (see Section 3.4).

3.2 Finslerian Geometrization of Hamiltonian Dynamics

There is another natural way, which makes use of Finsler spaces, of rephrasing in geometric terms Hamilton's least-action principle. Below we outline some basic ideas in a coordinate-dependent formulation of this geometric setting of a variational problem. Details can be found in the classic book by H. Rund [118] or, in modern language, in [121].

Consider a differentiable manifold M and a vector X tangent to M at a point P, that is, $X \in T_P M$. The norm $\|X\|$ of this vector—in analogy with its definition in a Euclidean space with curvilinear coordinates—will be given by the modulus of some function $F(x^j, X^j)$ of the coordinates x^j of the point P and of the components X^j of the vector X. This function, to properly define the norm of a vector, has to satisfy the following conditions: (i) $F(x^j, X^j)$ has to be smooth; (ii) $F(x^j, X^j)$ has to remain invariant under generic coordinate changes on M; (iii) $F(x^j, X^j)$ has to be positively homogeneous of first degree in the variables X^j, i.e., $F(x^j, \lambda X^j) = |\lambda| F(x^j, X^j)$ for any $\lambda > 0$.

Condition (i) means that the norm of a vector does not depend on the choice of coordinates on M, and condition (iii) means that, for any $\lambda > 0$, the norm $\|\lambda X\|$ of the vector whose components are λX^j is given by $\|\lambda X\| = \lambda |F(x^j, X^j)|$.

Then consider a curve $c(t)$ on M, represented in local coordinates by $x^j = x^j(t)$. The infinitesimal displacement ds along $c(t)$ is $ds = F(x^j, dx^j)$ with $dx^j = \dot{x}^j dt$; note that ds is a proper arc length if $F(x^j, dx^j) > 0$. The arc length ℓ_{AB} of the piece of $c(t)$ between the endpoints A and B is

$$\ell_{AB} = \int_A^B F(x^j, dx^j) = \int_{t_A}^{t_B} F(x^j, \dot{x}^j) \, dt \ ,$$

and, as can be easily verified, this arc length is invariant for arbitrary twice-differentiable and invertible time reparametrizations $t \rightarrow \tau(t)$ with $\dot{\tau} > 0$.

Euler's theorem on homogeneous functions states that $\dot{x}^j \partial_{\dot{x}^j} F(x, \dot{x}) = F(x, \dot{x})$, where $\partial_{\dot{x}^j} \equiv \partial/\partial \dot{x}^j$, and thus also $\dot{x}^h \partial^2_{\dot{x}^j \dot{x}^h} F(x, \dot{x}) = 0$, which entails

$$\det[\partial^2_{\dot{x}^j \dot{x}^h} F(x, \dot{x})] = 0 .$$

Now, from

$$\frac{1}{2} \frac{\partial F^2(x, \dot{x})}{\partial \dot{x}^j} = F(x, \dot{x}) \frac{\partial F(x, \dot{x})}{\partial \dot{x}^j}$$

we get

$$\frac{1}{2} \frac{\partial^2 F^2(x, \dot{x})}{\partial \dot{x}^j \partial \dot{x}^h} = \frac{\partial F(x, \dot{x})}{\partial \dot{x}^j} \frac{\partial F(x, \dot{x})}{\partial \dot{x}^h} + F(x, \dot{x}) \frac{\partial^2 F^2(x, \dot{x})}{\partial \dot{x}^j \partial \dot{x}^h} ,$$

and using the above-mentioned consequences of Euler's theorem for homogeneous functions, it follows that

$$F^2(x, \dot{x}) = \frac{1}{2} \frac{\partial^2 F^2(x, \dot{x})}{\partial \dot{x}^j \partial \dot{x}^h} \dot{x}^j \dot{x}^h . \tag{3.39}$$

Putting

$$g_{jh}(x, \dot{x}) = \frac{1}{2} \frac{\partial^2 F^2(x, \dot{x})}{\partial \dot{x}^j \partial \dot{x}^h} , \tag{3.40}$$

equation (3.39) is rewritten as

$$F^2(x, \dot{x}) = g_{jh}(x, \dot{x}) \dot{x}^j \dot{x}^h . \tag{3.41}$$

The norm $\|X\|$ of a vector $X \in T_P M$ is then

$$\|X\| = \|g_{jh}(x, X) X^j X^h\|^{1/2} ,$$

and with the components dx^j of an infinitesimal displacement at the point P, the line element ds reads

$$ds^2 = g_{jh}(x, dx) dx^j dx^h .$$

The quantities $g_{jh}(x, \dot{x})$ are the components of a $(0, 2)$-type symmetric tensor, as can be easily verified by explicitly computing how they transform under a nonsingular coordinate transformation $\bar{x}^j = \bar{x}^j(x^h)$. If (i) $F(x, \dot{x}) \geq 0$, (ii) $F(x, \dot{x}) = 0$ iff $\dot{x}^1 = \dot{x}^2 = \cdots = \dot{x}^N = 0$, and (iii) the rank of $\partial^2 F / \partial \dot{x}^j \partial \dot{x}^h$ is $N - 1$, then the manifold M is said to be a Finsler space, or a Finslerian manifold.

In order to get a regular variational problem from Hamilton's least-action principle, the Lagrangian function $L(q^i, \dot{q}^i)$ has to satisfy the conditions given above. In order to satisfy the request of invariance for time reparametrization, the function $L(q^i, \dot{q}^i)$ has to be homogeneous of degree one in the velocities, that is, $L(q^i, \lambda \dot{q}^i) = \lambda L(q^i, \dot{q}^i)$, $\lambda > 0$. This condition is not so stringent as it may appear. In fact, by adding a supplementary dimension, to be later identified with the physical time, it is possible to define a parametrically invariant extension of the initial Lagrangian on the space $\mathcal{M} = M \times \mathbb{R}$ as follows [122]:

$$\Lambda(q^a, q'^a) = L(q^i, q'^i/q'^0)q'^0 \ , \quad a = 0, 1, \ldots, N; \quad i = 1, \ldots, N \ , \quad (3.42)$$

where the parametric representation $q^a(u)$ is expressed as a function of an arbitrary parameter u. We have put $q'^a = dq^a/du$ and $dq^i/dt = q'^i/q'^0$. The function $\Lambda(q^a, q'^a)$ in (3.42) is homogeneous of degree one in the velocities. Then taking u as integration variable, the Hamiltonian action is given by

$$S = \int_{u_0}^{u_1} \Lambda(q^a, q'^a) du \qquad (3.43)$$

where the explicit expression of Λ reads as

$$\Lambda(q^a, q'^a) = \frac{1}{2} a_{ij} \frac{q'^i q'^j}{q'^0} - V(q) q'^0 \ .$$

In such a way, trajectories of a Hamiltonian system in configuration space-time are given by the extremals of the functional (3.43), and the formalism is invariant with respect to time reparametrizations.

To define a Finsler metric, the function $\Lambda(q^a, q'^a)$ has to be positive valued on the tangent bundle $T\mathcal{M}$. This is ensured by adding a convenient constant to the potential function, or by adding to the Lagrangian a suitable "gauge" function in the form of total time derivative [123], that is, $L(q, q') \to L(q, q') + (d\mathcal{G}(q)/dt)$. The metric tensor $g_{ab}(q^a, q'^a)$, defined through the metric function Λ as

$$g_{ab} = \frac{1}{2} \frac{\partial^2 \Lambda^2}{\partial q'^a \partial q'^b} \ ,$$

provides the manifold $\mathcal{M}$ with a Finslerian structure. For standard systems one obtains

$$g_{00} = \frac{1}{2} \frac{\partial^2 \Lambda^2}{\partial^2 (q'^0)^2} = 3T^2 + V^2 \ ,$$

$$g_{0i} = \frac{1}{2} \frac{\partial^2 \Lambda^2}{\partial q'^0 \partial q'^i} = -2T a_{ij} \frac{q'^j}{q'^0} \ , \qquad (3.44)$$

$$g_{ij} = \frac{1}{2} \frac{\partial^2 \Lambda^2}{\partial q'^i \partial q'^j} = a_{ih} a_{jk} q'^h q'^k (q'^0)^{-2} + a_{ij} (T - V) \ .$$

The geodesic equations are

$$\frac{d^2 q^a}{ds^2} + \gamma_{bc}^a (q, q') \frac{dq^b}{ds} \frac{q^c}{ds} = 0 \ , \qquad (3.45)$$

where s is the arc length, and

$$\gamma^a_{bc}(q, q') = \frac{1}{2} g^{ar} \left(\frac{\partial g_{cr}}{\partial q^b} + \frac{\partial g_{br}}{\partial q^r} - \frac{\partial g_{bc}}{\partial q^r} \right) . \tag{3.46}$$

These geodesic equations on $\mathcal{M}$, associated with the extended Lagrangian Λ, give

$$\frac{dp_i}{du} - q'^0 \frac{\partial L}{\partial q^i} = 0 \qquad i = 1, \dots, N ,$$

$$\frac{d}{du} \frac{\partial \Lambda}{\partial q'^0} = \frac{dH}{du} = 0 ,$$

where $p_i = \partial \Lambda / \partial q'^i$ and H is the Hamiltonian function associated with L.

The first N equations, written in terms of the initial parameter t (i.e., $q^0 = t$), are the equations of motion associated with the Lagrangian $L(q, \dot{q})$, while the zeroth equation is nothing but the energy conservation along any geodesic.

Different definitions can be given for the connection on a Finslerian manifold, hence of the covariant derivation, and consequently, different curvature tensors can be defined. However, there is an axact analogue in the Finsler setting of the canonical Levi-Civita connection on Riemannian manifolds: the Cartan connection. The Cartan connection is almost torsion free (in a sense that can be made precise), and it reduces to the Levi-Civita connection when the Finsler metric reduces to a Riemannian metric.

Finsler geometry has been successfully used to approach the study of Hamiltonian chaos in [123, 124] where dynamical systems with few degrees of freedom have been considered. A priori this could be a very useful geometric framework to tackle dynamical systems involving electromagnetic interactions.

3.3 Sasaki Lift on TM

It is natural to wonder whether a Riemannian description of Hamiltonian dynamics is possible that makes use, as ambient space, of the tangent bundle TM, the cotangent bundle T^*M (phase space), or of the constant-energy hypersurface $\Sigma_E \subset TM$, which is the manifold in which the trajectories of an autonomous Hamiltonian system are naturally defined. The answer is yes, and one possibility is to lift the Jacobi metric up to the tangent or cotangent bundle. As long as standard Hamiltonian systems, like those we are mainly interested in, are considered, the tangent and cotangent bundle can be interchanged, and we consider only TM, which for simplicity is also referred to as the phase space of our system. The particular lift of the Jacobi metric that we briefly consider is the so-called Sasaki—or diagonal—lift.

This metric can be defined as follows [125]: Given a Riemannian manifold (M, g) let us consider two vectors X and Y tangent to to TM at the point (q, v). Suppose that X and Y are tangent at $t = 0$ to the curves $\bar{\alpha}(t) = (\alpha(t), V(t))$ and $\bar{\beta}(t) = (\beta(t), W(t))$ respectively. Denote by $\nabla V/dt$ and $\nabla W/dt$ the covariant derivatives of the vector fields $V(t)$ and $W(t)$ along $\alpha(t)$ and $\beta(t)$. Then the Sasaki metric g_s on TM at (q, v) is defined by

$$g_s|_{(q,v)}(X, Y) = g(\dot{\alpha}(0), \dot{\beta}(0)) + g\left(\left.\frac{\nabla V}{dt}\right|_0, \left.\frac{\nabla W}{dt}\right|_0\right) . \tag{3.47}$$

This metric is perhaps the most natural metric on TM depending only on the Riemannian structure defined on M. It is worth noticing that the trajectories of the geodesic flow, i.e., the flow on TM given by the solutions of the geodesic equations on (M, g), are geodesics of (TM, g_s) [126, 127]; hence this metric is the natural tool to geometrize dynamics using the phase space as ambient manifold. Given an orthonormal frame $(e_1, \ldots, e_N)$ on an open set $U \subset M$, with $(q^1, \ldots, q^N)$ a local coordinate system on U, the natural coordinate system on $\pi^{-1}(U)$ is $(q^1, \ldots, q^N, v^1, \ldots, v^N)$, defined as follows:

$$q^i(q, v) = q^i , \quad v^i(q, v) = v^i , \quad (q, v) \in \pi^{-1}(U) ,$$

where $v = \sum_i v^i e_i(q)$. In this coordinate system the components of g_s are given by [128]

$$g_s = \begin{pmatrix} g_{ij} + g_{kl}\Gamma^k_{im}\Gamma^l_{jn}v^m v^n & g_{ik}\Gamma^k_{jl}v^l \\ g_{ik}\Gamma^k_{jl}v^l & g_{ij} \end{pmatrix} , \tag{3.48}$$

where the Γ's are the Christoffel symbols of the canonical connection associated with g. The reason why this metric is also referred to as the diagonal lift of g to TM is that it is possible to find a reference frame in which the metric is diagonal. A curve $\bar{\gamma} : I \subset \mathbf{R} \mapsto TM$, $t \mapsto (\gamma(t), V(t))$, is horizontal if the vector field $V(t)$ is parallel along $\gamma = \pi(\bar{\gamma})$. A vector on TM is horizontal if it is tangent to a horizontal curve, or vertical if it is tangent to a fiber. Let $\gamma : I \mapsto M$, $t \mapsto \gamma(t)$ be a curve through the point $q = \gamma(0)$. For each tangent vector $v \in T_q M$ there exists a unique horizontal curve $\gamma^H : I \mapsto TM$ through (q, v) that projects onto γ. This curve is defined by

$$\gamma^H(t) = (\gamma(t), V(t)) ,$$

where $V(t)$ is obtained from v by parallel transport along γ. The curve γ^H is called a horizontal lift of γ. The horizontal lift of a vector field X on M is the unique vector field X^H on TM that is horizontal and that projects onto X. Let us denote by Γ^i_j the quantities defined by[5]

$$\nabla_X e_j = \Gamma^i_j(X)e_i ; \tag{3.49}$$

---

[5] These $\Gamma^i_j(X)$ must not be confused with the Christoffel symbols Γ^i_{jk}; the two quantities are linked by the relation $\Gamma^i_{jk} = \langle \Gamma^i_j(e_i), e_k \rangle$.

then in terms of the local frame the horizontal lift is defined by

$$X^H = X - \Gamma^i_j(X)v^j \frac{\partial}{\partial v^i} , \qquad (3.50)$$

while the *vertical lift* of X is

$$X^V = X^i \frac{\partial}{\partial v^i} . \qquad (3.51)$$

Vertical and horizontal vectors are orthogonal with respect to g_s. For each pair of vectors X and Y on M, we have

$$g_s(X^H, Y^H) = g_s(X^V, Y^V) = g(X, Y) ,$$
$$g_s(X^H, Y^V) = 0 . \qquad (3.52)$$

Hence the frame $(e_1^H, \ldots, e_N^H, e_1^V, \ldots, e_N^V)$ is orthonormal and the components of the Sasaki metric in this frame become

$$g_s = \begin{pmatrix} g_{ij} & 0 \\ 0 & g_{ij} \end{pmatrix} . \qquad (3.53)$$

Let us remark that the metric induced on the fibers $\pi^{-1}(q)$ is Euclidean. This condition, together with the condition of orthogonality of the vertical and horizontal fields, uniquely determines the Sasaki metric.

3.4 Curvature of the Mechanical Manifolds

In this section we give some elementary formulae concerning basic curvature properties of the manifolds hitherto introduced to geometrize Hamiltonian flows.

Curvature of (M_E, g_J)

We have already observed that the Jacobi metric is a conformal deformation of the kinetic-energy metric, whose components are given by the kinetic-energy matrix a_{ij}. In the case of systems whose kinetic-energy matrix is diagonal, this means that the Jacobi metric is conformally flat (see Appendix B, Section B.3). This greatly simplifies the computation of curvatures. It is convenient to define then a symmetric tensor C whose components are [7]

$$C_{ij} = \frac{N-2}{4(E-V)^2} \left[2(E-V)\partial_i\partial_j V + 3\partial_i V \partial_j V - \frac{\delta_{ij}}{2}|\nabla V|^2 \right] , \qquad (3.54)$$

where V is the potential, E is the energy, and ∇ and $|\cdot|$ stand for the Euclidean gradient and norm, respectively. The curvature of (M_E, g_J) can be expressed through C. In fact, the components of the Riemann tensor are

$$R_{ijkm} = \frac{1}{N-2} [C_{jk}\delta_{im} - C_{jm}\delta_{ik} + C_{im}\delta_{jk} - C_{ik}\delta_{jm}] . \qquad (3.55)$$

By contraction of the first and third indices, we obtain the Ricci tensor, whose components are (see Appendix B)

$$R_{ij} = \frac{N-2}{4(E-V)^2} \left[2(E-V)\partial_i\partial_j V + 3\partial_i V\partial_j V \right]$$

$$+ \frac{\delta_{ij}}{4(E-V)^2} \left[2(E-V)\triangle V - (N-4)|\nabla V|^2 \right] , \qquad (3.56)$$

and by a further contraction we obtain the scalar curvature (see Appendix B)

$$\mathcal{R} = \frac{N-1}{4(E-V)^2} \left[2(E-V)\triangle V - (N-6)|\nabla V|^2 \right] . \qquad (3.57)$$

Curvature of $(M \times \mathbb{R}, G_e)$

The basic curvature properties of the Eisenhart metric G_e can be derived by means of the Riemann curvature tensor, which is found to have the non-vanishing components

$$R_{0i0j} = \frac{\partial_i\partial_j V}{(2V+2\eta)^2} - \frac{3(\partial_i V)(\partial_j V)}{(2V+2\eta)^3} , \qquad (3.58)$$

where $\partial_i \equiv \partial/\partial q^i$ and whence, after contraction, using $G^{00} = 2V + 2\eta$ the components of the Ricci tensor are found to be

$$R_{kj} = \frac{\partial_k\partial_j V}{(2V+2\eta)} - \frac{3(\partial_k V)(\partial_j V)}{(2V+2\eta)^2} ,$$

$$R_{00} = \frac{\triangle V}{(2V+2\eta)^2} - \frac{3(\nabla V)^2}{(2V+2\eta)^3} , \qquad (3.59)$$

where $\triangle V = \sum_{i=1}^{N} \partial^2 V/\partial q^{i\,2}$, and thus we find that the Ricci curvature at the point $q \in M \times \mathbb{R}$ and in the direction of the velocity vector $\dot{q}$ is

$$K_R(q,\dot{q}) = \triangle V + R_{ij}\dot{q}^i\dot{q}^j \qquad (3.60)$$

and the scalar curvature at $q \in M \times \mathbb{R}$ is

$$\mathcal{R}(q) = \frac{\triangle V}{(2V+2\eta)} - \frac{3(\nabla V)^2}{(2V+2\eta)} . \qquad (3.61)$$

Curvature of $(M \times \mathbb{R}^2, g_e)$

The curvature properties of the Eisenhart metric g_e are much simpler than those of the Jacobi metric and also simpler than those of the Eisenhart metric G_e, and this is obviously a great advantage from a computational point of view. The only nonvanishing components of the Riemann curvature tensor are

$$R_{0i0j} = \partial_i\partial_j V ; \qquad (3.62)$$

hence the Ricci tensor has only one nonzero component,

$$R_{00} = \Delta V , \tag{3.63}$$

so that the Ricci curvature is

$$K_R(q, \dot{q}) = R_{00} \dot{q}^0 \dot{q}^0 \equiv \Delta V ,$$

and the scalar curvature is identically vanishing,

$$\mathcal{R}(q) = 0 . \tag{3.64}$$

Curvature of (TM_E, g_s)

We already observed that the Sasaki metric is the most natural metric on the tangent bundle of a Riemannian manifold, since it depends only on the metric of M. Hence also the curvatures of (TM_E, g_s) can be expressed through the curvatures of (M, g_J) (see, e.g., [128]). Unfortunately, the expressions for the curvatures are very complicated. A "simple" expression has been established in [125] for the scalar curvature, which reads

$$\mathcal{R}_s(q, v) = \mathcal{R}(q) - \frac{1}{4} \sum_{i,j,k,l,m=1}^{N} R_{ijkm}(q) R_{ijlm}(q) v^k v^l . \tag{3.65}$$

The constant-energy hypersurface $\Sigma_E = \{(q, p) : \mathcal{H}(q, p) = E\}$ is a submanifold of TM after identifying p with the tangent vector v, which is possible in the case of the standard Hamiltonians that we are considering. In fact $\Sigma_E = T_1 M$, i.e., it is the set of all the pairs (q, v) such that the tangent vectors v have length 1 with respect to the Jacobi metric. Moreover, Σ_E is also a Riemannian submanifold of (TM_E, g_s), and the remarkable result is that it is a *totally geodesic* submanifold. In fact, since the trajectories of the geodesic flow on TM are geodesics of (TM_E, g_s) and by energy conservation these curves are constrained on Σ_E, every geodesic of TM that starts on Σ_E remains on Σ_E, and this is one of the definitions of a totally geodesic submanifold. The consequences are particularly important as long as the curvature properties are concerned, in fact the curvature of a totally geodesic submanifold N of M equipped with the induced metric is simply given by the restriction to N of the curvature of M. Hence measuring the curvature of (TM_E, g_s) at a point $(q, v) \in \Sigma_E$ we directly measure the curvature of Σ_E equipped with the metric induced by the immersion in (TM_E, g_s).

Curvature of $(M \times \mathbb{R})$ with a Finsler Metric

On a Finsler manifold, several plausible analogues of the covariant derivation on Riemannian manifolds can be given. The first useful definition concerns the so-called δ-derivative, which for a vector field X reads

$$\frac{\delta X^a}{\delta u} = \frac{d X^a}{du} + P^a_{bc}(q, q') X^b q^c , \tag{3.66}$$

where $P^a_{bc}(q, q') = \gamma^a_{bc}(q, q') - C^a_{bd}(q, q')\gamma^d_{rc}(q, q')q'^r$, with $\gamma^a_{bc}(q, q')$ defined in (3.46) and $C^a_{bc} = g^{ar}C_{rbc}$ with the coefficients C_{rbc} defined below. Then, by computing the commutator of a double δ-derivation one obtains the analogue of the Riemann–Christoffel curvature tensor, whose components, for a Finsler space, read [118]

$$K^a_{bcd}(q^r, q'^r) = \left(\frac{\partial\Gamma^{*a}_{bc}}{\partial q^d} - \frac{\partial\Gamma^{*a}_{bc}}{\partial q'^s}\frac{\partial G^s}{\partial q'^d}\right) - \left(\frac{\partial\Gamma^{*a}_{bd}}{\partial q^c} - \frac{\partial\Gamma^{*a}_{bd}}{\partial q'^s}\frac{\partial G^s}{\partial q'^c}\right)$$

$$+ \Gamma^{*a}_{sd}\Gamma^{*s}_{bc} - \Gamma^{*a}_{sc}\Gamma^{*s}_{bd} , \tag{3.67}$$

where

$$G^a = \frac{1}{2}\Gamma^a_{bd}\, q'^b q'^d , \tag{3.68}$$

$$\Gamma^{*c}_{db} = \Gamma^c_{db} - g^{ac}\left(C_{bas}\frac{\partial G^s}{\partial q'^d} + C_{das}\frac{\partial G^s}{\partial q'^b} - C_{dbs}\frac{\partial G^s}{\partial q'^a}\right) \tag{3.69}$$

and

$$C_{abd} = \frac{1}{2}\frac{\partial g_{ab}}{\partial q'^d} . \tag{3.70}$$

If we restrict to the Riemannian case $g = g(q^a)$, the symbols C_{abc} identically vanish, so that the coefficients Γ^{*a}_{bd} reduce to the usual Christoffel symbols Γ^a_{bd}, and the curvature tensor K^a_{bcd} reduces to the Riemann tensor. With the curvature tensor K^a_{bcd} we can define the analogue of sectional curvatures, of the Ricci curvature tensor and of the scalar curvature. Just to call the reader's attention to the different degrees of complexity of the geometries of the different frameworks reviewed in the present chapter, we give the analytic expression of the scalar curvature $\mathcal{R}(q, q')$ of an $(N+1)$-dimensional Finsler space:

$$\mathcal{R}(q, q') = \frac{K^a_{bca}(q, q')q'^b q'^c}{N\Lambda^2(q, q')(q'^0)^2} . \tag{3.71}$$

It is reasonable to think that the different degrees of complexity of the geometries of the different ambient manifolds have to correspond to different capabilities of "capturing" information on the dynamics.

3.5 Curvature and Stability of a Geodesic Flow

The possibility of identifying a Hamiltonian flow with a geodesic flow on a Riemannian manifold provides a real practical resource for studying the stability/instability properties (chaos) of Hamiltonian dynamics. In fact, the geometrization of dynamics allows one to borrow from geometry the link between the curvature features of a given manifold and the stability/instability properties of its geodesics.

Studying the instability of dynamics means determining the evolution of perturbations of a given trajectory, as has already been discussed in Section 2.2.2.

Let us now translate the stability problem into geometric language. By writing, in close analogy to what has been done above in the case of dynamical systems, a perturbed geodesic as

$$\tilde{q}^i(s) = q^i(s) + J^i(s) , \tag{3.72}$$

and then inserting this expression in the equation for the geodesics, one finds that the evolution of the perturbation vector J is given by the following equation:

$$\frac{\nabla^2 J^k}{ds^2} + R^k{}_{ijr}\frac{dq^i}{ds}J^j\frac{dq^r}{ds} = 0 , \tag{3.73}$$

where $R^k{}_{ijr}$ are the components of the Riemann–Christoffel curvature tensor (see Appendix B). Equation (3.73) is referred to as the Jacobi equation, and $(\nabla J^k/ds) = dJ^k/ds + \Gamma^k_{ij}(dq^i/ds)J^j$ is the covariant derivative of the tangent vector field J known as the Jacobi geodesic separation field. Since in the context of Riemannian geometry this equation was first studied by Levi-Civita, it is also often referred to as the equation of Jacobi and Levi-Civita. For a derivation we refer to Appendix B, Section B.4, where it is also shown that one can always assume that J is orthogonal to the velocity vector $\dot\gamma$ along the geodesic i.e.,

$$\langle J, \dot\gamma \rangle = 0 , \tag{3.74}$$

where $\langle \cdot, \cdot \rangle$ stands for the scalar product induced by the metric (see Appendix B). The remarkable fact is that the evolution of J—and then the stability or instability of the geodesic—is completely determined by the *curvature* of the manifold. Therefore, if the metric is associated with a physical system, as in the case of Jacobi or Eisenhart metrics, such an equation links the stability or instability of the trajectories to the curvature of the "mechanical" manifold. Let us begin by computing the left-hand side of (3.73).

From $(\nabla J^k/ds) = dJ^k/ds + \Gamma^k_{ij}(dq^i/ds)J^j$ we have

$$\frac{\nabla^2}{ds^2}J^k = \frac{d}{ds}\left(\frac{dJ^k}{ds} + \Gamma^k_{ij}\frac{dq^i}{ds}J^j\right) + \Gamma^k_{rt}\frac{dq^r}{ds}\left(\frac{dJ^t}{ds} + \Gamma^t_{ij}\frac{dq^i}{ds}J^j\right) ; \tag{3.75}$$

trivial algebra leads to

$$\frac{\nabla^2}{ds^2}J^k = \frac{d^2 J^k}{ds^2} + 2\Gamma^k_{ij}\frac{dq^i}{ds}\frac{dJ^j}{ds} + \left(\partial_r\Gamma^k_{ij} + \Gamma^k_{rt}\Gamma^t_{ij} - \Gamma^k_{tj}\Gamma^t_{ri}\right)\frac{dq^r}{ds}\frac{dq^i}{ds}J^j , \tag{3.76}$$

where $\partial_i \equiv \partial/\partial q^i$. Then, we use the expression for the components of the Riemann–Christoffel tensor to obtain

$$R^k{}_{ijr}\frac{dq^i}{ds}J^j\frac{dq^r}{ds} = \left(\Gamma^t_{ri}\Gamma^k_{jt} - \Gamma^t_{ji}\Gamma^k_{rt} + \partial_j\Gamma^k_{ri} - \partial_r\Gamma^k_{ji}\right)\frac{dq^r}{ds}J^j\frac{dq^i}{ds} , \tag{3.77}$$

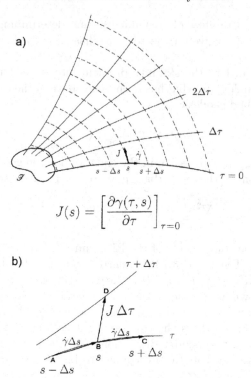

$$J(s) = \left[\frac{\partial \gamma(\tau, s)}{\partial \tau}\right]_{\tau=0}$$

Fig. 3.2. Jacobi vector field J. Figure (a) shows a congruence of geodesics $\gamma(\tau, s)$ issuing from a neighborhood $\mathcal{I}$. Each geodesic is labeled by a continuous parameter τ; on a reference geodesic $\gamma(0, s)$ the separation vector field $J(s)$ is defined by $J(s) = [\partial \gamma(\tau, s)/\partial \tau]_{\tau=0}$. Figure (b) shows that J locally measures the distance between close geodesics. From [129].

and by substituting (3.76) and (3.77) into (3.73) we finally get

$$\frac{d^2 J^k}{ds^2} + 2\Gamma^k_{ij}\frac{dq^i}{ds}\frac{dJ^j}{ds} + \left(\frac{\partial \Gamma^k_{ri}}{\partial q^j}\right)\frac{dq^r}{ds}\frac{dq^i}{ds}J^j = 0 , \qquad (3.78)$$

which has general validity *independently* of the metric of the ambient manifold.

Geodesic Spread Equation for the Jacobi Metric g_J

Let us now derive its explicit form in the case of the Jacobi metric. This metric is a conformal deformation of the pure kinetic energy metric, i.e., $(g_J)_{ij} = e^{-2f}a_{ij}$. Since we are mainly interested in studying standard Hamiltonian systems, $a_{ij} = \delta_{ij}$ is assumed. For a conformal metric $(g_J)_{ij} = e^{-2f}\delta_{ij}$ one readily obtains the following expression for the Christoffel coefficients: $\Gamma^k_{ij} = -\delta^k_j f_{,i} - \delta^k_i f_{,j} + \delta_{ij} f^{,k}$, where $f_{,i} = \partial_i f \equiv \partial f/\partial q^i$. Hence (3.78) is transformed into

$$\frac{d^2 J^k}{ds^2} - 2\frac{df}{ds}\frac{dJ^k}{ds} - 2\frac{dq^k}{ds}\frac{d}{ds}(f_{,j}J^j) + 2f_k\frac{dq^i}{ds}\delta_{ij}\frac{dJ^j}{ds} + f_{kj}J^j e^{2f} = 0 \ , \quad (3.79)$$

and using the relation $ds = e^{-2f}dt$, we can express it in terms of the physical time t instead of the proper time s:

$$\frac{d^2 J^k}{dt^2} + 2\left(f^{,k}\delta_{ij}\frac{dq^i}{dt} - f_{,j}\frac{dq^k}{dt}\right)\frac{dJ^j}{dt} + \left(f_{,kj}e^{-2f} - 2f_{,ji}\frac{dq^i}{dt}\frac{dq^k}{dt}\right)J^j = 0 \ ,$$

$$(3.80)$$

where $f_{,ij} = \partial^2_{ij}f$. Finally, since the Jacobi metric corresponds to $f = \frac{1}{2}\ln[1/2(E-V)]$, one obtains

$$f_{,i} = \frac{\partial_i V}{2(E-V)} \ , \tag{3.81}$$

$$f_{,ij} = \frac{\partial^2_{ij}V}{2(E-V)} + \frac{(\partial_i V)(\partial_j V)}{2(E-V)^2} \ , \tag{3.82}$$

$$e^{-2f} = 2(E-V) \ , \tag{3.83}$$

so that the final expression for the JLC equation for (M_E, g_J) is

$$\frac{d^2 J^k}{dt^2} + \frac{1}{E-V}\left(\partial_k V\delta_{ij}\frac{dq^i}{dt} - \partial_j V\frac{dq^k}{dt}\right)\frac{dJ^j}{dt} + [\partial^2_{kj}V]\, J^j$$

$$+ \frac{1}{E-V}\left[(\partial_k V)(\partial_j V) - \left(\partial^2_{ij}V + \frac{(\partial_i V)(\partial_j V)}{E-V}\right)\frac{dq^i}{dt}\frac{dq^k}{dt}\right]J^j = 0 \ .$$

$$(3.84)$$

This equation has been used in [130] to numerically work out a quantitative measure of the strength of chaos in a system with many degrees of freedom. The outcome of these computations has been found to be in excellent quantitative agreement with the outcome of standard numerical computations of the largest Lyapunov exponent for the same model.

Geodesic Spread Equation for the Eisenhart Metric G_e

Let us now give the explicit form of (3.78) in the case of $(M \times \mathbb{R}, G_e)$, the enlarged configuration space equipped with one of the Eisenhart metrics. Since the nonvanishing Christoffel coefficients are Γ^i_{00} and Γ^0_{0i}, then using the affine parametrization of the arc length with physical time, we obtain

$$\frac{d^2 J^k}{dt^2} + \frac{2(\partial_k V)}{2V+2\eta}\frac{dJ^0}{dt} + \left[\partial^2_{kj}V - \frac{4(\partial_k V)(\partial_j V)}{2V+2\eta}\right]J^j = 0 \ ,$$

$$(3.85)$$

$$\frac{d^2 J^0}{dt^2} - \frac{2(\partial_i V)\dot{q}^i}{2V+2\eta}\frac{dJ^0}{dt} - 2(\partial_i V)\frac{dJ^i}{dt} - \left[\partial^2_{ij}V - \frac{2(\partial_i V)(\partial_j V)}{2V+2\eta}\right]\dot{q}^i J^j = 0 \ ,$$

$$(3.86)$$

where the indexes i, j, k run from 1 to N. These equations have not yet been used to tackle Hamiltonian chaos, but they certainly deserve a thorough investigation.

Geodesic Spread Equation for the Eisenhart Metric g_e

Let us now give the explicit form of (3.78) in the case of $(M \times \mathbb{R}^2, g_e)$, the enlarged configuration space-time equipped with Eisenhart metric. One easily finds [131] that only the following Christoffel coefficients do not vanish: $\Gamma^i_{00} = (\partial V/\partial q_i)$ and $\Gamma^{N+1}_{0i} = (-\partial V/\partial q^i)$. Thus, the Jacobi equation (3.73) takes the form (we recall that the manifold now has dimension $N + 2$; all the indices $i, j, k, \ldots$ run from 1 to N)

$$\frac{\nabla^2 J^0}{ds^2} + R^0_{i0j} \frac{dq^i}{ds} J^0 \frac{dq^j}{ds} + R^0_{0ij} \frac{dq^0}{ds} J^i \frac{dq^j}{ds} = 0 , \tag{3.87}$$

$$\frac{\nabla^2 J^i}{ds^2} + R^i_{0j0} \left(\frac{dq^0}{ds}\right)^2 J^j + R^i_{00j} \frac{dq^0}{ds} J^0 \frac{dq^j}{ds} + R^i_{j00} \frac{dq^j}{ds} J^0 \frac{dq^0}{ds} = 0 , \tag{3.88}$$

$$\frac{\nabla^2 J^{N+1}}{ds^2} + R^{N+1}_{i0j} \frac{dq^i}{ds} J^0 \frac{dq^j}{ds} + R^{N+1}_{ij0} \frac{dq^i}{ds} J^j \frac{dq^0}{ds} = 0 , \tag{3.89}$$

where, for the sake of clarity, we have written out (3.5) separately for the 0, the $i = 1, \ldots, N$, and the $N + 1$ components, respectively. Since $\Gamma^0_{ij} = 0$ (see 3.32) we obtain, from the definition of covariant derivative (see Appendix B), $\nabla J^0/ds = dJ^0/ds$, and since $R^0_{ijk} = 0$ (see Section 3.4), we find that (3.87) becomes

$$\frac{d^2 J^0}{ds^2} = 0 , \tag{3.90}$$

so that J^0 does not accelerate, and without loss of generality, we can set $\frac{dJ^0}{ds}\Big|_{s=0} = J^0(0) = 0$. Combining the latter result with the definition of covariant derivative we obtain

$$\frac{\nabla J^i}{ds} = \frac{dJ^i}{ds} + \Gamma^i_{0k} \frac{dq^0}{ds} J^k , \tag{3.91}$$

and using $dq^0/ds = 0$ we finally get

$$\frac{\nabla^2 J^i}{ds^2} = \frac{d^2 J^i}{ds^2} , \tag{3.92}$$

so that (3.88) gives, for the projection in configuration space of the separation vector,

$$\frac{d^2 J^i}{ds^2} + \frac{\partial^2 V}{\partial q_i \partial q^k} \left(\frac{dq^0}{ds}\right)^2 J^k = 0 . \tag{3.93}$$

Equation (3.89) describes the evolution of J^{N+1}, which, however, does not contribute to the norm of J because $g_{N+1N+1} = 0$, so we can disregard it.

Along the physical geodesics of g_E, $ds^2 = (dq^0)^2 = dt^2$, so that (3.93) is exactly the usual tangent dynamics equation

$$\frac{d^2 J^i}{dt^2} + \frac{\partial^2 V}{\partial q_i \partial q^k} J^k = 0 , \qquad (3.94)$$

provided the identification $\xi = J$ is made.

It is a very interesting fact that the JLC equation (3.73) yields the usual tangent dynamics equation (2.108) for Hamiltonian flows, when explicitly worked out for the Eisenhart metric on $M \times \mathbb{R}^2$.

This circumstance allows one to make a direct link between the currently derived "experimental" data on Hamiltonian chaos, that is, between the numerical Lyapunov exponents discussed in Section 2.2.2, and the geometric treatment of a chaotic geodesic flow. This connection is a crucial point in the development of a geometric theory of Hamiltonian chaos without introducing any new definition of chaos in the geometric context.

Geodesic Spread Equation for a Finsler Space

In close analogy with Riemannian geometry, also in the case of Finsler geometry we can define a geodesic deviation equation. If we denote by ζ the vector field of geodesic separation, which locally measures the distance between pairs of points belonging to nearby geodesics at the same proper time s, the following equation is worked out for its evolution [118]:

$$\frac{\delta^2 \zeta^i}{\delta s^2} + K^i_{jhk}(q, q') \frac{dq'^j}{ds} \frac{dq'^h}{ds} \zeta^k = 0 , \qquad (3.95)$$

where the δ-derivation and the curvature tensor are those defined in Section 3.2. By introducing a normalized vector field X defined along a geodesic and having the same direction of ζ, there exists a scalar function $z(s)$ such that $\zeta^i = z X^i$, which can be shown to satisfy the equation

$$\frac{d^2 z}{ds^2} + z \left[R(q, q', X) - g_{ij}(q, q') \frac{\delta X^i}{\delta s} \frac{\delta X^j}{\delta s} \right] = 0 , \qquad (3.96)$$

where $R(q, q', X)$ is a sectional curvature defined as in the Riemannian case by replacing the Riemann curvature tensor with $K^i_{jhk}(q, q')$. In the two-dimensional case, the equation above simplifies to

$$\frac{d^2 z}{ds^2} + \mathcal{R}(q, q')z = 0 ,$$

where $\mathcal{R}(q, q')$ is the scalar curvature defined in (3.71), in strict analogy with the Riemannian case. This equation could be used to study the chaotic dynamics of 1.5-degrees-of-freedom Hamiltonian systems, that is, those described by a time-dependent Hamiltonian function with only a pair of conjugated $p-q$ variables. Also, Schur's theorem for constant-curvature Riemannian manifolds can be generalized to Finsler spaces, allowing, in principle, an extension to (3.96) of the quasi-isotropy assumption that will be developed in Chapter 5 for the Riemannian case.

Some successful applications of equation (3.95) to the study of Hamiltonian chaos in few degrees of freedom Hamiltonian systems are discussed in [123, 124, 132].

3.5.1 Concluding Remark

To summarize, we have shown that the dynamical trajectories of a Hamiltonian system of the form (1.1) can be seen as geodesics of the configuration space, or of at least three different enlargements of it,[6] or of the tangent bundle of configuration space. In each case one has to define an appropriate metric. The general relationship that holds between dynamical and geometrical quantities regardless of the precise choice of the metric can be sketched as follows:

$$
\begin{array}{ccc}
\text{dynamics} & & \text{geometry} \\[4pt]
\text{(time) } t & \leftrightarrow & s \ \text{(arc}-\text{length)} \\
\text{(potential energy) } V & \leftrightarrow & g \ \text{(metric)} \\
\text{(forces) } \partial V & \leftrightarrow & \Gamma \ \text{(Christoffel symbols)} \\
(\text{``curvature''of the potential}) \ \partial^2 V, (\partial V)^2 & \leftrightarrow & R \ \text{(curvature of the manifold)}
\end{array}
\tag{3.97}
$$

In the case of the Eisenhart metric, all these relations are extremely simple (maybe as simple as possible). In fact, the physical time t can be chosen as equal to the arc length s; the metric tensor g_e contains only the potential energy V, the nonvanishing Christoffel symbols Γ are equal to the forces ∂V; and the components of the Riemann curvature tensor R contain only the second derivatives $\partial_i \partial_j V$ of the potential energy.

We have also shown how the stability of the dynamical trajectories can be mapped onto the stability of the geodesics, which is completely determined by the curvature of the manifold. We have seen that in the case of the Eisenhart metric, as a consequence of its remarkably simple properties, also the relationship between the stability of the trajectories and the stability of the geodesics

---

[6] Notice that the space $M \times \mathbb{R}$ being equipped with the Eisenhart metric G_e is not the same as $M \times \mathbb{R}$ to be equipped with a Finsler metric. In the former case, the extra variable is related to the action, whereas in the latter case the extra variable coincides with the physical time.

becomes as simple as possible, i.e., the Jacobi equation (3.73) becomes identical to tangent dynamics equation (3.94) that is used to compute numerical Lyapunov exponents.

In Chapter 5 we shall see that the geometric framework, and in particular the link between curvature and stability, provides us with a natural explanation of the origin of the chaotic instability of Hamiltonian flows, thus overcoming all the difficulties and severe limitations of the traditional explanation based on homoclinic intersections.

Chapter 4

Integrability

4.1 Introduction

The problem of integrability in classical mechanics has been a seminal one. Motivated by celestial mechanics, it has stimulated a wealth of analytical methods and results. For example, as we have discussed in Chapter 2, the weaker requirement of only approximate integrability over finite times, or the existence of integrable regions in the phase space of a globally nonintegrable system, has led to the development of classical perturbation theory, with all its important achievements. However, deciding whether a given Hamiltonian system is globally integrable still remains a difficult task, for which a general constructive framework is lacking.

The topic of integrability is a vast one, and reviewing it is beyond the aim of the present monograph. For the sake of completeness, in this chapter we briefly discuss how the classical problem of integrability is rephrased in the Riemannian-geometric framework for the Hamiltonian dynamics introduced in Chapter 3.

In general, the existence of conservation laws, and of conserved quantities along the trajectories of a Hamiltonian system, is related to the existence of symmetries. The link is made by Noether's theorem [133]. A symmetry is seen as an invariance under the action of a group of transformations, and in the case of continuous symmetries, this can be related also to the existence of special vector fields: Killing vector fields on the mechanical manifold generating the transformations.

On a generic manifold M, a flow $\sigma : \mathbb{R} \times M \to M$ is generated by the ensemble of the integral curves of a vector field X on the manifold:

$$\frac{d}{dt}\sigma^\mu(t, x_0) = X^\mu(\sigma(t, x_0)) , \quad \mu \in \{1, \ldots, \dim(M)\} . \tag{4.1}$$

Given $t \in \mathbb{R}$, $\sigma(t, \cdot)$ is a diffeomorphism of M to itself. The summation in $\mathbb{R}$ endows σ with the structure of a commutative group

$$\sigma(t, \cdot) \circ \sigma(s, \cdot) = \sigma(t + s, \cdot); \tag{4.2}$$

such a group is called a one-parameter group of transformations $\sigma_t : M \rightarrow M$. Under the action of σ_ϵ, with ϵ infinitesimal, a point x of coordinates x^μ is transformed as

$$\sigma_\epsilon^\mu(x) = \sigma^\mu(\epsilon, x) = x^\mu + \epsilon X^\mu(x). \tag{4.3}$$

In this framework, the vector field X is called the infinitesimal generator of the transformation σ_ϵ. If two flows are given, $\sigma(t, x)$ and $\tau(t, x)$, generated by the vector fields X and Y respectively,

$$\frac{d\sigma^\mu(s, x)}{ds} = X^\mu(\sigma(s, x)) , \tag{4.4}$$

$$\frac{d\tau^\mu(t, x)}{dt} = Y^\mu(\tau(t, x)) , \tag{4.5}$$

the Lie derivative $\mathcal{L}_X Y$ of the vector field Y along the flow σ of X is defined by

$$(\mathcal{L}_X Y)(x) = \lim_{\epsilon \to 0} \frac{1}{\epsilon} [(\sigma_{-\epsilon})_*(Y(\sigma_\epsilon(x))) - Y(x)] , \tag{4.6}$$

where by $(\sigma_{-\epsilon})_* : T_{\sigma_\epsilon(x)} M \rightarrow T_x M$ we denote the derivative of $\sigma_{-\epsilon}$. This amounts to evaluating the variation of a vector field Y along a flow of σ, and this can also be extended to a tensor field A:

$$(\mathcal{L}_X A)(x) = \lim_{\epsilon \to 0} \frac{1}{\epsilon} [(\sigma_{-\epsilon})_*(A(\sigma_\epsilon(x))) - A(x)] . \tag{4.7}$$

On Riemannian manifolds (M, g), a special class of vector fields can be defined: Killing vector fields. A field X is such a vector field if

$$\mathcal{L}_X g = 0 . \tag{4.8}$$

It directly follows from (4.7) that a vector field is a Killing field iff the one-parameter group of transformations associated with it is an isometry.[1] This means that along the flow σ_t, geometry does not change, and therefore a Killing field represents an infinitesimal symmetry of the manifold. However, through Noetherian symmetries, and thus Killing vector fields, only a limited set of conservation laws can be accounted for. This is easily understood because only invariants that are linear functions of the momenta can be constructed by means of Killing vectors, while the energy, an invariant for any autonomous Hamiltonian system, is already a quadratic function of the momenta. The possibility of constructing invariants along a geodesic flow that are of higher order than linear in the momenta is related to the existence of Killing *tensor* fields on the mechanical manifolds [134–136].

In general, the components of any Killing tensor field on a mechanical manifold are solutions of a linear inhomogeneous system of first-order partial differential equations. Since the number of these equations always exceeds

---

[1] An isometry $f : M \rightarrow M$ is a diffeomorphism preserving distance: $f^* g = g$.

the number of the unknowns, the system is always *overdetermined*. The existence of Killing tensors thus requires *compatibility*. However, compatibility is generically very unusual, which suggests a possible explanation, at least of a qualitative kind, of the exceptionality of integrability with respect to nonintegrability.

4.2 Killing Vector Fields

On a Riemannian manifold, for any pair of vectors V and W, the following relation holds:

$$\frac{d}{ds}\langle V, W \rangle = \left\langle \frac{\nabla V}{ds}, W \right\rangle + \left\langle V, \frac{\nabla W}{ds} \right\rangle , \qquad (4.9)$$

where $\langle V, W \rangle = g_{ij} V^i W^j$ and ∇/ds is the covariant derivative along a curve $\gamma(s)$. If the curve $\gamma(s)$ is a geodesic, for a generic vector X we have

$$\frac{d}{ds}\langle X, \dot{\gamma} \rangle = \left\langle \frac{\nabla X}{ds}, \dot{\gamma} \right\rangle + \left\langle X, \frac{\nabla \dot{\gamma}}{ds} \right\rangle = \left\langle \frac{\nabla X}{ds}, \dot{\gamma} \right\rangle \equiv \langle \nabla_{\dot{\gamma}} X, \dot{\gamma} \rangle , \qquad (4.10)$$

where $(\nabla_{\dot{\gamma}} X)^i = \frac{dx^l}{ds}\frac{\partial X^i}{\partial x^l} + \Gamma^i_{jk}\frac{dx^j}{ds}X^k$, so that in components it reads

$$\frac{d}{ds}(X_i v^i) = v^i \nabla_i (X_j v^j) , \qquad (4.11)$$

where $v^i = dx^i/ds$; with $X_j v^i \nabla_i v^j = X_j \nabla_{\dot{\gamma}} \dot{\gamma}^j = 0$, because geodesics are autoparallel, this can obviously be rewritten as

$$\frac{d}{ds}(X_i v^i) = \frac{1}{2}v^j v^i (\nabla_i X_j + \nabla_j X_i) , \qquad (4.12)$$

telling that the vanishing of the left-hand side, i.e., the conservation of $X_i v^i$ along a geodesic, is guaranteed by the vanishing of the right-hand side, i.e.,

$$\nabla_{(i} X_{j)} \equiv \nabla_i X_j + \nabla_j X_i = 0 , \qquad i, j = 1, \ldots, \dim M_E. \qquad (4.13)$$

If such a field exists on a manifold, it is called a Killing vector field (KVF). Equation (4.13) is equivalent to $\mathcal{L}_X g = 0$. On the mechanical manifolds (M_E, g_J), the unit vector $\frac{dq^k}{ds}$ is proportional to the canonical momentum $p_k = \frac{\partial L}{\partial \dot{q}^k} = \dot{q}^k$, $(a_{ij} = \delta_{ij})$, and is tangent to a geodesic. The existence of a KVF X implies that the quantity, linear in the momenta,

$$J(q, p) = X_k(q)\frac{dq^k}{ds} = \frac{1}{\sqrt{2(E - V(q))}}X_k(q)\frac{dq^k}{dt} = \frac{1}{\sqrt{2W(q)}}\sum_{k=1}^{N} X_k(q)p_k$$

$$(4.14)$$

is a constant of motion along the geodesic flow. Thus, for an N-degrees-of-freedom Hamiltonian system, a physical conservation law involving a conserved quantity linear in the canonical momenta can always be related to a symmetry on the manifold (M_E, g_J) due to the action of a KVF on the manifold. These are conservation laws of Noetherian kind. Equation (4.13) is equivalent to the vanishing of the Poisson brackets

$$\{H, J\} = \sum_{i=1}^{N} \left(\frac{\partial H}{\partial q^i} \frac{\partial J}{\partial p_i} - \frac{\partial H}{\partial p_i} \frac{\partial J}{\partial q^i} \right) = 0 , \tag{4.15}$$

the standard definition of a constant of motion $J(q, p)$. In fact, a linear function of the momenta

$$J(q, p) = \sum_i C_i(q) p_i , \tag{4.16}$$

if conserved, can be associated with the vector of components

$$X_k = [E - V(q)] C_k(q). \tag{4.17}$$

The explicit expression of the system of equations (4.13) is obtained by writing in components the covariant derivatives associated with the connection coefficients (3.17), and it finally reads

$$[E - V(q)] \left[\frac{\partial C_i(q)}{\partial q^j} + \frac{\partial C_j(q)}{\partial q^i} \right] - \delta_{ij} \sum_{k=1}^{N} \frac{\partial V}{\partial q^k} C_k(q) = 0, \tag{4.18}$$

or equivalently

$$\frac{1}{2} \sum_{k=1}^{N} p_k^2 \left[\frac{\partial C_i(q)}{\partial q^j} + \frac{\partial C_j(q)}{\partial q^i} \right] - \delta_{ij} \sum_{k=1}^{N} \frac{\partial V}{\partial q^k} C_k(q) = 0, \tag{4.19}$$

which, according to the principle of polynomial identity, yields the following conditions for the coefficients $C_i(q)$:

$$\frac{\partial C_i(q)}{\partial q^j} + \frac{\partial C_j(q)}{\partial q^i} = 0 , \quad i \neq j , \quad i, j = 1, \dots, N ,$$

$$\frac{\partial C_i(q)}{\partial q^i} = 0 , \quad i = 1, \dots, N ,$$

$$\sum_{k=1}^{N} \frac{\partial V}{\partial q^k} C_k(q) = 0 . \tag{4.20}$$

One can easily check that the same conditions stem from (4.15). As an elementary example, we can give the explicit expression of the components of the Killing vector field associated with the conservation of the total momentum $P(q, p) = \sum_{k=1}^{N} p_k$.

In this case the coefficients are $C_i(q) = 1$, so that the momentum conservation can be geometrically related to the action of the vector field of components $X_i = E - V(q)$, $i = 1, \ldots, N$, on the mechanical manifold. At least this class of invariants has a geometric counterpart in a symmetry of (M_E, g_J).

However, in order to achieve a fully geometric rephrasing of integrability, we need something similar for any constant of motion. If a one-to-one correspondence is to exist between conserved physical quantities along a Hamiltonian flow and suitable symmetries of the mechanical manifolds (M_E, g_J), then *integrability* will be equivalent to the existence of a number of symmetries at least equal to the number of degrees of freedom ($=$ dim M_E).

If a Lie group G acts on the phase space manifold through completely canonical transformations, and there exists an associated *momentum mapping*,[2] then every Hamiltonian having G as a symmetry group, with respect to its action, admits the momentum mapping as a constant of motion [137]. These symmetries are usually referred to as *hidden symmetries*, because even though their existence is ensured by integrability, they are not easily recognizable.[3]

4.3 Killing Tensor Fields

Let us now extend what has been just presented about KVFs in an attempt trying to generalize the form of the conserved quantity along a geodesic flow from $J = X_i v^i$ to $J = K_{j_1 j_2 \ldots j_r} v^{j_1} v^{j_2} \cdots v^{j_r}$, with $K_{j_1 j_2 \ldots j_r}$ a tensor of rank r. Thus, we look for the conditions that entail

$$\frac{d}{ds}(K_{j_1 j_2 \ldots j_r} v^{j_1} v^{j_2} \cdots v^{j_r}) = v^j \nabla_j (K_{j_1 j_2 \ldots j_r} v^{j_1} v^{j_2} \cdots v^{j_r}) = 0 . \qquad (4.21)$$

In order to work out from this equation a condition for the existence of a suitable tensor $K_{j_1 j_2 \ldots j_r}$, which is called a *Killing tensor field* (KTF), let us first consider the rank-$2r$ tensor $K_{j_1 j_2 \ldots j_r} v^{i_1} v^{i_2} \cdots v^{i_r}$ and its covariant derivative along a geodesic, i.e.,

$$v^j \nabla_j (K_{j_1 j_2 \ldots j_r} v^{i_1} v^{i_2} \ldots v^{i_r})$$

$$= v^j \left(\frac{\partial K_{j_1 \ldots j_r}}{\partial x^j} - K_{l j_2 \ldots j_r} \Gamma^l_{j_1 j} - \cdots - K_{j_1 \ldots l} \Gamma^l_{j_r j} \right) v^{i_1} \cdots v^{i_r}$$

---

[2] This happens whenever this action corresponds to the lifting to the phase space of the action of a Lie group on the configuration space.

[3] An interesting account of these hidden symmetries can be found in [138], where it is surmised that integrable motions of N-degrees-of-freedom systems are the "shadows" of free motions in symmetric spaces (for example, Euclidean spaces $\mathbb{R}^n$, hyperspheres $\mathbb{S}^n$, hyperbolic spaces $\mathbb{H}^n$) of sufficiently large dimension $n > N$.

$$+K_{j_1\ldots j_r}\left(v^j\frac{\partial v^{i_1}}{\partial x^j}+\Gamma_{jl}^{i_1}v^lv^j\right)v^{i_2}\cdots v^{i_r}+\cdots$$

$$+K_{j_1\ldots j_r}v^{i_1}\cdots v^{i_{r-1}}\left(v^j\frac{\partial v^{i_r}}{\partial x^j}+\Gamma_{jl}^{i_r}v^lv^j\right)$$

$$=v^{i_1}v^{i_2}\cdots v^{i_r}v^j\nabla_jK_{j_1j_2\ldots j_r}\ ,\tag{4.22}$$

where we have again used $v^j\nabla_jv^{ik}=0$ along a geodesic, and a standard covariant differentiation formula B. Now, by contraction on the indices i_k and j_k, the rank-$2r$ tensor of the right-hand side of (4.22) provides a new expression for the right-hand side of (4.21), which reads

$$\frac{d}{ds}(K_{j_1j_2\ldots j_r}v^{j_1}v^{j_2}\cdots v^{j_r})=v^{j_1}v^{j_2}\cdots v^{j_r}v^j\nabla_{(j}K_{j_1j_2\ldots j_r)}\ ,\tag{4.23}$$

where $\nabla_{(j}K_{j_1j_2\ldots j_r)}=\nabla_jK_{j_1j_2\ldots j_r}+\nabla_{j_1}K_{jj_2\ldots j_r}+\cdots+\nabla_{j_r}K_{j_1j_2\ldots j_{r-1}j}$, as can be easily understood by rearranging the indices of the summations in the contraction of the $2r$-rank tensor in the last part of (4.22) (a direct check for the case $N=r=2$ is immediate). The vanishing of (4.23), entailing the conservation of $K_{j_1j_2\ldots j_r}v^{j_1}v^{j_2}\cdots v^{j_r}$ along a geodesic flow, is therefore guaranteed by the existence of a tensor field satisfying the conditions

$$\nabla_{(j}K_{j_1j_2\ldots j_r)}=0\ .\tag{4.24}$$

These equations generalize (4.13) and give the definition of a KTF on a Riemannian manifold. These N^{r+1} equations in $(N+r-1)!/r!(N-1)!$ unknown independent components[4] of the Killing tensor constitute an *over-determined* system of equations. Thus, a priori, we can expect that the existence of KTFs has to be rather exceptional.

If a KTF exists on a Riemannian manifold, then the scalar

$$K_{j_1j_2\ldots j_r}\frac{dq^{j_1}}{ds}\frac{dq^{j_2}}{ds}\cdots\frac{dq^{j_r}}{ds}\tag{4.25}$$

is a constant of motion for the geodesic flow on the same manifold.

Let us consider, as a generalization of the special case of rank one given by (4.16), the constant of motion

$$J(q,p)=\sum_{\{i_1,i_2,\ldots,i_N\}}C_{i_1i_2\ldots i_N}p_1^{i_1}p_2^{i_2}\cdots p_N^{i_N},\tag{4.26}$$

which, with the constraint $i_1+i_2+\cdots+i_N=r$, is a homogeneous polynomial of degree r. The index i_j denotes the power with which the momentum p_j contributes. If $r<N$ then necessarily some indices i_j must vanish. By repeating the procedure developed in the case $r=1$, and by identifying

---

[4] This number of independent components, i.e., the binomial coefficient $\binom{N+r-1}{r}$, is due to the totally symmetric character of Killing tensors.

$$J(q,p) \equiv K_{j_1 j_2 \ldots j_r} \frac{dq^{j_1}}{ds} \frac{dq^{j_2}}{ds} \cdots \frac{dq^{j_r}}{ds} , \qquad (4.27)$$

we get the relationship between the components of the Killing tensor of rank r and the coefficients $C_{i_1 i_2 \ldots i_N}$ of the invariant $J(q,p)$, that is,

$$K_{\underbrace{1 \ldots 1}_{i_1} \underbrace{2 \ldots 2}_{i_2} \ldots \underbrace{N \ldots N}_{i_N}} = 2^{r/2}[E - V(q)]^r C_{i_1 i_2 \ldots i_N} . \qquad (4.28)$$

With the only difference of a more tedious combinatorics, also in this case it turns out that the equations (4.24) are equivalent to the vanishing of the Poisson brackets of $J(q,p)$, that is,

$$\{H, J\} = 0 \iff \nabla_{(j} K_{j_1 j_2 \ldots j_r)} = 0 . \qquad (4.29)$$

Thus, the existence of Killing tensor fields satisfying (4.24) on a mechanical manifold (M, g_J) provides the rephrasing of integrability of Newtonian equations of motion, or equivalently, of standard Hamiltonian systems, within the Riemannian-geometric framework.

At first sight, it might appear too restrictive that prime integrals of motion have to be homogeneous functions of the components of p. However, as we shall discuss in the next section, the integrals of motion of the known integrable systems can actually be cast in this form. This is in particular the case of total energy, a quantity conserved by any autonomous Hamiltonian system.

4.4 Explicit KTFs of Known Integrable Systems

The first natural question to address concerns the existence of a KT field, on any mechanical manifold (M, g_J), to be associated with total energy conservation. Such a KT field actually exists and coincides with the metric tensor g_J. In fact, by definition it satisfies[5] (4.24).

One of the simplest case of integrable system is represented by a decoupled system described by a generic Hamiltonian

$$H = \sum_{i=1}^{N} \left[\frac{p_i^2}{2} + V_i(q_i) \right] = \sum_{i=1}^{N} H_i(q_i, p_i) \qquad (4.30)$$

for which all the energies E_i of the subsystems H_i, $i = 1, \ldots, N$, are conserved. On the associated mechanical manifold, N KT fields of rank 2 exist. They are given by

$$K_{jk}^{(i)} = \delta_{jk}\{V_i(q_i)[E - V(q)] + \delta_j^i[E - V(q)]^2\} . \qquad (4.31)$$

---

[5] A property of the canonical Levi-Civita connection, on which the covariant derivative is based, is just the vanishing of ∇g.

In fact, these tensor fields satisfy (4.24), which explicitly reads

$$
\nabla_k K^{(i)}_{lm} + \nabla_l K^{(i)}_{mk} + \nabla_m K^{(i)}_{kl}
$$
$$
= \frac{\partial K^{(i)}_{lm}}{\partial q^k} + \frac{\partial K^{(i)}_{mk}}{\partial q^l} + \frac{\partial K^{(i)}_{kl}}{\partial q^m} - 2\Gamma^j_{kl} K^{(i)}_{jm} - 2\Gamma^j_{km} K^{(i)}_{jl} - 2\Gamma^j_{lm} K^{(i)}_{jk} = 0 \ ,
$$

$$
(4.32)
$$

$$
k, l, m = 1, \ldots, N \ .
$$

The conserved quantities $J^{(i)}(q, p)$ are then obtained by saturation of the tensors $K^{(i)}$ with the velocities dq/ds:

$$
J^{(i)}(q, p) = \sum_{jk=1}^{N} K^{(i)}_{jk} \frac{dq^j}{ds} \frac{dq^k}{ds} = V_i(q_i) \frac{1}{E - V(q)} \sum_{k=1}^{N} \frac{p_k^2}{2} + \frac{p_i^2}{2} = E_i \ . \quad (4.33)
$$

This equation suggests that to require that the constants of motion be homogeneous polynomials of the momenta is not so restrictive as might appear. In fact, through the constant quantity

$$
\frac{1}{E - V(q)} \sum_{k=1}^{N} \frac{p_k^2}{2} = 1 \ , \quad (4.34)
$$

homogeneous of second degree in the momenta, any even-degree polynomial of the momenta can be made homogeneous. The possibility of inferring the existence of a conservation law from the existence of a KTF on (M, g_J) is thus extended to the constants of motion given by a sum of homogeneous polynomials whose degrees differ by an even integer,

$$
J(p, q) = P^{(r)}(p) + P^{(r-2)}(p) + \cdots
$$
$$
+ P^{(r-2n)}(p) \in C^\infty(q)[p_1, \ldots, p_N] \quad (4.35)
$$

$$
\mathrm{homdeg}\ P^s = s \ , \quad s = r, r - 2, \ldots, r - 2 \left[\frac{r}{2} \right] \ ,
$$

so that it can be recast in the homogeneous form

$$
J(p, q) = P^{(r)}(p) + P^{(r-2)}(p) \frac{1}{E - V(q)} \sum_{k=1}^{N} \frac{p_k^2}{2} + \cdots \quad (4.36)
$$

$$
+ P^{(r-2n)}(p) \left[\frac{1}{E - V(q)} \sum_{k=1}^{N} \frac{p_k^2}{2} \right]^n \ .
$$

4.4.1 Nontrivial Integrable Models

It is worth noting that the geodesic flow on an ellipsoid immersed in Euclidean three-dimensional space provides one of the simplest nontrivial examples of integrability. Besides the constant of motion obtained through the metric tensor (which corresponds to the energy for physical geodesic flows), the second constant of the motion is given by [139]

$$J^2 = c \sum_{i=1}^{3} (a^i)^{-2} \left(\frac{dx^i}{ds} \right)^2 ,$$

where c is a constant, a^i are half the major semiaxes, and x^i are the coordinates in the immersion space. According to what has been discussed above, this extra constant of motion has to correspond to a rank-2 KT.[6]

Nontrivial examples of nonlinear integrable Hamiltonian systems are provided by the following Hamiltonians:

$$H = \sum_{i=1}^{N} \left\{ \frac{p_i^2}{2} + \frac{a}{b} [e^{-b(q_{i+1} - q_i)} - 1] \right\} , \qquad (4.37)$$

known as the Toda model, which is integrable for any given pair of the constants a and b; and

$$H = \sum_{i=1}^{N} \frac{p_i^2}{2} + \frac{1}{2} \left(\sum_{i=1}^{N} q_i^2 \right)^2 - \sum_{i=1}^{N} \lambda_i q_i^2 , \qquad (4.38)$$

which is completely integrable for any $\lambda_1, \ldots, \lambda_N$ [140]. Recursive formulas are available for all the constants of motion of the Toda model at any N [141]; and also for the second Hamiltonian, the exact form of first integrals is known [140]. In both cases, the first integrals are polynomials of given parity of the momenta so that on the basis of what we have said above, each invariant $J^{(i)}$, $i = 1, \ldots, N$ can be derived from a KTF on (M, g_J). Thus, integrability of these systems admits a Riemannian-geometric interpretation.

Let us mention here another remarkable example of integrability that seems to demand a generalization of this Riemannian approach. It concerns a one-parameter family of Hamiltonian deformations of the Kepler problem leading to nonsymplectomorphic systems. Such deformations represent the motion of a charged particle in the field of a magnetic monopole with a special choice of the potential [142]. The components of a Runge–Lenz vector Poisson commute with the Hamiltonian and are quadratically dependent on the velocity. In order to associate with a geodesic flow the trajectories of a system subject to velocity-dependent forces, as is the case of the deformed Kepler models, the use of Finsler manifolds is necessary [118, 122], and thus

---

[6] Notice that the set of variables x^i is here redundant because of the algebraic equation defining the ellipsoid. In this case one has to consider the metric on the surface induced from $\mathbb{R}^3$, which, in contrast to the Jacobi metric on the mechanical manifolds, is not conformally flat.

a rephrasing of integrability through KT fields on Finsler manifolds could be necessary. To the best of our knowledge this is still an open problem.[7]

4.4.2 The Special Case of the $N = 2$ Toda Model

Let us consider the special case of a two-degrees-of-freedom Toda model described by the integrable Hamiltonian[8]

$$H = \frac{1}{2}(p_x^2 + p_y^2) + \frac{1}{24}\left[e^{2y+2\sqrt{3}x} + e^{2y-2\sqrt{3}x} + e^{-4y}\right] - \frac{1}{8} . \tag{4.39}$$

From what is already reported in the literature [141], we know that a third-order polynomial of the momenta has to be found eventually. Therefore, we look for a rank-3 KT satisfying

$$\nabla_i K_{jkl} + \nabla_j K_{ikl} + \nabla_k K_{ijl} + \nabla_l K_{ijk} = 0 , \qquad i, j, k, l = 1, 2 , \tag{4.40}$$

where, associating the label 1 to x and the label 2 to y,

$$\{(i, j, k, l)\} = \{(1, 1, 1, 1); (1, 1, 1, 2); (1, 1, 2, 2); (1, 2, 2, 2); (2, 2, 2, 2)\} .$$

The computation of the Christoffel coefficients according to (3.17) yields

$$\Gamma_{11}^1 = \frac{-\partial_x V}{2[E - V(x, y)]} , \quad \Gamma_{22}^1 = \frac{\partial_x V}{2[E - V(x, y)]} , \quad \Gamma_{11}^2 = \frac{\partial_y V}{2[E - V(x, y)]} ,$$

$$\Gamma_{22}^2 = \frac{-\partial_y V}{2[E - V(x, y)]} , \quad \Gamma_{12}^1 = \frac{-\partial_y V}{2[E - V(x, y)]} , \quad \Gamma_{12}^2 = \frac{-\partial_x V}{2[E - V(x, y)]} . \tag{4.41}$$

From (4.40) we get the system

$$\nabla_1 K_{111} = 0 ,$$

$$\nabla_1 K_{122} + \nabla_2 K_{112} = 0 ,$$

$$\nabla_2 K_{111} + 3\nabla_1 K_{211} = 0 ,$$

$$\nabla_1 K_{222} + 3\nabla_2 K_{122} = 0 ,$$

$$\nabla_2 K_{222} = 0 , \tag{4.42}$$

---

[7] The Killing–vector equations in Finsler spaces can be found in [118]. More recently these equations are studied in [143], where it is argued that Killing vectors in Finsler spaces can yield invariants of higher order than linear in the momenta.

[8] This is derived from an $N = 3$ Hamiltonian (4.37) by means of two canonical transformations of variables removing translational invariance; see, for example, [49]; the third-order expansion of this new Hamiltonian yields the Hénon–Heiles model of (4.46) with $C = D = 1$.

whence

$$\partial_x K_{111} - 3\Gamma^1_{11}K_{111} - 3\Gamma^2_{11}K_{211} = 0 \ ,$$

$$\partial_x K_{122} + \partial_y K_{211} - \Gamma^1_{11}K_{122} - \Gamma^2_{11}K_{222} - 4\Gamma^1_{12}K_{112}$$
$$-4\Gamma^2_{11}K_{212} - \Gamma^1_{22}K_{111} - \Gamma^2_{22}K_{211} = 0 \ ,$$

$$\partial_y K_{111} + 3\partial_x K_{211} - 6\Gamma^1_{12}K_{111} - 6\Gamma^2_{12}K_{112}$$
$$-6\Gamma^1_{11}K_{211} - 6\Gamma^2_{11}K_{212} = 0 \ ,$$

$$\partial_x K_{222} + 3\partial_y K_{122} - 6\Gamma^1_{21}K_{122} - 6\Gamma^2_{21}K_{222}$$
$$-6\Gamma^1_{22}K_{112} - 6\Gamma^2_{22}K_{212} = 0 \ ,$$

$$\partial_y K_{222} - 3\Gamma^1_{22}K_{122} - 3\Gamma^2_{22}K_{222} = 0 \ , \tag{4.43}$$

with the Christoffel coefficients given by (4.41), where one has to replace $V(x,y)$ with the potential part of the Hamiltonian (4.39) and $\partial_x V$, $\partial_y V$ with its derivatives. The general method of solving a linear inhomogeneous system of first-order partial differential equations in more than one dependent variables can be found in [144]. However, finding the explicit solution to the system of equations (4.43) is much facilitated because we already know a priori that this system is compatible and thus admits a solution, and we also have strong hints about the solution itself because the general form of the integrals of the Toda model is known [141]. The KTF, besides the metric tensor, for the model (4.39) is eventually found to have the components [145, 146]

$$K_{111} = 2(E-V)^2[3\partial_y V + 4(E-V)] \ ,$$
$$= 8(E-V)^3 + \frac{1}{2}(E-V)^2[e^{2y-2\sqrt{3}x} + e^{2y+2\sqrt{3}x} - 2e^{-4y}] \ ,$$

$$K_{122} = 2(E-V)^2[\partial_y V - 4(E-V)] \ ,$$
$$= -24(E-V)^3 + \frac{1}{2}(E-V)^2[e^{2y-2\sqrt{3}x} + e^{2y+2\sqrt{3}x} - 2e^{-4y}] \ ,$$

$$K_{112} = -2(E-V)^2\partial_x V = \frac{\sqrt{3}}{6}(E-V)^2(e^{2y+2\sqrt{3}x} - e^{2y-2\sqrt{3}x}) \ ,$$

$$K_{222} = -6(E-V)^2\partial_x V = \frac{\sqrt{3}}{2}(E-V)^2(e^{2y+2\sqrt{3}x} - e^{2y-2\sqrt{3}x}) \ , \tag{4.44}$$

as can be easily checked by substituting them into (4.43). Hence, the second constant of motion, besides energy, is given by

$$J(x,y,p_x,p_y)$$
$$= K_{ijk}\frac{dq^i}{ds}\frac{dq^j}{ds}\frac{dq^k}{ds} = K_{ijk}\frac{dq^i}{dt}\frac{dq^j}{dt}\frac{dq^k}{dt}\frac{1}{2\sqrt{2}[E-V(x,y)]^3}$$
$$= \frac{1}{2\sqrt{2}[E-V(x,y)]^3}(K_{111}p_x^3 + 3K_{122}p_xp_y^2 + 3K_{112}p_x^2p_y + K_{222}p_y^3)$$
$$= 8p_x(p_x^2 - 3p_y^2) + (p_x+\sqrt{3}p_y)e^{2y-2\sqrt{3}x} - 2p_xe^{-4y} + (p_x - \sqrt{3}p_y)e^{2y+2\sqrt{3}x} \ ,$$
$$\tag{4.45}$$

which coincides with the expression already reported in the literature [49] for the Hamiltonian (4.39).

4.4.3 The Generalized Hénon–Heiles Model

Let us now consider the two-degrees-of-freedom system described by the Hamiltonian

$$H = \frac{1}{2}(p_x^2 + p_y^2) + \frac{1}{2}(x^2 + y^2) + Dx^2 y - \frac{1}{3}Cy^3 . \qquad (4.46)$$

This model, originally derived to describe the motion of a test star in an axisymmetric galactic mean gravitational field, provided some of the first numerical evidence of the chaotic transition in nonlinear Hamiltonian systems [48]. Hénon and Heiles considered the case $C = D = 1$. The existence of a chaotic layer in the phase space of this model means lack of *global* integrability. However, by means of the Painlevé method, it has been shown [147] that for special choices of the parameters C and D this system is globally integrable. Let us now tackle integrability of this model from the viewpoint of the existence of KT fields on the manifold (M, g_J). We first begin with the equations for a Killing vector field. By means of (4.20) we look for possible coefficients $C_1(x, y)$, $C_2(x, y)$, thus obtaining

$$C_1 = C_1(y), \quad C_2 = C_2(x) ,$$

$$\frac{dC_1(y)}{dy} + \frac{dC_2(x)}{dx} = 0 , \qquad (4.47)$$

$$x(1 + 2Dy)C_1(y) + (y + Dx^2 - Cy^2)C_2(x) = 0 .$$

From the second equation of (4.47) it follows that

$$\frac{dC_1(y)}{dy} = -\frac{dC_2(x)}{dx} = \text{cost.} , \qquad (4.48)$$

whence, denoting the constant by α, the possible expressions for $C_1(y)$ and $C_2(x)$ are only of the form $C_1(y) = -\alpha y + \beta$, $C_2(x) = \alpha x + \gamma$, which, after substitution into the last equation of (4.47), implies

$$(x + 2Dxy)(-\alpha y + \beta) + (y + Dx^2 - Cy^2)(\alpha x + \gamma) = 0, \qquad (4.49)$$

which has a non-trivial solution only for $C = D = 0$. On the other hand, for these values of the parameters the potential simplifies to $V(x, y) = \frac{1}{2}x^2 + \frac{1}{2}y^2$, whence the existence of the Killing vector field X of components $X_1 = y$ and $X_2 = -x$, which is due to the invariance under rotations in the xy plane.

Let us now consider the case of a rank-2 KTF. Equations (4.40) become

$$\nabla_i K_{jk} + \nabla_j K_{ik} + \nabla_k K_{ij} = 0 , \qquad i, j, k = 1, 2 , \qquad (4.50)$$

where, associating again the label 1 to x and the label 2 to y, $\{(i,j,k)\} = \{(1,1,1);(1,1,2);(1,2,2);(2,2,2)\}$. The Christoffel coefficients are still given by (4.41), where we have to use the potential part of Hamiltonian (4.46) so that $\partial_x V(x,y) = x + 2Dxy$ and $\partial_y V(x,y) = y + Dx^2 - Cy^2$. The KTF equations are then

$$\nabla_1 K_{11} = 0 \ ,$$
$$2\nabla_1 K_{12} + \nabla_2 K_{11} = 0 \ ,$$
$$\nabla_1 K_{22} + 2\nabla_2 K_{12} = 0 \ ,$$
$$\nabla_2 K_{22} = 0 \ , \qquad\qquad (4.51)$$

whence

$$\partial_x K_{11} - 2\Gamma^1_{11} K_{11} - 2\Gamma^2_{11} K_{21} = 0 \ ,$$
$$2\partial_x K_{12} + \partial_y K_{11} - 4\Gamma^1_{12} K_{11} - (4\Gamma^2_{12} + 2\Gamma^1_{11})K_{12} - 2\Gamma^2_{11} K_{22} = 0 \ ,$$
$$\partial_x K_{22} + 2\partial_y K_{12} - 2\Gamma^1_{22} K_{11} - (4\Gamma^1_{12} + 2\Gamma^2_{22})K_{12} - 4\Gamma^2_{12} K_{22} = 0 \ ,$$
$$\partial_y K_{22} - 2\Gamma^1_{22} K_{12} - 2\Gamma^2_{22} K_{22} = 0 \ .$$
$$\qquad\qquad (4.52)$$

Since the Hamiltonian (4.46) is not integrable for a generic choice of the parameters C and D, we can reasonably expect that the generic property of the above *overdetermined* system of equations is *incompatibility*, i.e., only the trivial solution $K_{ij} = 0$ exists for the overwhelming majority of the pairs (C, D). However, the existence of special choices of C and D for which the Hamiltonian is integrable suggests that this overdetermined system can be *compatible* in special cases. For example, when $D = 0$ the variables x and y in (4.46) are decoupled, and thus two KT fields of rank 2 exist according to (4.31).

A non trivial solution for the system (4.52) must exist at least for the pair $(C = -6, D = 1)$. In fact, in this case the modified Hénon–Heiles model is known to be integrable [147]. An explicit solution for the system (4.52) is eventually found to be given by [145, 146]

$$K_{11} = (3 - 4y)(E - V(x,y))^2 + x^2(x^2 + 4y^2 + 4y + 3)(E - V(x,y)) \ ,$$
$$K_{12} = 2x(E - V(x,y)) \ ,$$
$$K_{22} = \frac{1}{2}(x^2 + 4y^2 + 4y + 3)(E - V(x,y)) \ . \qquad\qquad (4.53)$$

The associated constant of motion is therefore

$$J(x,y,p_x,p_y) = \frac{1}{(E - V(x,y))^2}(K_{11}p_x^2 + 2K_{12}p_xp_y + K_{22}p_y^2)$$
$$= x^4 + 4x^2y^2 - p_x^2 y + 4p_xp_yx + 4x^2y + 3p_x^2 + 3x^2. \quad (4.54)$$

This expression is identical to that reported in [147], worked out for the same values of C and D with a completely different method based on the Painlevé property.[9]

4.5 Open Problems

Let us now summarize the meaning of the results presented above and point out some open problems.

- Besides qualitative and quantitative descriptions of chaos, within the framework of Riemannian geometrization of Newtonian mechanics, also *integrability* has its own place. The idea of associating KTFs with integrability has been essentially developed in the context of classical general relativity; see, for example, [148–150] and references quoted therein. That Killing tensors generate "hidden" symmetries associated with constants of the motion in classical Newtonian mechanics has been considered in [135, 136], and, more recently, in [151]. In particular, the integrability conditions for quadratic invariants were obtained in [151].
- The reduction of the problem of integrability of a given Hamiltonian system to the existence of suitable KTFs on (M_E, g_J) offers several points of interest; in particular, we have seen that the system of equations in the unknown components of a KTF of a preassigned rank is overdetermined. Thus at a qualitative level, integrability seems a rather exceptional property, and the larger N, the "more exceptional" it seems to be, because of the rapidly growing mismatch between the number of unknowns and the number of equations. In principle, the existence of compatibility conditions for systems of linear first-order partial differential equations could allow one to decide about integrability prior to any explicit attempt at solving the equations for the components of a KTF. Even better, there are geometric constraints to the existence of KTFs. Early results in this sense are reported in [152], so that it seems possible, at least in some cases, to devise purely geometric criteria of *nonintegrability*. For example, hyperbolicity of compact manifolds excludes [152] the existence of KTFs, and this is consistent with the property of geodesic flows on compact hyperbolic manifolds of being strongly chaotic (Anosov flows).
- In general, we lack a criterion to restrict the search for KTFs to a small interval of ranks, and this constitutes a practical difficulty. Nevertheless, since the involution of two invariants translates into the vanishing of special brackets—the Schouten brackets [150]—between the corresponding Killing tensors, a shortcut to proving integrability, for a large class of systems satisfying the conditions of the Poincaré–Fermi theorem (see Chapter 2), might be to find *only one* KTF of vanishing Schouten brackets with

---

[9] This result, worked out in Chapter 2 of [145], was independently found also in [143] following a different computational strategy.

the metric tensor. In fact, for quasi-integrable systems with $N \geq 3$, the Poincaré–Fermi theorem states that generically only energy is conserved. Thus if another constant of motion is known to exist (apart from Noetherian ones such as angular momentum), then the system must be integrable and in fact there must be N constants of motion.

- Unlike Killing vectors, which are associated with Noetherian symmetries and conservation laws, Killing tensors no longer have a simple geometric interpretation [149, 153]. Therefore the associated symmetries are non-Noetherian and hidden.

The Riemannian-geometric approach to integrability deserves further attention and investigation. In fact, among the other reasons of interest, by considering, for example, the standard Hénon–Heiles model ($C = D = 1$), we might wonder whether the regular regions of phase space correspond to a *local* satisfaction of the compatibility conditions of the system (4.52), which would lead to a better understanding of the relationship between geometry and stability of Newtonian mechanics. Moreover, we could imagine that by suitably defining *weak* and *strong violations* of these compatibility conditions, we could better understand the geometric origin of *weak* and *strong chaos* in Hamiltonian dynamics (see Chapter 2), and perhaps this might even suggest a starting point to developing a "geometric perturbation theory" complementary to the more standard canonical perturbation theory.

Chapter 5

Geometry and Chaos

The purpose of the present chapter is to describe in some detail how it is possibile, using the Jacobi–Levi-Civita equation for geodesic spread as the main tool, to reach a twofold objective: first, to obtain a deeper understanding of the origin of chaos in Hamiltonian systems, and second, to obtain quantitative information on the "strength" of chaos in these systems.

5.1 Geometric Approach to Chaotic Dynamics

A physical theory should provide a conceptual framework for modeling and understanding—at least at a qualitative level—the observed features of the system that is the object of the theory, and should also have a predictive content, i.e., should provide quantitative tools able to compute, at least approximately, the outcomes of the experiments (whether laboratory experiments or numerical experiments performed on a computer). According to these requirements, a satisfactory theory of deterministic chaos is certainly still lacking. In fact, in both aspects the current theoretical approaches to chaos have some problems, especially if we consider the case of conservative flows, i.e., of the dynamics of conservative systems of ordinary differential equations.

To explain the origin of chaos in conservative dynamics one usually invokes the existence of homoclinic intersections of perturbed separatrices. In order to quantify the degree of instability of a trajectory or of a system we must instead resort to the notion of Lyapunov exponents. The Lyapunov exponents are defined as asymptotic quantities, so that their relation to local properties of phase space is far from evident; nonetheless they provide the natural measure of the degree of chaos, measuring the typical time scales over which a trajectory loses the memory of its initial conditions. A rigorous definition of the existence of chaotic regions in the phase space of a system, based on the detection of homoclinic intersections, does not provide any quantitative tool to measure chaos; on the other hand, Lyapunov exponents allow a very precise measure of chaos but give no information at all on the origin of such chaotic behavior. From a conceptual point of view this situation is far from being satisfactory, not to speak of the fact that the practical application of the methods based on homoclinic intersections become extremely difficult when the number of degrees of freedom is greater than two [54, 93]. From the predictive point of view the situation is even worse, for no analytic method at all exists to compute Lyapunov exponents, at least in the case of flows of physical relevance. It is worth noticing that in [154], it has been shown that one could build, in principle, a field-theoretic framework to compute Lyapunov exponents, but the practical application of such methods is still unclear. Needless to say, all the tools belonging to canonical perturbation theory, that have undergone remarkable developments in the last years [155], can hardly be used to compute quantities such as Lyapunov exponents since in this framework one can describe only the regular, i.e., nonchaotic, features of phase space.

The geometric approach to dynamical instability allows a unification of the method to measure chaos with the explanation of its origin. In fact, the evolution of the field J given by the Jacobi equation (3.73) contains all the information needed to compute Lyapunov exponents, and makes us also recognize in the curvature properties of the ambient manifold the origin of chaotic dynamics.

Obviously, this approach encounters also some practical difficulties. For instance, the only case in which it is possible to rigorously prove that some definite curvature properties imply chaos in the geodesic flow is the case of compact manifolds whose curvature is everywhere negative. In this case every point of the manifold is hyperbolic: In a sense, this is the opposite limit to the integrable case. Though abstract and unphysical, such systems can help intuition. In a geodesic flow on a compact negatively curved manifold, the negative curvature forces nearby geodesics to locally separate exponentially, while the compactness ensures that such a separation does not reduce to a trivial "explosion" of the system and obliges the geodesics to fold. The joint action of stretching and folding of sets of initial conditions in phase space is the essential ingredient of deterministic chaos.

Krylov tried to apply this framework to explain the origin of mixing in physical dynamical systems. Unfortunately, for many systems in which chaos is detected, the curvatures are found to be mainly positive, and there are examples, for instance the Hénon–Heiles system, see (5.6), geometrized with the Jacobi metric and the Fermi–Pasta–Ulam model, see (5.49), geometrized with the Eisenhart metric, where curvatures are always positive even in the presence of fully developed chaos. Hence even positive curvature must be able to produce chaos.

Only recently has an example been found of a compact surface with positive curvature where the presence of chaotic regions coexisting with regular ones can be rigorously proved [156], and this provides mathematical support for the available numerical evidence that negative curvature is not necessary at all to have chaos in a geodesic flow [86, 129, 131]. What then *is* the crucial feature of the curvature that is required to produce chaos? There is not yet a definite answer—at least on rigorous grounds—to this question. Nevertheless, it is sure that if positive, curvature must be nonconstant in order to originate instability, and we shall see that the *curvature fluctuations* along the geodesics can be responsible for the appearance of an instability through a parametric-instability mechanism.

The advantages of the geometric approach to chaos are not only conceptual: also on predictive grounds this framework proves very useful. For starting from the Jacobi equation, it is possible to obtain an effective stability equation that allows one to obtain an analytic estimate of the largest Lyapunov exponent in the thermodynamic limit [86, 157]. Such an estimate turns out to be in very good agreement with the results of numerical simulations for a number of systems (see Section 5.4). In order to understand the derivation of such an effective stability equation, let us investigate in greater detail the relation between stability and curvature that was introduced in the last chapter.

5.2 Geometric Origin of Hamiltonian Chaos

Let us consider an N-dimensional Riemannian (or pseudo-Riemannian) manifold (M, g) and a local coordinate frame with coordinates $(q^1, \ldots, q^N)$.

We already observed that the evolution of the Jacobi field J, that contains all the information on the stability of the geodesic flow, is completely determined by the curvature tensor R through the Jacobi equation (3.73). Unfortunately the number of independent components of the tensor R is $\mathcal{O}(N^4)$—even if this number can be considerably reduced by symmetry considerations—so that (3.73) becomes rather untractable already at fairly small N values.

Nevertheless, there is a particular case in which the Jacobi equation has a remarkably simple form: the case of *isotropic* manifolds, where (3.73) becomes

$$\frac{\nabla^2 J^i}{ds^2} + K J^i = 0 \ , \tag{5.1}$$

where K is the constant sectional curvature of the manifold (see Appendix B). Choosing a geodesic frame, i.e., an orthonormal frame parallel transported along the geodesic, covariant derivatives become ordinary derivatives, i.e., $\nabla/ds \equiv d/ds$, so that the solution of (5.1), with initial conditions $J(0) = 0$ and $dJ(0)/ds = w(0)$, is

$$J(s) = \begin{cases} \frac{w(s)}{\sqrt{K}} \sin\left(\sqrt{K}\, s\right) & (K > 0) \ ; \\[2mm] s\, w(s) & (K = 0) \ ; \\[2mm] \frac{w(s)}{\sqrt{-K}} \sinh\left(\sqrt{-K}\, s\right) & (K < 0) \ . \end{cases} \tag{5.2}$$

The geodesic flow is unstable only if $K < 0$, and in this case the instability exponent is just $\sqrt{-K}$.

As long as the curvatures are negative, the geodesic flow is unstable even if the manifold is no longer isotropic, and by means of the so-called comparison theorems (mainly Rauch's theorem; see Appendix B) it is possible to prove that the instability exponent is greater than or equal to $(-\max_M(K))^{1/2}$ [3]. However, no exact results of general validity have yet been found for the dynamics of geodesic flows on manifolds whose curvature is neither constant nor everywhere negative.

Equation (5.1) is valid only if K is constant. Nevertheless, in the case in which $\dim M = 2$ (surfaces), the Jacobi equation—again written in a geodesic reference frame for the sake of simplicity—takes a form very close to that for isotropic manifolds,

$$\frac{d^2 J}{ds^2} + K(s)\, J = 0 \,, \tag{5.3}$$

where

$$K(s) = \frac{1}{2}\mathcal{R}(s) \tag{5.4}$$

and, in contrast to (5.1), it is no longer a constant. We let $\mathcal{R}(s)$ denote the scalar curvature of the manifold at the point $P = \gamma(s)$ (see Appendix B). If the geodesics are unstable, (5.3) has exponentially growing solutions. As far as we know [158], the solutions of (5.3) can exhibit an exponentially growing envelope in two cases:

(i) the curvature $K(s)$ takes negative values;
(ii) the curvature $K(s)$, though mainly or even exclusively positive, fluctuates in such a way that it triggers a sort of *parametric instability* mechanism.

In the first case, the mechanism that is at the origin of the instability of the geodesics is the one usually considered in ergodic theory [3]. But in the second case a new mechanism of instability that does not require the presence of negatively curved regions on the manifold shows up, that is, the curvature-variations along any given geodesic make it unstable.

Let us now turn to physics, i.e., to the case of a mechanical manifold. In the case of the Jacobi metric with $N = 2$ the scalar curvature written in standard (Lagrangian) coordinates reads as

$$\mathcal{R} = \frac{(\nabla V)^2}{(E - V)^3} + \frac{\Delta V}{(E - V)^2} \,, \tag{5.5}$$

where ∇ and Δ stand respectively for the Euclidean gradient and Laplacian operators. Hence we can have $\mathcal{R} < 0$ only if $\Delta V < 0$, i.e., for stable physical potentials, when the potential has inflection points. In these cases Krylov's idea can work—even if in the high-dimensional case this becomes very complicated—and we may have dynamical chaos induced by negative curvatures of the manifold. Indeed, Krylov was mainly concerned with weakly nonideal gases, or in general dilute systems, where for typical interatomic interactions $\Delta V < 0$ so that the curvatures can be negative (see Krylov's PhD thesis in [2]).

An example in which, though chaos is present, curvatures are positive, is provided by the Hénon–Heiles model. It is described by the Hamiltonian [48]

$$H = \frac{1}{2}\left(p_x^2 + p_y^2\right) + \frac{1}{2}\left(x^2 + y^2\right) + x^2 y - \frac{1}{3}y^3 \ . \tag{5.6}$$

Introduced in an astrophysical context, as mentioned in Section 4.4.3, it can also be regarded as a model of a triatomic molecule (after one has used translational symmetry to eliminate the center-of-mass coordinate) [159]. The Hénon–Heiles model is a cornerstone in the study of Hamiltonian chaos: it was the first physical model for which chaos was found and where a transition from a mainly regular to a mainly chaotic phase space was identified under a variation of the energy. In this model, (5.3) is exact, but $\mathcal{R} > 0$ everywhere. Hence chaos in this system cannot come from any negative curvature in the associated mechanical manifold $(\mathcal{M}_E, g_J)$. As we shall see later on (see, e.g., Section 5.4), the absence of negative curvatures in the associated mechanical manifolds is not a peculiarity of this model, for it is shared with many systems of interest for field theory and condensed-matter physics that have chaotic trajectories. In particular, all the systems that in the low-energy limit behave as a collection of harmonic oscillators do belong to this class.

In these cases the second of the previously discussed instability mechanisms, the one mentioned in item (ii), comes in: curvature fluctuations may induce chaos through parametric instability. The latter is a well-known feature of differential equations whose parameters are time-dependent. The classical example (see, e.g., Arnold's book [133]) is the mathematical swing, i.e., a pendulum, initially at rest, whose length is modulated in time. If the modulation contains frequencies resonating with the free pendulum's fundamental frequency, the stable equilibrium position becomes unstable and the swing starts to oscillate with growing amplitude. In (5.3), $\sqrt{K(s)}$ and s play the roles of a frequency and a time, respectively, so that this equation can be thought of as the equation of motion of a harmonic oscillator with time-dependent frequency, often referred to as a (generalized) Hill's equation [160]. By expanding $K(s)$ in a Fourier series we get

$$K(s) = K_0 + \sum_{n=1}^{\infty} \left[a_n \cos(n\omega s) + b_n \sin(n\omega s)\right] \ , \tag{5.7}$$

where $\omega = 2\pi/L$ and L is the length of the geodesic. The presence of resonances between the average frequency $\sqrt{K_0}$ and the frequency in some term in the expansion (5.7) eventually forces an exponential growth of the solutions of the equation. In the simplest case, in which only one coefficient of the series (5.7), say a_1, is nonvanishing, the equation is called the Mathieu equation, and it is possible to compute analytically both the bounds of the instability regions in the parameter space and the actual value of the characteristic exponent [160]. In contrast to the Mathieu case, in the general case, where a large number of coefficients of the Fourier decomposition of $K(s)$ are nonzero, it is much more difficult to do something similar. Hence there is not yet any rigorous proof of the fact that this kind of parametric instability is the mechanism that produces chaos in Hamiltonian dynamical systems—in the two-degrees-of-freedom case or in the general case—and this still remains a conjecture. Nevertheless, such a conjecture is strongly supported by at least two facts. First of all, in recent papers [161, 162] it has been shown that the solutions of the Jacobi equation (5.3) for the Hénon–Heiles model and for a model of quartic coupled oscillators show an oscillatory behavior with an exponentially growing envelope—which is precisely what one expects from parametric instability—in the chaotic regions, while the oscillations are bounded in the regular regions. Second, also in high-dimensional flows the components of the Jacobi field J oscillate with an exponentially growing amplitude as long as the system is non-integrable, whereas they exhibit only bounded oscillations for integrable systems. Moreover, in the high-dimensional case (i.e., for systems with a large number of degrees of freedom) it is possible to establish a *quantitative* link between the largest Lyapunov exponent and the curvature fluctuations. In fact, as we shall see in the following, in the high-dimensional case it is possibile to write down, under suitable approximations, an effective stability equation that looks very similar to (5.3), but where the squared frequency $K(s)$ is a stochastic process, and, through this equation, it is possible to give an analytical estimate of the largest Lyapunov exponent. Since from now on we are going to consider only the largest Lyapunov exponent, the latter will be referred to as just the Lyapunov exponent.

5.3 Effective Stability Equation in the High-Dimensional Case

Let us now study the problem of the stability of the geodesics in manifolds whose dimension N is large: according to the correspondence between geometry and dynamics introduced in Section 3, we are considering a system with a large number N of degrees of freedom.

Our starting point is the Jacobi equation (3.73). Our aim is to derive from it an effective stability equation that no longer depends on the dynamics, i.e., on the evolution of the particular geodesic that we are following, but only on the average curvature properties of the underlying manifold. To do that, we need some assumptions and approximations that are not valid in general but that are very reasonable in the case of large-N mechanical manifolds.[1] For

---

[1] As we shall see in the following, taken alone these assumptions lack something when the topology of the mechanical manifolds is non-trivial.

the sake of clarity, we first summarize the assumptions and approximations leading to our final result, and later on we discuss them more thoroughly. Further details can be found in the papers where this approach was originally put forward [86, 157].

0. We assume that the evolution of a generic geodesic is chaotic. This assumption is reasonable in the case of a manifold whose geodesics are the trajectories of a generic Hamiltonian system with a large number of degrees of freedom N, for in this case the overwhelming majority of the trajectories will be chaotic (see Chapter 2, Section 2.2).

1. We assume that the manifold is *quasi-isotropic*. Loosely speaking, this assumption means that the manifold can be regarded somehow as a locally deformed constant-curvature manifold. However, we will give this assumption a precise formulation later, in (5.16). This approximation allows us to get rid of the dependence of the Jacobi equation (3.73) on the full Riemann curvature tensor by replacing it with an effective sectional curvature $\mathcal{K}(s)$ along the geodesic; moreover, the Jacobi equation becomes diagonal.

2. To get rid of the dependence of the effective sectional curvature $\mathcal{K}(s)$ on the dynamics, i.e., on the evolution of the geodesic, we model $\mathcal{K}(s)$ with a stochastic process. This assumption is motivated by assumption 0 above. Moreover, as long as we consider a high-dimensional mechanical manifold associated with a Hamiltonian flow with N degrees of freedom and we are eventually interested in taking the thermodynamic limit $N \to \infty$, the sectional curvature is formed by adding up many independent terms, so that invoking a central-limit-theorem-like argument, $\mathcal{K}(s)$ is expected to behave, in first approximation, like a Gaussian stochastic process.

3. We assume that the statistics of the effective sectional curvature $\mathcal{K}$ are the same as those of the Ricci curvature K_{R}, which is a suitably averaged sectional curvature (see Appendix B, Section B.3.1). Such an assumption is consistent with assumption 1 above, for in a constant-curvature manifold the sectional curvature equals the Ricci curvature times a constant, and this allows us to compute the mean and the variance of the stochastic process introduced in assumption 2 in terms of the average and the variance of K_{R} along a generic geodesic.

4. The last step, which completely decouples the problem of the stability of the geodesics from the evolution of the geodesics themselves, consists in replacing the (proper) time averages of the Ricci curvature with static averages computed with a suitable probability measure μ. If the manifold is a mechanical manifold, the natural choice for μ is the microcanonical measure. Again this assumption is reasonable if Assumption 0 is valid.

After these steps, we end up with an effective stability equation that no longer depends on the evolution of the geodesics, but only on the average and fluctuations of the Ricci curvature of the manifold.

Let us now discuss more thoroughly the above-sketched procedure. For that, it is convenient to introduce the Weyl projective tensor W, whose components are given by [163]:

$$W^i{}_{jkl} = R^i{}_{jkl} - \frac{1}{N-1}(R_{jl}\delta^i_k - R_{jk}\delta^i_l) \, , \qquad (5.8)$$

where $R_{ij} = R^m{}_{imj}$ are the components of the Ricci curvature tensor (see Appendix B). Weyl's projective tensor measures the deviation from isotropy of a given manifold, since it vanishes identically for an isotropic manifold. Then we can rewrite the Jacobi equation (3.73) in the following form [86]:

$$\frac{\nabla^2 J^i}{ds^2} + \frac{1}{N-1}R_{jk}\frac{dq^j}{ds}\frac{dq^k}{ds}J^i - \frac{1}{N-1}R_{jk}\frac{dq^j}{ds}J^k\frac{dq^i}{ds} + W^i{}_{jkl}\frac{dq^j}{ds}J^k\frac{dq^l}{ds} = 0 \, .$$
$$(5.9)$$

For an isotropic manifold, the third term in (5.9) vanishes because $R_{jk} = K\,g_{jk}$ (see Appendix A), so that $R_{jk}\dot{q}^j J^k = K\,\langle\dot{\gamma}, J\rangle$, and $\langle\dot{\gamma}, J\rangle = 0$ (see 3.74). Thus, for an isotropic manifold, (5.9) collapses to (5.1). In fact, the second term is nothing but $K\,J^i$. When the manifold is not isotropic, we see that (5.9) retains the structure of (5.1) up to its second term, since the coefficient of J^i is still a scalar. This coefficient has now the meaning of a sectional curvature averaged, at any given point, over the $N-1$ independent directions orthogonal to $\dot{\gamma}$, the velocity vector of the geodesic. However, such a mean sectional curvature is no longer constant along the geodesic $\gamma(s)$, and is just the Ricci curvature $K_{\rm R}$ divided by $N-1$ (see Appendix B). The fourth term of (5.9) accounts for the local degree of anisotropy of the ambient manifold.

Let us now rewrite (5.9) as

$$\frac{\nabla^2 J^i}{ds^2} + k_R(s)\,J^i - r^i_j(s)\,J^j + w^i_j(s)\,J^j = 0 \, , \qquad (5.10)$$

where, to ease the notation, we have put

$$k_R(s) = \frac{K_{\rm R}}{N-1} = \frac{1}{N-1}R_{jk}\frac{dq^j}{ds}\frac{dq^k}{ds} \, ; \qquad (5.11)$$

$$r^i_j(s) = \frac{1}{N-1}R_{jk}\frac{dq^k}{ds}\frac{dq^i}{ds} \, ; \qquad (5.12)$$

$$w^i_j(s) = W^i{}_{kjl}\frac{dq^k}{ds}\frac{dq^l}{ds} \, . \qquad (5.13)$$

Now let us formulate our assumption 1, namely, that the manifold is quasi-isotropic, in a more precise way. To do that, we recall that (see Appendix B, Section (B.3.2)) if and only if the manifold is isotropic, i.e., has constant curvature, the Riemann curvature tensor and the Ricci tensor can be written in the remarkably simple forms

$$R_{ijkl} = K\,(g_{ik}g_{jl} - g_{il}g_{jk}) \qquad (5.14)$$

and

$$R_{ij} = K \, g_{ij} \, , \tag{5.15}$$

where K is a scalar constant, the sectional curvature of the manifold. The precise formulation of assumption 1 is now that along a generic geodesic the Riemann curvature tensor and the Ricci tensor retain the same functional form as in the case (5.14), i.e., that

$$R_{ijkl} \approx \mathcal{K}(s) \, (g_{ik}g_{jl} - g_{il}g_{jk}) \tag{5.16}$$

and

$$R_{ij} \approx \mathcal{K}(s) \, g_{ij} \, , \tag{5.17}$$

where $\mathcal{K}(s)$, which is no longer a constant, is an effective sectional curvature. In the general case we are not able to give a rigorous explicit expression for $\mathcal{K}(s)$, because the functional dependence postulated in (5.17) holds only for constant-curvature manifolds. However, the effective curvature $\mathcal{K}(s)$ is expected to be essentially the sectional curvature $K(\dot{\gamma}, J)$ (see Appendix B, Section (B.3.1)) measured along the geodesic in the directions of the velocity vector $\dot{\gamma} = dq/ds$ and of the Jacobi vector J.

Combining (5.12) and (5.17), and recalling that the vector J is orthogonal to the velocity of the geodesic, i.e., $g_{ij} \frac{dq^i}{ds} J^j = 0$, we find that the third term in (5.10), $-r^i_j J^j$, vanishes as in the isotropic case. Now we combine (5.8) and (5.16) to obtain

$$W^i{}_{jkl} \approx \mathcal{K}(s)(\delta^i_j g_{kl} - \delta^i_l g_{kj}) - \frac{1}{N-1}(R_{jl}\delta^i_k - R_{jk}\delta^i_l) \, , \tag{5.18}$$

so that (5.13) can be rewritten as

$$w^i_j \approx \mathcal{K}(s)\delta^i_j - k_R(s)\delta^i_j - \mathcal{K}(s)\frac{N-2}{N-1}\frac{dq^i}{ds}g_{kj}\frac{dq^k}{ds} \, , \tag{5.19}$$

where we have used the definition of k_R given in (5.11) and the approximation (5.17) for the Ricci tensor. Let us now insert (5.19) into (5.10): the last term of (5.19) vanishes after having been multiplied by J^j and summed over j, because J and dq/ds are orthogonal, and the term $k_R(s)J^i$ is canceled by the term $-k_R(s)J^i$ coming from (5.19), so that (5.10) is finally rewritten as

$$\frac{\nabla^2 J^i}{ds^2} + \mathcal{K}(s) \, J^i = 0 \, , \tag{5.20}$$

and, with respect to a geodesic frame, it becomes

$$\frac{d^2 J^i}{ds^2} + \mathcal{K}(s) \, J^i = 0 \, . \tag{5.21}$$

Being a scalar quantity, the value of $\mathcal{K}(s)$ is independent of the coordinate system. Equation (5.21) is now diagonal. However, in order to use it, we should know the values of $\mathcal{K}(s)$ along the geodesic. Here, assumptions 2 and 3 come into play: we replace $\mathcal{K}(s)$ with a stochastic Gaussian process, and we assume that its probability distribution is the same as that of the Ricci curvature,

$$\mathcal{P}(\mathcal{K}) \approx \mathcal{P}(K_R) \ . \tag{5.22}$$

Such an assumption is consistent with our assumption 1, because for an isotropic manifold the sectional curvature is identical to the Ricci curvature divided by $N - 1$, so that if the manifold is quasi-isotropic, it is natural to assume that the probability distributions of the sectional curvature and of the Ricci curvature are similar. Moreover, such an assumption is also the only easy one, because we are able to compute, under some further assumptions, the probability distribution of K_R, but we do not know anything about $\mathcal{K}$.

To be consistent with the definition of the sectional and the Ricci curvatures (see Appendix B, Section (B.3.1)), the following relations are assumed to hold for the first two cumulants of (5.22):

$$\langle \mathcal{K}(s) \rangle_s = \frac{1}{N-1} \langle K_R(s) \rangle_s \equiv \langle k_R(s) \rangle_s \ , \tag{5.23}$$

$$\langle [\mathcal{K}(s) - \overline{\mathcal{K}}]^2 \rangle_s = \frac{1}{N-1} \langle [K_R(s) - \langle K_R \rangle_s]^2 \rangle_s \equiv \langle \delta^2 k_R \rangle_s \ , \tag{5.24}$$

where $\langle \cdot \rangle_s$ stands for a proper-time average along a geodesic $\gamma(s)$. A priori, the probability distributions (5.22) are unknown. will not be Gaussian, i.e., other cumulants in addition to the first two will be nonvanishing. However, since for a large-N system K_R is obtained by summing a large number of randomly varying independent terms, it is reasonable to invoke a central limit theorem argument to assume that both $\mathcal{P}(\mathcal{K})$ and $\mathcal{P}(K_R)$ are Gaussian distributions.

Our approximation for the effective sectional curvature $\mathcal{K}(s)$ is then the stochastic process

$$\mathcal{K}(s) \approx \langle k_R(s) \rangle_s + \langle \delta^2 k_R \rangle_s^{1/2} \, \eta(s) \ , \tag{5.25}$$

where $\eta(s)$ is a random Gaussian process with zero mean and unit variance.

Finally, in order to completely decouple the stability equation from the dynamics, we use assumption 4 and we replace time averages with static averages computed with a suitable measure μ. If the manifold is a mechanical manifold, the geodesics are the natural motions of the systems, and a natural choice for μ is then the microcanonical ensemble, so that (5.25) becomes

$$\mathcal{K}(s) \approx \langle k_R(s) \rangle_\mu + \langle \delta^2 k_R \rangle_\mu^{1/2} \, \eta(s) \ . \tag{5.26}$$

Our final effective (in)stability equation is then

$$\frac{d^2 \psi}{ds^2} + \langle k_R \rangle_\mu \, \psi + \langle \delta^2 k_R \rangle_\mu^{1/2} \, \eta(s) \, \psi = 0 \ , \tag{5.27}$$

where ψ stands for *any* of the components of J, since all of them now obey the *same* effective equation of motion.

Equation (5.27) implies that if the manifold is a mechanical manifold, the growth rate of ψ gives the dynamical instability exponent in our Riemannian framework. Equation (5.27) is a scalar equation that, *independently of the knowledge of the dynamics*, provides a measure of the degree of instability of the dynamics itself through the s-dependence of $\psi(s)$. The peculiar properties of a given Hamiltonian system enter (5.27) only through the global geometric properties $\langle k_{\mathrm{R}} \rangle_\mu$ and $\langle \delta^2 k_{\mathrm{R}} \rangle_\mu$ of the ambient Riemannian manifold (whose geodesics are natural motions) and are sufficient, as long as our assumptions 1–4 hold, to determine the average degree of chaoticity of the dynamics. Moreover, $\langle k_{\mathrm{R}} \rangle_\mu$ and $\langle \delta^2 k_{\mathrm{R}} \rangle_\mu$ are microcanonical averages, so that they are functions of the energy E of the system, or of the energy per degree of freedom $\varepsilon = E/N$, which is the relevant parameter as $N \to \infty$. Thus from (5.27) we can obtain the energy dependence of the geometric instability exponent.

Within the validity of our assumptions 1–4, transforming the Jacobi equation (3.73) into (5.27), the original complexity of the Jacobi equation has been considerably reduced: from a tensor equation we have obtained an effective scalar equation formally representing the equation of motion of a stochastic oscillator. In fact, (5.27), with a self-evident notation, is of the form

$$\frac{d^2\psi}{ds^2} + k(s)\,\psi = 0 \ , \tag{5.28}$$

where $k(s)$, the squared frequency, is a Gaussian stochastic process.

Moreover, such an equation admits a very suggestive geometric interpretation: since it is a scalar equation, i.e., it is formally the Jacobi equation on a 2-dimensional manifold whose Gaussian curvature is given, along a geodesic, by the random process $k(s)$ and can be regarded as an "effective" low-dimensional manifold approximating the "true" high-dimensional manifold where the dynamics of the geodesic flow takes place. This is the real geometrical content of our quasi-isotropy hypothesis. Hence the average global curvature properties $\langle k_{\mathrm{R}} \rangle_\mu$ and $\langle \delta^2 k_{\mathrm{R}} \rangle_\mu$, in addition to being the ingredients for a geometric computation of the instability exponent, convey also information on the geometric structure of this effective manifold. Thus we expect that it will be possible to gain some insight into the global properties of the dynamics by simply studying the behavior of these average curvature properties as the energy is varied.

5.3.1 A Geometric Formula for the Lyapunov Exponent

Let us now study the properties of the solutions of (5.28) in order to obtain an analytic estimate for the Lyapunov exponent. The derivation of the stochastic oscillator equation does not depend on a particular choice of the metric; within the approximations discussed above, (5.28) holds regardless of the choice of the metric. However, to make explicit the connection between the solutions of (5.28) and the stability of a dynamical system, one has to choose a particular metric; in the case of Hamiltonian systems of the form (1.1), the choice of the Eisenhart metric is the simplest one.

For this reason, we shall from now on restrict ourselves to standard Hamiltonian systems with a diagonal kinetic energy matrix, i.e., $a_{ij} = \delta_{ij}$, choosing as ambient manifold for the geometrization of the dynamics the enlarged configuration space-time equipped with the Eisenhart metric (3.28). The case of the Jacobi metric is discussed in [86, 164].

The fact that the Jacobi–Levi-Civita equation on $(M \times \mathbb{R}^2, g_e)$ coincides with the standard tangent dynamics equation (2.108), or, equivalently, (3.94), clarifies the relationship between the geometric description of the instability of a geodesic flow and the conventional numerical description of dynamical instability. We stress that from a formal viewpoint this is a peculiarity of the Eisenhart metric; nevertheless, the physical content of this result is valid independently of the metric used, as long as the identification between trajectories and geodesics holds true. Indeed, in papers [130, 161, 162] it has been found that using the Jacobi metric, (M_E, g_J), the instability growth rates of the solutions of the Jacobi equation (3.84) and of those of the tangent dynamics equation (3.94)—which in this case are two distinct equations—are the same at any given energy.

Let us now come to the computation of the instability growth rates of the solutions of the effective stochastic equation (5.28).

Along a physical geodesic, $(dq^0)^2 = dt^2 = ds^2$. Thus, we replace the arc length s along the geodesic with the physical time t, and the stochastic oscillator equation (5.28) can be written

$$\frac{d^2\psi}{dt^2} + k(t)\,\psi = 0 \ . \tag{5.29}$$

The Ricci curvature along a geodesic depends only on the coordinates and not on the velocities and for the Eisenhart metric reads

$$K_R(q) = \triangle V \ , \tag{5.30}$$

so that the mean and variance of $k(t)$ are given by

$$k_0 \equiv \langle k_R \rangle_\mu = \frac{1}{N} \langle \triangle V \rangle_\mu \ , \tag{5.31}$$

$$\sigma_k^2 \equiv \langle \delta^2 k_R \rangle_\mu = \frac{1}{N} \left(\langle (\triangle V)^2 \rangle_\mu - \langle \triangle V \rangle_\mu^2 \right) \ , \tag{5.32}$$

where μ is an invariant measure for the dynamics to be identified with the microcanonical measure. Since we are considering systems with large N—eventually taking the limit $N \to \infty$—we replaced $(N - 1)$ with N in (5.31) and (5.32).

The process $k(t)$ is not completely defined unless its time correlation function,

$$\Gamma_k(t_1, t_2) = \langle k(t_1) k(t_2) \rangle - \langle k(t_1) \rangle \langle k(t_2) \rangle \ , \tag{5.33}$$

is given. The simplest choice is to assume that $k(t)$ is a stationary random process which can be approximated[2] by a δ-correlated process, so that

$$\Gamma_k(t_1, t_2) = \Gamma_k(|t_2 - t_1|) = \Gamma_k(t) = \tau \sigma_k^2 \delta(t) , \tag{5.34}$$

where τ is the characteristic correlation time scale of the process.

Before we can actually solve (5.29), we have then to give an explicit expression for τ. To do that, first we will show how two independent characteristic correlation time scales, which will be referred to as τ_1 and τ_2, respectively, can be defined; then we will estimate τ by combining these two time scales.

A first time scale, which we will refer to as τ_1, is associated with the time needed to cover the average distance between two successive *conjugate points* along a geodesic. Conjugate points are the points where the Jacobi vector field vanishes (see Appendix B). As long as the curvature is positive and its fluctuations are small compared to the average, two nearby geodesics will remain close to each other until a conjugate point is reached. At each crossing of a conjugate point the Jacobi vector field increases as if the geodesics were kicked (this is what happens when parametric instability is active). Thus the average distance between conjugate points provides a meaningful estimate of the lower bound of the correlation time scale. It can be proved that if the sectional curvature K is bounded as $0 < L \leq K \leq H$, then the distance d between two successive conjugate points is bounded by $d > \frac{\pi}{2\sqrt{H}}$ (see Appendix B). The upper bound H of the curvature can then be approximated in our framework by

$$H \approx k_0 + \sigma_k , \tag{5.35}$$

so that we can define τ_1 as (remember that $dt = ds$)

$$\tau_1 = d_1 = \frac{\pi}{2\sqrt{k_0 + \sigma_k}} . \tag{5.36}$$

This time scale is expected to be the most relevant only as long as the curvature is positive and the fluctuations are small, compared to the average.

Another time scale, referred to as τ_2, is related to the local curvature fluctuations. These will be felt on a length scale of order at least $l = 1/\sqrt{\sigma_k}$ (the average fluctuation of curvature radius). The scale l is expected to be the relevant one when the fluctuations are of the same order of magnitude as the average curvature. Locally, the metric of a manifold can be approximated by [126]

$$g_{ik} \approx \delta_{ik} - \frac{1}{6} R_{ikjl} u^i u^k , \tag{5.37}$$

where the u^i are the components of the displacements from the point around which we are approximating the metric. When the sectional curvature is positive (negative), lengths and time intervals on a scale l are enlarged (shortened)

---

[2] Actually, $k(t)$ models $\mathcal{K}(t)$ which is a smooth function, thus $k(t)$ can be modeled by a random process if it represents a sampling of $\mathcal{K}(t)$ at time steps larger than some suitable correlation time scale τ.

by a factor $(l^2 K/6)$, so that the period $\frac{2\pi}{\sqrt{k_0}}$ has a fluctuation amplitude d_2 given by $d_2 = \frac{l^2 K}{6} \frac{2\pi}{\sqrt{k_0}}$; replacing K by the most probable value k_0, one gets

$$\tau_2 = d_2 = \frac{l^2 k_0}{6} \frac{2\pi}{\sqrt{k_0}} \approx \frac{k_0^{1/2}}{\sigma_k} . \tag{5.38}$$

Finally, τ in (5.34) is obtained by combining τ_1 with τ_2 as follows:

$$\tau^{-1} = \tau_1^{-1} + \tau_2^{-1} . \tag{5.39}$$

The present definition of τ is obviously by no means a direct consequence of any theoretical result, but only a rough, physically based estimate. Such an estimate might well be improved independently of the general geometric framework.

Now that all the quantities entering (5.29) have been fully defined, we can proceed to compute the instability growth rate of a generic solution of (5.29). Whenever $k(t)$ has a nonvanishing stochastic component, any solution $\psi(t)$ has an exponentially growing envelope [165] whose growth rate provides a measure of the degree of instability. How can one relate such a growth rate to the Lyapunov exponent of the physical system? Let us recall that, for a standard Hamiltonian system of the form (1.1), the Lyapunov exponent can be computed as the following limit (see (2.108) in Section 2.2.3):

$$\lambda = \lim_{t \to \infty} \frac{1}{2t} \log \frac{\xi_1^2(t) + \cdots + \xi_N^2(t) + \dot{\xi}_1^2(t) + \cdots + \dot{\xi}_N^2(t)}{\xi_1^2(0) + \cdots + \xi_N^2(0) + \dot{\xi}_1^2(0) + \cdots + \dot{\xi}_N^2(0)} , \tag{5.40}$$

where the ξ's are the components of the tangent vector, i.e., of the perturbation of a reference trajectory, which obey the tangent dynamics equation (2.108). In the case of the Eisenhart metric, each component of the Jacobi vector field J can be identified with the corresponding component of the tangent vector ξ; moreover, ψ in (5.29) stands for any of the components of J, which obey the same effective equation. Thus, (5.40) becomes

$$\lambda = \lim_{t \to \infty} \frac{1}{2t} \log \frac{\psi^2(t) + \dot{\psi}^2(t)}{\psi^2(0) + \dot{\psi}^2(0)} , \tag{5.41}$$

where $\psi(t)$ is a solution of (5.29). Equation (5.41) is our estimate for the (largest) Lyapunov exponent.

As a stochastic differential equation, the solutions of (5.29) are properly defined after an averaging over the realizations of the stochastic process: referring to such an averaging as $\langle \bullet \rangle$, we rewrite (5.41) as

$$\lambda = \lim_{t \to \infty} \frac{1}{2t} \log \frac{\langle \psi^2(t) \rangle + \langle \dot{\psi}^2(t) \rangle}{\langle \psi^2(0) \rangle + \langle \dot{\psi}^2(0) \rangle} . \tag{5.42}$$

The evolution of $\langle \psi^2 \rangle$, $\langle \dot{\psi}^2 \rangle$, and $\langle \psi \dot{\psi} \rangle$, i.e., of the vector of the second moments of ψ, obeys the following equation, which can be derived by means of a technique developed by van Kampen and sketched in Section 5.6.

$$\frac{d}{dt} \begin{pmatrix} \langle \psi^2 \rangle \\ \langle \dot{\psi}^2 \rangle \\ \langle \psi \dot{\psi} \rangle \end{pmatrix} = \begin{pmatrix} 0 & 0 & 2 \\ \sigma_k^2 \tau & 0 & -2k_0 \\ -k_0 & 1 & 0 \end{pmatrix} \begin{pmatrix} \langle \psi^2 \rangle \\ \langle \dot{\psi}^2 \rangle \\ \langle \psi \dot{\psi} \rangle \end{pmatrix} , \tag{5.43}$$

where k_0 and σ_k are the mean and the variance of $k(t)$, defined in (5.31) and (5.32), respectively. Equation (5.43) can be solved by diagonalizing the matrix on the right-hand side of (5.43). The result for the evolution of $\langle \psi^2 \rangle + \langle \dot{\psi}^2 \rangle$ is

$$\langle \psi^2(t) \rangle + \langle \dot{\psi}^2(t) \rangle = \left(\langle \psi^2(0) \rangle + \langle \dot{\psi}^2(0) \rangle \right) \exp(\alpha t) , \tag{5.44}$$

where α is the only real eigenvalue of the matrix. According to (5.42), the Lyapunov exponent is given by $\lambda = \alpha/2$, so that by computing explicitly α, one then obtains the final expression

$$\lambda(k_0, \sigma_k, \tau) = \frac{1}{2} \left(\Lambda - \frac{4k_0}{3\Lambda} \right) , \tag{5.45}$$

$$\Lambda = \left(\sigma_k^2 \tau + \sqrt{\left(\frac{4k_0}{3} \right)^3 + \sigma_k^4 \tau^2} \right)^{1/3} . \tag{5.46}$$

All the quantities k_0, σ_k and $\tau(k_0, \sigma_k)$ can be computed as *static* averages, as functions of the energy per degree of freedom, ε (see (5.31) and (5.32)). Therefore—within the limits of validity of the assumptions made above— equation (5.46) provide an approximate analytic formula to compute the largest Lyapunov exponent independently of the numerical integration of the dynamics and of the tangent dynamics.

Let us remark that expanding (5.46) in the limit $\sigma_k \ll k_0$, one finds that

$$\lambda \propto \sigma_k^2 , \tag{5.47}$$

which shows how close the relation is between curvature fluctuations and dynamical instability.

5.4 Some Applications

Let us now discuss briefly the results of the application of the geometric techniques described up to this point to some Hamiltonian models. In particular, we shall consider three cases: a chain of coupled nonlinear oscillators (the FPU β model, introduced in Section 2.3.1), a chain of coupled rotators (the 1D XY model), and the mean-field XY model. The reason for the choice of these three particular models is that they allow fully analytic calculations and are well suited to show advantages and limitations of the theory. The geometric theory developed above has already been applied to many other cases, some of which will be addressed in Chapter 6. For other applications we refer to the literature: in particular, a model of a homopolymer chain has been studied in [88], a model of a three-dimensional Lennard-Jones crystal has been studied in [166], and a classical lattice gauge theory has been considered in [167].

The systems we now consider are 1D models with nearest-neighbor interactions whose Hamiltonians H have the standard form (1.1) with

$$V = \sum_{i=1}^{N} v(q_i - q_{i-1}) \ . \tag{5.48}$$

The interaction potentials are, respectively,

$$v(q_i - q_{i-1}) = \frac{1}{2}(q_i - q_{i-1})^2 + \frac{u}{4}(q_i - q_{i-1})^4 \ , \quad \text{(FPU } \beta \text{ model)} \tag{5.49}$$

$$v(x) = -J\cos(q_i - q_{i-1}) \ , \quad \text{(1D } XY \text{ model)} \tag{5.50}$$

while the third model is a long-range interaction one, described by the potential

$$V = \frac{J}{2N} \sum_{i,j=1}^{N} [1 - \cos(q_i - q_{i-1})] \quad \text{(mean-field } XY \text{ model)} \ . \tag{5.51}$$

In (5.49) we use u instead of the customary β in order to avoid confusion with the inverse temperature β. We assume $u > 0$.

As we have seen in the preceding section, to compute the largest Lyapunov exponent by means of the geometric formula (5.45), we need to compute the average and the root mean square (r.m.s.) fluctuations of the Ricci curvature of the mechanical manifold. These are to be computed as microcanonical ensemble averages. These microcanonical quantities can be computed starting from the *canonical* partition function, which can be calculated exactly for an infinite chain, i.e., $N \to \infty$, for all the three models defined above.

The average and fluctuations, within the microcanonical ensemble, of any observable function $f(q)$ can be computed as follows, in terms of the corresponding quantities in the canonical ensemble. The canonical configurational partition function $Z(\beta)$ is given by

$$Z(\beta) = \int dq\, e^{-\beta V(q)} \ , \tag{5.52}$$

where $dq = \prod_{i=1}^{N} dq_i$. The canonical average $\langle f \rangle_{\text{can}}$ of the observable f can be computed as

$$\langle f \rangle_{\text{can}} = [Z(\beta)]^{-1} \int dq\, f(q)\, e^{-\beta V(q)} \ . \tag{5.53}$$

From this average, we can obtain the microcanonical average of f, $\langle f \rangle_\mu$, in the following (implicit) parametric form [108].

$$\left.\begin{aligned} \langle f \rangle_\mu(\beta) &= \langle f \rangle_{\text{can}}(\beta) \\[2mm] \varepsilon(\beta) &= \frac{1}{2\beta} - \frac{1}{N}\frac{\partial}{\partial\beta}[\log Z(\beta)] \end{aligned}\right\} \;\rightarrow\; \langle f \rangle_\mu(\varepsilon)\,. \tag{5.54}$$

Note that (5.54) is strictly valid only in the thermodynamic limit; at finite N, $\langle f \rangle_\mu(\beta) = \langle f \rangle_{\text{can}}(\beta) + \mathcal{O}(\frac{1}{N})$.

In contrast to the computation of $\langle f \rangle$, which is insensitive to the choice of the probability measure in the $N \to \infty$ limit, computing the fluctuations of f, that is, $\langle \delta^2 f \rangle = \frac{1}{N}\langle (f - \langle f \rangle)^2 \rangle$, by means of the canonical or microcanonical ensembles yields different results. The relationship between the canonical, i.e., computed with the Gibbsian weight $e^{-\beta H}$, and the microcanonical fluctuations is given by the Lebowitz–Percus–Verlet formula [108]

$$\langle \delta^2 f \rangle_\mu(\varepsilon) = \langle \delta^2 f \rangle_{\text{can}}(\beta) - \frac{\beta^2}{c_V}\left[\frac{\partial \langle f \rangle_{\text{can}}(\beta)}{\partial\beta}\right]^2, \tag{5.55}$$

where

$$c_V = -\frac{\beta^2}{N}\frac{\partial \langle H \rangle_{\text{can}}}{\partial\beta} \tag{5.56}$$

is the specific heat at constant volume. Thus (5.55) also reads

$$\langle \delta^2 f \rangle_\mu(\varepsilon) = \langle \delta^2 f \rangle_{\text{can}}(\beta) + \left(\frac{\partial \langle \varepsilon \rangle_{\text{can}}}{\partial\beta}\right)^{-1}\left[\frac{\partial \langle f \rangle_{\text{can}}(\beta)}{\partial\beta}\right]^2, \tag{5.57}$$

and $\beta = \beta(\varepsilon)$ is given in implicit form by the second equation in (5.54).

The average k_0 and the fluctuations σ_k of the Ricci curvature per degree of freedom are then obtained by replacing f with the explicit expression for Ricci curvature, which, according to the definition given in (5.30), is

$$K_R(q) = \sum_{i=1}^N \frac{\partial^2}{\partial q_i^2}\, v(q_i - q_{i-1})\,, \tag{5.58}$$

in (5.54) and (5.55), respectively.

5.4.1 FPU β Model

For the FPU β model, the dynamical observable that corresponds to the Ricci curvature reads, according to (5.58),

$$K_R = 2N + 6u \sum_{i=1}^N (q_{i+1} - q_i)^2\,. \tag{5.59}$$

Note that K_R is always positive and that this is also true for the sectional curvature along a physical geodesic. Computing the microcanonical average of K_R according to (5.54) we find that in the thermodynamic limit, $k_0(\varepsilon)$ is implicitly given by (the details are reported in [86])

$$\left.\begin{aligned}
\langle k_R \rangle_{\text{can}}(\theta) &= 2 + \frac{3}{\theta} \frac{D_{-3/2}(\theta)}{D_{-1/2}(\theta)} \\[2ex]
\varepsilon(\theta) &= \frac{1}{8\sigma}\left[\frac{3}{\theta^2} + \frac{1}{\theta}\frac{D_{-3/2}(\theta)}{D_{-1/2}(\theta)}\right]
\end{aligned}\right\} \rightarrow k_0(\varepsilon)\,, \qquad (5.60)$$

where the D_ν are parabolic cylinder functions [160] and θ is a parameter proportional to β, so that $\theta \in [0,+\infty)$.

The fluctuations of K_R

$$\sigma_k^2(\varepsilon) = \frac{1}{N}\langle \delta^2 K_R \rangle_\mu(\varepsilon) = \frac{1}{N}\langle (K_R - \langle K_R \rangle)^2 \rangle_\mu\,. \qquad (5.61)$$

are computed as follows.

According to (5.55), first the canonical fluctuations,

$$\langle \delta^2 k_R \rangle_{\text{can}}(\beta) = \frac{1}{N}\langle (K_R - \langle K_R \rangle)^2 \rangle_{\text{can}}(\beta)\,,$$

have to be computed and then a correction term must be added. The canonical fluctuations are given by [86]

$$\langle \delta^2 k_R \rangle_{\text{can}}(\theta) = \frac{9}{\theta^2}\left\{2 - 2\theta \frac{D_{-3/2}(\theta)}{D_{-1/2}(\theta)} - \left[\frac{D_{-3/2}(\theta)}{D_{-1/2}(\theta)}\right]^2\right\}\,, \qquad (5.62)$$

and, by adding the correction term, the final result for the microcanonical fluctuations of the Ricci curvature is

$$\left.\begin{aligned}
\langle \delta^2 k_R \rangle_\mu(\theta) &= \langle \delta^2 k_R \rangle_{\text{can}}(\theta) - \frac{\beta^2}{c_V(\theta)}\left(\frac{\partial \langle k_R \rangle_{\text{can}}(\theta)}{\partial \beta}\right)^2 \\[2ex]
\varepsilon(\theta) &= \frac{1}{8\mu}\left[\frac{3}{\theta^2} + \frac{1}{\theta}\frac{D_{-3/2}(\theta)}{D_{-1/2}(\theta)}\right]
\end{aligned}\right\} \rightarrow \sigma_k^2(\varepsilon)\,, \qquad (5.63)$$

where $\langle \delta^2 k_R \rangle_{\text{can}}(\theta)$ is given by (5.62), $\partial \langle k_R \rangle(\theta)/\partial \beta$ is given by

$$\frac{\partial \langle k_R \rangle(\theta)}{\partial \beta} = \frac{3}{8\mu\theta^3}\frac{\theta D_{-3/2}^2(\theta) + 2(\theta^2 - 1)D_{-1/2}(\theta)D_{-3/2}(\theta) - 2\theta D_{-1/2}^2(\theta)}{D_{-1/2}^2(\theta)}\,, \qquad (5.64)$$

and the specific heat per particle c_V is found to be [86, 168]

$$\begin{aligned}
c_V(\theta) = \frac{1}{16 D_{-1/2}^2(\theta)}\Big\{&(12 + 2\theta^2)D_{-1/2}^2(\theta) + 2\theta D_{-1/2}(\theta)D_{-3/2}(\theta) \\
&- \theta^2 D_{-3/2}(\theta)\left[2\theta D_{-1/2}(\theta) + D_{-3/2}(\theta)\right]\Big\}\,. \qquad (5.65)
\end{aligned}$$

The microcanonical averages and fluctuations computed in (5.60) and (5.63) are compared in Figures 5.1 and 5.2 with their corresponding time averages computed along numerically simulated trajectories of the FPU β model with the potential (5.49) for $N = 128$ and $N = 512$ with $u = 0.1$. Though the microcanonical averages have to be computed in the thermodynamic limit, the agreement between time and ensemble averages is excellent already at $N = 128$.

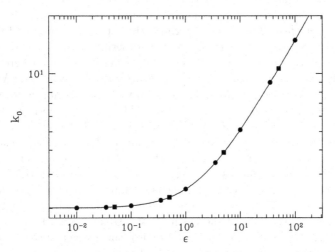

Fig. 5.1. Average Ricci curvature (Eisenhart metric) per degree of freedom, k_0, vs. energy density ε for the FPU β model. The continuous line is the analytic computation according to (5.60); circles and squares are time averages obtained by numerical simulations with $N = 128$ and $N = 512$ respectively; $u = 0.1$. From [131].

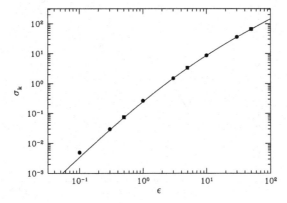

Fig. 5.2. Fluctuations of the Ricci curvature (Eisenhart metric), σ_k, vs. energy density ε for the FPU β model. Symbols and parameters as in Figure 5.1; the continuous line now refers to (5.63). From [131].

As we have seen in Chapter 2, Section 2.3.1, the FPU β model, as many other Hamiltonian dynamical systems at large-N, undergo a transition between weak and strong chaos. In particular, in the FPU β model, a weakly chaotic regime is found for specific energies smaller than $\varepsilon_c \approx 0.1/u$ [78, 79, 131]. Although in the weakly chaotic regime the dynamics is chaotic (i.e., the Lyapunov exponent is positive, though small), mixing is very slow, as witnessed by the existence of a rather long memory of the initial conditions, i.e., of long relaxation times if the initial conditions are far from equilibrium. For ε larger than ε_c the dynamics is strongly chaotic and relaxations to equilibrium are fast. The precise origin of these phenomena is still to be understood. However, the geometric approach described here is able to provide a suggestive interpretation [131, 169]. Let us consider Figure 5.3, where the ratio of the fluctuations and the average curvature σ_k/k_0 is reported. As $\varepsilon \to 0$, $\sigma_k \ll k_0$, so that the manifold looks essentially like a constant-curvature manifold with only small curvature fluctuations. This situation corresponds to the weakly chaotic dynamical regime. However, since ε is larger than ε_c, σ_k/k_0 tends to saturate toward a value of order unity, thus indicating that in the high-energy (strongly chaotic) regime the curvature fluctuations are of the same order of magnitude as the average curvature, so that the system no longer "feels" the isotropic (and integrable) limit. Hence the geometric approach can give a hint toward understanding, at least qualitatively, the origin of weak and strong chaos in the Fermi–Pasta–Ulam model. From the knowledge of $k_0(\varepsilon)$ and $\sigma_k(\varepsilon)$, via (5.45) and (5.46), the geometric theory allows us to make a quantitative prediction for the Lyapunov exponent as a function of the specific energy ε. The analytic result is shown in Figure 5.4 and is

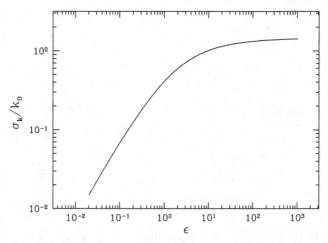

Fig. 5.3. Fluctuations of the Ricci curvature (Eisenhart metric) divided by the average curvature, σ_k/k_0, vs. energy density ε for the FPU β model.

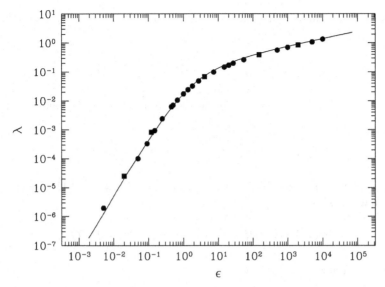

Fig. 5.4. Lyapunov exponent, λ, vs. energy density ε for the FPU β model with $u = 0.1$. The continuous line is the theoretical computation according to (5.46), while the circles and squares are the results of numerical simulations with N respectively equal to 256 and 2000. From [86].

compared with numerical simulations made for different values of N, for the FPU β case in a wide range of energy densities — more than six orders of magnitude [86, 157]. The agreement between theory and simulations is strikingly good, which confirms the existence of a nonempty validity domain of the simplifying assumptions that we had to introduce on physical grounds to capture some of the essentials of the configuration space geometry.

5.4.2 The Role of Unstable Periodic Orbits

Let us now make a digression about an interesting peculiarity of a special class of phase space trajectories: unstable periodic orbits (UPO). Surprisingly, this kind of orbits make a sort of "importance sampling" of the geometry of the mechanical manifolds. Along some of these trajectories, which can be analytically computed in the case of the FPU β model, one can also compute the time averages of the Ricci curvature and of its fluctuations. These averages are used to obtain a good estimate of the largest Lyapunov exponent. In the following, we give the details of this computation. The model is described by the potential function in (5.49), and thus by the Hamiltonian (2.115), with the nonlinear coupling constant β replaced by u. The linear terms of this Hamiltonian can be diagonalized by introducing suitable harmonic normal coordinates. The latter are obtained by means of a canonical linear transformation [170]. Denoting the normal coordinates and momenta by Q_k and P_k for $k = 0, \ldots, N-1$, the transformation is given by

$$Q_k(t) = \sum_{n=1}^{N} S_{kn} q_k(t) , \quad P_k(t) = \sum_{n=1}^{N} S_{kn} p_k(t) , \tag{5.66}$$

where $k = 0, \ldots, N-1$, and S_{kn} is the orthogonal matrix [78] whose elements are

$$S_{kn} = \frac{1}{\sqrt{N}} \left[\sin \left(\frac{2\pi kn}{N} \right) + \cos \left(\frac{2\pi kn}{N} \right) \right] , \tag{5.67}$$

$n = 1, \ldots, N$ and $k = 0, \ldots, N-1$. The full Hamiltonian (2.115) in the new coordinates reads

$$H(Q, P) = \frac{1}{2} P_0^2 + \frac{1}{2} \sum_{i=1}^{N-1} \left(P_i^2 + \omega_i^2 Q_i^2 \right) + H_1(Q) , \tag{5.68}$$

where the anharmonic term is

$$H_1(Q) = \frac{u}{8N} \sum_{i,j,k,l=1}^{N-1} \omega_i \omega_j \omega_k \omega_l C_{ijkl} Q_i Q_j Q_k Q_l . \tag{5.69}$$

The $\omega_k = 2 \sin(\pi k / N)$, for $k \in \{1, \ldots, N-1\}$, are the normal frequencies for the harmonic case ($\mu = 0$), being $\omega_k = \omega_{N-k}$. By defining

$$\Delta_r = \begin{cases} (-1)^m & \text{for } r = mN \text{ with } m \in \mathbb{Z} , \\ 0 & \text{otherwise} , \end{cases} \tag{5.70}$$

the integer-valued coupling coefficients C_{ijkl} are explicitly given by

$$C_{ijkl} = -\Delta_{i+j+k+l} + \Delta_{i+j-k-l} + \Delta_{i-j+k-l} + \Delta_{i-j-k+l} . \tag{5.71}$$

By eliminating the motion of the center of mass (which corresponds to the zero index), we now easily get the equations of motion for the remaining $N-1$ degrees of freedom, which, at the second order, read as

$$\ddot{Q}_r = -\omega_r^2 Q_r - \frac{u\omega_r}{2N} \sum_{j,k,l=1}^{N-1} \omega_j \omega_k \omega_l C_{rjkl} Q_j Q_k Q_l , \tag{5.72}$$

for $r = 1, \ldots, N-1$.

As shown in [170], the equations of motion (5.72) admit some exact periodic solutions that can be explicitly expressed in closed analytical form. The simplest ones, consisting of one mode, have only one excited mode, which we denote by the index e, and thus are characterized by $Q_j(t) \equiv 0$ for $j \neq e$. The solitary modes are found by setting $C_{reee} = 0 \; \forall r \in \{1, \ldots, N-1\}$ with $r \neq e$; it is easily verified that this condition is satisfied for

$$e = \frac{N}{4}; \frac{N}{3}; \frac{N}{2}; \frac{2N}{3}; \frac{3N}{4} . \tag{5.73}$$

Thus, for solutions with initial conditions $Q_j = 0$ and $\dot{Q}_j = 0$ for $j \neq e$, the whole system (5.72) reduces to a one-degree-of-freedom (and thus integrable) system described by the equation of motion

$$\ddot{Q}_e = -\omega_e^2 Q_e - \frac{u\omega_e^4 C_{eeee}}{2N} Q_e^3 , \tag{5.74}$$

where $C_{eeee} = 4, 4, 3, 3, 2$ for $e = N/4, 3N/4, N/3, 2N/3, N/2$, respectively. The harmonic frequencies of the modes (5.73) are $\omega_e = \sqrt{2}, \sqrt{2}, \sqrt{3}, \sqrt{3}, 2$ for $e = N/4, 3N/4, N/3, 2N/3, N/2$, respectively. In order to simplify the notation, in the following, let us set $\hat{C}_e = C_{eeee}$.

The general solution of (5.74) is a Jacobi elliptic cosine

$$Q_e(t) = A \, \text{cn} \left[\Omega_e(t - t_0), k \right] , \tag{5.75}$$

where the free parameters (modal) amplitude A and time origin t_0 are fixed by the initial conditions. The frequency Ω_e and the modulus k of Jacobi elliptic cosine function [160, 171] depend on A as follows:

$$\Omega_e = \omega_e \sqrt{1 + \delta_e A^2} , \quad k = \sqrt{\frac{\delta_e A^2}{2(1 + \delta_e A^2)}} , \tag{5.76}$$

with $\delta_e = u\omega_e^2 \hat{C}_e/(2N)$. This kind of solution is periodic, and its oscillation period T_e depends on the amplitude A, since it is given in terms of the complete elliptic integral of the first kind $\mathbf{K}(k)$ [160, 171] and in terms of Ω_e by

$$T_e = \frac{4\mathbf{K}(k)}{\Omega_e} . \tag{5.77}$$

The modal amplitude A is one-to-one related to the energy density $\epsilon = E/N$. In fact, computing the total energy (5.68) on the one mode solution $Q_j(t) \equiv \delta_{je} Q_e(t)$, one obtains

$$\epsilon N = \frac{1}{2} \left(P_e^2 + \omega_e^2 Q_e^2 \right) + \frac{u}{8N} \omega_e^4 \hat{C}_e Q_e^4 . \tag{5.78}$$

Since at $t = t_0$ the coordinates are $(Q_e(t_0), P_e(t_0)) = (A, 0)$, by solving the previous equation for A we get

$$A = \left[2N \left(\frac{\sqrt{1 + 2u\epsilon\hat{C}_e} - 1}{u\omega_e^2 \hat{C}_e} \right) \right]^{1/2} . \tag{5.79}$$

This relation allows us to express all the parameters of the solution (5.75) in terms of the more physically relevant parameter ϵ. The period T_e is

$$T_e = \frac{4\mathbf{K}(k)}{\omega_e (1 + 2u\epsilon\hat{C}_e)^{1/4}} , \tag{5.80}$$

where $k = k(\epsilon)$ can be found from (5.76) and (5.79).

In terms of the standard coordinates, the one mode solutions are

$$q_n(t) = \frac{1}{\sqrt{N}} Q_e(t) \left[\sin\left(\frac{2\pi ne}{N}\right) + \cos\left(\frac{2\pi ne}{N}\right) \right] , \tag{5.81}$$

where e is one of the values listed in (5.73).

The Ricci curvature along a periodic trajectory, obtained by substituting (5.81) into (5.59), is

$$k_R(t) = 2 + \frac{6u}{N} \omega_e^2 Q_e^2(t) , \tag{5.82}$$

and its time average $\overline{k}_R$ is

$$\overline{k}_R = 2 + \frac{6u}{N} \omega_e^2 \overline{Q^2}_e . \tag{5.83}$$

After simple algebra, using standard properties of the elliptic functions, one obtains

$$\overline{Q^2}_e = \frac{1}{T_e} \int_{t_0}^{T_e + t_0} dt \, Q_e^2 = \frac{A^2}{\mathbf{K} k^2} \left(\mathbf{E} + (k^2 - 1)\mathbf{K} \right) . \tag{5.84}$$

The time-averaged Ricci curvature is then

$$\overline{k}_R = 2 + \frac{12}{\mathbf{K} k^2 \hat{C}_e} \left[\sqrt{1 + 2u\epsilon\hat{C}_e} - 1 \right] \left[\mathbf{E} + (k^2 - 1)\mathbf{K} \right] , \tag{5.85}$$

where $\mathbf{K}$ and $\mathbf{E}$ are the complete elliptic integrals of the first and second kinds respectively, [160, 171] both depending on the modulus k, which, from (5.76) and (5.79), is determined by the energy density ϵ:

$$k^2 = \frac{1}{2} \left(1 - \frac{1}{\sqrt{1 + 2u\epsilon\hat{C}_e}} \right) . \tag{5.86}$$

Now, using (5.85) and (5.86), and the tabulated values for $\mathbf{E}$ and $\mathbf{K}$, $\overline{k}_R$ is given as a function of the energy density ϵ. In Figure 5.5 a comparison is made between $\overline{k}_R$ versus ϵ, worked out for the one mode solutions under consideration, and $\langle k_R \rangle_{\mu_E}$ versus ϵ, the microcanonical average Ricci curvature analytically computed in the preceding section. By replacing the microcanonical averages with time averages in the expression

$$\langle \delta^2 K_R \rangle_\mu = \langle (K_R - \langle K_R \rangle_\mu)^2 \rangle_\mu = (N-1)^2 \left[\langle (k_R)^2 \rangle_\mu - (\langle k_R \rangle_\mu)^2 \right] , \tag{5.87}$$

using (5.83), and after some trivial algebra, we get

$$\overline{\delta^2 k_R} = \frac{36u^2 \omega_e^4}{N^2} \left[\overline{Q^4}_e - \overline{Q^2}_e \, \overline{Q^2}_e \right] . \tag{5.88}$$

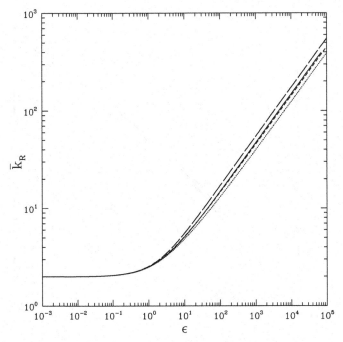

Fig. 5.5. $\overline{k}_R$ versus ϵ, worked out by means of the three single-mode solutions identified by the values of e listed in (5.73) (dotted, dashed, and long-dashed lines refer to $e = N/4, 3N/4$, $e = N/3, 2N/3$, and $e = N/2$, respectively), is compared with $\langle k_R \rangle_{\mu_E}$ computed in [86] (continuous line). The agreement is very good on a broad range of values of energy density ϵ. From [172].

The new term

$$\overline{Q_e^4} = \frac{A^4}{T_e} \int_0^{T_e} dt \; \mathrm{cn}^4(\Omega_e t, k) = \frac{A^4}{4\mathbf{K}} \int_0^{4\mathbf{K}} d\theta \; \mathrm{cn}^4(\theta, k)$$

can be computed by resorting to standard properties of the elliptic functions, and the result is

$$\overline{Q_e^4} = \frac{A^4}{3\mathbf{K}k^4} \left[\mathbf{K}(2 - 5k^2 + 3k^4) + 2\mathbf{E}(2k^2 - 1) \right] . \tag{5.89}$$

Finally, (5.89) and (5.84) in (5.88) yield

$$\overline{\delta^2 k_R} = \frac{192 \left[(k^2 - 1) + 2(2 - k^2)\frac{\mathbf{E}}{\mathbf{K}} - 3\left(\frac{\mathbf{E}}{\mathbf{K}}\right)^2 \right]}{(1 - 2k^2)^2 \hat{C}_e^2} . \tag{5.90}$$

From (5.86) and making use of the tabulated values for $\mathbf{E}$ and $\mathbf{K}$, equation (5.90) provides the mean squaredfluctuations of the Ricci curvature as a

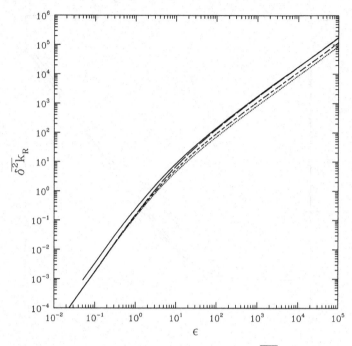

Fig. 5.6. In this figure we report three curves for $\overline{\delta^2 k_R}$ versus ϵ computed by integrating the curvature fluctuations along the three single mode solutions considered in the present section (dotted, dashed, and long-dashed lines refer to $e = N/4, 3N/4$, $e = N/3, 2N/3$ and $e = N/2$, respectively), and a comparison is made with the same quantity computed in [86] (continuous line). Also in this case the agreement is very good. From [172].

function of ϵ. In Figure 5.6, the time averages of the Ricci curvature fluctuations $\overline{\delta^2 k_R}$ versus the energy density ϵ, are compared to the microcanonical averages $\langle \delta^2 k_R \rangle_{\mu_E}$ versus ϵ analytically computed in the preceding section. The agreement is very good, thus confirming from a completely new point of view that unstable periodic orbits are special tools for dynamical systems analysis; in this case, the Ricci curvature of the enlarged configuration space $(M \times \mathbb{R}, g_e)$, and its fluctuations, are surprisingly well sampled by UPOs because time averages computed along them are very close to microcanonical averages performed on the whole configuration space. By replacing $\langle K_R \rangle_\mu$ and $\langle \delta^2 K_R \rangle_\mu$ with $\overline{k_R}$ and $\overline{\delta^2 k_R}$, and by inserting (5.85), (5.86) and (5.90) into the analytic formula (5.45), the largest Lyapunov exponent can be computed as a function of the energy density ϵ.

Figure 5.7 shows that the overall agreement between these analytic results, those reported in the preceding section, and the results obtained by numerical integration of the tangent dynamics is very good, especially at high energy density. At low energy density the discrepancy does not exceed, at worst,

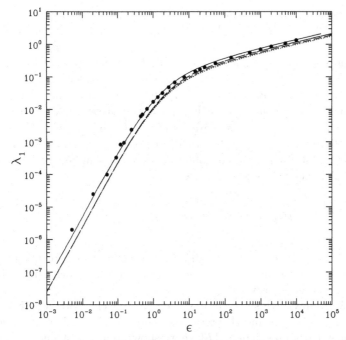

Fig. 5.7. This figure shows the largest Lyapunov exponent λ_1 obtained by integrating the Ricci curvature and its fluctuations along the three single mode solutions considered in the present section, plotted vs. ϵ. Dotted, dashed and long-dashed lines refer to $e = N/4, 3N/4$, $e = N/3, 2N/3$ and $e = N/2$, respectively. Continuous line refers to the Lyapunov exponent computed in [86]. The full circles are the values for λ_1 computed by numerical integration. The agreement is again very good on a broad range of ϵ values. From [172].

a factor of 2 on a range of many decades of energy density and with the use of only *one* unstable periodic orbit.

The theoretically interesting fact is that some geometric quantities, proper to the configuration-space manifold, are efficiently sampled even by means of a few unstable periodic orbits, as is shown in Figures 5.5 and 5.6.

Thus unstable periodic orbits can be used to compute Lyapunov exponents, and even though explicit computations have been hitherto performed only for one specific model, this method could be of more general validity to remove the ergodic assumption made to work out the effective instability equation (5.27). While chaotic trajectories can never be known analytically, in principle UPOs sometimes can, and efficiently do the same job.

5.4.3 1D *XY* Model

If the canonical coordinates q_i and p_i are given the meaning of angular coordinates and momenta, the 1D *XY* model, whose potential energy is given in

(5.50), describes a linear chain of N rotators constrained to rotate in a plane and coupled by a nearest-neighbor interaction. This model can be formally obtained by restricting the classical Heisenberg model with $O(2)$ symmetry to one spatial dimension. The potential energy of the $O(2)$ Heisenberg model is $V = -J \sum_{\langle i,j \rangle} \mathbf{s}_i \cdot \mathbf{s}_j$, where the sum is extended only over nearest-neighbor pairs, J is the coupling constant, and each $\mathbf{s}_i$ has unit modulus and rotates in the plane. To each "spin" $\mathbf{s}_i = (\cos q_i, \sin q_i)$, the velocity $\dot{\mathbf{s}}_i = (-\dot{q}_i \sin q_i, \dot{q}_i \cos q_i)$ is associated, so that $H = \sum_{i=1}^{N} \frac{1}{2}\dot{\mathbf{s}}_i^2 - J \sum_{\langle i,j \rangle} \mathbf{s}_i \cdot \mathbf{s}_j$.

This Hamiltonian system has two integrable limits. In the small-energy limit it represents a chain of harmonic oscillators, as can be seen by expanding the potential energy in a power series,

$$H(p,q) \approx \sum_{i=1}^{N} \left\{ \frac{p_i^2}{2} + J(q_{i+1} - q_i)^2 - 1 \right\} , \tag{5.91}$$

where $p_i = \dot{q}_i$, whereas in the high-energy limit a system of freely rotating objects is found, because the kinetic energy becomes much larger than the bounded potential energy.

The dynamics of this system has been extensively studied [145, 168, 173]. Numerical simulations and theoretical arguments independent of the geometric approach (see in particular [173]) have shown that also in this system there exist weakly and strongly chaotic dynamical regimes. It has been found that the dynamics is weakly chaotic in the low- and high-energy density regions, close to the two integrable limits. In contrast, fully developed chaos is found in the intermediate-energy region.

According to (5.58), the expression for the Ricci curvature K_R, computed with the Eisenhart metric, is

$$K_R(q) = 2J \sum_{i=1}^{N} \cos(q_{i+1} - q_i). \tag{5.92}$$

We note that for this model a relation exists between the potential energy V and Ricci curvature K_R:

$$V(q) = JN - \frac{K_R(q)}{2} . \tag{5.93}$$

The average Ricci curvature can be again expressed by implicit formulas (see [86] for details)

$$\left. \begin{aligned} \langle k_R \rangle_\mu (\beta) &= 2J \frac{I_0(\beta J)}{I_1(\beta J)} \\[2ex] \varepsilon(\beta) &= \frac{1}{2\beta} + J \left(1 - \frac{I_1(\beta J)}{I_0(\beta J)} \right) \end{aligned} \right\} \rightarrow k_0(\varepsilon) , \tag{5.94}$$

where the I_ν's are modified Bessel functions of index ν [160]. The fluctuations are given by the implicit equations

$$
\left.
\begin{aligned}
&\langle \delta^2 k_R \rangle (\beta) \\
&= \frac{4J}{\beta} \frac{\beta J I_0^2(\beta J) - I_0(\beta J) I_1(\beta J) - \beta J I_1^2(\beta J)}{I_0^2(\beta J)\left[1 + 2(\beta J)^2\right] - 2\beta J I_1(\beta J) I_0(\beta J) - 2\left[\beta J I_1(\beta J)\right]^2} \\
&\varepsilon(\beta) = \frac{1}{2\beta} + J\left[1 - \frac{I_1(\beta J)}{I_0(\beta J)}\right]
\end{aligned}
\right\} \to \sigma_k^2(\varepsilon).
$$

$$(5.95)$$

In Figures 5.8 and 5.9 a comparison between analytical and numerical results is provided for the average Ricci curvature and its fluctuations. The agreement between ensemble and time averages is again very good. Looking at Figures 5.8 and 5.9, we realize that the low-energy weakly chaotic region has the same geometric properties as the corresponding region of the FPU model, as expected, since the two low-energy integrable limits are the same. However, in the high-energy weakly chaotic region the fluctuations are far from being small with respect to the average curvature. The average curvature $k_0(\varepsilon)$ vanishes as $\varepsilon \to \infty$. In this case the weakly chaotic dynamics seems related to the fact that the manifold $(M \times \mathbb{R}^2, g_e)$ looks almost flat along the physical geodesics. The bounds of the two weakly chaotic regions, as estimated in [173], coincide with the values of ε where the asymptotic behavior of k (low-energy region) and σ_k (high-energy region) set in. Moreover, the case of the coupled rotators

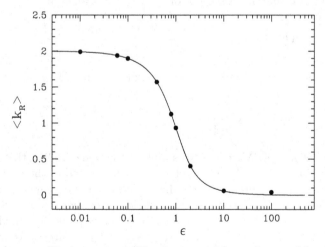

Fig. 5.8. Average Ricci curvature (Eisenhart metric) per degree of freedom k_0 vs. specific energy ε for the coupled rotators model. The continuous line is the result of an analytic computation according to (5.94); the full circles are time averages obtained by numerical simulations with $N = 150$; $J = 1$. From [86].

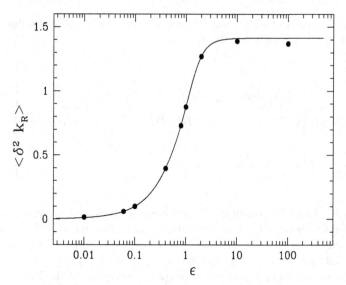

Fig. 5.9. Fluctuation of the Ricci curvature (Eisenhart metric), σ_k^2, vs. specific energy ε for the coupled rotators model. Symbols and parameters as in Figure 5.8; the continuous line now refers to (5.95). From [86].

is very different from the FPU case, since the sectional curvature $K(s)$ along a geodesic can take negative values. The probability $P(\varepsilon)$ that $K(s) < 0$ can be analytically estimated in the following simple way. The explicit expression for the sectional curvature $K(\dot{\gamma}, \xi)$, relative to the plane spanned by the velocity vector $\dot{\gamma} = dq/dt$ and a generic vector $\xi \perp \dot{\gamma}$, is (see Appendix B)

$$K(\dot{\gamma}, \xi) = R_{0i0k} \frac{dq^0}{dt} \frac{\xi^i}{\|\xi\|} \frac{dq^0}{dt} \frac{\xi^k}{\|\xi\|} \equiv \frac{\partial^2 V}{\partial q^i \partial q^k} \frac{\xi^i \xi^k}{\|\xi\|^2} , \qquad (5.96)$$

so that computing $\partial^2 V / \partial q^i \partial q^k$ using the explicit form of $V(q)$ given in (5.50), we get

$$K(\dot{\gamma}, \xi) = \frac{J}{\|\xi\|^2} \sum_{i=1}^{N} \cos(q_{i+1} - q_i) \left[\xi^{i+1} - \xi^i\right]^2 \qquad (5.97)$$

for the 1D XY model. We realize, by simple inspection of (5.97), that the probability of finding $K < 0$ along a geodesic must be related to the probability of finding an angular difference larger than $\frac{\pi}{2}$ between two nearest-neighboring rotators. From (5.97) we see that N orthogonal directions of the vector ξ exist such that the sectional curvatures—relative to the N planes spanned by these vectors together with $\dot{\gamma}$—are just $\cos(q_{i+1} - q_i)$. These directions are defined by the unit vectors of components $(1, 0, \ldots, 0), (0, 1, 0, \ldots, 0), \ldots, (0, \ldots, 0, 1)$. Hence the probability $P(\varepsilon)$ of occurrence of a negative value of a cosine is used to estimate the probability of occurrence of negative sectional curvatures along the geodesics. This probability function, calculated using the Boltzmann weight, has the following simple expression [86, 145]:

$$\left.\begin{array}{l} P(\beta) = \dfrac{\int_{-\pi}^{\pi} \Theta(-\cos x)e^{\beta J \cos x}dx}{\int_{-\pi}^{\pi} e^{\beta J \cos x}dx} = \dfrac{\int_{\frac{\pi}{2}}^{\frac{3\pi}{2}} e^{\beta J \cos x}dx}{2\pi I_0(\beta J)} \\[4mm] \varepsilon(\beta) = \dfrac{1}{2\beta} + J\left[1 - \dfrac{I_1(\beta J)}{I_0(\beta J)}\right] \end{array}\right\} \rightarrow P(\varepsilon)\,. \qquad (5.98)$$

where $\Theta(x)$ is the Heaviside unit step function and I_0 the modified Bessel function of index 0. The function $P(\varepsilon)$ is plotted in 5.10. We see that in the strongly chaotic region such a probability starts to increase rapidly from a very small value, while it approaches an asymptotic value $P(\varepsilon) \sim 1/2$ when the system enters its high-energy weakly chaotic region.

When the sectional curvatures are positive,[3] chaos is produced by curvature fluctuations; hence we expect chaos to be weak as long as $\sigma_k/k_0 \ll 1$, and to become strong when $\sigma_k \approx k_0$. In contrast, when $K(s)$ can assume both positive and negative values, the situation is much more complicated, for there are now two different and independent sources of chaos: negative curvature, which directly induces a divergence of nearby geodesics, and the bumpiness of the ambient manifold, which induces such a divergence via parametric instability. The results for the coupled-rotators model suggest that as long as the negative curvatures are "few," they do not dramatically change the

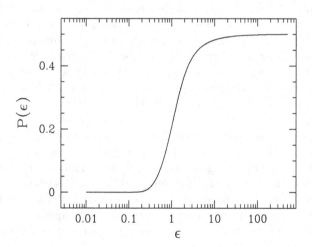

Fig. 5.10. Estimate of the probability $P(\varepsilon)$ of occurrence of negative sectional curvatures in the 1D XY model according to (5.98); $J = 1$. From [86].

---

[3] The sectional curvature is always strictly positive in the FPU β model, see Section 5.5.1; in the 1D XY model, in the low-energy region, negative sectional curvatures can occur, but have a very small probability.

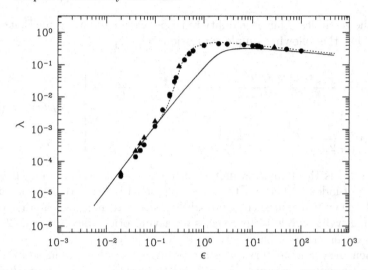

Fig. 5.11. Lyapunov exponent λ vs. energy density ε for the 1D XY model with $J = 1$. The continuous line is the theoretical computation according to (5.46), while full circles, squares and triangles are the results of numerical simulations with N, respectively, equal to 150, 1000, and 1500. The dotted line is the theoretical result where the value of k entering (5.46) has been corrected according to (5.100) with $\alpha = 150$. From [86].

picture, and may strengthen the parametrically generated chaos, while when their occurrence is equally likely as the occurrence of positive curvatures, the two mechanisms of chaos seem to inhibit each other and chaos becomes weak.[4]

Such a qualitative picture is consistent with the result of the geometric computation of λ for the coupled-rotator model. The result of the application of (5.46) to this model is plotted in Figure 5.11 (solid line). There is agreement between analytic and numeric values of the Lyapunov exponent only at low- and high-energy densities. As in the FPU case, at low energy, in the quasi-harmonic limit, we obtain $\lambda(\varepsilon) \propto \varepsilon^2$. At high energy, $\lambda(\varepsilon) \propto \varepsilon^{-1/6}$; here $\lambda(\varepsilon)$ is a decreasing function of ε because for $\varepsilon \to \infty$ the systems is integrable.

However, in the intermediate-energy range our theoretical prediction underestimates the actual degree of chaos of the dynamics. This energy range coincides with the region of fully developed (strong) chaos. According to the above discussion the origin of the underestimation can be found in the fact that the role of the negative curvatures, which appears to strengthen chaos in this energy range, is not correctly taken into account. The sectional

---

[4] The fact that the two mechanisms, when comparable, can inhibit rather than strengthen each other can be considered a "proof" of the fact that their nature is intrinsically different. A similar situation is found also in some billiard systems, where there are two mechanisms that can produce chaos: (i) defocusing, due to positively curved boundaries, and (ii) divergence of the trajectories due to scatterings with negatively curved boundaries [174].

curvature $K(s)$, whose expression is given by (5.97), can take negative values with nonvanishing probability regardless of the value of ε, whereas as long as $\varepsilon < J$, this possibility is lost in the replacement of K by the Ricci curvature, due to the constraint (5.93), which implies that at each point of the manifold,

$$k_{\mathrm{R}}(\varepsilon) \geq 2(J - \varepsilon). \tag{5.99}$$

Thus our approximation fails to account for the presence of negative sectional curvatures at values of ε smaller than J. In (5.97) the cosines have different and variable weights, $(\xi^{i+1} - \xi^i)^2$, which make it in principle possible to find somewhere along a geodesic a $K < 0$ even with only one negative cosine. This is not the case for k_{R} where all the cosines have the same weight.

Let us now show how the theoretical results can be improved. Our strategy is to modify the model for $K(s)$ in some *effective* way that takes into account the just-mentioned difficulty of $k_{\mathrm{R}}(s)$ to adequately model $K(s)$. This will be achieved by suitably "renormalizing" k_0 or σ_k to obtain an "improved" Gaussian process that can better model the behavior of the sectional curvature. Since our "bare" Gaussian model underestimates negative sectional curvatures in the strongly chaotic region, the simplest way to renormalize the Gaussian process is to shift the peak of the distribution $\mathcal{P}(K_{\mathrm{R}})$ toward the negative axis to make the average smaller. This can easily be done by the following rescaling of the average curvature k_0:

$$k_0 = \langle k_{\mathrm{R}}(\varepsilon) \rangle \rightarrow \frac{\langle k_{\mathrm{R}}(\varepsilon) \rangle}{1 + \alpha P(\varepsilon)}. \tag{5.100}$$

This correction has no influence either when $P(\varepsilon) \approx 0$ (below $\varepsilon \approx 0.2$) or when $P(\varepsilon) \approx 1/2$ (because in that case $\langle k_{\mathrm{R}}(\varepsilon) \rangle \rightarrow 0$). The simple correction (5.100) makes use of the information we have obtained analytically, i.e., of the $P(\varepsilon)$ given in (5.98), and is sufficient to obtain an excellent agreement of the theoretical prediction with the numerical data over the whole range of energies, as shown in 5.11. The parameter α in (5.100) is a free parameter, and its value is determined so as to obtain the best agreement between numerical and theoretical data. The result shown in 5.11 (dotted line) is obtained with $\alpha = 150$, but also very different values of α, up to $\alpha = 1000$, yield a good result, i.e., no particularly fine tuning of α is necessary to obtain a very good agreement between theory and numerical experiment.

5.4.4 Mean-Field XY Model

Now we consider the mean-field XY model, defined by the Hamiltonian [110]

$$H(p, \varphi) = \sum_{i=1}^{N} \frac{p_i^2}{2} + \frac{J}{2N} \sum_{i,j=1}^{N} [1 - \cos(\varphi_i - \varphi_j)] . \tag{5.101}$$

Here $\varphi_i \in [0, 2\pi]$ is the rotation angle of the ith rotator. For this model the canonical partition function can be explicitly computed (for the details see Chapter 10) and its configurational part reads

$$Z_c = \frac{1}{\pi} \frac{N}{2\beta J} \int_{-\infty}^{\infty} \int_{-\infty}^{\infty} d\mathbf{z} \, \exp\left[-N\psi(z, J)\right] ,$$

where $\psi(z, J) = z^2/2\beta J - \ln(2\pi I_0(z)) + \beta J/2$, with I_0 the modified Bessel function and z the modulus of a vector $\mathbf{z} \in \mathbb{R}^2$. In the limit $N \to \infty$, the saddle-point method prescribes that one solve the equation $\partial\psi/\partial z = 0$ in order to estimate the above integral, that is,

$$\frac{z}{\beta J} - \frac{I_1(z)}{I_0(z)} = 0 . \tag{5.102}$$

For $\beta J < 2$, $z = 0$ is the solution corresponding to the minimal free energy and to a vanishing magnetization. For $\beta J > 2$, the solution $\bar{z}$ is a value of z that is a function of β. Correspondingly, the magnetization

$$|\mathbf{M}| = \frac{I_1(\bar{z})}{I_0(\bar{z})} \tag{5.103}$$

bifurcates at the critical value $\beta J = 2$, and since from a simple calculation one obtains

$$V(q) = \frac{JN}{2}(1 - |\mathbf{M}|^2) ,$$

the critical energy density is $\varepsilon_c = 3J/4$.

Coming to the computation of the largest Lyapunov exponent, consider first that the canonical configurational partition function, worked out by means of the saddle-point method, reads

$$Z_c(\beta) \approx (2\pi)^N \frac{N\bar{z}}{\beta J} \exp[-N\psi(\bar{z}, \beta)] \sqrt{2\pi} \, [N\partial_\beta\psi(\bar{z}, \beta)]^{-1/2} .$$

The Ricci curvature, using the Eisenhart metric, is

$$k_R(q) = \frac{1}{N-1} \sum_{i=1}^{N} \frac{\partial^2 V(q)}{\partial q_i^2} = J - \frac{2}{N-1}V(q) \equiv J|\mathbf{M}|^2 . \tag{5.104}$$

In the large-N limit, $\langle k_R \rangle_{\mathrm{can}}$ and $\langle k_R \rangle_\mu$ coincide, so that from

$$\langle k_R \rangle_{\mathrm{can}} = J + \frac{2}{N}\partial_\beta \log Z_c \approx J - 2\partial_\beta[\psi(\bar{z}(\beta), \beta)]$$

$$= J - 2\frac{d\bar{z}}{d\beta}\partial_z\psi\Big|_{\bar{z}} - 2\beta\partial_\beta\psi|_{\bar{z}}$$

$$= J - 2\left(\frac{J}{2} - \frac{\bar{z}^2}{2\beta^2 J}\right) \tag{5.105}$$

we get

$$\left.\begin{array}{l} \langle k_{\mathrm{R}}\rangle_{\mathrm{can}}(\beta) = \dfrac{[\overline{z}(\beta)]^2}{2\beta^2 J} \\[4mm] \varepsilon(\beta) = \dfrac{1}{2\beta} - \dfrac{1}{N}\dfrac{\partial}{\partial\beta}\log Z_c(\beta) \end{array}\right\} \ \to k_0(\varepsilon) . \qquad (5.106)$$

We have seen above that up to an $\mathcal{O}(1/N)$ correction, it is $k_{\mathrm{R}} = J|\mathbf{M}|^2$. Thus $\langle k_{\mathrm{R}}\rangle_\mu(\varepsilon)$ must display the the same bifurcation, away from zero, at ε_c.

Then the computation of $\langle \delta^2 k_{\mathrm{R}}\rangle_\mu$ proceeds according to (5.57), and noting that $\partial_\beta \langle k_{\mathrm{R}}\rangle_{\mathrm{can}} = \frac{1}{2}\langle \delta^2 K_R\rangle_{\mathrm{can}}$, one finds that

$$\langle \delta^2 k_{\mathrm{R}}\rangle_\mu(\beta) = \langle \delta^2 K_{\mathrm{R}}\rangle_{\mathrm{can}}\left(1 + \frac{\beta^2}{2}\langle \delta^2 K_{\mathrm{R}}\rangle_{\mathrm{can}}\right)^{-1} , \qquad (5.107)$$

where we have

$$\langle \delta^2 K_{\mathrm{R}}\rangle_{\mathrm{can}} = \frac{4}{N}\partial_\beta^2 \log Z_c \approx \frac{4\overline{z}}{\beta^2}J\left(\partial_\beta \overline{z} - \frac{\overline{z}}{\beta}\right) ,$$

whence, together with $\varepsilon(\beta)$, we obtain $\sigma_k^2(\varepsilon)$ in parametric form.

The ε-dependence of these geometric averages are shown in Figures 5.12 and 6.11. Then, introducing these geometric quantities into the analytic formulas (5.45) and (5.46), we get finally $\lambda(\varepsilon)$, which is reported in Figure 5.13.

Note that this computation holds for $\varepsilon < \varepsilon_c$, the condition of existence of a nonvanishing solution $\overline{z}(\beta)$ of the consistency equation (5.102).

Above the critical energy ε_c, one gets

$$Z_c(\beta) \approx (2\pi)^N e^{-N\beta J/2}\left(1 - \frac{\beta J}{2}\right)^{-1} .$$

Since for $\varepsilon > \varepsilon_c$, $|\mathbf{M}|^2 \sim \mathcal{O}(N^{-1})$, using $k_{\mathrm{R}} = J|\mathbf{M}|^2 - J/N + \mathcal{O}(N^{-2})$ we have

$$\langle \delta^2 k_{\mathrm{R}}\rangle_\mu = \frac{\beta J^2}{N(2 - \beta J)} + \mathcal{O}(N^{-2}) ,$$

that is, in the $N \to \infty$ limit, $\langle \delta^2 k_{\mathrm{R}}\rangle_\mu$ vanishes. At the same time,

$$\langle \delta^2 K_{\mathrm{R}}\rangle_{\mathrm{can}} = \frac{4}{N}\frac{\partial^2}{\partial\beta^2}\log Z_c = \frac{4J^2}{N}(2 - \beta J)^{-2} = \mathcal{O}(N^{-1}) ,$$

that is, $\langle \delta^2 K_{\mathrm{R}}\rangle_{\mathrm{can}}$ also vanishes in the $N \to \infty$ limit. The correction term, since $\varepsilon(\beta) \approx 1/(2\beta) + J/2$, is $\mathcal{O}(N^{-2})$, whence $\langle \delta^2 k_{\mathrm{R}}\rangle_\mu = \langle \delta^2 K_{\mathrm{R}}\rangle_{\mathrm{can}} + \mathcal{O}(N^{-2}) = \mathcal{O}(N^{-1})$, thus also $\langle \delta^2 k_{\mathrm{R}}\rangle_\mu$ vanishes as $\mathcal{O}(N^{-1})$ in the $N \to \infty$ limit.

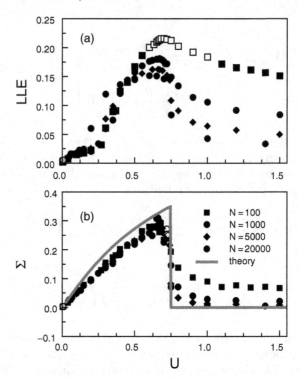

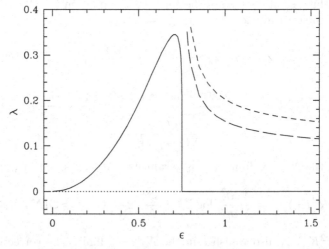

Fig. 5.12. Largest Lyapunov exponent (a), and half the variance of Ricci curvature fluctuations (b), computed numerically for the mean-field XY model. From [224].

Fig. 5.13. Largest Lyapunov exponent vs. ε, analytically computed for the mean-field XY model (solid curve). The curves above the transition are finite-N results for $N = 80$ and $N = 200$: here $\lambda \propto N^{-1/3}$.

Using again formulas (5.45) and (5.46), where we replace the leading-order terms of $\langle k_{\mathrm{R}} \rangle_\mu$ and $\langle \delta^2 k_{\mathrm{R}} \rangle_\mu$, we finally obtain

$$\lambda \approx \frac{4^{1/3} J \sqrt{\beta J}}{(2 - \beta J)^{3/2}} \, N^{-1/3} \; ,$$

and this N-scaling of λ at $\varepsilon > \varepsilon_{\mathrm{c}}$, checks very well with the numerical findings reported in the literature [176]. The overall qualitative agreement below and above ε_{c} is reasonably good, but the quantitative agreement at $\varepsilon < \varepsilon_{\mathrm{c}}$ is less good, because there is a factor 2 missing in the peak level just below ε. As we shall discuss in Section 5.5.1, we can reasonably doubt that this model satisfies the quasi-isotropy assumption. In fact, its configuration space has a highly nontrivial topology, as we shall see in Chapter 10, and this should be taken somehow into account, also with some phenomenological assumption, as was the case of the 1D XY model.

5.5 Some Remarks

Let us now comment about some points of the material presented so far. In particular, we would like to clarify the meaning of some of the approximations made and to draw the attention of the reader to some of the problems that are still open. A better understanding of these points could lead, in our opinion, to a considerable improvement of the theory, which is still developing and can by no means be considered a "closed" subject.

What has been presented in this chapter has a conceptual implication that goes far beyond the development of a method to analytically compute Lyapunov exponents. Rather, the strikingly good agreement between analytic and numeric Lyapunov exponents—obtained at the price of a restriction of the domain of applicability[5] of the analytic expression worked out for λ—has three main implications:

(i) the *local* geometry of mechanical manifolds contains all the relevant information about (in)stability of Hamiltonian dynamics;
(ii) once a good model for the local source of instability of the dynamics is provided, then a statistical-mechanical-like treatment of the average degree of instability of the dynamics can be worked out, in the sense that we do not need a detailed knowledge of the dynamics, but by computing *total* geometric quantities (that is, averages of local quantities computed on the whole manifold), we obtain a very good estimate of the average strength of chaos;

---

[5] Actually, the simplifying assumptions made to work out an effective (in)stability equation from the full geodesic deviation equation might be inadequate for many systems, as will be explained in the next Section 5.5.1.

(iii) due to the variational formulation of newtonian dynamics, the Riemannian-geometric framework a priori seems—and actually seemed in the past—the natural framework to investigate the instability of Hamiltonian dynamics, although no evidence at all was available to confirm such an idea until the above-mentioned development took place. It is now evident that the efforts to improve the theory by expanding its domain of applicability are worthwhile.

We must warn the reader, though, against "blind" applications of formula (5.45), i.e., without any idea about the satisfaction, by the Hamiltonian model under investigation, of the conditions under which it has been derived.

From a more technical point of view, one of the central results we have presented so far is the possibility of deriving, from the Jacobi equation, a *scalar* equation (equation (5.21)) describing the evolution of the Jacobi field J for a geodesic spread on a manifold. We would like to stress that such a result, though approximate, applies to a wide class of Hamiltonian systems. In fact, the only hypothesis needed to get such an equation is the quasi-isotropy hypothesis, i.e., the assumption that $R_{ijkl} \approx \mathcal{K}(s)(g_{ik}g_{jl} - g_{il}g_{jk})$. Loosely speaking, such an assumption means that, locally, the manifold can be regarded as isotropic, i.e., there is a neighborhood of each point where the curvature can be considered constant. This does not imply at all that there are only small-scale fluctuations. There can be fluctuations of curvature on many scales, provided that they are finite and there is a cutoff at a certain point. The only case in which such an assumption will surely *not* hold is that of curvature fluctuations over *all* scales. This might happen when the mechanical manifold undergoes topological changes.

Other approximations come into play when one actually wants to model $\mathcal{K}(s)$ along a geodesic with a stochastic process. It is true that replacing the sectional curvature by the Ricci curvature requires that the fluctuations be not only finite, but also small. Moreover, we use global averages to define the stochastic process, and here it is crucial that the fluctuations not extend over too large scales. Thus (5.28) has a less general validity than (5.21). A way to improve the theory might be to try to replace the sectional curvature with some quantity related also to the gradient of the Ricci curvature, in order to make the replacement of sectional curvature less sensitive to the large-scale variations of the Ricci curvature.

To get an explicit solution of (5.21), in addition to the quasi-isotropy assumption, we made further restrictions through the following steps:

(i) using the Eisenhart metric on the enlarged configuration space-time $(M \times \mathbb{R}^2, g_e)$;
(ii) considering standard systems where the kinetic energy does not depend on the q's;
(iii) estimating the characteristic correlation time τ of the curvature fluctuations.

As to item (iii), we have already remarked that our estimate given in (5.39) is by no means a consequence of any theoretical result, but only a reasonable estimate that could surely be improved.

As to item (ii), the case of a more general kinetic energy matrix $a_{ij} \neq \delta_{ij}$, though not conceptually different, is indeed different in practice, and the same final result is not expected to hold in that case.

Finally, item (i) should not reduce significantly the generality of the result. In fact, considering the Eisenhart metric only makes the calculations feasible, and in principle nothing should change if one were able to solve (5.28) in the case of the Jacobi metric (see the discussion in [130] and the numerical results in [161, 162]). However, Eisenhart and Jacobi metrics are *equivalent* for what concerns the computation of the average instability of the dynamics [130], but they might *not* be *equivalent* for other developments of the theory, this in view of the fact that (M_E, g_J) is a manifold that has better mathematical properties than $(M \times \mathbb{R}^2, g_e)$: (M_E, g_J) is a proper Riemannian manifold, it is compact, all of its geodesics are in one-to-one correspondence with mechanical trajectories, its scalar curvature does not identically vanish as is the case of $(M \times \mathbb{R}^2, g_e)$, and it can be naturally lifted to the tangent bundle where the associated geodesic flow on the submanifolds of constant energy coincides with the phase space trajectories. Other geometric frameworks that are worth considering are the enlarged configuration space with the Eisenhart metric $(M \times \mathbb{R}, G_e)$ (see Section 3.1.2), and the configuration space-time $(M \times \mathbb{R})$ with a Finsler metric (see Section 3.2). The former framework has never been considered hitherto, neither for numeric nor for analytic computations, and the latter has been successfully exploited only for numerical computations of Hamiltonian flows with two degrees-of-freedom [123, 124].

5.5.1 Beyond Quasi-Isotropy: Chaos and Nontrivial Topology

A significant improvement and generalization of the theory is expected by taking into account the role of nontrivial topology of configuration space. As is repeatedly stated throughout the present book, nontrivial topology of configuration space and of its submanifolds corresponds to the existence of critical points of the potential V: a generic situation indeed, which affects the (in)stability properties of the dynamics (see the beginning of Chapter 8).

In what follows we briefly draw the reader's attention to an interesting difference among the three models for which we have provided an analytic computation of the largest Lyapunov exponent. This difference consists in the absence of critical points in the case of the FPU β model and, in contrast, in the existence of a large number of critical points in the case of both the 1D and mean-field XY models. This difference is closely related to the different capabilities of the effective instability equation (5.27) to predict the good values of the Lyapunov exponent.

Let us begin by considering the FPU β model, described by the potential function of the Hamiltonian (2.115), and let us look for its critical points, i.e., those points where $\nabla V = 0$. For $i = 1, \ldots, N$ we have to solve

$$\nabla_i V = -(q_{i+1} - q_i) + (q_i - q_{i-1}) - u(q_{i+1} - q_i)^3 + (q_i - q_{i-1})^3 = 0 .$$

Put $\xi_i = (q_i - q_{i-1})$. It is

$$\nabla_i V = \xi_i(1 + u\xi_i^2) - \xi_{i+1}(1 + u\xi_{i+1}^2) = 0 ,$$

and the solutions are $\xi_i = 0$, $i = 1, \ldots, N$, and $\xi_i = \xi_{i+1}$, $i = 1, \ldots, N$. The first solution means that for any i, i' it is $q_i = q_{i'}$, that is, all the particles are displaced from their equilibria by the same amount. The second solution means that for $i = 1, \ldots, N$ it is $(q_{i+1} - q_i) = (q_i - q_{i-1})$, which has the same meaning of the previous solution, and both solutions entail that either for fixed boundary conditions or for periodic boundary conditions plus fixed center of mass, the only critical point of the potential corresponds to the equilibrium solution of lowest potential energy, that is, $q_i = 0$, $i = 1, \ldots, N$. At higher potential energies there are no other critical points. As a consequence, all the level sets $\Sigma_v = V^{-1}(v)$ and the regions bounded by them, $M_v = \{q \in \mathbb{R}^N | V(q) \leq v\}$, have the same (trivial) topology.

Thus the mechanical manifolds associated with the FPU β model, both (M_E, g_J) and $(M \times \mathbb{R}^2, g_E)$, which are defined on configuration space, are topologically trivial. At the same time, we have also seen in Section 5.4.1 that the quasi-isotropy assumption, which led to the effective (in)stability equation (5.27), works very well for the FPU β model and thus seems a posteriori well justified.

Another interesting example, for which the assumption of quasi-isotropy seems to work very well because of trivial topology of configuration space, is provided in [177]. The model, introduced to describe DNA denaturation, and defined by the Hamiltonian

$$H = \sum_n \left[\frac{m}{2}\dot{y}_n^2 + \frac{K}{2}(y_n - y_{n+1})^2 + D(e^{-ay_n} - 1)^2 \right] ,$$

corresponds to a topologically trivial configuration space below the unbinding transition energy [178], that is, where the dynamics is non-trivial, the largest Lyapunov exponent is non-vanishing, and its analytic prediction through (5.45) and (5.46) is in excellent agreement with the numerical findings.

In contrast, in the case of the one-dimensional chain of rotators, or 1D XY model, the quasi-isotropy assumption led to the underestimation of λ in an intermediate energy range, where chaos is strong. An excess of negative sectional curvatures with respect to what is predicted by the quasi-isotropy assumption is responsible for the observed enhancement of chaos. On the other hand, as we show in Section 10.4.2, there is a large number of critical points in the configuration space of this model (see Figure 10.8) and as we have discussed in the introduction and at the beginning of Chapter 8, the neighborhoods of critical points of the potential are *"scatterers"* of the trajectories, which enhance chaos by adding to parametric instability another instability mechanism. Necessarily these two facts, excess of negative sectional curvatures and existence of many neighborhoods of critical points, are the same phenomenon seen from different viewpoints. Also, the mean-field XY model, as is largely discussed in Chapter 10, has a huge number of critical points, but apparently they work in the opposite way with respect to the 1D XY model. In fact, with respect to the prediction based only on the quasi-isotropy assumption, the presence of critical points weakens chaos instead of strengthening it. This is not necessarily surprising; for example, the hyperbolic critical point of the Hamiltonian of a reversed pendulum can lose its unstable character by a fast parametric modulation (a fast oscillation of its pivotal point). In other words, the interplay between the two instability mechanisms must be nontrivial and, according to the model under consideration, has to be able either to enhance or to weaken chaos.

Now, since the existence of critical points—and correspondingly of a nontrivial topology of configuration space—should be a rather common feature of physical potentials, in order to further develop the Riemannian theory of Hamiltonian chaos beyond the restrictive assumption of quasi-isotropy, we have to take into account the role of nontrivial topology by generalizing the instability equation (1.10) in the form of a vector effective equation, which might read

$$\frac{d^2}{ds^2}\begin{pmatrix}\psi\\\phi\end{pmatrix} + \begin{pmatrix}\kappa(s) & \alpha\\\beta & \gamma\end{pmatrix}\begin{pmatrix}\psi\\\phi\end{pmatrix} = 0 \,, \qquad (5.108)$$

where α, β, γ are functions, to be worked out by the future developments of the theory, accounting for the relative frequency of encounters of neighborhoods of critical points, for the average number of unstable directions, and for the mentioned interplay between the two instability mechanisms. Then the instability exponent from (5.108) would be the average growth rate of $\|\psi^2\| + \|\dot\psi^2\| + \|\phi^2\| + \|\dot\phi^2\|$.

How to develop the theory so as to transform the wishful thinking represented by (5.108) into a real theoretical tool, is an open problem of great relevance within the present context, and it certainly deserves much attention and effort.

5.6 A Technical Remark on the Stochastic Oscillator Equation

In the following we will briefly describe how to cope with the stochastic oscillator problem that we encountered in Section 5.3.1. The discussion closely follows van Kampen [165], where all the details can be found.

A stochastic differential equation can be put in the general form

$$F(x, \dot{x}, \ddot{x}, \ldots, \Omega) = 0, \tag{5.109}$$

where F is an assigned function and the variable Ω is a random process defined by a mean, a standard deviation, and an autocorrelation function. A function $\xi(\Omega)$ is a solution of this equation, if $\forall \Omega \; F(\xi(\Omega), \Omega) = 0$. If equation (5.109) is linear of order n, it is written as

$$\dot{\mathbf{u}} = \mathbf{A}(t, \Omega)\mathbf{u} , \tag{5.110}$$

where

$$\mathbf{u} = \begin{pmatrix} u_1 \\ u_2 \\ u_3 \\ \vdots \\ u_n \end{pmatrix} = \begin{pmatrix} x \\ \dot{x} \\ \ddot{x} \\ \vdots \\ x^{(n)} \end{pmatrix} , \tag{5.111}$$

and $\mathbf{A}$ is an $n \times n$ matrix whose elements $A_{\mu\nu}(t)$ depend randomly on time.

For the purposes of the present chapter, we are interested in the evolution of the quantities $u_\nu u_\mu$, rather than of the u_μ's themselves. The products $u_\mu u_\nu$ obey the differential equation

$$\frac{d}{dt}(u_\nu u_\mu) = \sum_{k,\lambda} \tilde{A}_{\nu\mu,k\lambda}(t)(u_k u_\lambda) , \tag{5.112}$$

where

$$\tilde{A}_{\nu\mu,k\lambda} = A_{\nu k}\delta_{\mu\lambda} + \delta_{\nu k}A_{\mu\lambda} . \tag{5.113}$$

However, both (5.110) and (5.112) have exactly the same form and can be solved using the same procedure, thus, let us first consider such a procedure in general. In the following formulas, $\mathbf{u}$ refers to a vector whose components are either the u_μ's or the $u_\mu u_\nu$'s, and $\mathbf{A}$ denotes either the matrix $\mathbf{A}$ in (5.110) or the matrix $\tilde{\mathbf{A}}$ whose elements are given by (5.113). Then we apply this procedure to the case of the stochastic harmonic oscillator.

Now, solving a linear stochastic differential equation means determining the evolution of the average of $\mathbf{u}(t)$, $\langle \mathbf{u}(t) \rangle$, where the average is carried over all the realizations of the process. Let us consider the matrix $\mathbf{A}$ as the sum

$$\mathbf{A}(t, \Omega) = \mathbf{A}_0(t) + \alpha \mathbf{A}_1(t, \Omega) , \tag{5.114}$$

where the first term is Ω-independent and the second one is randomly fluctuating with zero mean. Let us also assume that $\mathbf{A}_0$ is time-independent. If the parameter α which determines the fluctuation amplitude, is small, we can treat (5.110) by means of a perturbation expansion. It is convenient to use the interaction representation, so that we put

$$\mathbf{u}(t) = \exp(\mathbf{A}_0 t)\mathbf{v}(t) \tag{5.115}$$

and

$$\mathbf{A}_1(t) = \exp(\mathbf{A}_0 t)\mathbf{v}(t)\exp(-\mathbf{A}_0 t) . \tag{5.116}$$

Formally one is then led to a Dyson expansion for the solution $\mathbf{v}(t)$. Then, going back to the previous variables and averaging, the second-order approximation gives

$$\frac{d}{dt}\langle\mathbf{u}(t)\rangle = \left\{\mathbf{A}_0 + \alpha^2 \int_0^{+\infty} \langle\mathbf{A}_1(t)\exp(\mathbf{A}_0\tau)\mathbf{A}_1(t-\tau)\rangle \exp(-\mathbf{A}_0\tau)d\tau\right\}\langle\mathbf{u}(t)\rangle . \tag{5.117}$$

Let us remark that if the stochastic process Ω is Gaussian, (5.117) is more than a second-order approximation: it is exact. In fact, the Dyson series can be written in compact form as

$$\langle\mathbf{u}(t)\rangle = T\left[\left\langle\exp\left(\int_0^t \mathbf{A}(t')dt'\right)\right\rangle\right]\langle\mathbf{u}(0)\rangle , \tag{5.118}$$

where $T[\cdots]$ stands for a time-ordered product. According to Wick's procedure we can rewrite (5.118) as a cumulant expansion, and when the cumulants of higher than the second order vanish (as in the case of a Gaussian process) one can easily show that (5.117) is exact.

We now apply this general approach to the case of interest for the main text, i.e., to the stochastic harmonic oscillator equation, which is the second-order linear stochastic differential equation given by

$$\ddot{x} + \Omega(t)\,x = 0 , \tag{5.119}$$

where $\Omega(t)$ is the random squared frequency $\Omega = \Omega_0 + \sigma_\Omega\eta(t)$, where Ω_0 is the mean of $\Omega(t)$, σ_Ω is the amplitude of the fluctuations, and $\eta(t)$ is a stochastic process with zero mean. In this case, (5.110) has the form

$$\frac{d}{dt}\begin{pmatrix} x \\ \dot{x} \end{pmatrix} = \begin{pmatrix} 0 & 1 \\ -\Omega & 0 \end{pmatrix}\begin{pmatrix} x \\ \dot{x} \end{pmatrix} . \tag{5.120}$$

In particular, we are interested in obtaining the averaged equation of motion for the second moments. Using (5.113) and (5.120), one finds that (5.112) becomes

$$\frac{d}{dt}\begin{pmatrix} x^2 \\ \dot{x}^2 \\ x\dot{x} \end{pmatrix} = \begin{pmatrix} 0 & 0 & 2 \\ 0 & 0 & -2\Omega \\ -\Omega & 1 & 0 \end{pmatrix}\begin{pmatrix} x^2 \\ \dot{x}^2 \\ x\dot{x} \end{pmatrix} = \mathbf{A}\begin{pmatrix} x^2 \\ \dot{x}^2 \\ x\dot{x} \end{pmatrix} . \tag{5.121}$$

As in (5.114), the matrix $\mathbf{A}$ splits into

$$\mathbf{A}(t) = \mathbf{A}_0 + \sigma_\Omega\eta(t)\mathbf{A}_1 = \begin{pmatrix} 0 & 0 & 2 \\ 0 & 0 & -2\Omega_0 \\ -\Omega_0 & 1 & 0 \end{pmatrix} + \sigma_\Omega\eta(t)\begin{pmatrix} 0 & 0 & 0 \\ 0 & 0 & -2 \\ -1 & 0 & 0 \end{pmatrix} , \tag{5.122}$$

so that the equation for the averages becomes

$$\frac{d}{dt}\begin{pmatrix} \langle x^2 \rangle \\ \langle \dot{x}^2 \rangle \\ \langle x\dot{x} \rangle \end{pmatrix} = \left\{ \mathbf{A}_0 + \sigma_\Omega^2 \int_0^{+\infty} \langle \eta(t)\eta(t-t') \rangle \mathbf{B}(t')dt' \right\} \begin{pmatrix} \langle x^2 \rangle \\ \langle \dot{x}^2 \rangle \\ \langle x\dot{x} \rangle \end{pmatrix}, \quad (5.123)$$

where $\mathbf{B}(t) = \mathbf{A}_1 \exp(\mathbf{A}_0 t)\mathbf{A}_1 \exp(-\mathbf{A}_0 t)$.

When the process $\eta(t)$ is Gaussian and δ-correlated, (5.123) is exact, and the integral can be computed explicitly: writing $\langle \eta(t)\eta(t-t') \rangle = \tau\delta(t')$, where τ is the correlation time scale of the random process, we obtain

$$\frac{d}{dt}\begin{pmatrix} \langle x^2 \rangle \\ \langle \dot{x}^2 \rangle \\ \langle x\dot{x} \rangle \end{pmatrix} = \left\{ \mathbf{A}_0 + \frac{\sigma_\Omega^2 \tau}{2}\mathbf{B}(0) \right\} \begin{pmatrix} \langle x^2 \rangle \\ \langle \dot{x}^2 \rangle \\ \langle x\dot{x} \rangle \end{pmatrix}. \quad (5.124)$$

From the definition of $\mathbf{B}(t)$ it follows then that $\mathbf{B}(0) = \mathbf{A}_1^2$, and by an easy calculations we obtain

$$\mathbf{A}_0 + \sigma_\Omega^2 \tau \mathbf{A}_1^2 = \begin{pmatrix} 0 & 0 & 2 \\ \sigma_\Omega^2 \tau & 0 & -2\Omega_0 \\ -\Omega_0 & 1 & 0 \end{pmatrix}, \quad (5.125)$$

which is the result used in Section 5.3.1.

Chapter 6

Geometry of Chaos and Phase Transitions

In the previous chapters we have shown how simple concepts belonging to classical differential geometry can be successfully used as tools to build a geometric theory of chaotic Hamiltonian dynamics. Such a theory is able to describe the instability of the dynamics in classical systems consisting of a large number N of mutually interacting particles, by relating these properties to the average and the fluctuations of the curvature of the configuration space. Such a relation is made quantitative through (5.45), which provides an approximate analytical estimate of the largest Lyapunov exponent in terms of the above-mentioned geometric quantities, and which compares very well with the outcome of numerical simulations in a number of cases, three of which have been discussed in detail in Chapter 5.

The macroscopic properties of large-N Hamiltonian systems can be understood by means of the traditional methods of statistical mechanics. One of the most striking phenomena that may occur in such systems is that when the external parameters (e.g., either the temperature or the energy) are varied until some critical value is reached, the macroscopic thermodynamic quantities may suddenly and even discontinuously change, so that though the microscopic interactions are the same above and below the critical value of the parameters, their macroscopic properties may be completely different. Such phenomena are referred to as phase transitions (see Chapter 2). In statistical mechanics, phase transitions are explained as true mathematical singularities that occur in the thermodynamic functions in the limit $N \to \infty$, the so-called thermodynamic limit. Such singularities come from the fact that the equilibrium probability distribution in configuration space, which in the canonical ensemble is the Boltzmann weight

$$\varrho_{\text{can}}(q_1, \ldots, q_N) = \frac{1}{Z} \exp\left[-\beta V(q_1, \ldots, q_N)\right] , \qquad (6.1)$$

where $\beta = 1/k_B T$, V is the potential energy, and $Z = \int dq\, e^{-\beta V(q)}$ is the configurational partition function, can itself develop singularities in the thermodynamic limit.

189

The statistical-mechanical theory of phase transitions is one of the most elaborate and successful physical theories now at hand, and at least as far as continuous phase transitions are concerned, also quantitative predictions can be made, with the aid of renormalization-group techniques, that are in very good agreement with laboratory experiments and numerical simulations. We are not going to discuss this here, referring the reader to the vast literature on the subject [109, 179–183].

However, the origin of the possibility of describing Hamiltonian systems via equilibrium statistical mechanics is the chaotic properties underlying the dynamics. In fact, though it is not possible to prove that generic Hamiltonian systems of the form (1.1) are ergodic and mixing, the fact that the trajectories are mostly chaotic (i.e., for the overwhelmingly majority of the trajectories positive Lyapunov exponents are found) means that such systems can be considered ergodic and mixing for all practical purposes.

The observation that chaos is at the origin of the statistical behavior of Hamiltonian systems and that chaotic dynamics can be described by means of the geometric methods described above naturally leads to the follwing two questions:

1. Is there any specific behavior of the largest Lyapunov exponent when the system undergoes a phase transition?
2. What are the geometric properties of the configuration-space manifold in the presence of a phase transition?

The aim of the present section is to discuss these two questions. We shall first give a concise phenomenological description that follows from numerical experiments, and then we shall report some analytically computed geometric quantities in the case of the mean-field XY model. From the discussion of the questions above and from the (at least partial) answers that we find, we are led to put forward a topological hypothesis about phase transitions, which will be discussed in Chapter 7.

6.1 Chaotic Dynamics and Phase Transitions

In order to look for an answer to question 1 above, we now review the numerical results that have been obtained up until now for various Hamiltonian dynamical systems that show a phase transition when considered as statistical-mechanical models for macroscopic systems in thermal equilibrium.

The first attempt to look for a chaotic-dynamic counterpart of an equilibrium phase transition is in the work by Butera and Caravati (BC) [91]. BC considered a two-dimensional XY model, i.e., a Hamiltonian dynamical system described by

$$H = \sum_{i,j=1}^{n} \left\{ \frac{1}{2}p_{i,j}^2 + J\left[2 - \cos(q_{i+1,j} - q_{i,j}) - \cos(q_{i,j+1} - q_{i,j})\right] \right\}, \quad (6.2)$$

where the q_{ij}'s are angles, i and j label the sites of a square lattice, and the sum runs over all the nearest-neighbor sites. This model is the two-dimensional version of the one studied in Section 5.4.3. As the temperature is decreased, such a system undergoes a peculiar phase transition (referred to as the Berežinskij–Kosterlitz–Thouless, or BKT, transition) from a disordered phase to a quasi-ordered phase, where, though no true long-range order is present, correlation functions decay as power laws, as occurs at a critical point [109]. Since there are no singularities in the finite-order derivatives of the free energy, the BKT transition is sometimes classified as an "infinite-order" phase transition. BC computed the Lyapunov exponent λ as a function of the temperature, and found that $\lambda(T)$ followed a rather smooth pattern; however, in a region around the transition, the dependence of λ on T changed from a steeply increasing function to a much less steep one.

BC's pioneering paper was for a long time the only example of a study of this kind. However, very recently there has been a renewed interest in the study of the behavior of Lyapunov exponents in systems undergoing phase transitions, and a number of papers have appeared [87,89,90,92,111–114,116, 175,184–187]. In [92,105], the two-dimensional XY model has been reconsidered, together with the three-dimensional case, defined by the Hamiltonian

$$H = \sum_{i,j,k=1}^{n} \left\{ \frac{1}{2}p_{i,j,k}^2 + J\left[3 - \cos(q_{i+1,j,k} - q_{i,j,k}) - \cos(q_{i,j+1,k} - q_{i,j,k}) \right. \right.$$
$$\left. \left. - \cos(q_{i,j,k+1} - q_{i,j,k}) \right] \right\} . \tag{6.3}$$

We remark that in three spatial dimensions the XY model undergoes a standard continuous (second-order) phase transition accompanied by the breaking of the $O(2)$ symmetry of the potential energy (6.3). The behavior of the largest Lyapunov exponent λ as a function of the temperature T is shown in Figures 6.1 and 6.2. The behavior found for the two-dimensional case confirms the BC results. The three-dimensional case shows a similar behavior, but the change of the shape of the $\lambda(T)$ function near T_c is somehow sharper than in the previous case.

Dellago and Posch [113] considered an extended XY model, whose potential energy is

$$V = 2 - 2 \sum_{\langle i,j \rangle} \cos\left(\frac{q_i - q_j}{2} \right)^{p^2} , \tag{6.4}$$

which includes the standard XY model (6.2) for $p^2 = 2$. On a two-dimensional lattice the transition, which is a continuous BKT transition for $p^2 = 2$, becomes a discontinuous transition for $p^2 = 100$. The results for the Lyapunov exponent λ show that for any considered value of p^2 there is a change in the shape of $\lambda(T)$ close to the critical temperature.

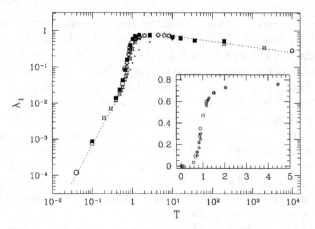

Fig. 6.1. Lyapunov exponent λ vs. the temperature T for the two-dimensional XY model defined in 2.121: the circles refer to a 10×10, the squares to a 40×40, the triangles to a 50×50, and the stars to a 100×100 lattice. The critical temperature of the BKT transition is $T_c \approx 0.95$. From [92].

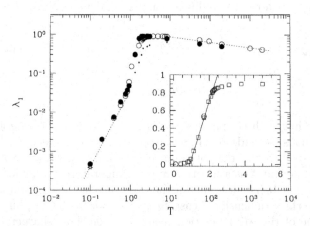

Fig. 6.2. Lyapunov exponent λ vs. the temperature T for the three-dimensional XY model, defined in 2.122, numerically computed on an $N = 10 \times 10 \times 10$ lattice (solid circles) and on an $N = 15 \times 15 \times 15$ lattice (solid squares). The critical temperature of the phase transition is $T_c \approx 2.15$. From [92].

One of the systems that have received considerable attention in this framework is the so-called lattice φ^4 model, i.e., a system with a Hamiltonian of the form (1.1) and a potential energy given by

$$V = \frac{J}{2} \sum_{\langle i,j \rangle} (\varphi_i - \varphi_j)^2 + \sum_i \left[-\frac{r^2}{2} \varphi_i^2 + \frac{u}{4!} \varphi_i^4 \right] , \qquad (6.5)$$

where the φ's are scalar variables, $\varphi_i \in [-\infty, +\infty]$ defined on the sites of a d-dimensional lattice, and r^2 and u are positive parameters. The lattice φ^4 model has a phase transition at a finite temperature provided that $d > 1$. The existence of such a transition, which belongs to the universality class of the d-dimensional Ising model, can be proved by renormalization-group arguments (see, e.g., [179, 188]). The cases $d = 2$ and $d = 3$ have been considered in [89] and [90], respectively. Moreover, in [90] also some vector versions of this model have been considered, namely, systems with potential energy given by

$$V = \frac{J}{2} \sum_{\langle i,j \rangle} \sum_{\alpha} (\varphi_i^\alpha - \varphi_j^\alpha)^2 + \sum_i \left\{ -\frac{r^2}{2} \sum_\alpha (\varphi_i^\alpha)^2 + \frac{u}{4!} \left[\sum_\alpha (\varphi_i^\alpha)^2 \right]^2 \right\} , \quad (6.6)$$

where α runs from 1 to n, with the components of the vectors labeled $\varphi_i = (\varphi_i^1, \ldots, \varphi_i^n)$. The potential energy (6.6) is $O(n)$-invariant; in the case $n = 1$ we recover the scalar model (6.5). Figures 6.3 and 6.4 show the behavior of λ in the φ^4 model, in two and three dimensions, respectively. Again we see that the Lyapunov exponent is sensitive to the presence of the transition, and that the shape of $\lambda(T)$ close to the transition is highly model-dependent. Moreover, such a shape can be significantly different within the same model as its parameters are varied. For instance, in the φ^4 model, λ either can be a monotonically increasing function of T or can display a maximum close to T_c, depending on the values of r^2 and u [89].

The Lyapunov exponents of systems undergoing phase transitions of the solid–liquid type have been recently determined numerically: Dellago and

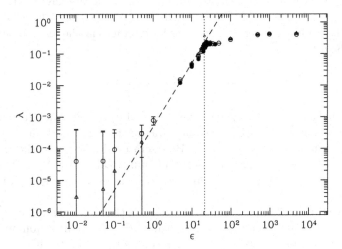

Fig. 6.3. Lyapunov exponent λ vs. the energy per particle ε, numerically computed for the two-dimensional $O(1)$ φ^4 model, with $N = 100$ (solid circles), $N = 400$ (open circles), $N = 900$ (solid triangles), and $N = 2500$ (open triangles). The critical energy is marked by a vertical dotted line, and the dashed line is the power law ε^2. From [89].

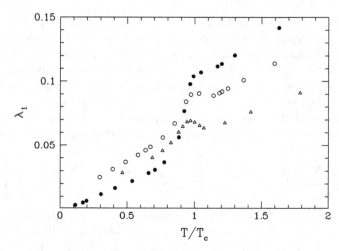

Fig. 6.4. Lyapunov exponent λ_1 vs. the temperature T for the three-dimensional φ^4 model. Full circles correspond to the $O(1)$ (scalar) case, open circles to the $O(2)$ case, and open triangles to the $O(4)$ case. From [90].

Posch (DP) considered, in two dimensions, a system of hard disks [111], a Lorentz-gas-like model, and a Lennard-Jones fluid [112], and, in three dimensions, a system of hard spheres [114]. DP found that in all these systems the Lyapunov exponent is sensitive to the phase transition, and again the shape of λ is different for different models, the common feature being that λ attains a maximum close, if not at, the transition. Similar results have been obtained by Mehra and Ramaswamy [186]. Bonasera et al. [184] considered a classical model of an atomic cluster, whose particles interact via phenomenological pair potentials of the form

$$v(r) = a\,e^{-\left(\frac{br}{\sigma}\right)} - c\left(\frac{\sigma}{r}\right)^6 , \qquad (6.7)$$

and of a nuclear cluster, with nucleons interacting via Yukawa pair potentials. Such systems undergo a so-called multifragmentation transition at a critical (model-dependent) temperature T_c. Bonasera et al. computed the Lyapunov exponents of these systems by means of numerical simulations at different temperatures. The resulting $\lambda(T)$ of both systems develops a sharp maximum close to T_c.

An interesting example of a phase transition that is not characterized by an order parameter is provided by the so-called Θ-transition in homopolymers. This is a transition between a globular configuration and a filamentary (or swollen) one of an atomic chain free to move in space. Traditionally signaled by a change of the scaling with N of the so-called gyration radius of the polymer, this transition is, however, well signaled by a peculiar energy pattern of the largest Lyapunov exponent [88, 189].

The model Hamiltonian for a so-called Lennard–Jones homopolymer composed of $N + 1$ beads connected by springs is [190–192]

$$H = \sum_{i=0}^{N} \frac{p_i^2}{2} + \sum_{i=1}^{N} \sum_{j<i} \left[\delta_{i,j+1} f_{\text{spr}}(r_{ij}) + \left(\frac{\sigma}{r_{ij}} \right)^{12} - \eta_{ij} \left(\frac{\sigma}{r_{ij}} \right)^{6} \right] , \qquad (6.8)$$

where the p_i are the canonical momenta, and a configuration of the chain is defined by the positions $\{q_0, \ldots, q_N\}$ of the beads in D-dimensional continuous space. The $r_{ij} = |\mathbf{r}_{ij}| = |\mathbf{q}_i - \mathbf{q}_j|$ are the interparticle distances.

The expression $f_{\text{spr}}(r_{ij})$, representing spring-like anharmonic interactions between neighboring beads in the homopolymer chain, is given by

$$f_{\text{spr}}(r_{ij}) = \frac{a}{2}(r_{ij} - r_0)^2 + \frac{b}{4}(r_{ij} - r_0)^4 , \qquad (6.9)$$

where r_0 is the equilibrium distance between nearest neighbors along the chain. From the numerical simulation of the dynamics of this model one obtains $\lambda_1(\epsilon)$. This is reported in Figure 6.5. A sharp change in the slope of $\lambda_1(\epsilon)$ takes place at the same value of the energy density which has been estimated to be the transition energy ϵ_θ through the standard analysis based on the scaling with N of the gyration radius. Even though the Θ-transition is only properly defined in the thermodynamic limit $N \to \infty$, numerical calculations of the Lyapunov exponent for $N = 25$ and $N = 50$ already appear to indicate the existence of the transition between a globular configuration and a filamentary one.

The numerical evidence that we have reviewed above, clearly shows that the Lyapunov exponent of a Hamiltonian dynamical system is sensitive to the

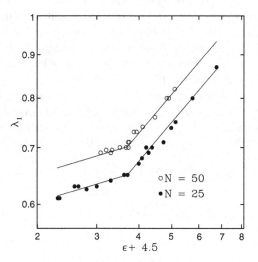

Fig. 6.5. Two-dimensional Lennard-Jones homopolymer. Cross-over between two different behaviors of the Lyapunov exponent λ_1 vs. the energy density ϵ at the Θ-transition point. From [88].

presence of a phase transition. However, the interpretation of the observed behavior as it now stands is difficult, because each model behaves differently and the behavior of λ close to the transition does not apparently exhibit any universal feature: on the other hand, there is no reason for this to happen, and the shape of $\lambda(T)$ can depend on the values of the parameters of the model. However, the qualitative behavior of $\lambda(T)$ appears to detect whether the transition is accompanied by a symmetry breaking, as in the case of the XY model: the shape of $\lambda(T)$ in two dimensions, where there is *not* any breaking of the $O(2)$ symmetry of the potential energy below the BKT transition temperature, is more "rounded" with respect to the corresponding shape of $\lambda(T)$ in the three-dimensional case, where the phase transition is accompanied by a symmetry breaking. In the latter case the "knee" of the $\lambda(T)$ curve is sharper. The difference between the two models is better appreciated by looking at the insets of Figures 6.1 and 6.2. As it will become clearer in the following, and, in particular, at the beginning of Chapter 8, we can anticipate that the largest Lyapunov exponent seems a "good" probing observable for the distribution of the critical points of the potential function in configurations space. Thus, the energy-pattern of λ should detect sudden changes of the topology of configuration-space submanifolds. And these sudden changes have no reason to occur in a model-independent way.

6.2 Curvature and Phase Transitions

In Chapter 5 we have seen that the origin of chaos in Hamiltonian mechanics can be understood from a geometrical point of view, and that the Lyapunov exponents are closely related to a geometric quantity, i.e., to the fluctuations of the Ricci curvature of the configuration space. Thus, it is natural to investigate whether such a geometric observable also has some peculiar behavior close to the phase transition. As we shall see in the following, the fluctuations of the curvature do indeed have such a peculiar behavior, which, in turn, suggests a topological intepretation of the phase transition itself.

The Ricci curvature along a geodesic of the enlarged configuration space-time equipped with the Eisenhart metric, which we denoted by K_R in the previous chapters, is given by the Laplacian of the potential energy; see (5.30). In the case of the XY model we obtain, as already shown in Section 5.4.3,

$$K_R = 2N - 2V = 2 \sum_{\langle i,j \rangle} \cos(q_i - q_j) \ . \tag{6.10}$$

The root mean square fluctuation of K_R divided by the number of degrees of freedom N, i.e.,

$$\sigma_k = \left(\frac{1}{N} \langle K_R^2 \rangle - \langle K_R \rangle^2 \right)^{1/2} , \tag{6.11}$$

is plotted in Figures 6.6 and 6.7 for the 2D and 3D cases, respectively.

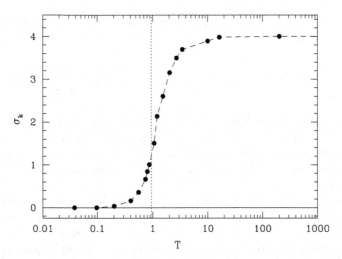

Fig. 6.6. Fluctuations of the Ricci curvature (Eisenhart metric), $\sigma_k(T)$, vs. the temperature T for the two-dimensional XY model. The solid circles are numerical values obtained for a 40×40 lattice; the dashed line is only a guide to the eye. The critical temperature of the BKT transition is $T_c \approx 0.95$ and is marked by a dotted line. From [92].

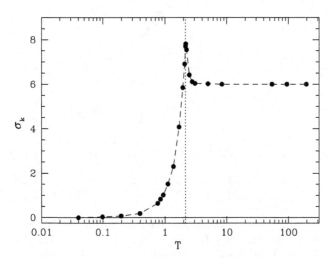

Fig. 6.7. Like Figure 6.6, for the three-dimensional XY model. Here $N = 10 \times 10 \times 10$, and the critical temperature of the phase transition is $T_c \approx 2.15$. From [92].

In the case of the φ^4 model with $O(n)$ symmetry, the Ricci curvature K_R is given by [89, 90]

$$K_R = \sum_{\alpha=1}^{n} \sum_{i=1}^{N} \frac{\partial^2 V}{\partial (\varphi_i^\alpha)^2} = Nn(2Jd - r^2) + \lambda(n+2) \sum_{\alpha=1}^{n} \sum_{i=1}^{N} (\varphi_i^\alpha)^2 \ . \qquad (6.12)$$

The root mean squared fluctuation of K_R, σ_k, is plotted against the energy per degree of freedom, ε, in the case of the two-dimensional $O(1)$ φ^4 model in Figure 6.8, and against the temperature T in the case of the two-dimensional $O(2)$ φ^4 model in Figure 6.9, and for the three-dimensional $O(n)$ φ^4 models in Figure 6.10.

Looking at Figures 6.6 and 6.10, one can clearly see that when a symmetry-breaking phase transition occurs, a cusplike ("singular") behavior of the curvature fluctuations is found at the phase transition point (Figures 6.7, 6.8 and 6.10), while when only a BKT transition is present, no cusplike pattern is observed[1] (Figures 6.6 and 6.9). We can summarize these results by saying that in general, curvature fluctuations always show a cusplike behavior when a continuous symmetry-breaking phase transition is present, and, within numerical accuracy, the cusp occurs at the critical temperature. No counterexamples have yet been found to this general rule.

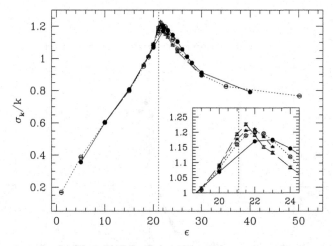

Fig. 6.8. Root mean square fluctuation of the Ricci curvature (Eisenhart metric) σ_k, divided by the average curvature k_0, numerically computed for the two-dimensional $O(1)$ φ^4 model. The inset shows a magnification of the region close to the transition. Symbols as in Figure 6.3. From [89].

---

[1] Although the cusplike behavior is lost, indeed some change of behavior is still visible in Figures 6.6 and 6.9 close to the critical temperature, so that a BKT transition appears as "intermediate" between the absence of a phase transition and the presence of a symmetry-breaking phase transition.

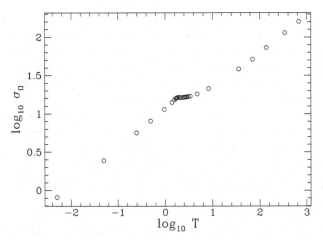

Fig. 6.9. Curvature fluctuations σ_Ω (notation of [90]) vs. the temperature T for the two-dimensional $O(2)$ φ^4 model, numerically computed on a square lattice of $30{\times}30$ sites. The critical temperature T_c of the BKT transition is located at $T_c \approx 1.5$. From [90].

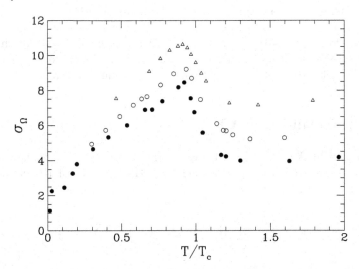

Fig. 6.10. Curvature fluctuations σ_Ω (notation of [90]) vs. the temperature T for the three-dimensional φ^4 model. Full circles correspond to the $O(1)$ (scalar) case, open circles to the $O(2)$ case, and open triangles to the $O(4)$ case. From numerical simulations performed on an $8 \times 8 \times 8$ cubic lattice, reported in [90].

The fact that the Lyapunov exponent is sensitive to the phase transition can now be understood, in the light of the fact that, as shown in the preceding chapters, chaos can be described geometrically, and under suitable assumptions, the Lyapunov exponent is closely related to the fluctuations of the

Ricci curvature[2] (see (5.47)). Thus, it is not surprising that also the energy (or temperature) pattern of the fluctuations of the Ricci curvature, $\sigma_k(\varepsilon)$ (or $\sigma_k(T)$), *is* a good observable to detect a phase transition. Both $\sigma_k(\varepsilon)$ and $\sigma_k(T)$ exhibit a clearly peculiar ("cuspy") pattern when a symmetry-breaking phase transition is present. In order to compare how $\lambda(T)$ and $\sigma_k(T)$ detect a phase transition, one has to look at Figures 6.4 and 6.10, where $\lambda(T)$ and $\sigma_k(T)$ are reported, respectively, for the $O(n)$ φ^4 models in 3D for $n = 1, 2, 4$. Likewise, Figures 6.3 and 6.8 allow one to make the comparison in the case of the 2D $O(1)$ φ^4 model, and Figures 6.2 and 6.7 for the 3D XY-model.

6.2.1 Geometric Estimate of the Lyapunov Exponent

At this point, it is worthwhile to point out that we can apply the geometric formula (5.46) for the Lyapunov exponent to estimate λ for all these models, since both k_0 and σ_k have been numerically computed. As shown in [89, 90, 92], one finds that in general, although the qualitative behavior of the Lyapunov exponent is well reproduced, the quantitative agreement between the values of λ extracted from the numerical simulations and those obtained applying (5.46) is *not* good, in a neighborhood of the phase transition.

However, this is to be expected, because among the assumptions under which the formula (5.46) was derived there was the hypothesis that the fluctuations of the curvature should be not too large, and this does not seem to be the case close to a phase transition, as we have just shown.[3]

6.3 The Mean-Field XY Model

The mean-field XY model [110] describes a system of N equally coupled planar classical rotators. It is defined by a Hamiltonian of the class (1.1), where the potential energy is

$$V(\varphi) = \frac{J}{2N} \sum_{i,j=1}^{N} [1 - \cos(\varphi_i - \varphi_j)] - h \sum_{i=1}^{N} \cos \varphi_i \ . \tag{6.13}$$

Here $\varphi_i \in [0, 2\pi]$ is the rotation angle of the ith rotator and h is an external field. Defining at each site i a classical spin vector $\mathbf{s}_i = (\cos \varphi_i, \sin \varphi_i)$, the model describes a planar (XY) Heisenberg system with interactions of equal

---

[2] The assumption of quasi-isotropy is somewhat restrictive because it seems to be adequate mainly when the mechanical manifolds are topologically trivial, however, as we shall see in the next chapter, also nontrivial topology brings about curvature fluctuations of its own.

[3] The results of the formula (5.46) can be improved using procedures that are specific to the model under consideration and that we are not going to describe here (see [92] for the XY case and [90] for the φ^4 case).

strength among all the spins. We consider only the ferromagnetic case $J > 0$; for the sake of simplicity, we set $J = 1$. The equilibrium statistical mechanics of this system is exactly described, in the thermodynamic limit, by mean-field theory [110]. In the limit $h \to 0$, the system has a continuous phase transition, with classical critical exponents, at $T_c = 1/2$, or $\varepsilon_c = 3/4$, where $\varepsilon = E/N$ is the energy per particle.

The Lyapunov exponent λ of this system is extremely sensitive to the phase transition. In fact, according to numerical simulations reported in [175, 176, 187, 193], $\lambda(\varepsilon)$ is positive for $0 < \varepsilon < \varepsilon_c$, shows a sharp maximum immediately below the critical energy, and drops to zero at ε_c in the thermodynamic limit, where it remains zero in the whole region $\varepsilon > \varepsilon_c$, which corresponds to the thermodynamic disordered phase. In fact, in this phase the system is integrable, reducing to an assembly of uncoupled rotators. These results are valid in the thermodynamic limit $N \to \infty$ in the sense that they have been obtained by estimating the infinite-N limit of finite-N numerical simulations [175, 187]: in the whole region $\varepsilon > \varepsilon_c$ the Lyapunov exponent, numerically computed for systems with different numbers of particles N, behaves as $\lambda \propto N^{-1/3}$, so that it extrapolates to zero at $N \to \infty$.

These results have received a theoretical confirmation in a work based on the application of the geometric techniques described in the preceding chapter [87]. The analytic computation of $\langle k_R \rangle$ and $\langle \delta^2 k_R \rangle$ in the thermodynamic limit for the mean-field XY model, show that these quantities indeed have a singular behavior at ε_c (see Figure 6.11). Using these quantities and (5.46), the analytical estimate for $\lambda(\varepsilon)$ is obtained. This is reported in Figure 5.13; it is remarkable that also the behavior $\lambda \propto N^{-1/3}$ at $\varepsilon > \varepsilon_c$ has been extracted from this theoretical calculation (see Section 5.4.4). This result gives a theoretical confirmation to the qualitative behavior of the Lyapunov exponent extrapolated from the numerical simulations. Moreover, these analytical

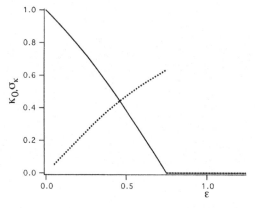

Fig. 6.11. Mean-field XY model: analytic expression for the microcanonical averages of the Ricci curvature (solid curve) and of its fluctuations (dotted curve). From [87].

results are in good quantitative agreement with numerical results reported in [176,187], also close to the phase transition and at variance with the cases of the nearest-neighbor XY and φ^4 models considered earlier. A tentative explanation of why the application of the geometric formula (5.46) gives such good quantitative results in the present case can be that the mean-field character of the model prevents the curvature fluctuations from being too wild.

Chapter 7

Topological Hypothesis on the Origin of Phase Transitions

In the previous chapter we have reported results of numerical simulations for the fluctuations of observables of a geometric nature (e.g., configuration-space curvature fluctuations) related to the Riemannian geometrization of the dynamics in configuration space.[1] These quantities have been computed, using time averages, for many different models undergoing continuous phase transitions, namely φ^4 lattice models with discrete and continuous symmetries and XY models. In particular, when plotted as a function of either the temperature or the energy, the fluctuations of the curvature have an apparently singular behavior at the transition point. Moreover, we have seen that the presence of a singularity in the statistical-mechanical fluctuations of the curvature at the transition point has been proved analytically for the mean-field XY model.

The aim of the present chapter is to try to understand on a deeper level the origin of this peculiar behaviour. In Section 7.1, we will show, using abstract geometric toy models, that a singular behavior in the fluctuations of the curvature of a Riemannian manifold can be associated with a change in the *topology* of the manifold itself. By "change of topology" we mean the following. Let us consider a surface $\mathcal{S}_\varepsilon$ that depends on a parameter ε in such a way that upon varying the parameter, the surface is continuously deformed: as long as the different deformed surfaces can be mapped *smoothly* one onto another,[2] the topology does not change; however, the topology changes if there is a critical value of the parameter, say ε_c, such that the surface $\mathcal{S}_{\varepsilon > \varepsilon_c}$ can no longer be mapped smoothly onto $\mathcal{S}_{\varepsilon < \varepsilon_c}$.

The observation that a singularity in the curvature fluctuations of a Riemannian manifold, of the same type as those observed numerically at phase transitions, can be associated with a change in the topology of the manifold leads us to conjecture that it is just this mechanism that could be the basis

---

[1] More precisely, we considered the enlarged configuration space-time, endowed with the Eisenhart metric.

[2] The different surfaces are then said to be diffeomorphic to each other (see Appendix A).

of thermodynamic phase transitions. Such a conjecture was originally put forward in [92] as follows: a thermodynamic transition might be related to a change in the topology of the configuration space, and the observed singularities in the statistical-mechanical equilibrium measure and in the thermodynamic observables at the phase transition might be interpreted as a "shadow" of this major topological change that happens at a more basic level. We will refer to this conjecture as the *topological hypothesis (TH)*.

In a first part of this chapter, we report about the logical path that, through heuristic and indirect evidence, led us to formulate the TH. In the second part of this chapter, a first *direct* numerical support to the validity of the TH is given for a specific model.

7.1 From Geometry to Topology: Abstract Geometric Models

Let us now describe how a singular behavior of the curvature fluctuations of a manifold can be put in correspondence with a change in the topology of the manifold itself. For the sake of clarity, we shall first discuss a simple example concerning two-dimensional surfaces [90,92], and then we will generalize it to the case of N-dimensional hypersurfaces [194,195].

The simple geometric model we are going to describe concerns surfaces of revolution. A surface of revolution $\mathbb{S} \in \mathbb{R}^3$ is obtained by revolving the graph of a function f around one of the axes of a Cartesian plane, and can be defined, in parametric form, as follows [196]:

$$\mathbb{S}(u,v) \equiv (x(u,v), y(u,v), z(u,v)) = (a(u)\cos v, a(u)\sin v, b(u)) , \qquad (7.1)$$

where either $a(u) = f(u)$ and $b(u) = u$, if the graph of f is revolved around the vertical axis, or $a(u) = u$ and $b(u) = f(u)$, if the graph is revolved around the horizontal axis; in both cases, u and v are local coordinates on the surface $\mathbb{S}$: $v \in [0, 2\pi]$ and u belongs to the domain of definition of the function f.

Let us consider now in particular the two families of surfaces of revolution defined as

$$\mathcal{F}_\varepsilon = (f_\varepsilon(u)\cos v, f_\varepsilon(u)\sin v, u) \qquad (7.2)$$

and

$$\mathcal{G}_\varepsilon = (u\cos v, u\sin v, f_\varepsilon(u)) , \qquad (7.3)$$

where

$$f_\varepsilon(u) = \pm\sqrt{\varepsilon + u^2 - u^4} , \qquad \varepsilon \in [\varepsilon_{min}, +\infty) , \qquad (7.4)$$

and $\varepsilon_{min} = -\frac{1}{4}$. Some cases are shown in Figure 7.1.

There exists for both families of surfaces a critical value of ε, $\varepsilon_c = 0$, corresponding to a change in the *topology* of the surfaces: the manifolds $\mathcal{F}_\varepsilon$ are diffeomorphic to a torus $\mathbb{T}^2$ for $\varepsilon < 0$ and to a sphere $\mathbb{S}^2$ for $\varepsilon > 0$; the manifolds $\mathcal{G}_\varepsilon$ are diffeomorphic to *two* spheres for $\varepsilon < 0$ and to one sphere

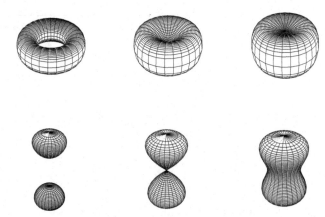

Fig. 7.1. Some representatives of the two families of surfaces $\mathcal{F}_\varepsilon$ and $\mathcal{G}_\varepsilon$ defined in (7.2) and (7.3) respectively. Each family is divided into two subfamilies by the critical surface corresponding to $\varepsilon_c = 0$ (middle members in the picture). Members of the same subfamily are diffeomorphic, whereas the two subfamilies are not diffeomorphic to each other. From [90].

for $\varepsilon > 0$. The Euler–Poincaré characteristic (see Appendix A) is $\chi(\mathcal{F}_\varepsilon) = 0$ if $\varepsilon < 0$, and $\chi(\mathcal{F}_\varepsilon) = 2$ otherwise, while $\chi(\mathcal{G}_\varepsilon)$ is 4 or 2 for ε negative or positive, respectively.

We now turn to the definition and the calculation of the curvature fluctuations on these surfaces. Let M belong to one of the two families; its Gaussian curvature K is [196]

$$K = \frac{a'(a''b' - b'a'')}{a(b'^2 + a'^2)^2} , \tag{7.5}$$

where $a(u)$ and $b(u)$ are the coefficients of (7.1), and primes denote differentiation with respect to u. The fluctuations of K can be then defined as

$$\sigma_K^2 = \langle K^2 \rangle - \langle K \rangle^2 = A^{-1} \int_M K^2 \, dS - \left(A^{-1} \int_M K \, dS \right)^2 , \tag{7.6}$$

where A is the area of M and dS is the invariant surface element. Both families of surfaces exhibit a singular behavior in σ_K as $\varepsilon \to \varepsilon_c$, as shown in Figure 7.2, in spite of their different curvature properties on average.[3]

We are now going to show that the result we have just obtained for two-dimensional surfaces has a much more general validity: a *generic* topology change in an n-dimensional manifold is accompanied by a singularity in its curvature fluctuations [194]. In order to do that, we have to make use of some concepts belonging to Morse theory, which will also be used in Section 7.4 below; the basic elementary concepts of Morse theory are sketched in

---

[3] For instance, $\langle K \rangle(\varepsilon) = 0$ for $\mathcal{F}_\varepsilon$ as $\varepsilon < 0$, while for $\mathcal{G}_\varepsilon$ the same average curvature is positive and diverges as $\varepsilon \to 0$.

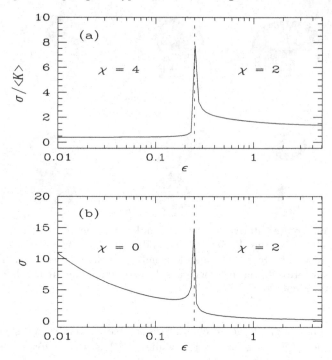

Fig. 7.2. The fluctuation σ of the Gaussian curvature of the surfaces $\mathcal{F}_\varepsilon$ and $\mathcal{G}_\varepsilon$ is plotted vs. ε; σ is defined in (7.6), and ε is shifted by $\varepsilon_{\min} = 0.25$ for reasons of clarity of presentation. (a) refers to $\mathcal{G}_\varepsilon$ and (b) refers to $\mathcal{F}_\varepsilon$. The cusps appear at $\varepsilon = 0$, where the topological transition takes place for both $\mathcal{F}_\varepsilon$ and $\mathcal{G}_\varepsilon$. From [90].

Appendix C, where also references to the literature are given. We consider then a hypersurface of $\mathbb{R}^N$ which is the u-level set of a function f defined in $\mathbb{R}^N$, i.e., a submanifold of $\mathbb{R}^N$ of dimension $n = N - 1$ defined by the equation

$$f(x_1, \ldots, x_N) = u ; \tag{7.7}$$

such a hypersurface can then be referred to as $f^{-1}(u)$. Let us now assume[4] that f is a *Morse function*, i.e., such that its critical points (i.e., the points of $\mathbb{R}^N$ where the differential df vanishes) are isolated. One of the most important results of Morse theory is that the topology of the hypersurfaces $f^{-1}(u)$ can change *only* by crossing a critical level $f^{-1}(u_c)$, i.e., a level set containing at least one critical point of f. This means that a generic change in the topology of the hypersurfaces can be associated with critical points of f. Now, the hypersurfaces $f^{-1}(u)$ can be given a Riemannian metric in a standard way [197], and it is possible to analyze the behavior of the curvature fluctuations in a neighborhood of a critical point. Let us assume, for the sake of simplicity,

---

[4] This is not a strong assumption: in fact, it can be shown that Morse functions are generic (see Appendix C).

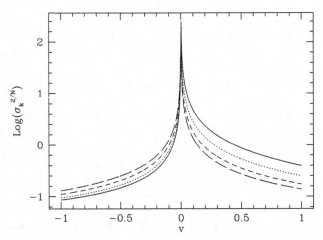

Fig. 7.3. Fluctuations of the Gauss curvature of a hypersurface $f^{-1}(u)$ of $\mathbb{R}^N$ vs. u close to a critical point. Here $\sigma_k^{2/N}$ is reported because it has the same dimensions of the scalar curvature. Also $\dim(f^{-1}(u)) = 100$, and the Morse indexes are $1, 15, 33, 48$, represented by solid, dotted, dashed, and long-dashed lines respectively. From [194].

that this critical point is located at $x_0 = 0$ and belongs to the level $u_c = 0$. Any Morse function can be parametrized, in the neighborhood of a x_0, by means of the so-called Morse chart, i.e., a system of local coordinates $\{y_i\}$ such that $f(y) = f(x_0) - \sum_{i=1}^k y_i^2 + \sum_{i=k+1}^N y_i^2$ (k is the Morse index of the critical point). Then standard formulas for the Gauss curvature K of hypersurfaces of $\mathbb{R}^N$ [197] can be used to compute explicitly the fluctuations of the curvature, σ_K, of the level set $f^{-1}(u)$. Numerical results for the curvature fluctuations are reported in Figure 7.3 and show that also at high dimension σ_K^2 develops a sharp, singular peak as the critical surface is approached (for computational details are reported in [195]).

7.2 Topology Changes in Configuration Space and Phase Transitions

As we have discussed in Chapter 6, the curvature fluctuations of the configuration space exhibit cusplike patterns in the presence of a second-order phase transition. A truly cuspy pattern, i.e., an analytic discontinuity, is mathematically proven in the case of mean-field XY model. In Section 7.1, we have shown that singular patterns in the fluctuations of the curvature of a Riemannian manifold can be seen as consequences of the presence of a topology change. Hence, we are led to the topological hypothesis (TH), i.e., to conjecture that at least continuous, symmetry-breaking phase transitions are associated with topology changes in the configuration space of the system.[5]

However, an important question arises, in that the fluctuations of the curvature considered in Chapter 6 have been obtained as time averages, computed along the dynamical trajectories of the Hamiltonian systems under investigation (or as statistical averages computed analytically, as in the case of the mean-field XY model). Now, time averages of geometric observables are usually found to be in excellent agreement with ensemble averages [86, 89, 90, 92, 131], so that one could argue that the above-mentioned singular-like patterns of the fluctuations of geometric observables are simply the precursors of truly singular patterns due to the fact that the measures of all the statistical ensembles tend to become singular in the limit $N \to \infty$ when a phase transition is present. In other words, geometric observables, like any other "honest" observable, already at finite N would feel the eventually singular character of the statistical measures, i.e., of the probability distribution functions of the statistical-mechanical ensembles. If this were the correct explanation, we could not attribute the cusplike patterns of the curvature fluctuations to any special geometric features of configuration space, and the cusp-like patterns observed in the numerical simulations could not be considered as (indirect) confirmations of the TH.

In order to elucidate this important point, three different paths have been followed: (i) *purely geometric* information about certain submanifolds of configuration space has been worked out *independently* of the statistical measures in the case of the two-dimensional φ^4 model, and the results lend *indirect* support to the TH [194]; (ii) a *direct* numerical confirmation of the TH has been given in [198] by means of the computation of a topologic invariant, the Euler characteristic, in the case of a 2D lattice φ^4 model; (iii) a *direct* analytic confirmation of the TH has been found in the particular case of the mean-field XY model [199] and of a trigonometric model with k-body interactions. We report on items (i) and (ii) in this chapter and (iii) in Chapter 10.

7.3 Indirect Numerical Investigations of the Topology of Configuration Space

In order to separate the singular effects due to the singular character of statistical measures at a phase transition from the singular effects due to some topological transition in configuration space, the first natural step is to consider again σ_K^2 as an observable, and to integrate it on suitable submanifolds of configuration space by means of a geometric measure, i.e., by means of a measure that has nothing to do with statistical ensemble measures.

Consider as ambient space the N-dimensional configuration space M of a Hamiltonian system with N degrees of freedom, which, when $N \to \infty$,

---

[5] As we shall see in the following chapters, also first-order phase transitions are necessarily driven by topological changes.

undergoes a phase transition at a certain finite temperature T_c (or critical energy per degree of freedom ε_c), and let $V(\varphi)$ be its potential energy.

Then the relevant geometrical objects are the submanifolds of M defined by

$$M_u = V^{-1}(-\infty, u] = \{\varphi \in M : V(\varphi) \le u\} \,, \qquad (7.8)$$

i.e., each M_u is the set $\{\varphi_i\}_{i=1}^N$ such that the potential energy does not exceed a given value u. As u is increased from $-\infty$ to $+\infty$, this family covers successively the whole manifold M. All the submanifolds M_u can be given a Riemannian metric g whose choice is largely arbitrary. On all these manifolds (M_u, g) there is a standard invariant volume measure:

$$d\eta = \sqrt{\det(g)} \, d\varphi_1 \cdots d\varphi_N \,, \qquad (7.9)$$

which has nothing to do with statistical measures. Let us finally define the hypersurfaces Σ_u as the u-level sets of V, i.e.,

$$\Sigma_u = V^{-1}(u) \,, \qquad (7.10)$$

which are nothing but the boundaries of the submanifolds M_u.

According to the discussion reported in Section 7.1, an indirect way to study the presence of topology changes in the family $\{(M_u, g)\}$ is to look at the behavior of the fluctuations of the Gaussian curvature, σ_K^2, defined as

$$\sigma_K^2 = \langle K_G^2 \rangle_{\Sigma_u} - \langle K_G \rangle_{\Sigma_u}^2 \,, \qquad (7.11)$$

where $\langle \cdot \rangle$ stands for integration over the surface Σ_u, as a function of u. The presence of cusplike singularities of σ_K^2 for some critical value of u, u_c, would eventually signal the presence of a topology change of the family $\{(M_u, g)\}$ at u_c [194]. Such an indirect geometric probing of the presence of critical points seems an expedient way to probe the possible topology changes of the manifolds (M_u, g). In fact, the properties of the manifolds M_u are closely related to those of the hypersurfaces $\{\Sigma_u\}_{u \le u_c}$, as can be inferred from the equation

$$\int_{M_u} f \, d\eta = \int_0^u dv \int_{\Sigma_v} f|_{\Sigma_v} d\omega / \|\nabla V\| \,, \qquad (7.12)$$

where $d\omega$ is the induced measure[6] on Σ_u and f a generic function [200]. From Morse theory (see Appendix C) we know that the surface Σ_{u_c} defined by $V = u_c$ is a degenerate quadric, so that in its vicinity some of the principal curvatures [197] of the surfaces $\Sigma_{u \approx u_c}$ tend to diverge.[7] Such a divergence

---

[6] If a surface is parametrically defined through the equations $x^i = x^i(z^1, \ldots, z^k)$, $i = 1, \ldots, N$, then the metric g_{ij} induced on the surface is given by $g_{ij}(z^1, \ldots, z^k) = \sum_{n=1}^N \frac{\partial x^n}{\partial z^i} \frac{\partial x^n}{\partial z^j}$. See Appendix B.

[7] The principal curvatures are the inverse of the curvature radii measured, at any given point of a surface, in suitable directions. At a Morse critical point some of these curvature radii vanish.

is generically detected by any function of the principal curvatures, and thus for practical computational reasons, instead of the Gauss curvature (which is the product of all the principal curvatures) we shall consider the total second variation of the *scalar* curvature $\mathcal{R}$ (i.e., the sum of all the possible products of two principal curvatures) of the manifolds (M_u, g), according to the definition

$$\sigma_{\mathcal{R}}^2(u) = [\text{Vol}(M_u)]^{-1} \int_{M_u} d\eta \left[\mathcal{R} - [\text{Vol}(M_u)]^{-1} \int_{M_u} d\eta \, \mathcal{R} \right]^2 \quad (7.13)$$

with $\mathcal{R} = g^{kj} R^l_{klj}$, where R^l_{kij} are the components of the Riemann curvature tensor [see Appendix B] and $\text{Vol}(M_u) = \int_{M_u} d\eta$. The subsets M_u of configuration space are given the structure of Riemannian manifolds (M_u, g) by endowing all of them with the *same* metric tensor g. However, the choice of the metric g is arbitrary in view of probing possible effects of the topology on the geometry of these manifolds.

What has been hitherto discussed now requires the choice of a model to perform a numerical investigation. A good candidate is represented by the so-called φ^4 model on a d-dimensional lattice $\mathbb{Z}^d$ with $d = 1, 2$, described by the potential function

$$V = \sum_{i \in \mathbb{Z}^d} \left(-\frac{\mu^2}{2} \varphi_i^2 + \frac{\lambda}{4} \varphi_i^4 \right) + \sum_{\langle ik \rangle \in \mathbb{Z}^d} \frac{1}{2} J(\varphi_i - \varphi_k)^2 , \quad (7.14)$$

where $\langle ik \rangle$ stands for nearest-neighbor sites. This system has a discrete $\mathbb{Z}_2$-symmetry and short-range interactions; therefore, in $d = 1$ there is no phase transition whereas in $d = 2$ there is a symmetry-breaking transition, at a finite temperature, of the same universality class of the 2D Ising model. In [194], three different types of metrics have been considered for this model, i.e.,

(i) $g_{\mu\nu}^{(1)} = [A - V(\varphi)]\delta_{\mu\nu}$, i.e., a conformal deformation (Section B.3.2) of the Euclidean flat metric $\delta_{\mu\nu}$, where $A > 0$ is an arbitrary constant chosen large enough to be sure that in the relevant interval of values of u the determinant of the metric is always positive;

(ii) $g_{\mu\nu}^{(2)}$ and $g_{\mu\nu}^{(3)}$ are generic metrics (no longer conformal deformations of the flat metric) defined by

$$(g_{\mu\nu}^{(k)}) = \begin{pmatrix} f^{(k)} & 0 & 1 \\ 0 & \mathbb{I} & 0 \\ 1 & 0 & 1 \end{pmatrix} , \quad k = 2, 3 , \quad (7.15)$$

where $\mathbb{I}$ is the $(N-2)$-dimensional identity matrix, $g^{(2)}$ is obtained by setting $f^{(2)} = \frac{1}{N} \sum_{\alpha \in \mathbb{Z}^d} \varphi_\alpha^4 + A$, and $g^{(3)}$ by setting $f^{(3)} = \frac{1}{N} \sum_{\alpha \in \mathbb{Z}^d} \varphi_\alpha^6 + A$, with $A > 0$, and α labels the N lattice sites of a linear chain ($d = 1$) or of a square lattice ($d = 2$, $N = n \times n$).

These choices are completely arbitrary, however, and only if metrics of very simple form are chosen are both analytical and numerical computations feasible also for rather large values of N. Thus the first metric has been chosen diagonal, and the other two metrics concentrate in only one matrix element all the nontrivial geometric information. Moreover, the first metric still contains a reference to the physical potential, whereas the other two define metric structures that are completely independent of the physical potential and contain only monomials of powers sufficiently high that they do not vanish after two successive derivatives have been taken (needed to compute curvatures). The topology of the subsets of points M_u and Σ_u of $\mathbb{R}^N$ is already determined (though well concealed) by the definitions of (7.8) and (7.10); the task is to "capture" some information about their topology through a mathematical object or structure, defined on these sets of points, that is capable of mirroring the variations of topology through the u-pattern of an analytic function. This idea follows the philosophy of standard mathematical theories of differential topology. For example, within Morse theory, the information about topology is extracted through the critical points of any function—defined on a given manifold—satisfying some conditions (necessary to be a good Morse function, see Appendix C), or, within cohomology theory, topology is probed through vector spaces of differential forms (the de Rham cohomology vector spaces, see Appendix A) "attached" to a given manifold. Provided that good mathematical quantities are chosen as topology-variation detectors, arbitrary Riemannian metric structures could work as well.

For the above-defined metrics $g^{(k)}$, $k = 1, 2, 3$, simple algebra leads from the definition of the scalar curvature (see Appendix B) to the following explicit expressions:

$$\mathcal{R}^{(1)} = (N - 1) \left[\frac{\Delta V}{(A - V)^2} - \frac{\|\nabla V\|^2}{(A - V)^3} \left(\frac{N}{4} - \frac{3}{2} \right) \right] , \tag{7.16}$$

$$\mathcal{R}^{(k)} = \frac{1}{(f^{(k)} - 1)} \left[\frac{\|\tilde{\nabla} f^{(k)}\|^2}{2(f^{(k)} - 1)} - \tilde{\Delta} f^{(k)} \right] , \quad k = 2, 3 , \tag{7.17}$$

where ∇ and Δ are the Euclidean gradient and Laplacian respectively, and $\tilde{\nabla}$ and $\tilde{\Delta}$ lack the derivative $\partial/\partial\varphi_\alpha$ with $\alpha = 1$ in the $d = 1$ case, and lack the derivative $\partial/\partial\varphi_\alpha$ with $\alpha = (1, 1)$ in the $d = 2$ case.

The numerical computation of the geometric integrals in (7.13) is worked out by means of a Monte Carlo algorithm [169, 195] to sample the geometric measure $d\eta$ by means of an "importance sampling" algorithm suitably modified (see Section 7.6.1).

In Figures 7.4 and 7.5, $\sigma_{\mathcal{R}}^2(\overline{u})$, where $\overline{u} = u/N$, are given for the one- and two-dimensional cases obtained for two different lattice sizes with $g^{(1)}$ (Figure 7.4), and at given lattice size with $g^{(2,3)}$ (Figure 7.5). Peaks of $\sigma_{\mathcal{R}}^2(\overline{u})$ appear at a certain value $\overline{u}_c = u_c/N$, of $\overline{u}$ in the two-dimensional case, whereas only smooth patterns are found in the one-dimensional case, where no phase transition is present.

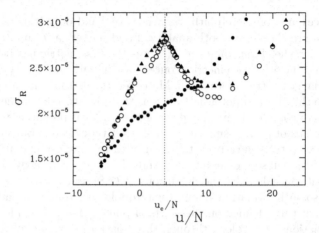

Fig. 7.4. Variance of the scalar curvature of M_u vs. u/N computed with the metric $g^{(1)}$. Full circles correspond to the 1D φ^4 model with $N = 400$. Open circles refer to the 2D φ^4 model with $N = 20 \times 20$ lattice sites, and full triangles refer to 40×40 lattice sites (whose values are rescaled for graphical reasons). From [194].

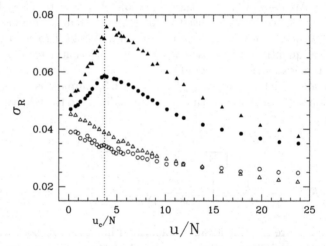

Fig. 7.5. $\sigma^2_{\mathcal{R}}(\overline{u})$ of M_u vs. u/N computed for the φ^4 model with metric $g^{(2)}$ in 1D, $N = 400$ (open triangles); metric $g^{(2)}$ in 2D, $N = 20 \times 20$ (full triangles); metric $g^{(3)}$ in 1D, $N = 400$ (open circles); metric $g^{(3)}$ in 2D, $N = 20 \times 20$ (full circles). From [194].

According to the discussion above, these peaks can be considered as indirect evidence of the presence of a topology transition in the manifolds M_u at $u = u_c$ in the case of the two-dimensional φ^4 model. It is in particular the persistence of cusplike patterns of $\sigma^2_{\mathcal{R}}(\overline{u})$ independently of the metric chosen that lends credence to the idea that this actually reflects a topological transition. Now we want to argue that the topological transition occurring at

u_c/N is related to a *thermodynamic phase transition* that occurs in the φ^4 model. In order to do that, in [194] the average potential energy per particle

$$\frac{u(T)}{N} = \frac{\langle V \rangle}{N} \qquad (7.18)$$

has been numerically computed, as a function of T, by means of both Monte Carlo averaging with the canonical configurational measure and Hamiltonian dynamics. In the latter case the temperature T is given by the average kinetic energy per degree of freedom, and u is obtained as the time average. Figure 7.6 shows a perfect agreement between time and ensemble averages. The phase transition point is well visible at $\overline{u}_c = u_c/N \approx 3.75$. Looking at Figures 7.4 and 7.5, we realize that within the numerical accuracy, the critical value of the potential energy per particle u_c/N where the topological change occurs equals the statistical-mechanical average value of the potential energy at the phase transition. At this point the doubt, formulated at the beginning of this chapter, about the possible nongeometrical origin of the "singular" cusplike patterns of $\sigma_{\mathcal{R}}^2(\overline{u})$ has been dissipated. These results have been found *independently* of statistical-mechanical measures and of their singular character in the presence of a phase transition. These results are also *independent*—at least to the limited extent of the three metric tensors reported above—of the geometric structure given to the family $\{M_u\}$. Thus they seem most likely to have their origin at a deeper level than the geometric one, i.e. at the topologic level. Hence the observed phenomenology strongly hints that some *major change in the topology of the configuration-space submanifolds $\{M_u\}$* occurs when a second-order phase transition takes place.

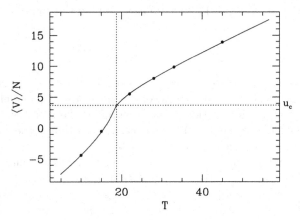

Fig. 7.6. Average potential energy vs. temperature for the 2D lattice φ^4 model with $O(1)$ symmetry. Lattice size $N = 20 \times 20$. The solid line is made out of 200 points obtained as time averages. Full circles represent Monte Carlo estimates of canonical ensemble averages. The dotted lines locate the phase transition. From [194].

7.4 Topological Origin of the Phase Transition in the Mean-Field XY Model

Until now we have not yet given any *direct* analytic evidence of the validity of the TH. Let us now consider again the mean-field XY model (6.13). As we shall see in detail in Chapter 10, for this model we can analytically compute both its thermodynamics and a topological invariant (the Euler–Poincaré characteristic) of the submanifolds M_v of its configuration space. Hence, it is possible to show analytically that a particular topological change in the configuration space is related to the thermodynamic phase transition. However, in this chapter we begin by discussing a first simplified approach to the model [199] giving evidence of the topological transition in a space of collective variables.

Let us consider again, as was already done in Section 7.3, the family M_v of submanifolds of the configuration space defined in (7.8); now the potential energy per degree of freedom is that of the mean-field XY model, i.e.,

$$\mathcal{V}(\varphi) = \frac{V(\varphi)}{N} = \frac{J}{2N^2} \sum_{i,j=1}^{N} [1 - \cos(\varphi_i - \varphi_j)] - \frac{h}{N} \sum_{i=1}^{N} \cos \varphi_i \,, \qquad (7.19)$$

where $\varphi_i \in [0, 2\pi]$. Such a function can be considered a Morse function on M, so that, according to Morse theory (see Appendix C), all these manifolds have the same topology until a critical level $\mathcal{V}^{-1}(v_c)$ is crossed, where the topology of M_v changes.

A change in the topology of M_v can occur only when v passes through a critical value of $\mathcal{V}$. Thus in order to detect topological changes in M_v we have to find the critical values of $\mathcal{V}$, which means solving the equations

$$\frac{\partial \mathcal{V}(\varphi)}{\partial \varphi_i} = 0 \,, \qquad i = 1, \ldots, N \,. \qquad (7.20)$$

For a general potential energy function $\mathcal{V}$, the solution of the (7.20) would be a formidable task [202], but in the case of the mean-field XY model, the mean-field character of the interaction greatly simplifies the analysis, allowing an analytical treatment of the (7.20); moreover, a projection of the configuration space onto a two-dimensional plane is possible.

We recall that in the limit $h \to 0$, the system has a continuous phase transition, with classical critical exponents, at $T_c = 1/2$, or $\varepsilon_c = 3/4$, where $\varepsilon = E/N$ is the energy per particle. Let us now show that this phase transition has its foundation in a basic topological change that occurs in the configuration space M of the system. To begin with, note that since $\mathcal{V}(\varphi)$ is bounded, $-h \leq \mathcal{V}(\varphi) \leq 1/2 + h^2/2$, the manifold is empty as long as $v < -h$, and when v increases beyond $1/2 + h^2/2$ no changes in its topology can occur, so that the manifold M_v remains the same for any $v > 1/2 + h^2/2$, and is then an N-torus. To detect topological changes we have to solve (7.20). To this end it is useful to define the magnetization vector, i.e., the collective spin vector $\mathbf{m} = \frac{1}{N} \sum_{i=1}^{N} \mathbf{s}_i$, which as a function of the angles is given by

$$\mathbf{m} = (m_x, m_y) = \left(\frac{1}{N} \sum_{i=1}^{N} \cos \varphi_i, \frac{1}{N} \sum_{i=1}^{N} \sin \varphi_i \right) . \tag{7.21}$$

Due to the mean-field character of the model, the potential energy (6.13) can be expressed as a function of $\mathbf{m}$ alone (remember that $J = 1$), so that the potential energy per particle reads

$$\mathcal{V}(\varphi) = \mathcal{V}(m_x, m_y) = \frac{1}{2}(1 - m_x^2 - m_y^2) - h\, m_x . \tag{7.22}$$

This allows us to write (7.20) in the form $(i = 1, \ldots, N)$

$$(m_x + h) \sin \varphi_i - m_y \cos \varphi_i = 0 . \tag{7.23}$$

Now we can solve these equations and find all the critical values of $\mathcal{V}$. The solutions of (7.23) can be grouped into three classes:

(i) The minimal energy configuration $\varphi_i = 0\ \forall i$, with a critical value $v = v_0 = -h$, which tends to 0 as $h \to 0$. In this case, $m_x^2 + m_y^2 = 1$.

(ii) Configurations such that $m_y = 0, \sin \varphi_i = 0\ \forall i$. These are the configurations in which φ_i equals either 0 or π; i.e., we have again $\varphi_i = 0\ \forall i$, but also the N configurations with $\varphi_k = \pi$ and $\varphi_i = 0\ \forall i \neq k$, as well as the $N(N-1)$ configurations with 2 angles equal to π and all the others equal to 0, and so on, up to the configuration with $\varphi_i = \pi\ \forall i$. The critical values corresponding to these critical points depend only on the number of π's, n_π, so that $v(n_\pi) = \frac{1}{2}[1 - \frac{1}{N^2}(N - 2n_\pi)^2] - \frac{h}{N}(N - 2n_\pi)$. We see that the largest critical value is, for N even, $v(n_\pi = N/2) = 1/2$ and that the number of critical points corresponding to it is $\mathcal{O}(2^N)$.

(iii) Configurations such that $m_x = -h$ and $m_y = 0$, which correspond to the critical value $v_c = 1/2 + h^2/2$, which tends to $1/2$ as $h \to 0$. The number of these configurations grows with N not slower than $N!$ [199].

Configurations (i) are the absolute minima of $\mathcal{V}$, (iii) are the absolute maxima, and (ii) are all the other stationary configurations of $\mathcal{V}$.

Since for $v < v_0$ the manifold is empty, the topological change that occurs at v_0 is the one corresponding to the "birth" of the manifold from the empty set; subsequently there are many topological changes at values $v(n_\pi) \in (v_0, 1/2]$ till at v_c there is a final topological change that corresponds to the "completion" of the manifold. We remark that the number of critical values in the interval $[v_0, 1/2]$ grows with N and that eventually the set of these critical values becomes dense in the limit $N \to \infty$. However, the critical value v_c remains isolated from other critical values also in that limit. We observe that it is necessary to consider a nonzero external field h in order that $\mathcal{V}$ be a Morse function, because if $h = 0$ all the critical points of classes (i) and (ii)

are degenerate, in which case topological changes do not necessarily occur.[8]
This degeneracy is due to the $O(2)$-invariance of the potential energy in the
absence of an external field. To be sure, for $h \neq 0$, $\mathcal{V}$ may not be a Morse
function on the whole of M either, but only on M_v with $v < v_c$, because
the critical points of class (iii) may also be degenerate, so that v_c does not
necessarily correspond to a topological change. However, this difficulty could
be dealt with by using that the potential energy can be written in terms of
the collective variables m_x and m_y, as in (7.22). This implies that we consider
the system of N spins projected onto the two-dimensional configuration space
of the collective spin variables. According to the definition (7.21) of $\mathbf{m}$, the
accessible configuration space is now not the whole plane, but only the disk

$$D = \{(m_x, m_y) : m_x^2 + m_y^2 \leq 1\} . \tag{7.24}$$

Thus we want to study the topology of the submanifolds

$$D_v = \{(m_x, m_y) \in D : \mathcal{V}(m_x, m_y) \leq v\} . \tag{7.25}$$

The sequence of topological transformations undergone by D_v can now be
very simply determined in the limit $h \to 0$ (see Figure 7.7), as follows. As
long as $v < 0$, D_v is the empty set. The first topological change occurs at
$v = v_0 = 0$, where the manifold appears as the circle $m_x^2 + m_y^2 = 1$, i.e., the
boundary ∂D of D. Then as v grows D_v is given by the conditions

$$1 - 2v \leq m_x^2 + m_y^2 \leq 1 , \tag{7.26}$$

i.e., it is the ring with a hole centered at $(0,0)$ (punctured disk) contained
between two circles of radii 1 and $\sqrt{2v}$. As v continues to grow, the hole
shrinks and is eventually completely filled when $v = v_c = 1/2$, where the
second topological change occurs (see Section A.5 on the fundamental group).
In this coarse-grained two-dimensional description in D, all the topological

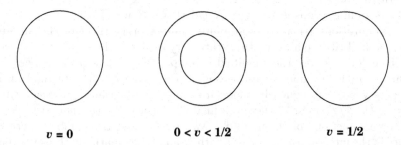

$$v = 0 \qquad\qquad 0 < v < 1/2 \qquad\qquad v = 1/2$$

Fig. 7.7. The sequence of topological changes undergone by the manifolds D_v with
increasing v in the limit $h \to 0$.

---

[8] It would also be possible to avoid this problem by considering an improved version
of Morse theory, referred to as *equivariant Morse theory* [203].

changes that occur in M between $v = 0$ and $v = 1/2$ disappear, and only
the two topological changes corresponding to the extrema of $\mathcal{V}$, occurring at
$v = v_0$ and $v = v_c$, survive. This means that the topological change at v_c
should be present also in the full N-dimensional configuration space, so that
the degeneracies mentioned above for the critical points of class (iii) should
not prevent a topological change.

Now we want to argue that the topological change occurring at v_c is related
to the thermodynamic phase transition of the mean-field XY model. Since the
Hamiltonian is of the standard form (1.1), the temperature T, the energy per
particle ε, and the average potential energy per particle $\bar{u} = \langle \mathcal{V} \rangle$ obey, in the
thermodynamic limit, the following equation:

$$\varepsilon = \frac{T}{2} + \bar{u}(T) , \qquad (7.27)$$

where we have set Boltzmann's constant equal to 1. Substituting the values
of the critical energy per particle $\varepsilon_c = 3/4$ and of the critical temperature $T_c = 1/2$ we get $\bar{u}_c = \bar{u}(T_c) = 1/2$, so that the critical value of the
potential energy per particle v_c where the last topological change occurs equals
the statistical-mechanical average value of the potential energy at the phase
transition,

$$v_c = \bar{u}_c . \qquad (7.28)$$

Thus although a topological change in M occurs at any N, and v_c is
independent of N, there is a connection of such a topological change and
a thermodynamic phase transition *only* in the limit $N \to \infty$, $h \to 0^+$, when
indeed a thermodynamic phase transition can be defined.

Since *not all* topological changes correspond to phase transitions, those
that do correspond remain to be determined to make the conjecture of [92]
more precise. In this context, we consider one example where there are topological changes very similar to the ones of our model but no phase transitions,
i.e., the one-dimensional XY model with nearest-neighbor interactions, whose
Hamiltonian is of the class (1.1) with interaction potential

$$V(\varphi) = \frac{1}{4} \sum_{i=1}^{N} [1 - \cos(\varphi_{i+1} - \varphi_i)] - h \sum_{i=1}^{N} \cos \varphi_i . \qquad (7.29)$$

In this case the configuration space M is still an N-torus, and using again the
potential energy per degree of freedom $\mathcal{V} = V/N$ as a Morse function, we can
see that also here there are many topological changes in the submanifolds M_v
as v is varied in the interval $[0, 1/2]$ (after taking $h \to 0^+$). However there are
critical points of the type $\varphi_j = \varphi_k = \varphi_l = \cdots = \pi$, $\varphi_i = 0$ $\forall i \neq j, k, l, \ldots$; in
contrast to the mean-field XY model, it is now no longer the number of π's
that determines the value of $\mathcal{V}$ at the critical point, but rather the number of
domain walls, n_d, i.e., the number of boundaries between "islands" of π's and
"islands" of 0's: $v(n_d) = n_d/2N$. Since $n_d \in [0, N]$, the critical values lie in the

same interval as in the case of the mean-field XY model; but now the maximum critical value $v = 1/2$, instead of corresponding to a huge number of critical points, which rapidly grows with N, corresponds to *only two* configurations with N domain walls, which are $\varphi_{2k} = 0$, $\varphi_{2k+1} = \pi$, with $k = 1, \ldots, N/2$, and the reversed one. There are also "spin-wave-like" critical points, i.e., such that $e^{i\theta_k} = \text{const } e^{2\pi i k n / N}$ with $n = 1, \ldots, N$ [204]; their critical energies are contained in the interval above but again there is not a critical value associated with a huge number of critical points.

Thus this example suggests the conjecture that a topological change in the configuration-space submanifolds M_v occurring at a critical value v_c is associated with a phase transition in the thermodynamic limit only if the number of critical points corresponding to the critical value v_c is sufficiently rapidly growing with N and makes a big jump at v_c. On the basis of the behavior of the mean-field XY model we expect then that such a growth should be at least exponential. Furthermore, a relevant feature appears to be that v_c remains an isolated critical value also in the limit $N \to \infty$: in the mean-field XY model this holds only if the thermodynamic limit is taken *before* the $h \to 0^+$ limit: this appears as a topological counterpart of the noncommutativity of the limits $h \to 0^+$ and $N \to \infty$ in order to get a phase transition in statistical mechanics.

The sequence of topological changes occurring with growing V makes the configuration space larger and larger, till at v_c the whole configuration space becomes fully accessible to the system through the last topological change. From a physical point of view, this corresponds to the appearance of more and more disordered configurations as T grows, which ultimately lead to the phase transition at T_c.

7.5 The Topological Hypothesis

Let us consider the canonical partition function Z_N for an N-degrees-of-freedom system described by a standard Hamiltonian $H(p,q) = \sum p^2/2 + V(q)$, where p and q are vectors. For these systems, after a trivial integration of the kinetic energy term, it reads

$$Z(\beta, N) = \int d^N p \, d^N q \, e^{-\beta H(p,q)} = \left(\frac{\pi}{\beta} \right)^{\frac{N}{2}} \int d^N q \, e^{-\beta V(q)} , \quad (7.30)$$

showing that its nontrivial part is the configurational partition function

$$Z_c(\beta, N) = \int_{\mathbb{R}^N} d^N q \, e^{-\beta V(q)} = \int_0^{+\infty} dv \, e^{-\beta v} \int_{\Sigma_v} \frac{d\sigma}{\|\nabla V\|} , \quad (7.31)$$

where a coarea formula [200] has been used to unfold the *structure integrals*

$$\Omega_N(v) \equiv \int_{\Sigma_v} \frac{d\sigma}{\|\nabla V\|} \ , \tag{7.32}$$

an infinite collection of integrals on the *equipotential hypersurfaces* Σ_v of the configuration space defined by $\Sigma_v \equiv \{q \in \mathbb{R}^N | V(q) = v\} \subset \mathbb{R}^N$, and $d\sigma$ is the volume form induced by the immersion of Σ_v in $\mathbb{R}^N$.

Equation (7.31) shows that, formally, Z_c is the Laplace transform of the structure integral.

Then, if we consider the *microcanonical* ensemble, the basic mathematical object is the phase space volume

$$\Omega(E) = \int_0^E d\eta \ \Omega^{(-)}(E - \eta) \int d^N p \ \delta \left(\sum_i \frac{1}{2} p_i^2 - \eta \right)$$

where

$$\Omega^{(-)}(E - \eta) = \int d^N q \ \Theta[V(q) - (E - \eta)] = \int_0^{E-\eta} dv \int_{\Sigma_v} \frac{d\sigma}{\|\nabla V\|} \ , \tag{7.33}$$

whence

$$\Omega(E) = \int_0^E d\eta \ \frac{(2\pi\eta)^{N/2}}{\eta \Gamma(\frac{N}{2})} \int_0^{E-\eta} dv \int_{\Sigma_v} \frac{d\sigma}{\|\nabla V\|} \ . \tag{7.34}$$

Here too, as in the above decomposition of $Z_c(\beta, N)$, the only nontrivial objects are the structure integrals (7.32).

Once the microscopic interaction potential $V(q)$ is given, the configuration space of the system is *automatically* foliated into the family $\{\Sigma_v\}_{v \in \mathbb{R}}$ of equipotential hypersurfaces *independently* of any statistical measure we may wish to use. Now, from standard statistical-mechanical arguments we know that the larger the number N of particles, the closer to some Σ_v are the microstates that significantly contribute to the statistical averages of thermodynamic observables. At large N, and at any given value of the inverse temperature β, the effective support of the canonical measure is narrowed very close to a single $\Sigma_v = \Sigma_{v(\beta_c)}$; similarly, in the microcanonical ensemble, the fluctuations of potential and kinetic energies tend to vanish at increasing N so that the effective contributions to $\Omega(E)$ come from a close neighborhood of a $\Sigma_v = \Sigma_{v(E_c)}$.

Now, the *topological hypothesis* consists in assuming that some suitable change of the topology of the $\{\Sigma_v\}$, occurring at some $v_c = v_c(\beta_c)$ (or $v_c = v_c(E_c)$), is the deep origin of the singular behavior of thermodynamic observables at a phase transition (by change of topology we mean that $\{\Sigma_v\}_{v<v_c}$ are *not diffeomorphic* to the $\{\Sigma_v\}_{v>v_c}$). In other words, the claim is that the canonical and microcanonical measures must "feel" a big and sudden change, if any, of the topology of the equipotential hypersurfaces of their underlying supports, the consequence being the appearance of the typi-cal signals of a phase transition, i.e., almost singular energy or temperature dependencies of the averages of appropriate observables. The larger N the narrower is the effective support of the measure and hence the sharper can be the mentioned signals. Eventually, in the $N \to \infty$ limit this sharpening will lead to nonanalyticity.

On the basis of this and what was found in [92, 194, 199], we formulate the TH as follows:

Topological Hypothesis: *The basic origin of a phase transition lies in a topological change of the support of the measure describing a system. This change of topology induces a change of the measure itself at the transition point.*

In other words, this hypothesis stipulates that some change of the topo-logy of the $\{\Sigma_v\}$, occurring at some $v_c = v_c(\beta_c)$, could be the origin of the singular behavior of thermodynamic observables at a phase transition rather than measure singularities, which in this view are induced from a deeper level where the topological changes take place.

Remark 7.1. As we shall see in the following chapters, topological changes of the manifolds Σ_v and M_v are associated to the existence of critical points of the potential function, i.e., points where $\nabla V = 0$. By looking at the defin-ition of the structure integral (7.32), one could naively infer that since the denominator $\|\nabla V\|$ vanishes at the critical points, entailing a divergence of the structure integral, the critical points, and thus topology, must be relevant to the divergence of thermodynamic observables. However, such a kind of rea-soning would be completely wrong and misleading. On large–N hypersurfaces the integration measure regularizes the structure integral also at the critical points so that the vanishing of the denominator does not entail any diver-gence of the structure integral (Consider, for example, that Σ_v is a large–N hypersphere, and write ∇V near a critical point as a quadratic form using the Morse chart). The way topology induces the appearance of thermodynamic singularities is by far more subtle, as will be clarified in the next chapters.

7.6 Direct Numerical Investigations of the Topology of Configuration Space

The first successful attempt at obtaining a direct evidence that topologi-cal changes are associated with phase transitions, and thus the first direct evidence supporting the topological hypothesis, is numerical.

Since the counterpart of a phase transition is expected to be a suitable breaking of diffeomorphicity among the surfaces Σ_v, it is appropriate to choose a *diffeomorphism invariant* to probe whether and how the topology of the Σ_v changes as a function of v. This is a very challenging task because one has to deal with high-dimensional manifolds. Fortunately, a topological invariant exists whose computation is feasible, yet demands a big effort. This is the *Euler characteristic*, a diffeomorphism invariant, expressing fundamental topological information (see Appendix A).

Let us recall how it is defined. First, let us consider that a basic way to analyze a geometrical object is to fragment it into other more familiar objects and then to examine how these pieces fit together. Take, for example, a surface Σ in Euclidean three-dimensional space. Slice Σ into pieces that are curved triangles (this is called a triangulation of the surface). Then count the number F of faces of the triangles, the number E of edges, and the number V of vertices on the tesselated surface. Now, no matter how we triangulate a compact surface Σ, $\chi(\Sigma) = F - E + V$ will always equal a constant that is characteristic of the surface and that is invariant under diffeomorphisms $\phi : \Sigma \to \Sigma'$. This is the Euler characteristic of Σ. At higher dimensions this can be again defined by using higher-dimensional generalizations of triangles (simplices) and by defining the Euler characteristic of the n-dimensional manifold Σ to be

$$\chi(\Sigma) = \sum_{k=0}^{n} (-1)^k (\# \text{ of "faces of dimension k"}). \tag{7.35}$$

In differential topology an equivalent definition of $\chi(\Sigma)$ is

$$\chi(\Sigma) = \sum_{k=0}^{n} (-1)^k b_k(\Sigma) , \tag{7.36}$$

where the numbers b_k, the Betti numbers of Σ, are diffeomorphism invariants (see Appendix A). While it would be hopeless to try to compute $\chi(\Sigma)$ from (7.36) in the case of nontrivial physical models at large dimension, there is a possibility given by a powerful theorem, the Gauss–Bonnet–Hopf theorem, that relates $\chi(\Sigma)$ to the total Gauss–Kronecker curvature of the manifold, that is,

$$\chi(\Sigma) = \gamma \int_{\Sigma} K_G \, d\sigma , \tag{7.37}$$

which is valid for even dimensional hypersurfaces of Euclidean spaces $\mathbb{R}^N$ [here $\dim(\Sigma) = n \equiv N - 1$], and where: $\gamma = 2/\mathrm{Vol}(\mathbb{S}^n_1)$ is twice the inverse of the volume of an n-dimensional sphere of unit radius; K_G is the Gauss–Kronecker curvature of the manifold; $d\sigma = \sqrt{\det(g)} dx^1 dx^2 \cdots dx^n$ is the invariant volume measure of Σ, and g is the Riemannian metric induced from $\mathbb{R}^N$. The definition and significance of the Gauss–Kronecker curvature are given in Chapter 8. The practical computation of K_G at any point $x \in \Sigma_v$ proceeds from the knowledge of a basis $\{\mathbf{v}_1, \ldots, \mathbf{v}_n\}$ for the tangent space of Σ_v at x, so that, using the directional derivatives $\nabla_{\mathbf{v}_i} V$, it is

$$K_G(x) = \frac{(-1)^n}{\|\nabla V\|^n} \left| \begin{pmatrix} \nabla_{\mathbf{v}_1} \nabla V \\ \vdots \\ \nabla_{\mathbf{v}_n} \nabla V \\ \nabla V \end{pmatrix} \right| \left| \begin{pmatrix} \mathbf{v}_1 \\ \vdots \\ \mathbf{v}_n \\ \nabla V \end{pmatrix} \right|^{-1} . \tag{7.38}$$

7.6.1 Monte Carlo Estimates of Geometric Integrals

In order to perform a numerical computation of the topological invariant given by the Gauss–Bonnet–Hopf formula (7.37), the numerical evaluation of multiple integrals in high-dimensional spaces is needed. To this end, the Monte Carlo–Metropolis method was devised a long time ago and apart from a large number of improvements and modifications of the basic scheme to fit it to different kinds of problems, the main feature remains unaltered: we can compute only densities and not the actual numerical value of the multiple integral under examination. The basic scheme consists in any algorithm capable of generating a Markov chain, in the high-dimensional space of interest, whose asymptotic probability density coincides with the measure of the integral to compute.

In particular, since we have to compute surface integrals $\int_{\Sigma_v} g\, d\sigma$, it is necessary to devise an efficient algorithm to generate a Markov chain on a hypersurface. In order to constrain a Markov chain generated with the standard "importance sampling" [205] on a given Σ_v, one has to adopt a projection algorithm.

Suppose that $x_k \in \Sigma_v$ is the point generated at the kth step of the Monte Carlo Markov chain, and that the updated point at the following step $\tilde{x}_{k+1} = x_k + \Delta x_k$ is not too far from the preceding one ($\|\Delta x_k\| \ll 1$ in convenient units). In general, it is $\tilde{x}_{k+1} \in \Sigma_{v+\Delta v}$; thus $\tilde{x}_{k+1}$ is projected on the tangent plane at x_k to Σ_v. The coordinates of the updated and projected configuration are thus the following:

$$x_{k+1} = x_k - \frac{\Delta v}{\|\nabla V\|^2} \cdot \nabla V \ , \tag{7.39}$$

where $\Delta v = (\tilde{x}_{k+1} - x_k)\cdot \nabla V$ is the difference in potential energy between the two configurations. The projection algorithm allows one to efficiently perform a random walk on a hypersurface Σ_v, provided that the new configurations proposed at each step are generated and/or preselected so as not to be too far from Σ_v.

Another important point concerns the measure $d\sigma$ entering the integral 7.37. This is the canonical volume form associated with the metric tensor $g_{\alpha\gamma}(v)$ of the hypersurface that is obtained by restricting the Euclidean metric of $\mathbb{R}^N$ to Σ_v. Thus we have

$$g_{\alpha\gamma}(v) = \delta_{\alpha\gamma} + \frac{\partial_\alpha V \partial_\gamma V}{\partial_N V \partial_N V} \ ,$$

where it is understood that $g_{\alpha\gamma}(v)$is a function of the point where it is computed. The volume form is

$$d\sigma = \prod_{i=1}^{N-1} dx_i \sqrt{|g(v)|} = \prod_{i=1}^{N-1} dx_i \frac{\|\nabla V\|}{|\partial_N V|} \ . \tag{7.40}$$

The Monte Carlo algorithm has to be defined to sample the geometric measure (7.40) through the standard "importance sampling" method. Let $\mathcal{O}(x)$ be the observable that we want to average with the measure $\sqrt{|g(v)|}$, and let $x \in \Sigma_v$ be an arbitrary initial condition. Then one proceeds with a random update of the coordinates of x so as to obtain a new configuration $\tilde{x} = (x + \Delta x) \in \Sigma_{\tilde{v}}$ not too far from x; then the coordinates of $\tilde{x}$ are modified by projecting them according to (7.39) on Σ_v. Then the following ratio of weights is computed:

$$\zeta = \frac{\sqrt{|g(v')|}}{\sqrt{|g(v)|}} \ . \tag{7.41}$$

If $\zeta > 1$, then the new proposed configuration is accepted; if $\zeta < 1$, then a random number $w \in [0,1]$ is generated; and if $w < \zeta$, again the new configuration is accepted. Otherwise, it is rejected and the old configuration is counted once again. The observable $\mathcal{O}(x)$ is averaged on the set of all the accepted configurations

By means of a Monte Carlo algorithm one can estimate only *averages* of observables, that is,

$$\langle \mathcal{O} \rangle_{\mathrm{MC}}(v) = \frac{\int_{\Sigma_v} d\sigma \mathcal{O}(x)}{\int_{\Sigma_v} d\sigma} \ , \tag{7.42}$$

while we are interested in evaluating the actual values of the integral (7.37). In order to do this, we should be able to estimate the volume that appears in the denominator of (7.42). Denote by

$$\omega(v) = \int_{\Sigma_v} d\sigma \tag{7.43}$$

the volume of interest, and notice that the following identity holds:

$$\frac{d}{dv} \log \omega(v) = \frac{\omega'(v)}{\omega(v)} \ , \tag{7.44}$$

where $\omega'(v)$ stands for the first derivative of the volume (7.43) with respect to v. In the absence of critical points, Federer's derivation formula (see Chapter 8) gives

$$\omega'(v) = \int_{\Sigma_v} \frac{d\sigma}{\|\nabla V\|} \nabla \left(\frac{\nabla V}{\|\nabla V\|} \right) \ . \tag{7.45}$$

However, since we tackle potentials that are good Morse functions, the number of critical values of these potentials is finite in any finite interval of potential energy values, so that, even in the presence of critical values, (7.45) can be safely used in Monte Carlo computations. In fact, the probability of numerically falling exactly on a critical level set is zero. Moreover, the volume $\omega(v)$ and its first derivative are regular (see Theorem 9.14 of the Chapter 9). By combining (7.44) and (7.45) we get

$$\frac{d}{dv} \log \omega(v) = \frac{\int_{\Sigma_v} \frac{d\sigma}{\|\nabla V\|} \nabla \left(\frac{\nabla V}{\|\nabla V\|} \right)}{\int_{\Sigma_v} d\sigma} , \qquad (7.46)$$

which is still in suitable form to be numerically computed through a Monte Carlo algorithm.

Let us introduce the quantity[9]

$$M_1 = \nabla \left(\frac{\nabla V}{\|\nabla V\|} \right) , \qquad (7.47)$$

which is proportional to the *mean curvature*, and integrate (7.46) to obtain

$$\omega(v) = \omega(v_0) \exp \left\{ \int_{v_0}^{v} dw \left\langle \frac{M_1}{\|\nabla V\|} \right\rangle_{\mathrm{MC}} \right\} , \qquad (7.48)$$

so that the potential energy dependence of the volume, $\omega(v)$, is determined apart from a constant $\omega(v_0)$, which, however, is the same for any value v. This last equation makes it possible to numerically estimate, by means of a Monte Carlo algorithm, the integral (7.37), with the only indeterminacy due to the unknown multiplicative constant $\omega(v_0)$.

7.6.2 Euler Characteristic for the Lattice ϕ^4 Model

Let us now consider the family of $\{\Sigma_v\}_{v \in \mathbb{R}}$ associated again with the lattice φ^4 model, described by the potential function (7.14) and show how things work in practice.

By computing $\chi(\Sigma_v)$ vs. v according to (7.37), one can probe whether and how the topology of the hypersurfaces Σ_v varies with v. A variation of the Euler characteristic signals a change of topology. However, the converse can be false. For example, odd-dimensional manifolds have vanishing Euler characteristic no matter what their topology is. But the Euler characteristic, as far as a numerical investigation of topology is concerned, seems to be "the only game in town." So, in order to make possible the numerical estimate of the variations of the Euler characteristic, we resort to the Monte Carlo algorithm described above. By means of a Monte Carlo scheme we can estimate only $\int_{\Sigma_v} K_G \, d\sigma / \int_{\Sigma_v} d\sigma$ rather than the total value (7.37) of K_G on Σ_v, hence the need for an estimate of $\omega(v) = \int_{\Sigma_v} d\sigma$ as a function of v. This is achieved by means of formula (7.48), which requires us to compute also the Monte Carlo average $\langle M_1 / \|\nabla V\| \rangle_{\mathrm{MC}}^{\Sigma_v}$. Thus the final outcome of these computations is the *relative* variation of the Euler characteristic as a function of v.

The computation of K_G at any point $x \in \Sigma_v$ proceeds by working out an orthogonal basis for the tangent space at x, orthogonal to $\boldsymbol{\xi} = \nabla V / \|\nabla V\|$,

---

[9] In mathematical textbooks, mean curvature is defined as $-\frac{1}{N} \nabla \left(\frac{\nabla V}{\|\nabla V\|} \right)$.

by means of a Gram–Schmidt orthogonalization procedure. Equation (7.38) is used to compute K_G at x.

In [198], K_G was computed at a number of points on each Σ_v varying between $1 \cdot 10^6$ and $3.5 \cdot 10^6$. The computations were performed for $\dim(\Sigma_v) = 48$ and $= 80$ (i.e., $N = 7 \times 7$, and 9×9) and with the choice $\lambda = 0.6$, $\mu^2 = 2$, $J = 1$ for the parameters of the potential.

In order to test the reliability of the numerical procedure to compute $\chi(\Sigma_v)$, the method is checked against a simplified form of the potential V in (7.14), i.e., with $\lambda = J = 0$, $\mu^2 = -1$. In this case the Σ_v are hyperspheres and therefore $\chi(\mathbb{S}_v^n) = 2$ for any even n. The integral $\int_{\Sigma_v} d\sigma$ is analytically known as a function of the radius $\sqrt{v}$. Therefore, the starting value $\omega(v_0) = \int_{\Sigma_{v_0}} d\sigma$ is known, and in this case we can compute the actual values of $\chi(\Sigma_v)$ instead of their relative variations only. In Figure 7.8 we report $\chi(\Sigma_v = \mathbb{S}_v^n)$ vs. v/N for $N = 5 \times 5$; the results are in agreement with the theoretical value within an error of few percent, a very good precision in view of the large variations of $\chi(\Sigma_v)$ that are found with the full expression (7.14) of V.

In Figure 7.9 we report the results for the 1D lattice, which is known not to undergo any phase transition. Apart from some numerical noise—here enhanced by the more complicated topology of the Σ_v when $\lambda, J \neq 0$—a monotonical (on average) decreasing pattern of $\chi(v/N)$ is found. Since the variation of $\chi(v/N)$ signals a topology change of the $\{\Sigma_v\}$, Figure 7.9 tells that a "smoothly" varying topology is not *sufficient* for the appearance of a phase transition. In fact, when the 2D lattice is considered, the pattern of $\chi(v/N)$ is very different: it displays a rather abrupt *change of the topology variation rate with v/N* at some v_c/N. This result is reported in Fig. 7.10 for a lattice of $N = 7 \times 7$ sites. The question is now whether the value v_c/N, at which $\chi(v/N)$ displays a cusp, has anything to do with the thermodynamic phase transition, i.e., we wonder whether the effective support of the canonical measure shrinks close to $\Sigma_{v \equiv v_c}$ just at $\beta \equiv 1/T_c$, the (inverse) critical temperature of the phase transition. The answer is in the affirmative. In fact, the numerical analysis, already discussed in this chapter, shows that

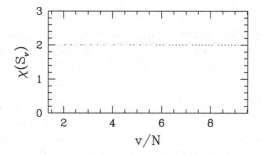

Fig. 7.8. Numerical computation of the Euler characteristic for 24-dimensional spheres. Here v is the squared radius. From [198].

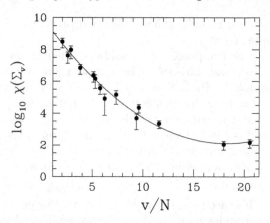

Fig. 7.9. 1D φ^4 model. Relative variations of the Euler characteristic of Σ_v vs. v/N (potential energy density). Lattice of $N = 1 \times 49$ sites. Full line is a guide to the eye. From [198].

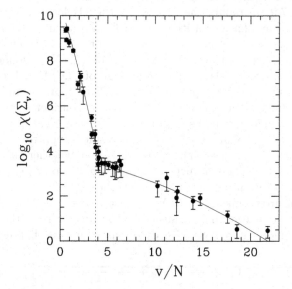

Fig. 7.10. 2D φ^4 model. Relative variations of the Euler characteristic of Σ_v vs. v/N (potential energy density). Lattice of $N = 7 \times 7$ sites. The vertical dotted line corresponds to the phase transition point. Full line is a guide to the eye. From [198].

with $\lambda = 0.6$, $\mu^2 = 2$, $J = 1$, the function $\frac{1}{N}\langle V \rangle(T)$ and its derivative signal the phase transition at $\frac{1}{N}\langle V \rangle \approx 3.75$, a value in very good agreement, within the numerical precision, with v_c/N, where the cusp of $\chi(v/N)$ shows up. We see that a sudden *"second-order variation"* of the topology of these hypersurfaces is the "suitable" topology change—mentioned at the beginning of the present section—that underlies the phase transition of the second kind in the lattice φ^4 model.

In conclusion, through the computation of the v-dependence of a topological invariant, the hypothesis of a deep connection between topological changes of the $\{\Sigma_v\}$ and phase transitions is given *direct* confirmation.

We emphasize, and clarify in the following chapters, that not all topological transitions lead to physical phase transitions, as is clearly shown by the results given above for the 1D version of the φ^4 model. Being certain that not every topological transition corresponds to a phase transition, it seems that on the basis of the above given results, a phase transition corresponds to a supercombination of many simultaneous elementary topological transitions taking place,[10] where many might mean at least exponentially growing with the number of degrees of freedom. It seems therefore more like a supertopologically constructed transition, as will be discussed in Chapter 10.

---

[10] With elementary topological transition we mean any change of topology associated with a single critical point, and thus with the attachment of the corresponding handle. See Appendix C.

Chapter 8

Geometry, Topology and Thermodynamics

In the preceding chapter we have seen that configuration-space topology is suspected to play a significant role in the emergence of phase transition phenomena. We have summarized all the clues in the form of a working hypothesis that we called the *topological hypothesis*. Then this has been given strong support by a direct numerical investigation of the topological changes of configuration space of 1D and 2D lattice φ^4 models. This conjecture stems from the peculiar energy density patterns of the largest Lyapunov exponent at phase transition points. In fact, Lyapunov exponents are closely related to configuration space geometry, which, in turn, can be strongly influenced by topology. However, there is another argument, independent of the Riemannian geometrization of Hamiltonian dynamics, that suggests how to make another link between Lyapunov exponents and topology.

We have already seen that the largest Lyapunov exponent λ_1, for a standard Hamiltonian system, is computed by solving the tangent dynamics equation

$$\frac{d^2 J_i}{dt^2} + \left(\frac{\partial^2 V}{\partial q^i \partial q^j}\right)_{q(t)} J^j = 0 , \tag{8.1}$$

where $q(t) = [q_1(t), \cdots, q_N(t)]$, and then

$$\lambda_1 = \lim_{t \to \infty} \frac{1}{2t} \log \left\{ \Sigma_{i=1}^N [\dot{J}_i^2(t) + J_i^2(t)] / \Sigma_{i=1}^N [\dot{J}_i^2(0) + J_i^2(0)] \right\} .$$

If there are critical points of V in configuration space, that is, points $q_c = [\bar{q}_1, \ldots, \bar{q}_N]$ such that $\nabla V(q)|_{q=q_c} = 0$, according to the *Morse lemma*, see Appendix C, in the neighborhood of any critical point q_c there always exists a coordinate system $\tilde{q}(t) = [\tilde{q}_1(t), \cdots, \tilde{q}_N(t)]$ for which

$$V(\tilde{q}) = V(q_c) - \tilde{q}_1^2 - \cdots - \tilde{q}_k^2 + \tilde{q}_{k+1}^2 + \cdots + \tilde{q}_N^2 , \tag{8.2}$$

where k is the index of the critical point, i.e. the number of negative eigenvalues of the Hessian of V. In the neighborhood of a critical point, (8.2) yields $\partial_{ij}^2 V = \pm \delta_{ij}$, which, substituted into (8.1), gives

229

$$\frac{d^2\mathbf{J}}{dt^2} + \begin{pmatrix} -1 & 0 & \cdots & 0 & 0 \\ 0 & -1 & \cdots & 0 & 0 \\ \vdots & \vdots & \ddots & \vdots & \vdots \\ 0 & 0 & \cdots & 1 & 0 \\ 0 & 0 & \cdots & 0 & 1 \end{pmatrix} \mathbf{J} = 0 \,,$$

where there are k unstable directions that contribute to the exponential growth of the norm of the tangent vector J. This means that the strength of dynamical chaos, measured by the largest Lyapunov exponent λ_1, is affected by the existence of critical points of V. In particular, let us consider the possibility of a sudden variation, with the potential energy v, of the number of critical points (or of their indexes) in configuration space at some value v_c. It is then reasonable to expect that the pattern of $\lambda_1(v)$—as well as that of $\lambda_1(E)$, since $v = v(E)$—will be consequently affected, thus displaying jumps or cusps or other "singular" patterns at v_c (this heuristic argument has been given evidence in the case of the XY mean-field model, see [24] and [206]).

On the other hand, consider a smooth function f, bounded below, such that $f : \mathbb{R}^N \to \mathbb{R}$. Its level sets $\Sigma_u = f^{-1}(u)$ are diffeomorphically transformed one into the other by the flow [23]

$$\frac{dx}{du} = -\frac{\nabla f}{\|\nabla f\|^2} \,,$$

where $x \in \mathbb{R}^N$, i.e., the points of a hypersurface Σ_{u_0} with $u_0 \in [a, b]$, are mapped by this flow to the points of another Σ_{u_1} with $u_1 \in [a, b]$, provided that ∇f never vanishes in the interval $[a, b]$. In other words, if in the interval $[a, b]$ the function f has no critical points, all the level sets $\Sigma_u = f^{-1}(u)$, with $u \in [a, b]$, have the same topology. Conversely, the appearance of critical points of f at some critical value u_c breaks the diffeomorphicity among the $\Sigma_{u<u_c}$ and $\Sigma_{u>u_c}$. This is illustrated by one of the simplest possible examples in Figure 8.1.

Within *Morse theory* a systematic study is developed of the relationship between topological properties of a manifold and the critical points of a suitable class of real-valued functions (Morse functions) defined on it. In particular, if $f \equiv V$, Morse theory tells us that the existence of critical points of V is associated with topological changes of the hypersurfaces $\{\Sigma_v\}_{v\in\mathbb{R}}$, provided that V is a good Morse function (that is, bounded below, with no vanishing eigenvalues of its Hessian matrix).

In conclusion, the existence of critical points of the potential V makes possible a conceptual link between dynamical chaos (measured through Lyapunov exponents) and configuration-space topology.

Taken alone, this argument would not be very strong, because we know that chaos stems also from curvature fluctuations. Thus unless we can disentangle the contribution due to curvature fluctuations from the contribution due to the "scattering" neighborhoods of critical points, we are left with

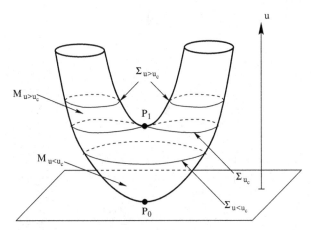

Fig. 8.1. The function f is here the height of a point of the bended cylinder with respect to the ground. In P_1 it is $df = 0$. The level sets below this critical point are circles, whereas above are the union of two circles. The manifolds $M_u = f^{-1}((-\infty, u])$ are disks for $u < u_c$ and cylinders for $u > u_c$.

the doubt of a real involvement of topology in shaping $\lambda(\varepsilon)$. However, taken together with the content of Chapter 7, this argument strengthens the topological hypothesis.

In the present chapter we make a further step forward by working out a quantitative connection between geometry and topology of the energy landscape in phase space, or of the potential energy landscape in configuration space, and thermodynamic entropy. In so doing, we begin to suspect that quite a bit of topological complexity of the energy landscape in phase space must be at the grounds of standard phase transitions, perhaps not so far (at least qualitatively) from what could be at the origin of the various transitions in complex systems such as glasses, spin-glasses, and proteins. In other words, we believe that phenomenologically very different kinds of phase transitions could perhaps fit in a unified topology-based framework.

Remarkably, the way of linking topology and thermodynamics, which is given in what follows, indicates the existence of a common ground where both *microscopic dynamics* and *macroscopic thermodynamics* are rooted.

Finally, the analytic links established between entropy and topological invariants of submanifolds of phase space and of configuration space through equations (8.31), (8.33), (8.35), (8.36) (9.133), and give precious hints for future investigations aiming at clarifying which kind of topological changes can entail a phase transition and of what kind. In the following chapter this problem is referred to as the *sufficiency conditions* for the necessity theorems therein discussed.

We begin with a sketchy presentation of some basic definitions of extrinsic curvatures of hypersurfaces of N-dimensional Euclidean spaces that are necessary for understanding some classic results that link analytic and topological

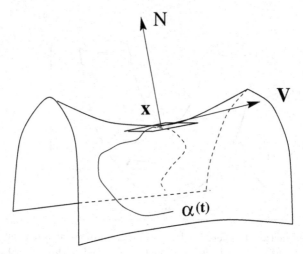

Fig. 8.2. Illustration of the ingredients necessary to construct the shape operator of a surface.

properties of differentiable manifolds, as is the case of the Gauss–Bonnet–Hopf, Chern–Lashof, and Pinkall theorems.

Then we proceed by giving the derivation of the link between entropy and Morse indexes (and Betti numbers) of the constant-energy hypersurfaces in phase space or of the phase space volumes enveloped by them. This derivation eventually relates entropy to topological invariants of configuration-space subsets bounded by equipotential hypersurfaces.

8.1 Extrinsic Curvatures of Hypersurfaces

Let us briefly sketch some basic concepts and definitions concerning the extrinsic geometry of hypersurfaces of a Euclidean space. The starting point is to study the way in which an n-surface Σ curves around in $\mathbb{R}^N$ by measuring the way the normal direction changes as we move from point to point on the surface. The rate of change of the normal direction $\mathbf{N}$ at a point $x \in \Sigma$ in direction $\mathbf{v}$ is described by the *shape operator* (sometimes also called Weingarten's map) $L_x(\mathbf{v}) = -\nabla_{\mathbf{v}}\mathbf{N} = -(\mathbf{v} \cdot \nabla)\mathbf{N}$, where $\mathbf{v}$ is a tangent vector at x and $\nabla_{\mathbf{v}}$ is the directional derivative; gradients and vectors are represented in $\mathbb{R}^N$. For the level sets of a regular function, as is the case of the constant-energy hypersurfaces in the phase space of Hamiltonian systems or of the equipotential hypersurfaces in configuration space, thus generically defined through a regular real-valued function f as $\Sigma_a := f^{-1}(a)$, the normal vector is $\mathbf{N} = \nabla f / \|\nabla f\|$. Let $\{\mathbf{e}_\mu\}_{\mu=1,\dots,N} = \{\mathbf{e}_1,\dots,\mathbf{e}_n,\mathbf{N}\}$, with $\mathbf{e}_\alpha \cdot \mathbf{e}_\beta = \delta_{\alpha,\beta}$. We denote by Greek subscripts, $\alpha = 1,\dots,N$, the components in the embedding space $\mathbb{R}^N$, and with Latin subscripts, $i = 1,\dots,n$, the components on a generic tangent space $T_x\Sigma_a$ at $x \in \Sigma_a$. We are interested in the case of codimension one, that is, $N = n + 1$.

From $\partial_\mu N_\alpha N_\alpha = 0 = 2N_\alpha\partial_\mu N_\alpha$ we see that for any $\mathbf{v}$, $\mathbf{N} \cdot L_x(\mathbf{v}) = -N_\alpha v_\mu \partial_\mu N_\alpha = 0$, which means that $L_x(\mathbf{v})$ projects on the tangent space $T_x\Sigma_a$.

Now the *principal curvatures* $\kappa_1, \ldots, \kappa_n$ of Σ_a at x are the eigenvalues of the shape operator restricted to $T_x\Sigma_a$. Define the matrix $\mathcal{L}_x$ to be the restriction of L_x to $T_x\Sigma_a$

$$\mathcal{L}_{ij}(x) = \mathbf{e}_i \cdot L_x(\mathbf{e}_j) = -(\mathbf{e}_i)_\alpha(\mathbf{e}_j)_\beta\partial_\beta N_\alpha \ ,$$

whence the *mean curvature* is defined as

$$M_1(x) = \frac{1}{n}\mathrm{Tr}^{(n)}\mathcal{L}_{ij}(x) = \frac{1}{n}\sum_{i=1}^{n}\kappa_i \tag{8.3}$$

and the *Gauss–Kronecker curvature* is defined as

$$K_G(x) = \det{}^{(n)}\mathcal{L}_{ij}(x) = \prod_{i=1}^{n}\kappa_i \ . \tag{8.4}$$

Notice that

$$\mathbf{N} \cdot L_x(\mathbf{e}_j) = -N_\alpha(\mathbf{e}_j)_\beta\partial_\beta N_\alpha = 0 \ ,$$
$$\mathbf{N} \cdot L_x(\mathbf{N}) = 0 \ ,$$
$$\mathbf{e}_i \cdot L_x(\mathbf{N}) = -(\mathbf{e}_i)_\alpha N_\beta\partial_\beta N_\alpha \neq 0 \ . \tag{8.5}$$

The explicit computation of the mean curvature M_1 proceeds from

$$M_1(x) = \frac{1}{n}\mathrm{Tr}^{(n)}\mathcal{L}_{ij}(x) = -\frac{1}{n}\sum_{i=1}^{n}(\mathbf{e}_i)_\alpha(\mathbf{e}_i)_\beta\partial_\beta N_\alpha \ . \tag{8.6}$$

Defining $A_{\mu\nu} = (\mathbf{e}_\mu)_\nu$, so that $AA^T = \mathbb{I}$, we have

$$\sum_{i=1}^{n}(\mathbf{e}_i)_\alpha(\mathbf{e}_i)_\beta = \delta_{\alpha\beta} - N_\alpha N_\beta$$

and thus

$$M_1(x) = -\frac{1}{n}(\delta_{\alpha\beta} - N_\alpha N_\beta)\partial_\beta N_\alpha = -\frac{1}{n}\partial_\alpha N_\alpha = -\frac{1}{n}\nabla \cdot \left(\frac{\nabla f}{\|\nabla f\|}\right) \ . \tag{8.7}$$

In order to compute the Gauss–Kronecker curvature, we define

$$B(\mathbf{v}, \mathbf{u}) = -v_\alpha w_\beta\partial_\beta N_\alpha + v_\alpha N_\alpha u_\beta N_\beta \ ,$$

where $\mathbf{v}, \mathbf{u} \in T_x\Sigma_a$, so that $\mathcal{L}_{ij}(x) = B(\mathbf{e}_i, \mathbf{e}_j)$, $B(\mathbf{e}_i, \mathbf{N}) \neq 0$, $B(\mathbf{N}, \mathbf{e}_j) = 0$, $B(\mathbf{N}, \mathbf{N}) = 1$. The matrix elements in the embedding space are $B_{\mu\nu} =$

$B(\mathbf{e}_\mu, \mathbf{e}_\nu)$, and we readily see that $\det^{(n+1)} B = \det^{(n)} L$. Having put $C_{\alpha\beta} = \partial_\beta N_\alpha$, we compute

$$B_{\mu\nu} = B(\mathbf{e}_\mu, \mathbf{e}_\nu) = -(\mathbf{e}_\mu)_\alpha (\mathbf{e}_\nu)_\beta \partial_\beta N_\alpha + \delta_{\mu,n+1}\delta_{\nu,n+1}$$
$$= A_{\mu\alpha} C_{\alpha\beta} A^T_{\beta\nu} + \delta_{\mu,n+1}\delta_{\nu,n+1} \tag{8.8}$$

whence

$$B = -ACA^T + \delta = A(-C + A^T \delta A)A^T \ ,$$

but $(A^T \delta A)_{\alpha\beta} = A_{\mu\alpha}\delta_{\mu,n+1}\delta_{\nu,n+1}A_{\nu\beta} = A_{n+1,\alpha}A_{n+1,\beta} = N_\alpha N_\beta$ and with $B = A(-C + N \otimes N)A^T$ we get $\det^{(n+1)} B = \det^{(n+1)}(-C + N \otimes N) = \det^n \mathcal{L}$, which is the Gauss–Kronecker curvature. Explicitly we obtain

$$K_G(x) = \det\left[-\partial_\beta \frac{\partial_\alpha f}{\|\nabla f\|} + \frac{\partial_\alpha f \partial_\beta f}{\|\nabla f\|^2} \right]$$
$$= \det\left[-\frac{\partial_\alpha \partial_\beta f}{\|\nabla f\|} + \frac{\partial_\alpha f \partial_\beta \partial_\mu f \partial_\mu f}{\|\nabla f\|^3} + \frac{\partial_\alpha f \partial_\beta f}{\|\nabla f\|^2} \right] . \tag{8.9}$$

This explicit expression for the Gauss–Kronecker curvature of a hypersurface $\Sigma_a = f^{-1}(a)$ is of prospective practical interest in the numerical computation of one among its topologic invariants, the Euler characteristic.

In fact, according to the Gauss-Bonnet–Hopf theorem, the following equation holds:

$$\int_{\Sigma_a} d\sigma \ K_G = \frac{1}{2}\mathrm{vol}(\mathbb{S}_1^{N-1}) \ \chi(\Sigma_a) \ , \tag{8.10}$$

where $\mathbb{S}_1^{N-1}$ is an $(N-1)$-dimensional hypersphere of unit radius, and $\chi(\Sigma_a)$ is the Euler characteristic of the level set Σ_a of the function f (hypersurface of $\mathbb{R}^N$). A slight modification of this formula yields the following nontrivial result due to Chern and Lashof [207]:

$$\int_{\Sigma_a} d\sigma \ |K_G| = \frac{1}{2}\mathrm{vol}(\mathbb{S}_1^{N-1}) \sum_{i=0}^{N-1} \mu_i(\Sigma_a) \geq \frac{1}{2}\mathrm{vol}(\mathbb{S}_1^{N-1}) \sum_{i=0}^{N-1} b_i(\Sigma_a) \ , \tag{8.11}$$

and $\mu_i(\Sigma_a)$ are the Morse indexes of Σ_a, which are defined as the number μ of critical points of index i on a given level set Σ_a; a critical point is a point where $\nabla f = 0$; the index i of a critical point is the number of negative eigenvalues of the Hessian of f computed at the critical point. The $b_i(\Sigma_a)$ are the Betti numbers of the hypersurfaces. The Betti numbers are fundamental *topological invariants* (see Appendix A) of differentiable manifolds; they are the diffeomorphism-invariant dimensions of suitable vector spaces (the de Rham cohomology spaces); thus they are integers (see Appendix A).

Another interesting connection between curvature and Betti numbers is provided by Pinkall's inequality [208]. Let M^n be a compact smooth manifold,

$\phi : M^n \to \mathbb{R}^m$ an immersion, and $N(\phi)$ the unit normal bundle of ϕ. For every $\xi \in N_p(\phi)$ there is a shape operator $A_\xi : T_p M^n \to T_p M^n$ whose eigenvalues $\kappa_1(\xi), \ldots, \kappa_n(\xi)$ are the principal curvatures at ξ. Denoting by

$$\sigma(A_\xi)^2 = \frac{1}{n^2} \sum_{i<j} (\kappa_i - \kappa_j)^2$$

the dispersion of these principal curvatures, then

$$\frac{1}{\text{vol}(\mathbb{S}^m)} \int_{N(\phi)} [\sigma(A_\xi)]^n \, d\xi \geq \sum_{i=1}^{n-1} \left(\frac{i}{n-i}\right)^{n/2-i} b_i(M^n) , \qquad (8.12)$$

where b_i are the Betti numbers of the manifold M^n.

8.1.1 Two Useful Derivation Formulas

In several places, throughout the present book we use the derivation formula that is derived below.

For a regular and measurable function $f(x)$, defined in a bounded subset M_s of $\mathbb{R}^N$, and given a regular function $\psi(x)$ defined in M_s, the following derivation formula holds:

$$\frac{d}{ds} \int_{M_s} f(x) \, d\mu(x) = \int_{\psi(x)=s} f(x) \frac{d\sigma}{\|\nabla \psi(x)\|} , \qquad (8.13)$$

where $d\mu(x)$ is a volume element of M_s, $d\sigma$ a surface element of the level set $\Sigma_s = \{x = (x_1, \ldots, x_N) \in \mathbb{R}^N | \psi(x) = s\}$, and

$$\|\nabla \psi(x)\| = \left[\sum_{i=1}^N (\partial_{x_i} \psi)^2\right]^{1/2} .$$

Then we put $\int_{M_s} f(x) \, d\mu(x) = \int_{M_s} f(x) \, d\sigma dn$, where dn is the infinitesimal element of the outward normal to $d\sigma$ whose projections are $dx_i = dn \cos(n, x_i)$, $1 \leq i \leq N$, so that an infinitesmal variation of the level value s can be expressed as

$$ds = \sum_{i=1}^N (\partial_{x_i} \psi) \, dx_i = dn \sum_{i=1}^N (\partial_{x_i} \psi) \cos(n, x_i) ,$$

and from the relation $\cos(n, x_i) = (\partial_{x_i} \psi)/\|\nabla \psi\|$ we get

$$ds = dn \sum_{i=1}^N \frac{(\partial_{x_i} \psi)^2}{\|\nabla \psi(x)\|} = dn \|\nabla \psi(x)\| .$$

Finally,

$$\int_{M_s} f(x)\, d\mu(x) = \int_{M_s} f(x)\, \frac{d\sigma\, ds}{\|\nabla\psi(x)\|} = \int_0^s d\eta \int_{\psi(x)=\eta} f(x)\, \frac{d\sigma}{\|\nabla\psi(x)\|} \;,$$

whence, by deriving both sides of the equation above with respect to s, the formula (8.13) immediately follows.

Both in the present and in the next chapter, we crucially resort to a differentiation formula [209] of geometric integrals on level sets with respect to the level, which generalizes the so-called Federer's derivation formula. This is given within the enunciation of a theorem in the proof of Lemma 9.20 in Section 9.3. Here we show how it is obtained.

Given two real-valued smooth functions, $g(x)$ and $\psi(x)$, defined on a bounded subset of $\mathbb{R}^N$, and given

$$F(s) = \int_{\psi(x)=s} g(x)\, d\mu^{N-1} \;,$$

where $d\mu^{N-1}$ is the measure on the level set (which can even be a Hausdorff measure), if there exists a constant $C > 0$ such that $\|\psi\| \geq C$, i.e., ψ has no critical points, then

$$\frac{d^k}{ds^k} \int_{\psi(x)=s} g(x)\, d\mu^{N-1} = \int_{\psi(x)=s} A^k\, g(x)\, d\mu^{N-1} \;, \tag{8.14}$$

where A^k stands for k iterates of the operator defined by

$$Ag = \nabla \cdot \left(g\frac{\nabla\psi}{\|\nabla\psi\|} \right) \frac{1}{\|\nabla\psi\|} \;.$$

Since $\|\nabla\psi\| \geq C > 0$ and ψ is smooth, the implicit function theorem implies that the level sets $\{\psi(x) = s\}$ are nested closed $(N-1)$-dimensional surfaces representable in local coordinates as the graph of a smooth function. Assume that $\nabla\psi$ points toward the inside of these surfaces. Then form the quotient

$$\frac{F(s+h) - F(s)}{h} = \frac{1}{h}\left\{ \int_{\psi(x)=s+h} g(x)\, d\mu^{N-1} - \int_{\psi(x)=s} g(x)\, d\mu^{N-1} \right\} \;.$$

Let $n(x) = \nabla\psi(x)/\|\nabla\psi(x)\|$, the unit inner normal to $\{\psi(x) = s\}$ at x, and since $g(x) = g(x)(\nabla\psi/\|\nabla\psi\|) \cdot n(x)$, we have

$$\frac{F(s+h) - F(s)}{h} \tag{8.15}$$

$$= \frac{1}{h} \int_{\psi(x)=s+h} g(x)\, \frac{\nabla\psi}{\|\nabla\psi\|} \cdot n\, d\mu^{N-1} - \frac{1}{h} \int_{\psi(x)=s} g(x)\, \frac{\nabla\psi}{\|\nabla\psi\|} \cdot n\, d\mu^{N-1}$$

$$= \frac{1}{h} \int_{\psi(x)=s+h} g(x)\, \frac{\nabla\psi}{\|\nabla\psi\|} \cdot n_e\, d\mu^{N-1} + \frac{1}{h} \int_{\psi(x)=s} g(x)\, \frac{\nabla\psi}{\|\nabla\psi\|} \cdot n_e\, d\mu^{N-1} \;,$$

where n_e is the unit normal that now points toward the exterior (only one is reversed in sign) of the annular region $\{s < \psi < s+h\}$, so that since $\{\psi = s\}$ are smooth closed surfaces and $g(\nabla\psi/\|\nabla\psi\|)$ is smooth, we can apply the divergence theorem to get

$$\frac{F(s+h) - F(s)}{h} = \frac{1}{h} \int_{s<\psi<s+h} \nabla \cdot \left(g \frac{\nabla\psi}{\|\nabla\psi\|} \right) dx .$$

Now rewrite the right hand-side of the equation above using the standard coarea formula, that is,

$$\frac{F(s+h) - F(s)}{h} = \frac{1}{h} \int_s^{s+h} dr \int_{\psi(x)=r} \nabla \cdot \left(g \frac{\nabla\psi}{\|\nabla\psi\|} \right) \frac{1}{\|\nabla\psi\|} d\mu^{N-1} , \quad (8.16)$$

which, in the limit $h \to 0$, obviously gives

$$\frac{dF}{ds}(s) = \int_{\psi(x)=s} \nabla \cdot \left(g \frac{\nabla\psi}{\|\nabla\psi\|} \right) \frac{1}{\|\nabla\psi\|} d\mu^{N-1} , \quad (8.17)$$

which is the Federer derivation formula.

8.2 Geometry, Topology and Thermodynamics

Consider a generic classical system described by a standard Hamiltonian

$$H = \frac{1}{2} \sum_{i=1}^n p_i^2 + V(q) , \quad (8.18)$$

where $q = (q_1, \ldots, q_n)$ and the symbols have standard meaning. We assume that $V(q)$ is such that H is a good Morse function (the reason is given in the remark at the end of this section). Then consider the microcanonical definition of entropy

$$S = \frac{k_B}{N} \log \Omega_N(E) , \quad (8.19)$$

where $N = 2n - 1$ and

$$\Omega_N(E) = \frac{1}{N!} \int_{\Sigma_E} \frac{d\sigma}{\|\nabla H\|}, \quad (8.20)$$

with $\|\nabla H\| = \{\sum_i p_i^2 + [\nabla_i V(q)]^2\}^{1/2}$. Here Σ_E is the constant-energy hypersurface in the $2n$-dimensional phase space Γ corresponding to the total energy E, that is, $\Sigma_E = \{(p_1, \ldots, p_n, q_1, \ldots, q_n) \in \Gamma | H(p,q) = E\}$.

The above given Federer's derivation formula now reads

$$\frac{d^k}{dE^k} \int_{\Sigma_E} \alpha \, d\sigma = \int_{\Sigma_E} A^k(\alpha) \, d\sigma , \quad (8.21)$$

where α is an integrable function and the operator A is

$$A(\alpha) = \frac{\nabla}{\|\nabla H\|} \cdot \left(\alpha \cdot \frac{\nabla H}{\|\nabla H\|} \right) .$$

Using this formula leads to the following result:

$$\frac{d\Omega_N(E)}{dE} = \frac{1}{N!} \int_{\Sigma_E} \frac{d\sigma}{\|\nabla H\|} \frac{M_1^\star}{\|\nabla H\|} + \mathcal{O}\left(\frac{1}{N} \right) , \qquad (8.22)$$

where $M_1^\star = \nabla(\nabla H/\|\nabla H\|)$, and $M_1^\star$ is directly proportional to the mean curvature M_1 of Σ_E seen as a submanifold of $\mathbb{R}^{2n}$ [126] according to the simple relation $M_1 = -M_1^\star/(2n-1)$. By integrating equation (8.22), we obtain

$$\Omega_N(E) = \frac{1}{N!} \int_0^E d\eta \int_{\Sigma_\eta} \frac{d\sigma}{\|\nabla H\|} \frac{M_1^\star}{\|\nabla H\|} = \frac{1}{N!} \int_{M_E} d\mu \frac{M_1^\star}{\|\nabla H\|} , \qquad (8.23)$$

and then, at large N, considering that the volume measure $d\mu$ concentrates on the boundary Σ_E, we write

$$\frac{1}{N!} \int_{M_E} d\mu \frac{M_1^\star}{\|\nabla H\|} \approx \frac{(\delta E)}{N!} \int_{\Sigma_E} \frac{d\sigma}{\|\nabla H\|} \frac{M_1^\star}{\|\nabla H\|}$$

$$\approx \frac{(\delta E)}{N!} \langle \|\nabla H\|^{-1} \rangle \int_{\Sigma_E} \frac{d\sigma}{\|\nabla H\|} M_1^\star , \qquad (8.24)$$

where in the last approximate replacement, we have used that $\|\nabla H\|$ is positive and only very weakly varying at large N.

By means of Hölder's inequality for integrals we get

$$\int_{\Sigma_E} \frac{d\sigma}{\|\nabla H\|} M_1^\star \leq \left(\int_{\Sigma_E} \frac{d\sigma}{\|\nabla H\|} |M_1^\star|^N \right)^{\frac{1}{N}} \left(\int_{\Sigma_E} \frac{d\sigma}{\|\nabla H\|} \right)^{\frac{N-1}{N}} ,$$

the sign of equality being better approached when $M_1^\star$ is everywhere positive. Hence, using (8.20), (8.23), and (8.24),

$$\Omega_N(E) \leq [\Omega_\nu(E)]^{\frac{N-1}{N}} \left(\frac{1}{N!} \int_{\Sigma_E} \frac{d\sigma}{\|\nabla H\|} |M_1^\star|^N \right)^{\frac{1}{N}} \frac{\delta E}{\langle \|\nabla H\| \rangle} , \qquad (8.25)$$

and introducing a suitable deformation factor $d(E)$ we can write

$$[\Omega_N(E)]^{\frac{1}{N}} = \frac{d(E)\delta E}{\langle \|\nabla H\| \rangle} \left(\frac{1}{N!} \int_{\Sigma_E} \frac{d\sigma}{\|\nabla H\|} |M_1^\star|^N \right)^{\frac{1}{N}} , \qquad (8.26)$$

so that

$$\Omega_N(E) = \frac{[d(E)]^N (\delta E)^N}{\langle \|\nabla H\| \rangle^{2n}} \frac{1}{N!} \int_{\Sigma_E} \frac{d\sigma}{\|\nabla H\|} |M_1^\star|^N . \qquad (8.27)$$

Let us now remember the expression of a multinomial expansion ($\rho \in \mathbb{N}$)

$$(x_1 + \cdots + x_\rho)^\rho = \sum_{\{n_i\}, \sum n_k = \rho} \frac{\rho!}{n_1! \cdots n_\rho!} \cdot x_1^{n_1} \cdots x_\rho^{n_\rho} \,, \tag{8.28}$$

and use it by identifying the x_i with $\kappa_1, \ldots, \kappa_N$, the principal curvatures of Σ_E. Put $\epsilon_0 = \text{sign}(\kappa_1 + \ldots + \kappa_N)$, so that $(\kappa_1 + \kappa_2 + \cdots + \kappa_N)^N = [\epsilon_0 \,|(\kappa_1 + \kappa_2 + \cdots + \kappa_N)| \,]^N = (\epsilon_0)^N |M_1^\star|^N$. Then, using (8.28), we can write $(M_1^\star)^N = (\kappa_1 + \kappa_2 + \cdots + \kappa_N)^N = N! \, K_G + \mathcal{R}(E)$, where $\mathcal{R}(E)$ contains all the terms of the expansion (8.28) but that one with $n_1 = n_2 = \ldots = n_\rho = 1$; K_G is the Gauss–Kronecker curvature of Σ_E, $K_G = \prod_{i=1}^{N} \kappa_i$. Now put $\epsilon = \text{sign}(K_G)$ and $\epsilon' = \text{sign}[\mathcal{R}(E)]$ so that it is

$$\left|(\epsilon_0)^N |M_1^\star|^N\right| = |M_1^\star|^N = \left| \, N! \, \epsilon \, |K_G| + \epsilon' \, |\mathcal{R}(E)| \, \right|$$

that can be rewritten as

$$|M_1^\star|^N = \left| \, N! \, |K_G| + \frac{\epsilon'}{\epsilon} |\mathcal{R}(E)| \, \right|$$

where $\epsilon = \epsilon(x, E)$ and $\epsilon' = \epsilon'(x, E)$—as well as $|K_G|$ and $|\mathcal{R}(E)|$—are functions of the point on Σ_E and of the energy. By replacing $\epsilon'(x, E)/\epsilon(x, E)$ by its average on Σ_E, let us denote it by $w(E)$, and by putting $\widetilde{\mathcal{R}}(E) = w(E)|\mathcal{R}(E)|$, the following approximate relation follows[1]

$$|M_1^\star|^N \approx N! \, |K_G| + \widetilde{\mathcal{R}}(E)$$

so that (8.27) now reads

$$\Omega_N(E) \approx \frac{[d(E)]^N (\delta E)^N}{\langle \|\nabla H\| \rangle^{2n}} \int_{\Sigma_E} d\sigma \left(|K_G| + \frac{\widetilde{\mathcal{R}}(E)}{N!} \right) . \tag{8.29}$$

Again we have used that $\|\nabla H\|$ is only very weakly varying at large N and that it is always positive.

---

[1] Here we are implicitly assuming either that $w(E)$ is positive, or that, if negative, $w(E)$ is sufficiently small so that $N!|K_G|$ is larger than $w(E)|\mathcal{R}(E)|$ practically everywhere. Otherwise we should consider $| \, N!(\epsilon/\epsilon')|K_G| + |\mathcal{R}(E)| \, |$ and proceed by taking into account both possibilities; however, for the aims of the present computation, this would not substantially change the final result. Just to give a rough idea of the reason why $w(E)$ could be supposed to be "sufficiently small", consider that $\int_{\Sigma_E} d\sigma \{N!\epsilon|K_G| + \epsilon'|\mathcal{R}(E)|\}$ has to be positive because it is proportional to the volume $\Omega_N(E)$, and that $\int_{\Sigma_E} d\sigma \epsilon|K_G| = 0$ because it is proportional to the Euler characteristic of Σ_E which is always vanishing since the Σ_E are odd-dimensional. Thus, a-priori, ϵ could assume the values $+1$ and -1 with almost the same frequency on Σ_E, whereas ϵ' would a-priori assume most frequently the value $+1$, thus the ratio ϵ'/ϵ could average on Σ_E to small values.

According to the Chern–Lashof theorem (8.11),

$$\int_{\Sigma_E} d\sigma \, |K_G| = \frac{1}{2} \text{vol}(\mathbb{S}_1^{2n-1}) \sum_{i=0}^{2n-1} \mu_i(\Sigma_E) \, , \tag{8.30}$$

where $\mu_i(\Sigma_E)$ are the Morse indexes of Σ_E.

Therefore, using the above relation in (8.29), the entropy per degree of freedom can be written as

$$S(E) = \frac{k_B}{N} \log \Omega_N(E)$$

$$\approx \frac{1}{N} \log \left[\text{vol}(\mathbb{S}_1^{2n-1}) \sum_{i=0}^{2n-1} \mu_i(\Sigma_E) + \int_{\Sigma_E} d\sigma \frac{\widetilde{\mathcal{R}}(E)}{N!} \right]$$

$$+ \frac{1}{N} \log \frac{[d(E)]^N (\delta E)^N}{\langle \|\nabla H\| \rangle^{2n}} \, . \tag{8.31}$$

The meaning of (8.31) is better understood if we consider that the Morse indexes $\mu_i(M)$ of a differentiable manifold M are related to the Betti numbers $b_i(M)$ of the same manifold by the inequalities

$$\mu_i(M) \geq b_i(M) \, . \tag{8.32}$$

At large dimension we can safely replace (8.32) with $\mu_i(M) \approx b_i(M)$ [206].

Equation (8.31), rewritten as

$$S(E) \approx \frac{k_B}{N} \log \left[\text{vol}(\mathbb{S}_1^{2n-1}) \sum_{i=0}^{2n-1} b_i(\Sigma_E) + \int_{\Sigma_E} d\sigma \frac{\widetilde{\mathcal{R}}(E)}{N!} \right]$$

$$+ \frac{1}{N} \log \frac{[d(E)]^N (\delta E)^N}{\langle \|\nabla H\| \rangle^{2n}} \, , \tag{8.33}$$

links topological properties of the *microscopic* phase space with the *macroscopic* thermodynamic potential $S(E)$.

In particular, even though the function $\widetilde{\mathcal{R}}(E)$ is unknown, sudden changes of the topology of the hypersurfaces Σ_E (reflected by the energy variation of $\sum b_i(\Sigma_E)$) necessarily affect the energy variation of the entropy.

Finally, we resort to the fact that at large N, the volume measure of Σ_E concentrates on $\Sigma_E^* = \mathbb{S}_{\langle 2K \rangle^{1/2}}^{n-1} \times M_{v=\langle V \rangle}$, where $\mathbb{S}_{\langle 2K \rangle^{1/2}}^{n-1} = \{(p_1, \ldots, p_n) | \sum p_i^2 = \langle 2K \rangle \}$ and $M_{v=\langle V \rangle} = \{(q_1, \ldots, q_n) | V(q) \leq \langle V \rangle \}$, so that Σ_E can be approximated by this product manifold, and we resort to the Kunneth formula (see Appendix A) for the Betti numbers of a product manifold $A \times B$, i.e.,

$$b_i(A \times B) = \sum_{j+k=i} b_j(A) b_k(B) \, , \tag{8.34}$$

which, applied to $\mathbb{S}^{n-1}_{\langle 2K \rangle^{1/2}} \times M_{\langle V \rangle}$, gives $b_i(\Sigma_E^\star) = b_0(\mathbb{S}^{n-1}_{\langle 2K \rangle^{1/2}})b_i(M_v) + b_{n-1}$
$(\mathbb{S}^{n-1}_{\langle 2K \rangle^{1/2}})b_{i-n+1}(M_v)$ since all the Betti numbers of a $(n-1)$-dimensional
hypersphere vanish except for b_0 and b_{n-1}, which are equal to 1 (see
Appendix A). This entails that $b_i(\Sigma_E^\star) = b_i(M_v)$ for $i = 1, \ldots, n-1$,
and $b_i(\Sigma_E^\star) = b_{i-n+1}(M_v)$ for $i = n, \ldots, 2n-1$.

Eventually we obtain

$$S(v) \approx \frac{k_B}{N} \log \left[vol(\mathbb{S}^{2n-1}_1) \left(b_0 + \sum_{i=1}^{n-1} 2b_i(M_v) + b_n \right) + \overline{\mathcal{R}}(E(v)) \right]$$
$$+ \frac{1}{N} \log \frac{[d(E(v))]^N (\delta E)^N}{\langle \| \nabla H \| \rangle^{2n}}, \tag{8.35}$$

where $\overline{\mathcal{R}}(E(v))$ stands for the integral on the product manifold of the remainder that appears in (8.33).

From this equation we see that a fundamental topological quantity, the sum of the Betti numbers of the submanifolds $M_v = \{(q_1, \ldots, q_n) \in \mathbb{R}^n | V(q) \leq v\}$ of configuration space, is related, although with some approximation, to the thermodynamic entropy of the system.

Note that we derived the equation above under the assumption that $\mu_i(M) \approx b_i(M)$, so we can also rewrite it as

$$S(v) \approx \frac{k_B}{N} \log \left[vol(\mathbb{S}^{2n-1}_1) \left(\mu_0 + \sum_{i=1}^{n-1} 2\mu_i(M_v) + \mu_n \right) + \overline{\mathcal{R}}(E(v)) \right] + r(E(v)),$$
$$\tag{8.36}$$

with an obvious meaning for $r(E(v))$.

An important remark. Simple inspection of the formulas in (8.36) and (8.35) shows an apparently serious problem: $S(v)$, which has to be smooth at any finite N, is a function of the sum of a real-valued function, $\widetilde{\mathcal{R}}(E(v))$, and an integer-valued function, the sum of Betti numbers or the Morse indexes, and thus it seems to have discontinuous jumps whenever the b_i or the μ_i jump. Moreover, we derived the above formulas using Federer's derivation formula, obviously—in view of the final result—with the implicit assumption of the existence of critical points of the Hamiltonian, thus contradicting the validity hypotheses of Federer's derivation formula. However, if the Hamiltonian is a Morse function (not a very restrictive condition at all) we know after Sard's theorem [210] that the ensemble of critical values, here of the energy, is a point set. Therefore, any finite interval of energy values is the union of a finite number of open sets where no critical energy value—and thus no critical point on the Σ_E—is present. On all these open sets, free of critical points, Federer's derivation formula can be legally applied. As a consequence, (8.36) and (8.35) are valid on the disjoint union of all these open sets of values for the energy. In other words, we have provided a piecewise approximation (with possibly a huge number of small discontinuous jumps) of a smooth function. In conclusion, from a rigorous standpoint we should define a smooth envelope of the expression given in (8.36), but in practice this would not change the situation very much, so one can use these formulas keeping in mind the present remark.

From (8.36) it appears intuitively reasonable that "mild" variations of the v-patterns of $\sum_{i=1}^{n-1} \mu_i(M_v)$ can exist in the absence of a phase transition. In other words, "mild" variations of the topology of the M_v can leave bounded the v-derivatives of the entropy. This is in qualitative agreement with the "monotonic" pattern of $\chi(\Sigma_v)$ numerically found for the 1D lattice φ^4 model and reported in Chapter 7. We can already surmise that some appropriate jump in the v-patterns of $\sum_{i=1}^{n-1} \mu_i(M_v)$—thus appropriate "strong" topological changes of the M_v with a suitable n-dependence—have to be sufficient to entail the appearance of a phase transition.

As we shall see in Chapter 10, there are models for which the $\mu_i(M_v)$ can be computed analytically, thus making of (8.36) a crucial constructive result to relate topology changes of configuration space with phase transitions. More generally, (8.36) has a great theoretical relevance since it relates thermodynamics with the *energy landscape* topology in configuration space, unifying the treatment of "simple" and complex systems. We shall comment more on this point in Chapter 11.

8.2.1 An Alternative Derivation

Let us anticipate here that another derivation of an analytic relationship between entropy and topology can be worked out following a different strategy that leads to an exact formula, proved under Theorem 9.2 in the following chapter. It is obtained again for standard Hamiltonian systems, starting from the following equivalent microcanonical definition of the entropy $S^{(-)}(E)$,

$$S^{(-)}(E) = \frac{k_B}{N} \log M(E, N) , \qquad (8.37)$$

where

$$M(E, N) = \text{vol}(M_E) = \frac{1}{N!} \int_{H(p,q) \leq E} d^n p \, d^n q, \qquad (8.38)$$

where $N = 2n$ and M_E is the subset of the $2n$-dimensional phase space Γ bounded by the constant-energy hypersurface Σ_E, that is $M_E = \{x \equiv (p_1, \ldots, p_n, q_1, \ldots, q_n) \in \mathbb{R}^{2n} | H(p, q) \leq E\}$.

By splitting the phase space volume into the union of suitably defined neighborhoods $\Gamma(x_c^{(i)})$ of the critical points $x_c^{(i)}$ of the Hamiltonian plus its complement, it is found that the thermodynamic entropy can be written in the form

$$S_N^{(-)}(E) = \frac{1}{N} \log \left[\int_{M_E \backslash \bigcup_{i=1}^{\mathcal{N}(E)} \Gamma(x_c^{(i)})} d^n p \, d^n q + \sum_{i=0}^{N} w_i \, \mu_i(M_E) + R(N, E) \right] ,$$

where the first term in square brackets is the phase space subset M_E minus the union of the mentioned neighborhoods of critical points, the second term is a weighed sum of the Morse indexes of M_E, and the third term, $R(N, E)$, is a smooth function. The clue point, which is exploited to derive such a formula, is the existence of a universal parametrization (through the Morse chart) of the level sets Σ_E, which holds in small neighborhoods of the critical points. Of course, given any function, here the phase space volume, we can always add and subtract to it another arbitrary function, but this is useless, unless we add some extra information. In the next chapter it is proved that the first term in square brackets cannot be at the origin of an unbound growth with N of any derivative of the entropy; thus in particular, singular behaviors of thermodynamic observables can be originated only by the second term of topological meaning.

As we shall see in the next chapter, this formula is in very good qualitative agreement with the one given in (8.36), and suggests that the term $\widetilde{\mathcal{R}}(E(v))$, which we called remainder and which enters (8.36), does not entail singular behaviors of thermodynamic observables, an interesting mutual support of the two results indeed.

Chapter 9

Phase Transitions and Topology:
Necessity Theorems

In the preceding chapters, we discussed the conceptual development that, starting from the Riemannian theory of Hamiltonian chaos, led us first to conjecture the involvement of topology in phase transition phenomena— formulating what we called the *topological hypothesis*—and then provided both indirect and direct numerical evidence of this conjecture. The present chapter contains a major leap forward: the rigorous proof that topological changes of equipotential hypersurfaces of configuration space—and of the regions of configuration space bounded by them—are a *necessary* condition for the appearance of thermodynamic phase transitions. This is obtained for a wide class of potential functions of physical relevance, and for first- and second-order phase transitions. However, long-range interactions, nonsmooth potentials, unbound configuration spaces, "exotic" and higher-order phase transitions, are not encompassed by the theorems given below and are still open problems deserving further work. For this reason, and mainly because we do not yet know precisely what kinds of topological changes entail a phase transition, we give in what follows the details of the proofs, making the presentation of the content of this chapter rather formal. We deem it useful to provide these details in order, we hope, to inspire and stimulate the interested reader to cope with these challenging tasks. On the other hand, the presentation is organized so as to facilitate the reader, if so inclined, to grasp only the main results and the ideas behind them, and to skip the details.

As we have recalled in Chapter 2, the central task of the mathematical theory of phase transitions has been to prove the loss of differentiability of the pressure function—or of other thermodynamic functions—with respect to temperature, or volume, or an external field. The first rigorous results of this kind are the exact solution of the 2D Ising model due to Onsager [211], and the Yang–Lee theorem [36,38] showing that despite the smoothness of the canonical and grand-canonical partition functions, in the $N \to \infty$ limit piecewise differentiability of pressure or other thermodynamic functions also becomes possible. In the canonical ensemble, after the introduction of the concept of a Gibbs measure for infinite systems by Dobrushin, Lanford, and Ruelle,

the phenomenon of phase transition is seen as the consequence of nonunique-
ness of a Gibbs measure for a given type of interaction among the particles of
a system [212,213]. In these standard approaches, a phase transition is seen as
stemming from singular properties of the statistical measures, whereas in the
present chapter, we show that these singularities are not "primitive" phenomena but are induced from a deeper level, that of configuration-space topology.
In other words, once the microscopic interaction potential is given, the information about the existence of a phase transition is already contained in the
topology of its level sets, *prior to and independently of* the definition of any
statistical measure.

Here we prove the topological hypothesis by proving the following theorems
[195, 214–217]:

Theorem 9.1 (Regularity under diffeomorphicity). *Let* $V_N(q_1, \ldots, q_N)$:
$\mathbb{R}^N \rightarrow \mathbb{R}$, *be a smooth, nonsingular, finite-range potential. Denote by*
$\Sigma_v := V_N^{-1}(v)$, $v \in \mathbb{R}$, *its* level sets, *or* equipotential hypersurfaces, *in*
configuration space.

Then let $\bar{v} = v/N$ *be the potential energy per degree of freedom.*

If for any pair of values $\bar{v}$ *and* $\bar{v}'$ *belonging to a given interval* $I_{\bar{v}} = [\bar{v}_0, \bar{v}_1]$
and for any $N > N_0$, *we have*

$$\Sigma_{N\bar{v}} \approx \Sigma_{N\bar{v}'}$$

that is, $\Sigma_{N\bar{v}}$ *is diffeomorphic to* $\Sigma_{N\bar{v}'}$, *then the sequence of the Helmholtz free*
energies $\{F_N(\beta)\}_{N \in \mathbb{N}}$—*where* $\beta = 1/T$ *(T is the temperature) and* $\beta \in I_\beta =$
$(\beta(\bar{v}_0), \beta(\bar{v}_1))$—*is uniformly convergent at least in* $C^2(I_\beta)$, *so that* $F_\infty \in C^2(I_\beta)$
and neither first- nor second-order phase transitions can occur in the (inverse)
temperature interval $(\beta(\bar{v}_0), \beta(\bar{v}_1))$.

This is our first theorem, formulated precisely in Section 9.2. In general,
given a model described by a smooth, nonsingular, finite-range potential,
it is a hard task to locate all its critical points and thus to ascertain
whether Theorem 9.1 actually applies to it. Moreover, the requirement of
the existence—at any N—of an energy density interval $[\bar{v}_0, \bar{v}_1]$ free of critical values seems rather strong. There are systems for which this condition is
satisfied, as is the case of the Peyrard–Bishop model defined in Section 10.6
(though this model has other problems). However, this does not seem to be
the generic case. In fact, consider, for example, the 1D XY model. As we
shall see in Chapter 10, there is a minimum energetic cost Δv to pass from
one critical value of the potential to the next one. Any configuration with
a random distribution of rotators with angles $\pm\pi$ ("spins" up and "spins"
down) corresponds to a critical point of the potential. Any "spin" flip from
such a configuration requires an energy Δv, and this is just the distance in
energy between two successive critical values of the potential. At increasing
N the quantity Δv remains the same and so does the length of an interval
$[v_c^j, v_c^{j+1}] \equiv [v_c^j, v_c^j + \Delta v]$, whereas $[\bar{v}_c^j, \bar{v}_c^{j+1}] \equiv [v_c^j/N, v_c^j/N]$ obviously shrinks
with N. This seems a generic situation for lattices and, in general, for short-
range interactions.

Therefore, if at large N the critical values of the potential densely crowd on the axis of potential energy values, then the claim that a phase transition point must correspond to a critical value of the potential is trivially satisfied and cannot be given the meaning that phase transitions have a topological origin. This notwithstanding, Theorem 9.1 is very useful and *crucial* to prove Theorem 9.2 which establishes that the occurrence of a phase transition is *necessarily* driven by topological changes in configuration space. To do this we have to consider what happens to the entropy when a critical value of the potential is crossed. Taking just *one* critical value v_c of the potential, and allowing an arbitrary growth with N of the number of critical points on Σ_{v_c}, one can see that it is the energy variation of the volume only in the vicinity of critical points that can entail an unbounded growth with N of the third- or fourth-order derivative of the entropy. In other words, the breaking of uniform convergence of the entropy in $\mathcal{C}^3$ or in $\mathcal{C}^2$ can be originated *only* by a topological change of the Σ_v or, equivalently, of the M_v. To rule out any role—in the breaking of uniform convergence—of the part of configuration space volume which is free of critical points, one resorts to Theorem 9.1.

Theorem 9.2 applies to all those systems whose potential is a good Morse function.[1] But are there systems with only one critical value in an interval $[\bar{v}_0, \bar{v}_1]$? At present we can conjecture that the result expressed by Theorem 9.2 extends at least to those potential functions for which the number of critical values $\bar{v}_c^j$ contained in $[\bar{v}_0, \bar{v}_1]$ grows at most linearly with N (thus encompassing a wide class of short-range interaction potentials).[2] The basic case of only one critical value has a great *conceptual* meaning: it allows a direct proof of the role of critical points. Once we have proved that phase transitions can stem only from the neighborhoods of critical points in the *ideal* case of one v_c in $[\bar{v}_0, \bar{v}_1]$, we can hardly imagine how the part of configuration space volume which is free of critical points could start playing any role when the number of critical values in the interval is let grow. This appears particularly reasonable considering the relation (9.109) between entropy and Morse indexes (see also below, Theorem 9.2). This relation is valid in general and is in very good qualitative agreement with a previous (approximate) derivation of a similar relation between entropy and Morse indexes reported in (8.36).[3] Now, forgetting for a moment Theorem 9.1, equation (9.109) suggests that

---

[1] Let us keep in mind that Morse functions are dense in the space of smooth functions bounded below. See Appendix C.

[2] This conjecture is based on the fundamental property of good Morse functions of having, at arbitrary finite N, a finite number of critical values and a finite number of isolated critical points on each critical level set, so that isolated neighborhoods of critical points can always be defined. Proving this conjecture seems only a technical point.

[3] In both relations (9.109) and (8.36) the topological term depends on the sum with constant coefficients of the Morse indexes of the M_v (with equal weights in (8.36), with a-priori different weights in (9.109)), and—very important—both topological terms have the same N-dependence.

big jumps, possibly steepening with N, in the v-pattern of the topological term $\sum_{i=0}^{N} A(N, i, \varepsilon_0) g_i \mu_i(M_v)$ should be *sufficient* to entail the appearance of a phase transition. The question that remains open concerns the volume term deprived of the neighborhoods of critical points. In principle it could also generate thermodynamic singularities independently of the topological term. However, this possibility is ruled out by resorting to Theorem 9.1, though in the special case of one critical value in an interval $[\bar{v}_0, \bar{v}_1]$. Again, it seems very hard to imagine how this could change by simply allowing the existence of more critical values. As a consequence, topology changes are also *necessary* for the existence of phase transitions.

Theorem 9.2, proved in Section 9.5, is enunciated as follows:

Theorem 9.2 (Entropy and Topology). *Let $V_N(q_1, \ldots, q_N) : \mathbb{R}^N \to \mathbb{R}$, be a smooth, nonsingular, finite-range potential. Denote by $M_v := V_N^{-1}((-\infty, v])$, $v \in \mathbb{R}$, the generic submanifold of configuration space bounded by Σ_v. Let $\{q_c^{(i)} \in \mathbb{R}^N\}_{i \in [1, \mathcal{N}(v)]}$ be the set of critical points of the potential, that is, such that $\nabla V_N(q_c^{(i)}) = 0$, and let $\mathcal{N}(v)$ be the number of critical points up to the potential energy value v. Let $\Gamma(q_c^{(i)}, \varepsilon_0)$ be pseudocylindrical neighborhoods of the critical points, and $\mu_i(M_v)$ the Morse indexes of M_v. Then there exist real numbers $A(N, i, \varepsilon_0)$, g_i and real smooth functions $B(N, i, v, \varepsilon_0)$ such that the following equation for the microcanonical configurational entropy $S_N^{(-)}(v) = (1/N) \log \int_{V(q) \leq v} d^N q$ holds:*

$$
S_N^{(-)}(v) = \frac{1}{N} \log \Bigg[\int_{M_v \setminus \bigcup_{i=1}^{\mathcal{N}(v)} \Gamma(q_c^{(i)}, \varepsilon_0)} d^N q + \sum_{i=0}^{N} A(N, i, \varepsilon_0) \, g_i \, \mu_i(M_{v-\varepsilon_0})
$$
$$
+ \sum_{n=1}^{\mathcal{N}_{cp}^{\nu(v)+1}} B(N, i(n), v - v_c^{\nu(v)}, \varepsilon_0) \Bigg]
$$

(details and appropriate definitions are given in Section 9.1). Moreover, an unbounded growth with N of one of the derivatives $|\partial^k S^{(-)}(v)/\partial v^k|$, for $k = 3, 4$, and thus the occurrence of a first- or second-order phase transition, can be entailed only by the topological term $\sum_{i=0}^{N} A(N, i, \varepsilon_0) \, g_i \, \mu_i(M_{v-\varepsilon_0})$.

Together, these two theorems imply that for a wide class of potentials that are good Morse functions, a first- or second-order phase transition can only be the consequence of a topological change of the submanifolds M_v (or equivalently of the Σ_v) of configuration space.

The converse is not true: topological changes are necessary but not sufficient for the occurrence of phase transitions. In the following chapter, the study of exactly solvable models provides some hints about the sufficiency conditions, but rigorous results of general validity are not yet available.

9.1 Basic Definitions

For a physical system $\mathcal{S}$ of n particles confined in a bounded subset Λ^d of $\mathbb{R}^d$, $d = 1, 2, 3$, and interacting through a real-valued potential function V_N defined on $(\Lambda^d)^{\times n}$, with $N = nd$, the *configurational microcanonical volume* $\Omega(v, N)$ is defined for any value v of the potential V_N as

$$\Omega(v, N) = \int_{(\Lambda^d)^{\times n}} dq_1 \cdots dq_N \ \delta[V_N(q_1, \ldots, q_N) - v] = \int_{\Sigma_v} \frac{d\sigma}{\|\nabla V_N\|} , \quad (9.1)$$

where $d\sigma$ is a surface element of $\Sigma_v := V_N^{-1}(v)$; in what follows, $\Omega(v, N)$ is also called *structure integral*. The norm $\|\nabla V_N\|$ is defined as $\|\nabla V_N\| = [\sum_{i=1}^{N} (\partial_{q_i} V_N)^2]^{1/2}$. The *configurational partition function* $Z_c(\beta, N)$ is defined as

$$Z_c(\beta, N) = \int_{(\Lambda^d)^{\times n}} dq_1 \ldots dq_N \ \exp[-\beta V_N(q_1, \ldots, q_N)]$$
$$= \int_0^\infty dv \ e^{-\beta v} \int_{\Sigma_v} \frac{d\sigma}{\|\nabla V_N\|} , \quad (9.2)$$

where the real parameter β has the physical meaning of an inverse temperature. Notice that the formal Laplace transform of the structure integral in the right hand-side of (9.2) stems from a coarea formula [201] that is of very general validity (it holds also for Hausdorff measurable sets).

Now we can define the configurational thermodynamic functions to be used in this chapter.

Definition 9.3. *Using the notation $\bar{v} = v/N$ for the value of the potential energy per particle, we introduce the following functions:*

- Configurational microcanonical entropy, relative to Σ_v. *For any $N \in \mathbb{N}$ and $\bar{v} \in \mathbb{R}$,*

$$S_N(\bar{v}) \equiv S_N(\bar{v}; V_N) = \frac{1}{N} \log \Omega(N\bar{v}, N) .$$

- Configurational canonical free energy. *For any $N \in \mathbb{N}$ and $\beta \in \mathbb{R}$,*

$$f_N(\beta) \equiv f_N(\beta; V_N) = \frac{1}{N} \log Z_c(\beta, N) .$$

- Configurational microcanonical entropy, relative to the volume bounded by Σ_v. *For any $N \in \mathbb{N}$ and $\bar{v} \in \mathbb{R}$,*

$$S_N^{(-)}(\bar{v}) \equiv S_N^{(-)}(\bar{v}; V_N) = \frac{1}{N} \log M(N\bar{v}, N) ,$$

where

$$M(v, N) = \int_{(\Lambda^d)^{\times n}} dq_1 \cdots dq_N \ \Theta[V_N(q_1, \ldots, q_N) - v] = \int_0^v d\eta \int_{\Sigma_\eta} \frac{d\sigma}{\|\nabla V_N\|} , \quad (9.3)$$

with $\Theta[\cdot]$ the Heaviside step function; $M(v, N)$ is the codimension-0 subset of configuration space enclosed by the equipotential hypersurface Σ_v. The representation of $M(v, N)$ given in the right hand-side stems from the already mentioned coarea formula in [201]. Moreover, $S_N^{(-)}(\bar{v})$ is related to the configurational canonical free energy, f_N, for any $N \in \mathbb{N}$ and $\bar{v} \in \mathbb{R}$, through the Legendre transform [98]

$$-f_N(\beta) = \inf_{\bar{v}}\{\beta \cdot \bar{v} - S_N^{(-)}(\bar{v})\}, \tag{9.4}$$

yielding, for any $N \in \mathbb{N}$ and $\beta \in \mathbb{R}$,

$$-f_N(\beta) = \beta \cdot \bar{v}_N - S_N^{(-)}(\bar{v}_N) \tag{9.5}$$

with, for any $N \in \mathbb{N}$ and $\bar{v} \in \mathbb{R}$,

$$\beta_N(\bar{v}) = \frac{\partial S_N^{(-)}}{\partial \bar{v}}(\bar{v}), \tag{9.6}$$

and the inverse relation, valid for any $N \in \mathbb{N}$ and $\beta \in \mathbb{R}$,

$$\bar{v}_N(\beta) = -\frac{\partial f_N}{\partial \beta}(\beta). \tag{9.7}$$

Finally, for a system described by a Hamiltonian function H of the kind $H = \sum_{i=1}^{N} p_i^2/2 + V_N(q_1, \ldots, q_N)$, the Helmholtz free energy is defined by

$$F_N(\beta; H) = -(N\beta)^{-1} \log \int d^N p \, d^N q \, \exp[-\beta H(p, q)], \tag{9.8}$$

whence

$$F_N(\beta; H) = -(2\beta)^{-1} \log(\pi/\beta) - f_N(\beta, V_N)/\beta \tag{9.9}$$

with its thermodynamic limit ($N \to \infty$ and $\text{vol}(\Lambda^d)/N = \text{const}$)

$$F_\infty(\beta) = \lim_{N \to \infty} F_N(\beta; H). \tag{9.10}$$

Definition 9.4 (First- and Second-Order Phase Transitions). *We say that a physical system S undergoes a phase transition if there exists a thermodynamic function that—in the thermodynamic limit ($N \to \infty$ and $\text{vol}(\Lambda^d)/N = \text{const}$)—is only piecewise analytic. In particular, if the first-order derivative of the Helmholtz free energy $F_\infty(\beta)$ is discontinuous at some point β_c, then we say that a first-order phase transition occurs. If the second-order derivative of the Helmholtz free energy $F_\infty(\beta)$ is discontinuous at some point β_c, then we say that a second-order phase transition occurs.*

Definition 9.5 (Standard potential, fluid case). *We say that an N-degrees-of-freedom potential V_N is a standard potential for a fluid if it is of the form*

$$V_N : \quad \mathcal{B}_N \subset \mathbb{R}^N \to \mathbb{R} ,$$

$$V_N(q) = \sum_{i \neq j=1}^{n} \Psi(\|\mathbf{q}_i - \mathbf{q}_j\|) + \sum_{i=1}^{n} U_\Lambda(\mathbf{q}_i) , \qquad (9.11)$$

where $\mathcal{B}_N$ is a compact subset of $\mathbb{R}^N$, $N = nd$, Ψ is a real-valued function of one variable such that additivity holds, and where U_Λ is any smoothed potential barrier to confine the particles in a finite volume Λ, that is,

$$U_\Lambda(\mathbf{q}) = \begin{cases} 0 & \text{if } \mathbf{q} \in \Lambda' , \\ +\infty & \text{if } \mathbf{q} \in \Lambda^c, \text{ complement in } \mathbb{R}^N , \\ \mathcal{C}^\infty & \text{function for } \mathbf{q} \in \Lambda \setminus \Lambda' , \end{cases} \qquad (9.12)$$

where $\Lambda' \subset \Lambda$ and Λ' is arbitrarily close to $\Lambda \subset \mathbb{R}^N$, closed and bounded. U_Λ is a confining potential in a limited spatial volume with the additional property that given two limited d-dimensional regions of space, Λ_1 and Λ_2, having in common a $(d-1)$-dimensional boundary, $U_{\Lambda_1} + U_{\Lambda_2} = U_{\Lambda_1+\Lambda_2}$. By additivity we mean what follows. Consider two systems $\mathcal{S}_1$ and $\mathcal{S}_2$, having $N_1 = n_1 d$ and $N_2 = n_2 d$ degrees of freedom, occupying volumes Λ_1^d and Λ_2^d, having potential energies v_1 and v_2, for any $(q_1, \ldots, q_{N_1}) \in (\Lambda_1^d)^{\times n_1}$ such that $V_{N_1}(q_1, \ldots, q_{N_1}) = v_1$, for any $(q_{N_1+1}, \ldots, q_{N_1+N_2}) \in (\Lambda_2^d)^{\times n_2}$ such that $V_{N_2}(q_{N_1+1}, \ldots, q_{N_1+N_2}) = v_2$, for $(q_1, \ldots, q_{N_1+N_2}) \in (\Lambda_1^d)^{\times n_1} \times (\Lambda_2^d)^{\times n_2}$ let $V_N(q_1, \ldots, q_{N_1+N_2}) = v$ be the potential energy v of the compound system $\mathcal{S} = \mathcal{S}_1 + \mathcal{S}_2$ that occupies the volume $\Lambda^d = \Lambda_1^d \cup \Lambda_2^d$ and contains $N = N_1 + N_2$ degrees of freedom. If

$$v(N_1 + N_2, \Lambda_1^d \cup \Lambda_2^d) = v_1(N_1, \Lambda_1^d) + v_2(N_2, \Lambda_2^d) + v'(N_1, N_2, \Lambda_1^d, \Lambda_2^d) , \quad (9.13)$$

where v' stands for the interaction energy between $\mathcal{S}_1$ and $\mathcal{S}_2$, and if $v'/v_1 \to 0$ and $v'/v_2 \to 0$ for $N \to \infty$, then V_N is additive. Moreover, at short distances, Ψ must be a repulsive potential so as to prevent the concentration of an arbitrary number of particles within small, finite volumes of any given size.

Definition 9.6 (Standard potential, lattice case). *We say that an N degrees-of-freedom potential V_N is a standard potential for a lattice if it is of the form*

$$V_N : \quad \mathcal{B}_N \subset \mathbb{R}^N \to \mathbb{R} ,$$

$$V_N(q) = \sum_{\underline{i},\underline{j} \in \mathcal{I} \subset \mathbb{N}^d} C_{\underline{ij}} \Psi(\|\mathbf{q}_{\underline{i}} - \mathbf{q}_{\underline{j}}\|) + \sum_{\underline{i} \in \mathcal{I} \subset \mathbb{N}^d} \Phi(\mathbf{q}_{\underline{i}}) , \qquad (9.14)$$

where $\mathcal{B}_N$ is a compact subset of $\mathbb{R}^N$. Denoting by $a_1, \ldots, a_d$ the lattice spacings, if $\underline{i} \in \mathbb{N}^d$, then $(i_1 a_1, \ldots, i_d a_d) \in \Lambda^d$. We denote by m the number of *lattice sites in each spatial direction, by $n = m^d$ the total number of lattice sites, by D the number of degrees of freedom on each site. Thus $\mathbf{q}_{\underline{i}} \in \mathbb{R}^D$ for any $\underline{i}$. The total number of degrees of freedom is $N = m^d D$. Having two systems made of $N = m^d D$ degrees of freedom, whose site indexes $i^{(1)}$ and $i^{(2)}$ run over $1 \leq i_1^{(1)}, \ldots, i_d^{(1)} \leq m$, and $1 \leq i_1^{(2)}, \ldots, i_d^{(2)} \leq m$, after gluing together the two systems through a common $(d-1)$-dimensional boundary, the new system has indexes i running over, for example, $1 \leq i_1 \leq 2m$ and $1 \leq i_2, \ldots, i_d \leq m$. If*

$$v(N + N, \Lambda_1^d \cup \Lambda_2^d) = v_1(N, \Lambda_1^d) + v_2(N, \Lambda_2^d) + v'(N, N, \Lambda_1^d, \Lambda_2^d) \qquad (9.15)$$

where v' stands for the interaction energy between the two systems and if $v'/v_1 \to 0$ and $v'/v_2 \to 0$ for $N \to \infty$, then V_N is additive.

Definition 9.7 (Short-range potential). *In defining a short-range potential, a distinction has to be made between lattice systems and fluid systems. Given a standard potential V_N on a lattice, we say that it is a short-range potential if the coefficients $C_{\underline{ij}}$ are such that for any $\underline{i}, \underline{j} \in \mathcal{I} \subset \mathbb{N}^d$, $C_{\underline{ij}} = 0$ iff $|\underline{i} - \underline{j}| > c$, with c is definitively constant for $N \to \infty$.*

Given a standard potential V_N for a fluid system, we say that it is a short-range potential if there exist $R_0 > 0$ and $\epsilon > 0$ such that for $\|\mathbf{q}\| > R_0$ we have $|\Psi(\|\mathbf{q}\|)| < \|\mathbf{q}\|^{-(d+1+\epsilon)}$, where $d = 1, 2, 3$ is the spatial dimension.

Definition 9.8 (Stable potential). *We say that a potential V_N is stable [98] if there exists $B \geq 0$ such that*

$$V_N(q_1, \ldots, q_N) \geq -NB \qquad (9.16)$$

for any $N > 0$ and $(q_1, \ldots, q_N) \in (\Lambda^d)^{\times n}$, or for $\mathbf{q}_{\underline{i}} \in \mathbb{R}^D$, $\underline{i} \in \mathcal{I} \subset \mathbb{N}^d$, $N = m^d D$, for lattices.

Definition 9.9 (Confining potential). *With the above definitions of standard potentials V_N, in the fluid case the potential is said to be confining in the sense that it contains U_Λ, which constrains the particles in a finite spatial volume, and in the lattice case the potential V_N contains an on-site potential such that at finite energy, $\|\mathbf{q}_{\underline{i}}\|$ is constrained in a compact set of values.*

Remark 9.10 (Compactness of equipotential hypersurfaces). From the previous definition it follows that for a confining potential, the equipotential hypersurfaces Σ_v are compact (because they are closed by definition and bounded in view of particle confinement).

Proposition 9.11 (Pointwise convergence). *Assume that V_N is a standard, confining, short-range, and stable potential. Assume also that there exists $N_0 \in \mathbb{N}$ such that $\bigcap_{N > N_0}^{\infty} \mathrm{dom}(S_N^{(-)})$ and $\bigcap_{N > N_0}^{\infty} \mathrm{dom}(S_N)$ are nonempty sets. Then the following pointwise limits exist almost everywhere:*

$$\lim_{N \longrightarrow \infty} S_N^{(-)}(\bar{v}) \equiv S_\infty^{(-)}(\bar{v}) \quad \text{for} \quad \bar{v} \in \bigcap_{N>N_0}^\infty \text{dom}(S_N^{(-)})$$

$$\lim_{N \longrightarrow \infty} S_N(\bar{v}) \equiv S_\infty(\bar{v}) \quad \text{for} \quad \bar{v} \in \bigcap_{N>N_0}^\infty \text{dom}(S_N)$$

and moreover,

$$S_\infty^{(-)}(\bar{v}) = S_\infty(\bar{v}) \quad \text{for} \quad \bar{v} \in \bigcap_{N>N_0}^\infty \text{dom}(S_N^{(-)}) \cap \bigcap_{N>N_0}^\infty \text{dom}(S_N) . \tag{9.17}$$

Proof. The existence of the thermodynamic limit for the sequences of functions $S_N^{(-)}$ and S_N, associated with a standard potential function V_N with short-range interactions, stable and confining is formally proved in [98], Sections 3.3 and 3.4. To prove that in the thermodynamic limit the two entropies $S_\infty^{(-)}$ and S_∞ are equal, we proceed from the definitions of $S_N^{(-)}$ and of $\beta_N(\bar{v})$, that is,

$$S_N^{(-)}(\bar{v}) = \frac{1}{N} \log M(N\bar{v}, N)$$

and

$$\beta_N(\bar{v}) = \frac{\partial S_N^{(-)}}{\partial \bar{v}}(\bar{v}) ,$$

noting that from the right hand-side of (9.3) we obtain

$$\frac{dM(N\bar{v}, N)}{d\bar{v}} = N\Omega(N\bar{v}, N) , \tag{9.18}$$

so that

$$\beta_N(\bar{v}) = \frac{1}{NM(N\bar{v}, N)} \frac{dM(N\bar{v}, N)}{d\bar{v}} = \frac{\Omega(N\bar{v}, N)}{M(N\bar{v}, N)} , \tag{9.19}$$

whence

$$\frac{1}{N} \log \Omega(\bar{v}N, N) = \frac{1}{N} \log M(\bar{v}N, N) + \frac{1}{N} \log \beta_N(\bar{v}) . \tag{9.20}$$

Because of the existence of the thermodynamic limit $\beta(\bar{v})$ of the sequence of functions $\beta_N(\bar{v})$ [see Proposition 9.13], for any given $\bar{v} \in \mathbb{R}$, we have

$$\lim_{N \to \infty} \frac{1}{N} \log \beta_N(\bar{v}) = 0 .$$

Thus, with $S_N(\bar{v}) = 1/N \log \Omega(\bar{v}N, N)$, in the thermodynamic limit, that is, in the limit $N \to \infty$ with $\text{vol}(\Lambda^d)/N = \text{const}$, for any $\bar{v} \in \mathbb{R}$, (9.20) implies

$$S_\infty(\bar{v}) = S_\infty^{(-)}(\bar{v}) . \tag{9.21}$$

$\square$

Remark 9.12 (Equivalent definitions of entropy). In [98] it is proved that the Legendre transform relating $S_N^{(-)}(\bar{v})$ to $f_N(\beta)$ still holds in the thermodynamic limit, that is, $S_\infty^{(-)}(\bar{v})$ and $f_\infty(\beta)$ are still related by a Legendre transform (see theorem 3.4.4 in [98]). Thus, by equation (9.21), $S(\bar{v})$ is related to $f_\infty(\beta)$ by the same Legendre transform.

Proposition 9.13 (Pointwise convergence). *Assume that V_N is a standard, confining, short-range, and stable potential. Assume also that there exists $N_0 \in \mathbb{N}$ such that $\bigcap_{N>N_0}^\infty \mathrm{dom}(f_N)$ and $\bigcap_{N>N_0}^\infty \mathrm{dom}(\beta_N)$ are nonempty. Then the following limits exist pointwise almost everywhere:*

$$\lim_{N \to \infty} f_N(\beta) \equiv f(\beta) , \quad \text{for} \quad \beta \in \bigcap_{N>N_0}^\infty \mathrm{dom}(f_N)$$

$$\lim_{N \to \infty} \beta_N(\bar{v}) \equiv \beta(\bar{v})) , \quad \text{for} \quad \bar{v} \in \bigcap_{N>N_0}^\infty \mathrm{dom}(\beta_N) . \tag{9.22}$$

Proof. See [98], Section 3.4.

Henceforth, we shall use V instead of V_N if no explicit reference the N-dependence of V is necessary.

9.2 Main Theorems: Theorem 1

In this section we prove the first of our two main theorems.

Theorem 9.14 (Regularity under diffeomorphicity). *Let V_N be a standard, smooth, confining, short-range potential bounded from below (Definitions 9.5, 9.7, 9.8, and 9.9):*

$$V_N : \quad \mathcal{B}_N \subset \mathbb{R}^N \to \mathbb{R} ,$$

$$V_N(q) = \sum_{\underline{i},\underline{j} \in \mathcal{I} \subset \mathbb{N}^d} C_{\underline{i}\underline{j}} \Psi(\|\mathbf{q}_{\underline{i}} - \mathbf{q}_{\underline{j}}\|) + \sum_{\underline{i} \in \mathcal{I} \subset \mathbb{N}^d} \Phi(\mathbf{q}_{\underline{i}}) . \tag{9.23}$$

Let (Ψ, Φ) be real-valued one-variable functions, let $\underline{i}, \underline{j}$ label interacting pairs of degrees of freedom within a short-range, and let $\{\bar{\Sigma}_v\}_{v \in \mathbb{R}}$ be the family of $(N-1)$-dimensional equipotential hypersurfaces $\Sigma_v := V_N^{-1}(v)$, $v \in \mathbb{R}$, of $\mathbb{R}^N$.

Let $\bar{v}_0, \bar{v}_1 \in \mathbb{R}$, $\bar{v}_0 < \bar{v}_1$. If there exists N_0 such that for any $N > N_0$ and for any $\bar{v}, \bar{v}' \in I_{\bar{v}} = [\bar{v}_0, \bar{v}_1]$,

$$\Sigma_{N\bar{v}} \text{ is } C^\infty - \text{diffeomorphic to } \Sigma_{N\bar{v}'} \tag{9.24}$$

(notation: $\Sigma_{N\bar{v}} \approx \Sigma_{N\bar{v}'}$), then the limit entropy $S(\bar{v})$ is of differentiability class $C^3(I_{\bar{v}})$, and, consequently, $\beta(\bar{v})$ belongs to $C^2(I_{\bar{v}})$, whence the limit Helmholtz free energy function F_∞ is in $C^2(\overset{\circ}{I}_\beta)$, where $\overset{\circ}{I}_\beta$ denotes the open interior of $\beta([\bar{v}_0, \bar{v}_1])$, so that the system described by V has neither first- nor second-order phase transitions in the inverse-temperature interval $\overset{\circ}{I}_\beta$.

The idea of the proof of Theorem 9.14 is the following. In order to prove that a *topological change* of the equipotential hypersurfaces Σ_v of configuration space is a *necessary* condition for a thermodynamic phase transition to occur, we shall prove the *equivalent proposition* that if any two hypersurfaces $\Sigma_{v(N)}$ and $\Sigma_{v'(N)}$ with $v(N), v'(N) \in (v_0(N), v_1(N))$ are *diffeomorphic* for all N, possibly greater than some finite N_0, then *no phase transition* can occur in the (inverse) temperature interval $[\lim_{N \to \infty} \beta(\bar{v}_0(N)), \lim_{N \to \infty} \beta(\bar{v}_1(N))]$. To this end we have to show that in the limit $N \to \infty$ and $\mathrm{vol}(\Lambda^d)/N = \mathrm{const}$, the Helmholtz free energy $F_\infty(\beta; H)$ is at least twice differentiable as a function of $\beta = 1/T$ in the interval $[\lim_{N \to \infty} \beta(\bar{v}_0(N)), \lim_{N \to \infty} \beta(\bar{v}_1(N))]$. For the standard Hamiltonian systems that we consider throughout this chapter, showing that $F_N(\beta) = -(2\beta)^{-1} \log(\pi/\beta) - f_N(\beta)/\beta$ is equivalent to showing that the sequence of configurational free energies $\{f_N(T; H)\}_{N \in \mathbb{N}_+}$ is *uniformly convergent* at least in $\mathcal{C}^2$, so that also $\{f_\infty(T; H)\} \in \mathcal{C}^2$.

We shall give the proof of Theorem 9.14 through the following lemmas, which are separately proven in subsequent sections.

Lemma 9.15 (Absence of critical points). *Let $f : M \to [a, b]$ a smooth map on a compact manifold M with boundary such that its Hessian is nondegenerate. Suppose $f(\partial M) = \{a, b\}$ and that for any $c, d \in [a, b]$, we have $f^{-1}(c) \approx f^{-1}(d)$, that is, all the level surfaces of f are diffeomorphic. Then f has no critical points, that is, $\|\nabla f\| \geq C > 0$ in $[a, b]$, where C is a constant.*

Proof. Since f is a good Morse function, let us consider the case of the existence of at least one critical value $c \in [a, b]$ such that $\nabla f = 0$ at some points of the level set $f^{-1}(c)$. The set of critical points $\sigma(c) = \{x_c^{i,k_i} \in f^{-1}(c) | (\nabla f)(x_c^{i,k_i}) = 0\}$ is a point set [210]; the index i labels the different critical points; and k_i is the Morse index of the ith critical point. By the "noncritical neck" theorem [210], we know that the level sets $f^{-1}(v)$ with $v \in [a, c - \varepsilon]$ and arbitrary $\varepsilon > 0$ are diffeomorphic because in the absence of critical points in the interval $[a, c - \varepsilon]$ for any $v, v' \in [a, c - \varepsilon]$, with arbitrary $\varepsilon > 0$, $f^{-1}(v)$ is a deformation retraction of $f^{-1}(v')$ through the flow associated with the vector field [23] $X = -\nabla f / \|\nabla f\|^2$. Now, in the neighborhood of each critical point x_c^{i,k_i}, the existence of the Morse chart [23] allows one to represent the function f as follows:

$$f(x) = f(x_c^{i,k_i}) - x_1^2 - \cdots - x_{k_i}^2 + x_{k_i+1}^2 + \cdots + x_n^2 , \qquad (9.25)$$

whence the degeneracy of the quadrics for $v = c$ entailing that the level set $f^{-1}(c)$ no longer qualifies as a differentiable manifold. Thus for any $v \in [a, c - \varepsilon]$ and arbitrary $\varepsilon > 0$, we have

$$f^{-1}(v) \not\approx f^{-1}(c) . \qquad (9.26)$$

In conclusion, if for any pair of values $v, v' \in [a, b]$ one has $f^{-1}(v') \approx f^{-1}(v)$, then no critical point of f can exist in the interval $[a, b]$. $\qquad \square$

Lemma 9.16 (Smoothness of the structure integral). *Let V_N be a standard, short-range, stable, and confining potential function bounded below. Let $\{\Sigma_v\}_{v \in \mathbb{R}}$ be the family of $(N-1)$-dimensional equipotential hypersurfaces $\Sigma_v := V_N^{-1}(v)$, $v \in \mathbb{R}$, of $\mathbb{R}^N$. Then*

$$\text{if for any } v, v' \in [v_0, v_1], \Sigma_v \approx \Sigma_{v'}, \text{ then } \Omega_N(v) \in \mathcal{C}^\infty(]v_0, v_1[).$$

Proof. The proof of this lemma is given in Section 9.3.

Lemma 9.17 (Uniform convergence). *Let U and U' be two open intervals of $\mathbb{R}$. Let h_N be a sequence of functions from U to U', differentiable on U, and let $h : U \longrightarrow U'$ be such that for any $x \in U$, $\lim_{N \to \infty} h_N(x) = h(x)$. If there exists $M \in \mathbb{R}$ such that for any $N \in \mathbb{N}$ and for any $a \in U$, $\left| \dfrac{dh_N}{dx}(a) \right| \leq M$, then h is continuous at a for any $a \in U$.*

Proof. From the assumption that for any $N \in \mathbb{N}$ and for any $a \in U$ we have $|h_N'(a)| \leq M$, and by the fundamental theorem of calculus, the set of functions $\{h_N\}_{N \in \mathbb{N}}$ is equilipschitzian and thus uniformly equicontinuous [218]. Then, by Ascoli's theorem on equicontinuous sets of applications [218], it follows that for any $a \in U$ the closure of the set of functions $\{h_N\}_{N \in \mathbb{N}}$ is equicontinuous, and thus the limit function h is continuous at a for any $a \in U$.   □

Lemma 9.18 (Uniform upper bounds). *Let V_N be a standard, short-range, stable, and confining potential function bounded below. Let $\{\Sigma_v\}_{v \in \mathbb{R}}$ be the family of $(N-1)$-dimensional equipotential hypersurfaces $\Sigma_v := V_N^{-1}(v)$, $v \in \mathbb{R}$, of $\mathbb{R}^N$. If*

$$\text{for any } N \text{ and } \bar{v}, \bar{v}' \in I_{\bar{v}} = [\bar{v}_0, \bar{v}_1], \text{ we have } \Sigma_{N\bar{v}} \approx \Sigma_{N\bar{v}'},$$

then

$$\sup_{N, \bar{v} \in I_{\bar{v}}} |S_N(\bar{v})| < \infty \quad and \quad \sup_{N, \bar{v} \in I_{\bar{v}}} \left| \frac{\partial^k S_N}{\partial \bar{v}^k}(\bar{v}) \right| < \infty, \quad k = 1, 2, 3, 4.$$

Proof. The proof of this lemma is given in Section 9.4.

Proof. (Theorem 9.14) Under the hypothesis that all the level surfaces of V_N are diffeomorphic in the interval $I_{\bar{v}}$, we know from Lemma 9.15 that there are no critical points of V_N in $I_{\bar{v}}$, i.e., there exists $C(N) > 0$ such that for any $N > N_0$,

$$\text{for } \bar{v} \in I_{\bar{v}}, \text{ and for any } x \in \Sigma_{N\bar{v}}, \quad \|\nabla V_N(x)\| \geq C > 0. \tag{9.27}$$

Therefore, the restriction of V_N,

$$\tilde{V}_N = V_{|V_N^{-1}(I_{N\bar{v}})} : V_N^{-1}(I_{N\bar{v}}) \subset B \rightarrow \mathbb{R} , \tag{9.28}$$

always defines a Morse function, since V_N is bounded below. Notice that

$$S_N(\bullet \, ; V_N)_{|\mathring{I}_{\bar{v}}} \equiv S_N(\bullet \, ; \tilde{V}_N)_{|\mathring{I}_{\bar{v}}} . \tag{9.29}$$

In what follows we shall drop the tilde and V_N will denote the above given restriction.

Now, since the condition (9.27) holds for the hypersurfaces $\{\Sigma_{N\bar{v}}\}_{\bar{v} \in \mathring{I}_{\bar{v}}}$, from Lemma 9.16 it follows that for any $N > N_0$, $\Omega(N\bar{v}, N)$ is actually in $C^{\infty}(\mathring{I}_{\bar{v}})$, where $\mathring{I}_{\bar{v}} = (\bar{v}_0, \bar{v}_1)$; this implies that for any $N > N_0$, S_N also belongs to $C^{\infty}(\mathring{I}_{\bar{v}})$.

While at any finite N, under the main assumption of the theorem, the entropy functions S_N are smooth, we do not know what happens in the $N \rightarrow \infty$ limit. To know the behavior at the limit, we have to prove the uniform convergence of the sequence $\{S_N\}_{N \in \mathbb{N}_+}$. Lemmas 9.17 and 9.18 prove exactly that this sequence is uniformly convergent at least in the space $C^3(\mathring{I}_{\bar{v}})$, so that we can conclude that also $S \in C^3(\mathring{I}_{\bar{v}})$.

Since $S = S^{(-)}$ in $I_{\bar{v}}$ (Proposition 9.11), also $S^{(-)}$ lies in $C^3(\mathring{I}_{\bar{v}})$ and β in $C^2(\mathring{I}_{\bar{v}})$. Moreover, by definition and existence of the uniform limit of $\{S_N\}_{N \in \mathbb{N}_+}$, for any $\bar{v} \in \mathring{I}_{\bar{v}}$ we can write

$$S(\bar{v}) = f(\beta(\bar{v})) + \beta(\bar{v}) \cdot \bar{v} , \tag{9.30}$$

which entails $f \in C^2(\beta(\mathring{I}_{\bar{v}})) \equiv C^2(\mathring{I}_{\beta})$.

Since the kinetic-energy term of the Hamiltonian describing the system S gives only a smooth contribution, the Helmholtz free energy F_{∞} also has differentiability class $C^2(\mathring{I}_{\beta})$. Hence we conclude that the system S undergoes neither first- nor second-order phase transitions in the inverse-temperature interval $\beta \in \mathring{I}_{\beta}$. $\qquad \square$

Corollary 9.19. *Under the hypotheses of Theorem 9.14, let $\{M_v\}_{v \in \mathbb{R}}$ be the family of the N-dimensional subsets $M_v := V_N^{-1}((-\infty, v])$, $v \in \mathbb{R}$, of $\mathbb{R}^N$. Let $\bar{v}_0, \bar{v}_1 \in \mathbb{R}$, $\bar{v}_0 < \bar{v}_1$. If there exists N_0 such that for any $N > N_0$ and for any $\bar{v}, \bar{v}' \in I_{\bar{v}} = [\bar{v}_0, \bar{v}_1]$,*

$$M_{N\bar{v}} \text{ is } C^{\infty} - \text{diffeomorphic to } M_{N\bar{v}'}, \tag{9.31}$$

then the limit entropy $S^{(-)}(\bar{v})$ is of differentiability class $C^3(I_{\bar{v}})$, and consequently, $\beta(\bar{v}) = \partial S^{(-)}/\partial \bar{v}$ belongs to $C^2(I_{\bar{v}})$, whence the limit Helmholtz free energy function F_{∞} is in $\in C^2(\mathring{I}_{\beta})$, where $\mathring{I}_{\beta}$ denotes the open interior of $\beta([\bar{v}_0, \bar{v}_1]))$, so that the system described by V has neither first- nor second-order phase transitions in the inverse-temperature interval $\mathring{I}_{\beta}$.

Proof. If for any $\bar{v}, \bar{v}' \in I_{\bar{v}} = [\bar{v}_0, \bar{v}_1]$ we have $M_{N\bar{v}} \approx M_{N\bar{v}'}$, then by Bott's "critical-neck theorem" [219], there are no critical points of V_N in the interval $[\bar{v}_0, \bar{v}_1]$. As a consequence of the absence of critical points in $[\bar{v}_0, \bar{v}_1]$, by the "noncritical neck theorem" [210], for any $\bar{v}, \bar{v}' \in I_{\bar{v}} = [\bar{v}_0, \bar{v}_1]$, $\Sigma_{N\bar{v}} \approx \Sigma_{N\bar{v}'}$. Now Theorem 9.14 implies $S(\bar{v}) \in \mathcal{C}^3(I_{\bar{v}})$, so that using Proposition 9.11 we have also $S^{(-)}(\bar{v}) \in \mathcal{C}^3(I_{\bar{v}})$. Then using equation (9.5) we have $f_\infty(\beta) \in \mathcal{C}^2(I_{\bar{v}})$ and thus $F_\infty \in \mathcal{C}^2(\overset{o}{I}_\beta)$, so that neither first- nor second-order phase transitions can occur in the inverse-temperature interval $\overset{o}{I}_\beta = (\partial S^{(-)}/\partial \bar{v}|_{\bar{v}=\bar{v}_0}, \partial S^{(-)}/\partial \bar{v}|_{\bar{v}=\bar{v}_1})$. $\qquad \square$

9.3 Proof of Lemma 2, Smoothness of the Structure Integral

We make use of the following lemma:

Lemma 9.20. *Let U be a bounded open subset of $\mathbb{R}^N$, let ψ be a Morse function defined on U, $\psi : U \subset \mathbb{R}^N \to \mathbb{R}$, and let $\mathcal{F} = \{\Sigma_v\}_v$ be the family of hypersurfaces defined as $\Sigma_v = \{x \in U | \psi(x) = v\}$. Then*

$$\text{if for any } v, v' \in [v_0, v_1], \ \Sigma_v \approx \Sigma_v',$$

$$\text{then for any } g \in \mathcal{C}^\infty(U), \ \int_{\Sigma_v} g \, d\sigma \text{ is } \mathcal{C}^\infty \text{ in }]v_0, v_1[.$$

Proof. To prove this lemma we need the following theorem [201, 209]:

Theorem 9.21 (Federer, Laurence). *Let $O \subset \mathbb{R}^p$ be a bounded open set. Let $\psi \in \mathcal{C}^{n+1}(\bar{O})$ be constant on each connected component of the boundary ∂O and $g \in \mathcal{C}^n(O)$.*

Define $O_{t,t'} = \{x \in O \mid t < \psi(x) < t'\}$ and $F(v) = \int_{\{\psi=v\}} g \, d\sigma^{p-1}$, where $d\sigma^{p-1}$ represents the Lebesgue measure of dimension $p-1$.

If $C > 0$ exists such that for any $x \in O_{t,t'}, \|\nabla\psi(x)\| \geq C$, then for any k such that $0 \leq k \leq n$, for any $v \in]t, t'[$, one has

$$\frac{d^k F}{dv^k}(v) = \int_{\{\psi=v\}} A^k g \, d\sigma^{p-1} . \tag{9.32}$$

with $Ag = \nabla \left(\frac{\nabla\psi}{\|\nabla\psi\|} g \right) \frac{1}{\|\nabla\psi\|}$.

By applying this theorem to the function ψ of Lemma 9.20 we have that if there exists a constant $C > 0$ such that *for any $x \in O_{v_0, v_1}$* we have $\|\nabla\psi(x)\| \geq C$, then

$$\frac{d^k F}{dv^k}(v) = \int_{\Sigma_v} A^k g \, d\sigma, \ \forall v \in]v_0, v_1[. \tag{9.33}$$

Now, under the hypothesis that *for any* $v, v' \in [v_0, v_1]$, $\Sigma_v \approx \Sigma_{v'}$, we know from Lemma 9.15, "absence of critical points," that this hypothesis is equivalent to the assumption that *for any* $v \in [v_0, v_1]$, Σ_v has no critical points. Hence there exists a constant $C > 0$ such that $\forall x \in O_{v_0, v_1}$ $\|\nabla \psi(x)\| \geq C$. Furthermore, since $\|\nabla \psi\|$ is strictly positive, A is a continuous operator on O_{v_0, v_1}. Thus, being Σ_v compact, $(d^k F / dv^k)$ is continuous on the interval $]v_0, v_1[$, $\forall k$, namely $\int_{\Sigma_v} g \, d\sigma \in C^\infty(]v_0, v_1[)$.

To conclude the proof of Lemma 9.16 we have to use Lemma 9.20 taking $\psi = V_N$ and $g = 1/\|\nabla V_N\|$, assuming that V_N is a Morse function and that $\|\nabla V_N\|$ is strictly positive (absence of critical points of V_N stemming from the hypothesis of diffeomorphicity of Theorem 9.14). $\square$

9.4 Proof of Lemma 9.18, Upper Bounds

The proof of Lemma 9.18 is split into two parts. In part A some preliminary results to be used in part B are given, and in part B the inequalities of Lemma 9.18 are proved.

The proof of Lemma 9.18 is the core of the proof of Theorem 9.14. Thus, since the proof of Lemma 9.18 is lengthy, in order to ease its reading we offer a summary of it.

Sketch of the Proof.

In order to prove Theorem 9.14), we have to show that the assumption of diffeomorphicity among the $\Sigma_{N\bar{v}}$ for $\bar{v} \in [\bar{v}_0, \bar{v}_1]$ entails that $S_\infty(\bar{v})$ is three times differentiable. By Ascoli's theorem [218], this is proved by showing that for $\bar{v} \in I_{\bar{v}} = [\bar{v}_0, \bar{v}_1]$ and *for any* N, the function $S_N(\bar{v})$ and its first four derivatives are uniformly bounded in N from above, that is, for any $N \in \mathbb{N}$ and $\bar{v} \in [\bar{v}_0, \bar{v}_1]$,

$$\sup |S_N(\bar{v})| < \infty, \quad \sup \left| \frac{\partial^k S_N}{\partial \bar{v}^k} \right| < \infty, \quad k = 1, \ldots, 4. \tag{9.34}$$

By *Definition 9.3* for the entropy, the first four derivatives of $S_N(\bar{v})$ are

$$\begin{aligned}
\partial_{\bar{v}} S_N &= (1/N)(dv/d\bar{v}) \Omega' / \Omega \, , \\
\partial_{\bar{v}}^2 S_N &= N[\Omega''/\Omega - (\Omega'/\Omega)^2] \, , \\
\partial_{\bar{v}}^3 S_N &= N^2[\Omega'''/\Omega - 3\Omega''\Omega'/\Omega^2 + 2(\Omega'/\Omega)^3] \, , \\
\partial_{\bar{v}}^4 S_N &= N^3[\Omega^{iv}/\Omega - 4\Omega'''\Omega'/\Omega^2 - 3(\Omega''/\Omega)^2 \\
&\quad + 12\Omega''(\Omega')^2/\Omega^3 - 6(\Omega'/\Omega)^4] \, ,
\end{aligned} \tag{9.35}$$

where the primes stand for derivations of $\Omega(v, N)$ with respect to $v = \bar{v}N$. In order to verify whether the conditions (9.34) are satisfied, we must be able to estimate the N-dependence of all the addenda in these expressions for the derivatives of S_N.

Since the assumption of diffeomorphicity of the $\Sigma_{N\bar{v}}$ is equivalent to the absence of critical points of the potential, we can use the derivation formula [201, 209]

$$\frac{d^k}{dv^k}\Omega_N(v) = \int_{\Sigma_v} \|\nabla V\| \, A^k \left(\frac{1}{\|\nabla V\|}\right) \frac{d\sigma}{\|\nabla V\|} \quad , \qquad (9.36)$$

where A^k stands for k iterations of the operator

$$A(\bullet) = \nabla \left(\frac{\nabla V}{\|\nabla V\|} \bullet\right) \frac{1}{\|\nabla V\|} \; .$$

A technically crucial step to prove the theorem is to use the above formula (9.36) to compute the derivatives of $\Omega(v, N)$. In fact, these are transformed into the surface integrals of explicitly computable combinations and powers of a few basic ingredients, such as $\|\nabla V\|$, $\partial V/\partial q_i$, $\partial^2 V/\partial q_i \partial q_j$, and $\partial^3 V/\partial q_i \partial q_j \partial q_k$.

The first uniform bound in (9.34), $|S_N(\bar{v})| < \infty$, is a simple consequence of the intensivity of $S_N(\bar{v})$.

To prove the boundedness of the first derivative of S_N, we compute its expression by means of the first of equations (9.35) and of Eq.(9.36), which reads

$$\frac{\partial S_N}{\partial \bar{v}} = \frac{1}{\Omega} \int_{\Sigma_{\bar{v}N}} \left[\frac{\Delta V}{\|\nabla V\|^2} - 2\frac{\sum_{i,j} \partial^i V \partial^2_{ij} V \partial^j V}{\|\nabla V\|^4}\right] \frac{d\sigma}{\|\nabla V\|} \quad , \qquad (9.37)$$

with $\partial_i V = \partial V/\partial q^i$ and $i,j = 1, \ldots, N$, whence (with an obvious meaning of $\langle \cdot \rangle_{\Sigma_v}$)

$$\left|\frac{\partial S_N}{\partial \bar{v}}\right| \leq \left\langle \frac{|\Delta V|}{\|\nabla V\|^2}\right\rangle_{\Sigma_v} + 2\left\langle \frac{\left|\sum_{i,j} \partial^i V \partial^2_{ij} V \partial^j V\right|}{\|\nabla V\|^4}\right\rangle_{\Sigma_v} . \qquad (9.38)$$

the right hand-side of this inequality—in the absence of critical points of the potential—can be bounded from above by (see Lemma 9.29)

$$\frac{\langle|\Delta V|\rangle_{\Sigma_v}}{\langle\|\nabla V\|^2\rangle_{\Sigma_v}} + O\left(\frac{1}{N}\right) + 2\frac{\left\langle\sum_{i,j=1}^{N}|\partial^i V \partial^2_{ij} V \partial^j V|\right\rangle_{\Sigma_v}}{\langle\|\nabla V\|^4\rangle_{\Sigma_v}} + O\left(\frac{1}{N^2}\right) . \quad (9.39)$$

Since we have assumed that V is smooth and bounded below, and using the argument put forward in Remark 9.28, we have $\langle|\Delta V|\rangle_{\Sigma_v} = \langle|\sum_{i=1}^{N} \partial^2_{ii} V|\rangle_{\Sigma_v} \leq N \max_i \langle|\partial^2_{ii} V|\rangle_{\Sigma_v}$, and since we have also assumed that V is a short-range potential, the number of nonvanishing matrix elements $\partial^2_{ij} V$ is $N(n_p + 1)$, where n_p is the number of neighboring particles in the interaction range of the potential. Thus $\langle|\partial^i V \partial^2_{ij} V \partial^j V|\rangle_{\Sigma_v} \leq N(n_p + 1) \max_{i,j}\langle|\partial^i V \partial^2_{ij} V \partial^j V|\rangle_{\Sigma_v}$.

Moreover, the following lower bounds exist for the denominators in the inequality (9.39): $\langle \|\nabla V\|^2 \rangle_{\Sigma_v} \geq N \, \min_i \langle (\partial_i V)^2 \rangle_{\Sigma_v}$, and $\langle \|\nabla V\|^4 \rangle_{\Sigma_v} \geq N^2 \, \min_{i,j} \langle (\partial_i V)^2 (\partial_j V)^2 \rangle_{\Sigma_v}$.

Finally, putting $m = \max_{i,j} \langle | \, \partial^i V \partial^2_{ij} V \partial^j V \, | \rangle_{\Sigma_v}$, $c_1 = \min_i \langle (\partial_i V)^2 \rangle_{\Sigma_v}$, and $c_2 = \min_{i,j} \langle (\partial_i V)^2 (\partial_j V)^2 \rangle_{\Sigma_v}$, by substituting in (9.39) the upper bounds for the numerators and the lower bounds for the denominators we obtain

$$\left| \frac{\partial S_N}{\partial \bar{v}} \right| \leq \frac{\max_i \langle | \, \partial^2_{ii} V \, | \rangle_{\Sigma_v}}{c_1} + O\left(\frac{1}{N} \right) + 2 \frac{n_p \, m}{c_2 N} + O\left(\frac{1}{N^2} \right) , \qquad (9.40)$$

which, in the limit $N \to \infty$, shows that the first derivative of the entropy is uniformly bounded by a finite constant. This first step proves that $S_\infty(\bar{v})$ is continuous.

The three further steps, concerning boundedness of the higher-order derivatives, involve similar arguments to be applied to a number of terms that is rapidly increasing with the order of the derivative. But many of these terms can be grouped in the form of the variance or higher moments of certain quantities, thus allowing the use of a powerful technical trick to compute their N-dependence. For example, using (9.36) in the expression for $\partial^2_{\bar{v}} S_N$, we get

$$\left| \frac{\partial^2 S_N}{\partial \bar{v}^2} \right| \leq N \left| \langle \alpha^2 \rangle_{\Sigma_v} - \langle \alpha \rangle^2_{\Sigma_v} \right| + N \left| \langle \underline{\psi}(V) \cdot \underline{\psi}(\alpha) \rangle_{\Sigma_v} \right| , \qquad (9.41)$$

where $\alpha = \|\nabla V\| \, A(1/\|\nabla V\|)$ and $\underline{\psi} = \nabla / \|\nabla V\|$. Now, it is possible to think of the scalar function α as if it were a random variable, so that the first term in the right hand-side of (9.41) would be its second moment. Such a possibility is related to the general validity of the Monte Carlo method for computing multiple integrals. In particular, since the Σ_v are smooth, closed (V is non-singular), without critical points, and representable as the union of suitable subsets of $\mathbb{R}^{N-1}$, the standard Monte Carlo method [220] is applicable to the computation of the averages $\langle \cdot \rangle_{\Sigma_v}$, which become sums of standard integrals in $\mathbb{R}^{N-1}$. This means that a random walk can be constructively defined on any Σ_v, which conveniently samples the desired measure on the surface (see Lemma 9.22). Along such a random walk, usually called a Monte Carlo Markov chain (MCMC), α and its powers behave as random variables whose "time" averages along the MCMC converge to the surface averages $\langle \cdot \rangle_{\Sigma_v}$. Notice that the actual computation of these surface averages goes beyond our aim; in fact, we do not need the numerical values, but only the N-dependences, of the upper bounds of the derivatives of the entropy. Therefore, all that we need is just knowing that in principle a suitable MCMC exists on each Σ_v. Now, the function α is the integrand in square brackets in (9.37), where the second term vanishes at large N, as is clear from (9.40). Therefore, at increasingly large N, the approximate expression $\alpha = \sum_{i=1}^{N} \partial^2_{ii} V / \|\nabla V\|^2$ tends to become exact. Then α is in the form of a sum function $\alpha = N^{-1} \sum_{i=1}^{N} a_i$ of terms $a_i = N \partial^2_{ii} V / \|\nabla V\|^2$, of $O(1)$ in N, which, along an MCMC, behave as independent random variables with probability densities $u_i(a_i)$, which we do not need to know explicitly. Then, after a classical ergodic theorem for sum functions, due to Khinchin [221], based on the central limit theorem of probability theory, α is a Gaussian-distributed random variable; since its variance decreases linearly with N, $\lim_{N \to \infty} N |\langle \alpha^2 \rangle_{\Sigma_v} - \langle \alpha \rangle_{\Sigma_v}^2| = \text{const} < \infty$.

Arguments similar to those above used for the first derivative of S_N lead to the result $\lim_{N \to \infty} N |\langle \psi(V) \cdot \psi(\alpha) \rangle_{\Sigma_v}| = \text{const} < \infty$, which, together with what has been just found for the variance of α, proves the uniform boundedness also of the second derivative of S_N under the hypothesis of diffeomorphicity of the Σ_v.

Similarly, but with increasingly tedious work, we can treat the third and fourth derivatives of the entropy. In fact, despite the large number of terms contained in their expressions, they again belong to only two different categories: those terms that can be grouped in the form of higher moments of the function α, and whose N-dependence is known from the above-mentioned theorem due to Khinchin and Lemma 9.25, and those terms whose N-dependence can be found by means of the same kind of estimates given above for $\partial_{\bar{v}} S_N$. Eventually, after lengthy but rather mechanical work, also the third and fourth derivatives of S_N are shown to be uniformly bounded as prescribed by (9.34), completing the proof of Theorem 9.14.

9.4.1 Part A

We begin by showing that on any $(N-1)$-dimensional hypersurface $\Sigma_{N\bar{v}} = V_N^{-1}(N\bar{v}) = \{X \in \mathbb{R}^N \mid V_N(X) = N\bar{v}\}$ of $\mathbb{R}^N$, we can define a homogeneous nonperiodic random Markov chain whose probability measure is the configurational microcanonical measure, namely $d\sigma/\|\nabla V_N\|$.

Notice that at any finite N and in the absence of critical points of the potential V_N (because of $\|\nabla V_N\| \geq C > 0$) the microcanonical measure is smooth. The microcanonical averages $\langle \ \rangle_{N,v}^{\mu c}$ are then equivalently computed as "time" averages along the previously mentioned Markov chains.

In the following, when no ambiguity is possible, for the sake of notation we shall drop the suffix N of V_N.

Lemma 9.22. *On each finite-dimensional level set $\Sigma_{N\bar{v}} = V^{-1}(N\bar{v})$ of a standard, smooth, confining, short-range potential V bounded below, and in the absence of critical points, there exists a random Markov chain of points $\{X_i \in \mathbb{R}^N\}_{i \in \mathbb{N}_+}$, constrained by the condition $V(X_i) = N\bar{v}$, which has*

$$d\mu = \frac{d\sigma}{\|\nabla V\|} \left(\int_{\Sigma_{N\bar{v}}} \frac{d\sigma}{\|\nabla V\|} \right)^{-1} \tag{9.42}$$

as its probability measure, so that for a smooth function $F : \mathbb{R}^N \to \mathbb{R}$ we have

$$\left(\int_{\Sigma_{N\bar{v}}} \frac{d\sigma}{\|\nabla V\|} \right)^{-1} \int_{\Sigma_{N\bar{v}}} \frac{d\sigma}{\|\nabla V\|} F = \lim_{n \to \infty} \frac{1}{n} \sum_{i=1}^{n} F(X_i) . \tag{9.43}$$

Proof. Sinc the level sets $\{\Sigma_{N\bar{v}}\}_{\bar{v}\in\mathbb{R}}$ are compact codimension-1 hypersurfaces of $\mathbb{R}^N$, there exists on each of them a partition of unity [197]. Thus, denoting by $\{U_i\}$, $1 \leq i \leq m$, an arbitrary finite covering of $\Sigma_{N\bar{v}}$ by means of domains of coordinates (for example by means of open balls), a set of smooth functions $\{\varphi_i\}$ exists, with $1 \geq \varphi_i \geq 0$ and $\sum_i \varphi_i = 1$, for any point of $\Sigma_{N\bar{v}}$. Since the hypersurfaces $\Sigma_{N\bar{v}}$ are compact and oriented, the partition of unity $\{\varphi_i\}$ on $\Sigma_{N\bar{v}}$, subordinate to a collection $\{U_i\}$ of one-to-one local parametrizations of $\Sigma_{N\bar{v}}$, allows one to represent the integral of a given smooth $(N-1)$-form ω as follows A.4.6:

$$\int_{\Sigma_{N\bar{v}}} \omega^{(N-1)} = \int_{\Sigma_{N\bar{v}}} \left(\sum_{i=1}^{m} \varphi_i(x)\right) \omega^{(N-1)}(x) = \sum_{i=1}^{m} \int_{U_i} \varphi_i \omega^{(N-1)}(x) \,.$$

Now we proceed constructively by showing how a Monte Carlo Markov chain (MCMC), having (9.42) as its probability measure, is constructed on a given $\Sigma_{N\bar{v}}$.

We consider sequences of random values $\{x_i : i \in \Lambda\}$, with Λ the finite set of indexes of the elements of the partition of unity on $\Sigma_{N\bar{v}}$, and $x_i = (x_i^1, \ldots, x_i^{N-1})$ the local coordinates with respect to U_i of an arbitrary representative point of the set U_i itself. Then we define the weight $\pi(i)$ of the ith element of the partition as

$$\pi(i) = \left(\sum_{k=1}^{m} \int_{U_k} \varphi_k \frac{d\sigma}{\|\nabla V\|}\right)^{-1} \int_{U_i} \varphi_i \frac{d\sigma}{\|\nabla V\|} \tag{9.44}$$

and the transition matrix elements [220]

$$p_{ij} = \min\left[1, \frac{\pi(j)}{\pi(i)}\right] \tag{9.45}$$

that satisfy the detailed balance equation $\pi(i)p_{ij} = \pi(j)p_{ji}$. Starting from an arbitrary element of the partition, labeled by i_0, and using the transition probability (9.45), we obtain a random Markov chain $\{i_0, i_1 \ldots, i_k, \ldots\}$ of indexes and, consequently, a random Markov chain of points $\{x_{i_0}, x_{i_1}, \ldots, x_{i_k}, \ldots\}$ on the hypersurface $\Sigma_{N\bar{v}}$. Now let $(x_P^1, \ldots, x_P^{N-1})$ be the local coordinates of a point P on $\Sigma_{N\bar{v}}$ and define a local reference frame as $\{\partial/\partial x_P^1, \ldots, \partial/\partial x_P^{N-1}, n(P)\}$, where $n(P)$ is the outward unit normal vector at P; through the point-dependent matrix that operates the change from this basis to the canonical basis $\{e_1, \ldots, e_N\}$ of $\mathbb{R}^N$ we can associate to the Markov chain $\{x_{i_0}, x_{i_1}, \ldots, x_{i_k}, \ldots\}$ an equivalent chain $\{X_{i_0}, X_{i_1}, \ldots, X_{i_k}, \ldots\}$ of points identified through their coordinates in $\mathbb{R}^N$ but still constrained to belong to the subset $V(X) = v$, that is, to $\Sigma_{N\bar{v}}$. By construction, this Monte Carlo Markov chain has the probability density (9.42) as its invariant probability measure [220]. Moreover, for smooth functions F, smooth potentials V, and in the absence of critical points, $F/\|\nabla V\|$ has a limited variation on each set U_i. Thus the partition of unity can be made as fine-grained as needed—keeping it finite—to make Lebesgue integration convergent; hence equation (9.43) follows.   □

In part B we shall need the N-dependence of the momenta, up to the fourth order, of the sum of a large number N of mutually independent random variables. These N-dependencies are worked out in what follows by using and extending some results due to Khinchin [221].

Definition 9.23. *Let us consider a sequence $\{\eta_k\}_{k=1,\ldots,N}$ of mutually independent random quantities with probability densities $\{u_k(x)\}_{k=1,\ldots,N}$. Let us denote by $a_k = \int x\, u_k(x)\, dx$ the mean of the kth quantity and by*

$$b_k = \int (x - a_k)^2\, u_k(x)\, dx\ , \qquad c_k = \int |x - a_k|^3\, u_k(x)\, dx\ ,$$

$$d_k = \int (x - a_k)^4\, u_k(x)\, dx\ , \qquad e_k = \int |x - a_k|^5\, u_k(x)\, dx\ ,$$

its higher moments.

Theorem 9.24 (Khinchin). *Let us consider a sequence $\{\eta_k\}_{k=1,\ldots,N}$ of mutually independent random quantities with probability densities $\{u_k(x)\}_{k=1,\ldots,N}$. Without any significant loss of generality we assume that the a_k are zero. Under the conditions of validity of the central limit theorem (see [221]), the probability density $U_N(x)$ of $s_N = \sum_{k=1}^{N} \eta_k$ is given by*

$$U_N(x) = \frac{1}{(2\pi B_N)^{\frac{1}{2}}} \exp\left[-\frac{x^2}{2B_N} \right] + \frac{S_N + T_N x}{B_N^{\frac{5}{2}}}$$

$$+ O\left(\frac{1 + |x|^3}{N^2} \right), \qquad \forall\, |x| < 2\log^2 N\ , \tag{9.46}$$

$$U_N(x) = \frac{1}{(2\pi B_N)^{\frac{1}{2}}} \exp\left[-\frac{x^2}{2B_N} \right] + O\left(\frac{1}{N} \right), \qquad \forall x \in \mathbb{R}\ , \tag{9.47}$$

where $B_N = \sum_{i=1}^{N} b_i$ and where S_N and T_N are independent of x such that $\lim_{N \to \infty} N^{-1} S_N$ and $\lim_{N \to \infty} N^{-1} T_N$ are finite values (allowed to vanish) and where $\log^2 N$ stands for $(\log N)^2$.

Lemma 9.25. *Consider a sequence $\{\eta_k\}_{k=1,\ldots,N}$ of zero-mean, mutually independent random variables with probability densities $\{u_k(x)\}_{k=1,\ldots,N}$. Denote by B'_N, C'_N and D'_N the second, third, and fourth moments respectively of $s'_N = \frac{1}{N} \sum_{k=1}^{N} \eta_k$, and by $K'_N = D'_N - 3B'^2_N$ the fourth cumulant of s'_N.*
If the random quantities satisfy the hypotheses of the central limit theorem, then

$$\begin{aligned}
&(i) \quad \lim_{N \to \infty} N\, B'_N = \text{const} < \infty\ , \\
&(ii) \quad \lim_{N \to \infty} N^2\, C'_N = 0\ , \\
&(iii) \quad \lim_{N \to \infty} N^3\, K'_N = 0\ ,
\end{aligned} \tag{9.48}$$

Proof. Assertion (i). Let $\tilde{B}_N$ be the second moment of $s_N = \sum_{k=1}^{N} \eta_k$. By the above reported Khinchin theorem, we have

$$
\tilde{B}_N = \int | x |^2 \, \tilde{U}_N(x) dx
$$

$$
= \frac{1}{(2\pi B_N)^{\frac{1}{2}}} \int | x |^2 \exp\left[-\frac{x^2}{2B_N}\right] dx + \int | x |^2 R_N(x) dx
$$

where $R_N(x)$ is a remainder of order $1/N$. The right hand-side of this equation is the second moment of the Gaussian distribution, which is just B_N. Then $\tilde{B}_N$ can be rewritten, using again Khinchin's theorem, as

$$
\lim_{N\to\infty} \tilde{B}_N = \lim_{N\to\infty} B_N + \lim_{N\to\infty} \int_{|x|<2\log^2 N} | x |^2 \, \frac{S_N + T_N x}{B_N^{\frac{5}{2}}}
$$

$$
= \lim_{N\to\infty} B_N + \lim_{N\to\infty} \int_{|x|<2\log^2 N} | x |^2 \, \frac{S_N}{B_N^{\frac{5}{2}}}
$$

$$
= \lim_{N\to\infty} B_N + \frac{2^4}{3} \lim_{N\to\infty} \frac{S_N \, \log^6 N}{B_N^{\frac{5}{2}}} \; ,
$$

Now let $U'_N(x)$ be the probability density of $s'_N = \frac{1}{N} \sum_{k=1}^{N} \eta_k$. Its second moment B'_N is equal to

$$
B'_N = \int | x |^2 \, U'_N(x) dx = \frac{1}{N^2} \, \tilde{B}_N \; ,
$$

and thus

$$
\lim_{N\to\infty} N B'_N = \lim_{N\to\infty} \frac{B_N}{N} + \frac{2^4}{3} \lim_{N\to\infty} \frac{S_N \, \log^6 N}{N \, B_N^{\frac{5}{2}}} \; . \tag{9.49}
$$

Since $\lim_{N\to\infty} N^{-1} B_N$ is a finite nonvanishing value and $\lim_{N\to\infty} N^{-1} S_N$ is a finite value, we conclude that

$$
\lim_{N\to\infty} N B'_N = \mathrm{const} < \infty \; . \tag{9.50}
$$

Proof. Assertion (ii). Let $\tilde{C}_N$ be the third moment of $s_N = \sum_{k=1}^{N} \eta_k$. By Khinchin's theorem we have

$$
\tilde{C}_N = \int | x |^3 \, \tilde{U}_N(x) dx
$$

$$
= \frac{1}{(2\pi B_N)^{\frac{1}{2}}} \int | x |^3 \exp\left[-\frac{x^2}{2B_N}\right] dx + \int | x |^3 R_N(x) dx
$$

where $R_N(x)$ is a remainder of order $1/N$. The first term of the right hand-side is identically vanishing because it is an odd moment of a Gaussian distribution. Thus $\tilde{C}_N$ can be rewritten, using again Khinchin's theorem, as

$$
\lim_{N\to\infty} \tilde{C}_N = \lim_{N\to\infty} \int_{|x|<2\log^2 N} |x|^3 \frac{S_N + T_N x}{B_N^{\frac{5}{2}}}
$$

$$
= \lim_{N\to\infty} \int_{|x|<2\log^2 N} |x|^3 \frac{S_N}{B_N^{\frac{5}{2}}} = 2^3 \lim_{N\to\infty} \frac{S_N \log^8 N}{B_N^{\frac{5}{2}}} .
$$

Now let $U'_N(x)$ be the probability density of $s'_N = \frac{1}{N}\sum_{k=1}^{N} \eta_k$. Its third moment C'_N is equal to

$$
C'_N = \int |x|^3 U'_N(x)dx = \frac{1}{N^3} \tilde{C}_N ,
$$

which leads to the conclusion

$$
\lim_{N\to\infty} N^2 C'_N = 2^3 \lim_{N\to\infty} \frac{S_N \log^8 N}{N B_N^{\frac{5}{2}}} = 0 . \tag{9.51}
$$

Proof. Assertion (iii). Let $\tilde{K}_N$ be the fourth cumulant of $s_N = \sum_{k=1}^{N} \eta_k$. we have

$$
\tilde{K}_N = \frac{1}{3} \int x^4 \tilde{U}_N(x)dx - \left(\int x^2 \tilde{U}_N(x)dx \right)^2 , \tag{9.52}
$$

which, using Khinchin theorem, can be written as

$$
\tilde{K}_N = \frac{1}{3} \int x^4 G_N(x)dx - \left(\int x^2 G_N(x)dx \right)^2
$$

$$
+ \frac{1}{3} \int x^4 R_N(x)dx - \left(\int x^2 R_N(x)dx \right)^2
$$

$$
- 2 \int x^2 R_N(x)dx \int x^2 G_N(x)dx
$$

where $G_N(x) = (2\pi B_N)^{-\frac{1}{2}} \exp\left[-x^2/(2B_N)\right]$ is a Gaussian probability distribution and $R_N(x)$ the remainder of order $1/N$.

The sum of the first two terms of the right hand-side of the equation above is the fourth cumulant of a Gaussian distribution, thus vanishing.

Again using Khinchin's theorem we can write

$$
\lim_{N\to\infty} \tilde{K}_N = \frac{1}{3} \lim_{N\to\infty} \int_{|x|<2\log^2 N} x^4 \frac{S_N + T_N x}{B_N^{\frac{5}{2}}} dx
$$

$$
- \lim_{N\to\infty} \left(\int_{|x|<2\log^2 N} x^2 \frac{S_N + T_N x}{B_N^{\frac{5}{2}}} dx \right)^2
$$

$$- \lim_{N \to \infty} \int_{|x| < 2 \log^2 N} x^2 \frac{S_N + T_N x}{B_N^{\frac{5}{2}}} dx \int x^2 G_N(x) dx$$

$$= \frac{2^6}{15} \lim_{N \to \infty} \frac{\log^{10} N \ S_N}{B_N^{\frac{5}{2}}} - \frac{2^8}{9} \lim_{N \to \infty} \frac{\log^{12} N \ S_N^2}{B_N^5}$$

$$- \frac{2^4}{3} \lim_{N \to \infty} \frac{\log^6 N \ S_N}{B_N^{\frac{5}{2}}} . \tag{9.53}$$

Knowing that $\lim_{N \to \infty} N^{-1} B_N$ is a finite nonvanishing value, that $\lim_{N \to \infty} N^{-1} S_N$ is a finite value, that $\int x^2 G_N(x) dx \equiv B_N$, and that

$$K_N' = \frac{1}{3} \int |x|^4 U_N'(x) dx - \left(\int |x|^2 U_N'(x) dx \right)^2 = \frac{1}{N^4} \tilde{K}_N ,$$

we conclude that

$$\lim_{N \to \infty} N^3 K_N' = \frac{2^6}{15} \lim_{N \to \infty} \frac{\log^{10} N \ S_N}{N \ B_N^{\frac{5}{2}}} - \frac{2^8}{9} \lim_{N \to \infty} \frac{\log^{12} N \ S_N^2}{N} B_N^5$$

$$- \frac{2^4}{3} \lim_{N \to \infty} \frac{\log^6 N \ S_N}{N \ B_N^{\frac{3}{2}}} = 0 .$$

This completes the proof of our Lemma 9.25. $\square$

Remark 9.26. If V_N is a standard, confining, short-range, and stable potential, at large N the entropy function $S_N(\bar{v}) = \frac{1}{N} \log \Omega(N\bar{v}, N)$ is an intensive quantity, that is,

$$S_{2N}(\bar{v}) \approx S_N(\bar{v}) . \tag{9.54}$$

This is the obvious consequence of the well-known fact that

$$S_N(\Lambda^d, \bar{v}) = S_{N_1}(\Lambda_1^d, \bar{v}) + S_{N_2}(\Lambda_2^d, \bar{v}) + \mathcal{O}\left(\frac{\log N}{N} , \right) \tag{9.55}$$

which is proved in textbooks [98] and which has also the important consequence summarized in the following remark.

Remark 9.27. A consequence of equation (9.55) is that

$$\Omega^{1/N}(N\bar{v}, N_1 + N_2, \Lambda_1^d \cup \Lambda_2^d) = \Omega^{1/N_1}(N_1\bar{v}, N_1, \Lambda_1^d) \ \Omega^{1/N_2}(N_2\bar{v}, N_2, \Lambda_2^d) \ \theta(N) , \tag{9.56}$$

where $\theta(N) = \mathcal{O}(N^{1/N}) \to 1$ for $N \to \infty$. For two identical subsystems the potential energy is equally shared among them, with vanishing relative fluctuations in the $N \to \infty$ limit.

Remark 9.28. In the hypotheses of Theorem 9.14, V contains only short-range interactions and its functional form does not change with N, i.e., the functions Ψ and Φ in Definitions 9.3 and 9.4 do not depend on N. In other words, we are tackling physically homogeneous systems, which, at any N, can be considered as the union of smaller and identical subsystems. At large N, if a system is partitioned into a number k of sufficiently large subsystems, then the generalization to k components of the factorization of configuration space given in Remark 9.27 holds. Therefore, the averages of functions of interacting variables belonging to a given block depend neither on the subsystems where they are computed (the potential functions are the same on each block after suitable relabeling of the variables) nor on the total number N of degrees of freedom.

Lemma 9.29. *Let $\{x_i\}_{i=1,\dots,N}$ and $\{y_i\}_{i=1,\dots,N}$ be two independent sets of mutually independent nonnegative random quantities. Define $X = \sum_{i=1}^{N} x_i$ and $Y = \sum_{i=1}^{N} y_i$. Let $Y > 0$ for any realization of the random variables $\{y_i\}_{i=1,\dots,N}$. Let $\langle X \rangle$, $\langle Y \rangle$ denote the averages over an arbitrarily large number of realizations of the sets of random variables $\{x_i\}_{i=1,\dots,N}$ and $\{y_i\}_{i=1,\dots,N}$, respectively.*

In the limit $N \to \infty$, we have

$$\left\langle \frac{X}{Y} \right\rangle = \frac{\langle X \rangle}{\langle Y \rangle} \ .$$

Proof. By Khinchin's theorem recalled below Definition 9.23, in the large-N limit, both X and Y are Gaussian distributed random variables. Setting $\delta X = X - \langle X \rangle$ and $\delta(1/Y) = 1/Y - \langle 1/Y \rangle$, we have

$$\left\langle \frac{X}{Y} \right\rangle = \langle X \rangle \left\langle \frac{1}{Y} \right\rangle + \left\langle \delta X \ \delta \left(\frac{1}{Y} \right) \right\rangle \ . \tag{9.57}$$

Moreover,

$$\left\langle \delta X \ \delta \left(\frac{1}{Y} \right) \right\rangle \leq \left\langle \delta Z \ \delta \left(\frac{1}{Z} \right) \right\rangle \ ,$$

where $Z = X$ if $\langle (\delta X)^2 \rangle \geq \langle [\delta(1/Y)]^2 \rangle$ and $Z = Y$ if $\langle (\delta Y)^2 \rangle \geq \langle (\delta X)^2 \rangle$, and

$$\left\langle \delta Z \ \delta \left(\frac{1}{Z} \right) \right\rangle = 1 - 2\langle Z \rangle \left\langle \frac{1}{Z} \right\rangle + \langle Z \rangle^2 \left\langle \frac{1}{Z^2} \right\rangle \ . \tag{9.58}$$

Now, for a Gaussian random variable Z such that $\langle Z \rangle > 0$, we have

$$\left\langle \frac{1}{Z} \right\rangle = \frac{1}{\langle Z \rangle} \left\langle \frac{1}{1 + (Z - \langle Z \rangle)/\langle Z \rangle} \right\rangle = \frac{1}{\langle Z \rangle} \left[1 + \frac{\langle (Z - \langle Z \rangle)^2 \rangle}{3 \langle Z \rangle^2} - \cdots \right] \ ,$$

where all the terms with odd powers in the series expansion of $1/(1 + \delta Z / \langle Z \rangle)$ vanish, and the even-power terms are powers of the quadratic term, which is $O(1/N)$, thus in the limit $N \to \infty$,

$$\left\langle \frac{1}{Z} \right\rangle = \frac{1}{\langle Z \rangle} \,. \tag{9.59}$$

Using (9.59) in (9.58) we get

$$\left\langle \delta X \; \delta \left(\frac{1}{Y} \right) \right\rangle \leq -1 + \frac{\langle Z \rangle^2}{\langle Z^2 \rangle} = O(1/N) \,,$$

which, used in (9.57) together with (9.59), leads to the final result. $\square$

9.4.2 Part B

This part is devoted to the proof of the existence of uniform upper bounds as affirmed in Lemma 9.18.

We shall prove that the *supremum* on N and on $\bar{v} \in I_{\bar{v}}$ exists of up to the fourth derivative of $S_N(\bar{v})$. The proof of the existence of $\sup_N$ will be given by showing that the functions considered have a finite value in the $N \to \infty$ limit for any $\bar{v} \in I_{\bar{v}}$. The existence of the *supremum* on $\bar{v}$ is then a consequence of compactness[4] of the set $I_{\bar{v}}$.

Remark 9.30. In what follows, a detailed proof is given for lattice potentials V_N. However, in the fluid case the only difference is that the number of particles interacting with a given one is not preassigned. For this reason, in the fluid case, the number of particles within the interaction range of any other particle has to be replaced by its average.

Proof of $\sup_{N, \bar{v} \in I_{\bar{v}}} \left| \frac{\partial S_N}{\partial \bar{v}} (\bar{v}) \right| < \infty$

By definition of S_N we have

$$\frac{\partial S_N}{\partial \bar{v}} (\bar{v}) = \frac{1}{N} \frac{\Omega'_N(v)}{\Omega_N(v)} \cdot \frac{dv}{d\bar{v}} = \frac{\Omega'_N(v)}{\Omega_N(v)} \,,$$

where $\Omega'_N(v)$ stands for the derivative of $\Omega_N(v)$ with respect to the potential energy value $v = N\bar{v}$.

The assumptions of our main theorem allow the use of the Federer–Laurence theorem enunciated in Section 9.3 and of the derivation formula given therein. Thus

$$\Omega'_N(v) = \int_{\Sigma_v} \|\nabla V\| A \left(\frac{1}{\|\nabla V\|} \right) \frac{d\sigma}{\|\nabla V\|} \,, \tag{9.60}$$

---

[4] Since at any finite N all these functions are C^∞, the *supremum* always exists for finite N.

whence

$$\frac{\partial S_N}{\partial \bar{v}}(\bar{v}) = \frac{\Omega'_N(v)}{\Omega_N(v)} = \langle \|\nabla V\| A(1/\|\nabla V\|) \rangle^{\mu c}_{N,v} , \qquad (9.61)$$

where $\langle \ \rangle^{\mu c}_{N,v}$ stands for the configurational microcanonical average performed on the equipotential hypersurface of level v.

Let us proceed to show that this derivative is bounded by a term that is independent of N.

To ease notations, we define

$$\chi \equiv \frac{1}{\|\nabla V\|} , \qquad (9.62)$$

so that (9.61) now reads

$$\frac{\partial S_N}{\partial \bar{v}}(\bar{v}) = \left\langle \frac{1}{\chi} A(\chi) \right\rangle^{\mu c}_{N,v} . \qquad (9.63)$$

We then have

$$\frac{1}{\chi} A(\chi) = \frac{\Delta V}{\|\nabla V\|^2} - 2 \frac{\sum_{i,j=1}^{N} \partial^i V \partial^2_{ij} V \partial^j V}{\|\nabla V\|^4} , \qquad (9.64)$$

and hence

$$\left| \frac{1}{\chi} A(\chi) \right| \leq \frac{|\Delta V|}{\|\nabla V\|^2} + 2 \frac{|\sum_{i,j=1}^{N} \partial^i V \partial^2_{ij} V \partial^j V|}{\|\nabla V\|^4} ,$$

where $\partial_i V = \partial V/\partial q^i$, q^i being the ith coordinate of configuration space $\mathbb{R}^N$.

In the absence of critical points of V we have $\|\nabla V\|^2 \geq C > 0$. Thus we can apply Lemma 9.29, where $Y > 0$ is required, to obtain

$$\left| \frac{\partial S_N}{\partial \bar{v}}(\bar{v}) \right| = \left| \left\langle \frac{1}{\chi} A(\chi) \right\rangle^{\mu c}_{N,v} \right| \leq \left\langle \left| \frac{1}{\chi} A(\chi) \right| \right\rangle^{\mu c}_{N,v}$$

$$\leq \left\langle \frac{|\Delta V|}{\|\nabla V\|^2} \right\rangle^{\mu c}_{N,v} + 2 \left\langle \frac{|\sum_{i,j=1}^{N} \partial^i V \partial^2_{ij} V \partial^j V|}{\|\nabla V\|^4} \right\rangle^{\mu c}_{N,v}$$

$$\leq \frac{\langle |\Delta V| \rangle^{\mu c}_{N,v}}{\langle \|\nabla V\|^2 \rangle^{\mu c}_{N,v}} + O\left(\frac{1}{N}\right)$$

$$+ 2 \frac{\left\langle \sum_{i,j=1}^{N} |\partial^i V \partial^2_{ij} V \partial^j V| \right\rangle^{\mu c}_{N,v}}{\langle \|\nabla V\|^4 \rangle^{\mu c}_{N,v}} + O\left(\frac{1}{N^2}\right) .$$

Consider now the term $\langle |\Delta V| \rangle^{\mu c}_{N,v}$. Since the potential V is assumed smooth and bounded below, one has

$$\langle\mid \Delta V\mid\rangle^{\mu c}_{N,v} = \left\langle\left|\left|\sum_{i=1}^{N}\partial^2_{ii}V\right|\right|\right\rangle^{\mu c}_{N,v} \leq \sum_{i=1}^{N}\langle\mid \partial^2_{ii}V\mid\rangle^{\mu c}_{N,v} \leq N\max_{i=1,\dots,N}\langle(\mid \partial^2_{ii}V\mid)\rangle^{\mu c}_{N,v}.$$

As a consequence of Remark 9.28, at large N (when the fluctuations of the averages are vanishingly small) $\max_{i=1,\dots,N}\langle\mid \partial^2_{ii}V\mid\rangle^{\mu c}_{N,v}$ does not depend on N. The same holds for $\langle\mid \partial^i V\partial^2_{ij}V\partial^j V\mid\rangle^{\mu c}_{N,v}$ and $\max_{i=1,\dots,N}\langle\mid \partial^i V\partial^2_{ij}V\partial^j V\mid\rangle^{\mu c}_{N,v}$. We set $m_1 = \max_{i=1,\dots,N}\langle\mid \partial^2_{ii}V\mid\rangle^{\mu c}_{N,v}$ and $m_2 = \max_{i,j=1,\dots,N}\langle\mid \partial^i V\partial^2_{ij}V\partial^j V\mid\rangle^{\mu c}_{N,v}$.

Let us now consider the terms $\langle\|\nabla V\|^{2n}\rangle^{\mu c}_{N,v}$ for $n = 1, 2$. One has

$$\langle\|\nabla V\|^2\rangle^{\mu c}_{N,v} = \left\langle\sum_{i=1}^{N}(\partial_i V)^2\right\rangle^{\mu c}_{N,v} = \sum_{i=1}^{N}\langle(\partial_i V)^2\rangle^{\mu c}_{N,v} \geq N\min_{i=1,\dots,N}\langle(\partial_i V)^2\rangle^{\mu c}_{N,v},$$

$$\langle\|\nabla V\|^4\rangle^{\mu c}_{N,v} = \left\langle\left[\sum_{i=1}^{N}(\partial_i V)^2\right]^2\right\rangle^{\mu c}_{N,v} = \sum_{i,j=1}^{N}\langle(\partial_i V)^2(\partial_j V)^2\rangle^{\mu c}_{N,v}$$

$$\geq N^2\min_{i,j=1,\dots,N}\langle(\partial_i V)^2(\partial_j V)^2\rangle^{\mu c}_{N,v}.$$

By setting

$$c_1 = \min_{i=1,\dots,N}\langle(\partial_i V)^2\rangle^{\mu c}_{N,v}$$

and

$$c_2 = \min_{i,j=1,\dots,N}\langle(\partial_i V)^2(\partial_j V)^2\rangle^{\mu c}_{N,v}$$

we can finally write

$$\left|\left\langle\frac{1}{\chi}A(\chi)\right\rangle^{\mu c}_{N,v}\right| \leq \frac{m_1}{c_1} + O\left(\frac{1}{N}\right) + 2\frac{n_p\,m_2}{c_2 N} + O\left(\frac{1}{N^2}\right), \tag{9.65}$$

where n_p is the number of nearest neighbors. It is evident that in the limit $N \to \infty$ the right hand-side of the equation above tends to the finite constant m_1/c_1.

The upper bound thus obtained ensures that $\sup_{N,\bar{v}\in I_{\bar{v}}}\left|\frac{\partial S_N}{\partial\bar{v}}(\bar{v})\right| < \infty$. $\square$

Remark 9.31. Notice that in the fluid case, the computation of quantities like $\langle(\partial_i V)^2\rangle^{\mu c}_{N,v}$ and $\langle\mid \partial^2_{ii}V\mid\rangle^{\mu c}_{N,v}$ involves an a priori unknown number of neighbors of the ith particle (we say that a particle is a neighbor of another one if the distance between the two particles is smaller than the interaction range of the potential). However, the requirement that V be repulsive at short distance, so that clusters of an arbitrary number of particles are forbidden, guarantees that each particle has a finite average number of neighbors. Thus, averaging quantities like the above-mentioned ones yields N independent values.

In order to extend to the fluid case the proofs of uniform boundedness of the derivatives of the entropy (one has to interpret n_p as the average number of neighbors of a given particle.

Remark 9.32. Notice that the above computations show that

$$\lim_{N \longrightarrow \infty} \left\langle \frac{A(\chi)}{\chi} \right\rangle_{N,v}^{\mu c} = \text{const} < \infty \ ,$$

which follows from the boundedness of $|\langle A(\chi)/\chi \rangle|$.

Proof of $\sup_{N, \bar{v} \in I_{\bar{v}}} \left| \frac{\partial^2 S_N}{\partial \bar{v}^2} (\bar{v}) \right| < \infty$

The second derivative of S_N can be rewritten in the form

$$\frac{\partial^2 S_N}{\partial \bar{v}^2} (\bar{v}) = N \cdot \left[\frac{\Omega''(v, N)}{\Omega_N(v)} - \left(\frac{\Omega_N'(v)}{\Omega_N(v)} \right)^2 \right] \ , \tag{9.66}$$

or, by using the same notation as before,

$$\frac{\partial^2 S_N}{\partial \bar{v}^2} (\bar{v}) = N \left\{ \left\langle \frac{1}{\chi} A^2(\chi) \right\rangle_{N,v}^{\mu c} - \left[\left\langle \frac{1}{\chi} A(\chi) \right\rangle_{N,v}^{\mu c} \right]^2 \right\} \ . \tag{9.67}$$

Again we are going to show that an upper bound, independent of N, exists also for this derivative. In order to make notation compact, we define

$$\underline{\psi} \equiv \frac{\nabla}{\|\nabla V\|}$$

$$\text{for any } h_1, h_2, \ \underline{\psi}(h_1) \cdot \underline{\psi}(h_2) = \sum_{i=1}^{N} \psi_i(h_1)\psi_i(h_2) \ ,$$

whence simple algebra yields

$$\underline{\psi}(V) \cdot \underline{\psi}(\chi) = \chi^2 M_1 - \chi^3 \Delta V \ , \tag{9.68}$$

$$\underline{\psi}^2(V) \equiv \underline{\psi}\left(\cdot \underline{\psi}(V) \right) = \frac{1}{\chi} \underline{\psi}(V) \cdot \underline{\psi}(\chi) + \chi^2 \Delta V \ , \tag{9.69}$$

$$\psi_i(\psi_j(V)) = \chi^2 \partial_{ij}^2 V - \chi^2 \psi_j(V) \sum_{k=1}^{N} \psi_k(V) \partial_{ik}^2 V \ , \tag{9.70}$$

$$\psi_i(\chi) = -\chi^3 \sum_{j=1}^{N} \partial_{ij}^2 V \psi_j(V) \ , \tag{9.71}$$

$$\psi_i(\psi_j(V)) = \chi^2 \partial_{ij}^2 V - \chi^2 \psi_j(V) \sum_{k=1}^{N} \psi_k(V) \partial_{ik}^2 V \ , \tag{9.72}$$

$$\psi_i(\partial_{jr}^2 V) = \chi \partial_{ijr}^3 V \ , \tag{9.73}$$

$$\psi_i(\partial_{jj}^2 V) = \chi \partial_{ijj}^3 V \ , \tag{9.74}$$

where $M_1 = \nabla(\nabla V/\|\nabla V\|) \equiv -N \cdot (mean\ curvature\ of\ \Sigma_v)$. With this notation we have

$$A^2(\chi) = A(A(\chi)) = A\left(\underline{\psi}(V) \cdot \underline{\psi}(\chi) + \chi^3 \Delta V\right)$$
$$= \frac{1}{\chi}(A(\chi))^2 + \chi\underline{\psi}(V) \cdot \underline{\psi}\left(\frac{A(\chi)}{\chi}\right), \tag{9.75}$$

and thus (9.67) now reads

$$\left|\frac{\partial^2 S_N}{\partial \bar{v}^2}(\bar{v})\right| \leq N \left|\left\langle\left[\frac{A(\chi)}{\chi}\right]^2\right\rangle_{N,v}^{\mu c} - \left[\left\langle\frac{A(\chi)}{\chi}\right\rangle_{N,v}^{\mu c}\right]^2\right|$$
$$+ N \left|\left\langle\underline{\psi}(V) \cdot \underline{\psi}\left(\frac{A(\chi)}{\chi}\right)\right\rangle_{N,v}^{\mu c}\right|. \tag{9.76}$$

Using the relations (9.68)–(9.74), the term $\frac{1}{\chi}A(\chi)$ is rewritten as

$$\frac{A(\chi)}{\chi} = \frac{1}{\chi}\underline{\psi}\left(\cdot\underline{\psi}(V)\chi\right) = \frac{2}{\chi}\underline{\psi}(V) \cdot \underline{\psi}(\chi) + \chi^2 \Delta V$$
$$= 2\chi M_1 - \chi^2 \Delta V$$
$$= \frac{\Delta V}{\|\nabla V\|^2} - 2\frac{\sum_{i,j=1}^{N} \partial^i V \partial_{ij}^2 V \partial^j V}{\|\nabla V\|^4}. \tag{9.77}$$

Now we consider the following inequalities:

$$\left|\left\langle\frac{\sum_{i,j=1}^{N} \partial^i V \partial_{ij}^2 V \partial^j V}{\|\nabla V\|^4}\right\rangle_{N,v}^{\mu c}\right| \leq \left\langle\frac{\left|\sum_{\langle i,j\rangle\ ;\ i,j=1}^{N} \partial^i V \partial_{ij}^2 V \partial^j V\right|}{\|\nabla V\|^4}\right\rangle_{N,v}^{\mu c}$$
$$\leq \frac{\sum_{\langle i,j\rangle\ ;\ i,j=1}^{N} \left\langle|\partial^i V \partial_{ij}^2 V \partial^j V|\right\rangle_{N,v}^{\mu c}}{\left\langle\|\nabla V\|^4\right\rangle_{N,v}^{\mu c}} + O\left(\frac{1}{N^2}\right)$$
$$\leq \frac{N\, n_p\, m_2}{c_2 N^2} + O\left(\frac{1}{N^2}\right), \tag{9.78}$$

where n_p is the number of nearest neighbors, and again

$$m_2 = \max_{i,j=1,\ldots,N} \left\langle| \partial^i V \partial_{ij}^2 V \partial^j V |\right\rangle_{N,v}^{\mu c}.$$

Since m_2 keeps a finite value for $\lim_{N\to\infty}$, the left hand-side of equation (9.78) vanishes in the $N \to \infty$ limit.

Thus, the larger N, the better the term $\frac{1}{\chi}A(\chi)$ is approximated by $\xi = \sum_{i=1}^{N} \partial_{ii}^2 V/\|\nabla V\|^2 = \sum_{i=1}^{N} \xi_i$, where $\xi_i = \partial_{ii}^2 V/\|\nabla V\|^2$. Here we resort to Lemma 9.22 and replace the microcanonical averages by "time" averages obtained along an ergodic stochastic process. Each term ξ_i, for any i, can then be considered as a stochastic process on the manifold Σ_v with a probability density $u_i(\xi_i)$. In the presence of short-range potentials, as prescribed in the hypotheses of our main theorem, and at large N, these processes are independent.

By simply writing $\xi = \sum_{i=1}^{N} \xi_i = 1/N \sum_{i=1}^{N} N\xi_i$, we are allowed to apply Lemma 9.25, which tells us that the the second moment B'_N of the distribution of ξ is such that $\lim_{N\to\infty} N\, B'_N = c < \infty$.

The first term of the right hand-side of (9.76) is the second moment of $\frac{1}{\chi}A(\chi)$ multiplied by N. This term, in the light of what we have just seen, remains finite in the $N \to \infty$ limit.

Then we consider the second term of the right hand-side of equation (9.76). This can be computed with simple algebra through the relations (9.68)–(9.74) to give

$$
\underline{\psi}(V) \cdot \underline{\psi}\left(\frac{A(\chi)}{\chi}\right) = 8\chi^4 \left(\langle \underline{\psi}(V); \underline{\psi}(V)\rangle\right)^2 - 4\chi^4 \langle \underline{\psi}(V)|\underline{\psi}(V)\rangle
$$

$$
-2\chi^4 \langle \underline{\psi}(V); \underline{\psi}(V)\rangle \Delta V + \chi^3 \sum_{i,j=1}^{N} \psi_i(V)\partial^3_{ijj}V
$$

$$
-2\chi^3 \sum_{i,j,k=1}^{N} \psi_i(V)\psi_j(V)\psi_k(V)\partial^3_{ijk}V \ , \tag{9.79}
$$

where

$$
\langle \underline{\psi}(V); \underline{\psi}(V)\rangle \equiv \frac{\sum_{i,j=1}^{N} \partial_i V \partial^2_{ij} V \partial_j V}{\|\nabla V\|^2} \ , \tag{9.80}
$$

$$
\langle \underline{\psi}(V)|\underline{\psi}(V)\rangle \equiv \frac{\sum_{i,j,k=1}^{N} \partial_i V \partial^2_{ij} V \partial^2_{jk} V \partial_k V}{\|\nabla V\|^2} \ , \tag{9.81}
$$

$$
\psi_i(V)\partial^3_{ijj}V \equiv \frac{\partial_i V \partial^3_{ijj} V}{\|\nabla V\|} \ , \tag{9.82}
$$

$$
\psi_i(V)\psi_j(V)\psi_k(V)\partial^3_{ijk}V \equiv \frac{\partial_i V \partial_j V \partial_k V \partial^3_{ijk} V}{\|\nabla V\|^3} \ . \tag{9.83}
$$

The same kind of computation developed for equations (9.78) gives

$$
N\left\langle \chi^4 \left(\langle \underline{\psi}(V); \underline{\psi}(V)\rangle\right)^2 \right\rangle_{N,v}^{\mu c} \leq \frac{N^3 n_p^2 m_4}{c_4 N^4} + O\left(\frac{1}{N^2}\right), \tag{9.84}
$$

$$
N\left\langle \chi^4 \langle \underline{\psi}(V)|\underline{\psi}(V)\rangle \right\rangle_{N,v}^{\mu c} \leq \frac{N^2 n_p^2 m_5}{c_3 N^3} + O\left(\frac{1}{N^2}\right), \tag{9.85}
$$

$$
N\left\langle \chi^4 \langle \underline{\psi}(V); \underline{\psi}(V)\rangle \Delta V \right\rangle_{N,v}^{\mu c} \leq \frac{N^3 n_p m_6}{c_3 N^3} + O\left(\frac{1}{N}\right), \tag{9.86}
$$

$$N \left\langle \chi^3 \sum_{i,j=1}^{N} \psi_i(V) \partial^3_{ijj} V \right\rangle^{\mu c}_{N,v} \leq \frac{N^2 n_p m_7}{c_2 N^2} + O\left(\frac{1}{N}\right), \quad (9.87)$$

$$N \left\langle \chi^3 \sum_{i,j,k=1}^{N} \psi_i(V)\psi_j(V)\psi_k(V) \partial^3_{ijk} V \right\rangle^{\mu c}_{N,v} \leq \frac{N^2 n_p^2 m_8}{c_3 N^3} + O\left(\frac{1}{N^2}\right), \quad (9.88)$$

where, resorting again to the argument of Remark 9.28, we have defined the following quantities independent of N:

$$m_4 = \max_{i,j,k,l=1,N} \left\langle (\partial_i V \partial^2_{ij} V \partial_j V)(\partial_k V \partial^2_{kl} V \partial_l V) \right\rangle^{\mu c}_{N,v} ,$$

$$m_5 = \max_{i,j,k=1,N} \left\langle \partial_i V \partial^2_{ij} V \partial^2_{jk} V \partial_k V \right\rangle^{\mu c}_{N,v} ,$$

$$m_6 = \max_{i,j,k=1,N} \left\langle (\partial_i V \partial^2_{ij} V \partial_j V)(\partial^2_{kk} V) \right\rangle^{\mu c}_{N,v} ,$$

$$m_7 = \max_{i,j=1,N} \left\langle \partial_i V \partial^3_{ijj} V \right\rangle^{\mu c}_{N,v} ,$$

$$m_8 = \max_{i,j,k=1,N} \left\langle (\partial_i V \partial_j V \partial_k V) \partial^3_{ijk} V \right\rangle^{\mu c}_{N,v} ,$$

and

$$c_3 = \min_{i_1,\dots,i_6=1,N} \left\langle (\partial_{i_1} V)^2 (\partial_{i_2} V)^2 \cdots (\partial_{i_6} V)^2 \right\rangle^{\mu c}_{N,v} ,$$

$$c_4 = \min_{i_1,\dots,i_8=1,N} \left\langle (\partial_{i_1} V)^2 (\partial_{i_2} V)^2 \cdots (\partial_{i_8} V)^2 \right\rangle^{\mu c}_{N,v} ,$$

so that the right hand-sides of (9.86) and (9.87) have finite limits for $N \to \infty$, while the right hand-sides of (9.84), (9.85), and (9.88) vanish in the limit $N \to \infty$.

In conclusion, since the ensemble of terms entering equation (9.76) is bounded above, we have $\sup_{N,\bar{v} \in I_{\bar{v}}} \left| \frac{\partial^2 S_N}{\partial \bar{v}^2}(\bar{v}) \right| < \infty$. $\qquad\square$

Remark 9.33. Notice that the above computations show that

$$\lim_{N \to \infty} N \left\langle \underline{\psi}(V) \cdot \underline{\psi}\left(\frac{A(\chi)}{\chi}\right) \right\rangle^{\mu c}_{N,v} = \text{const} < \infty .$$

Proof of $\sup_{N,\bar{v} \in I_{\bar{v}}} \left| \frac{\partial^3 S_N}{\partial \bar{v}^3}(\bar{v}) \right| < \infty$

The third derivative of S_N can be expressed as

$$\frac{\partial^3 S_N}{\partial \bar{v}^3}(\bar{v}) = N^2 \left\{ \frac{\Omega'''(v,N)}{\Omega(v,N)} - 3\frac{\Omega''(v,N)\Omega'(v,N)}{(\Omega(v,N))^2} + 2\left(\frac{\Omega'(v,N)}{\Omega(v,N)}\right)^3 \right\},$$

$$(9.89)$$

or, using Federer's operator A,

$$\frac{\partial^3 S_N}{\partial \bar{v}^3}(\bar{v}) = N^2 \left\{ \left\langle \frac{A^3(\chi)}{\chi} \right\rangle_{N,v}^{\mu c} - 3 \left\langle \frac{A^2(\chi)}{\chi} \right\rangle_{N,v}^{\mu c} \left\langle \frac{A(\chi)}{\chi} \right\rangle_{N,v}^{\mu c} + 2 \left(\left\langle \frac{A(\chi)}{\chi} \right\rangle_{N,v}^{\mu c} \right)^3 \right\}, \tag{9.90}$$

where

$$\frac{A^3(\chi)}{\chi} = \left(\frac{A(\chi)}{\chi} \right)^3 + 3 \frac{A(\chi)}{\chi} \, \underline{\psi}(V) \cdot \underline{\psi} \left(\frac{A(\chi)}{\chi} \right)$$
$$+ \underline{\psi}(V) \cdot \underline{\psi} \left(\underline{\psi}(V) \cdot \underline{\psi} \left(\frac{A(\chi)}{\chi} \right) \right) \tag{9.91}$$

$$\frac{A^2(\chi)}{\chi} = \left(\frac{A(\chi)}{\chi} \right)^2 + \underline{\psi}(V) \cdot \underline{\psi} \left(\frac{A(\chi)}{\chi} \right) \tag{9.92}$$

$$\frac{A(\chi)}{\chi} = \frac{2}{\chi} \underline{\psi}(V) \cdot \underline{\psi}(\chi) + \frac{\Delta V}{\|\nabla V\|^2} . \tag{9.93}$$

By substituting the expressions (9.91)–(9.93) into the right hand-side of equation (9.90), we get

$$\left| \frac{\partial^3 S_N}{\partial \bar{v}^3}(\bar{v}) \right|$$

$$\leq N^2 \left| \left\langle \underline{\psi}(V) \cdot \underline{\psi} \left(\underline{\psi}(V) \cdot \underline{\psi} \left(\frac{A(\chi)}{\chi} \right) \right) \right\rangle_{N,v}^{\mu c} \right|$$

$$+ 3N^2 \left| \left\langle \frac{A(\chi)}{\chi} \underline{\psi}(V) \cdot \underline{\psi} \left(\frac{A(\chi)}{\chi} \right) \right\rangle_{N,v}^{\mu c} \right.$$

$$\left. - \left\langle \frac{A(\chi)}{\chi} \right\rangle_{N,v}^{\mu c} \left\langle \underline{\psi}(V) \cdot \underline{\psi} \left(\frac{A(\chi)}{\chi} \right) \right\rangle_{N,v}^{\mu c} \right|$$

$$+ N^2 \left| \left\langle \left(\left(\frac{A(\chi)}{\chi} \right) - \left\langle \left(\frac{A(\chi)}{\chi} \right) \right\rangle_{N,v}^{\mu c} \right)^3 \right\rangle_{N,v}^{\mu c} \right| . \tag{9.94}$$

By explicitly expanding the first term of the right hand-side of (9.94), more than 30 terms are found. Nevertheless, these terms are similar or equal to those already encountered above, and consequently, their N-dependence can be similarly dominated as in the inequalities (9.84)–(9.88).

Consider now the second term of the right hand-side of equation (9.94). If we put

$$\mathcal{A} = \frac{A(\chi)}{\chi} , \qquad \mathcal{P} = \underline{\psi}(V) \cdot \underline{\psi} \left(\frac{A(\chi)}{\chi} \right) ,$$

using equations (9.64) and (9.79) we can write

$$\mathcal{A} = \sum_{i=1}^{N} a_i , \qquad \mathcal{P} = \sum_{j=1}^{N} p_j .$$

Then

$$\left\langle \frac{A(\chi)}{\chi}\, \underline{\psi}(V) \cdot \underline{\psi}\left(\frac{A(\chi)}{\chi}\right)\right\rangle_{N,v}^{\mu c} - \left\langle \frac{A(\chi)}{\chi}\right\rangle_{N,v}^{\mu c}\left\langle \underline{\psi}(V) \cdot \underline{\psi}\left(\frac{A(\chi)}{\chi}\right)\right\rangle_{N,v}^{\mu c}$$

$$= \langle \mathcal{A}\mathcal{P}\rangle_{N,v}^{\mu c} - \langle \mathcal{A}\rangle_{N,v}^{\mu c}\langle \mathcal{P}\rangle_{N,v}^{\mu c}$$

$$= \sum_{i,j=1}^{N}\left(\langle a_i p_j\rangle_{N,v}^{\mu c} - \langle a_i\rangle_{N,v}^{\mu c}\langle p_j\rangle_{N,v}^{\mu c}\right) . \tag{9.95}$$

Let us consider the terms in the last sum for which i and j label sites that are not nearest-neighbors.[5] The corresponding expressions of a_i and p_j have no common coordinate variables. Thus, when computing microcanonical averages through "time" averages along the random Markov chains of Lemma 9.22, we take advantage of the complete decorrelation of a_i and p_j so that

for any i,j such that $0 \leq i,j \leq N, \rangle i, j\langle, then \langle a_i p_j\rangle_{N,v}^{\mu c} - \langle a_i\rangle_{N,v}^{\mu c}\langle p_j\rangle_{N,v}^{\mu c} = 0$

(where $\rangle i, j\langle$ stands for i,j nonnearest neighbors), which simplifies equation (9.95) to

$$\langle \mathcal{A}\mathcal{P}\rangle_{N,v}^{\mu c} - \langle \mathcal{A}\rangle_{N,v}^{\mu c}\langle \mathcal{P}\rangle_{N,v}^{\mu c} = \sum_{\langle i,j\rangle}\left(\langle a_i p_j\rangle_{N,v}^{\mu c} - \langle a_i\rangle_{N,v}^{\mu c}\langle p_j\rangle_{N,v}^{\mu c}\right)$$

$$\leq N\, n_p \max_{\langle i,j\rangle}\left(\langle a_i p_j\rangle_{N,v}^{\mu c} - \langle a_i\rangle_{N,v}^{\mu c}\langle p_j\rangle_{N,v}^{\mu c}\right) .$$

Now, equations (9.65) and (9.84)–(9.88) imply

for any i,j such that $0 \leq i,j \leq N$, $\langle i,j\rangle$, $\lim\limits_{N\longrightarrow\infty} N^3\, \langle a_i p_j\rangle_{N,v}^{\mu c} < \infty$,

while equations (9.64) and (9.79) imply

for any i,j such that $0 \leq i,j \leq N$, $\langle i,j\rangle$, $\lim\limits_{N\longrightarrow\infty} N^3\, \langle a_i\rangle_{N,v}^{\mu c}\langle p_j\rangle_{N,v}^{\mu c} < \infty$,

where $\langle i,j\rangle$ stands for i,j nearest neighbors. Thus, the second term in the right hand-side of equation (9.94) is bounded independently of N in the limit $N \to \infty$.

The third term of the right hand-side of equation (9.94) is smaller than the third moment of the stochastic variable $A(\chi)/\chi$ (multiplied by N^2). As we have already seen, we can rewrite $A(\chi)/\chi = (1/N)\sum_{i=1}^{N} N\partial_{ii}^2 V/\|\nabla V\|^2$, to which Lemma 9.25 applies, thus ensuring that the third moment C_N' of the distribution of $A(\chi)/\chi$ is such that $\lim_{N\to\infty} N^2\, C_N' = 0$.

Finally we are left with a finite upper bound of the left hand-side of equation (9.94) in the $N \to \infty$ limit. □

---

[5] For simplicity we are here assuming that the configurational coordinates belong to a lattice, but such a restriction is not necessary. If our potential describes a fluid, replace "nearest neighbors" with "within the interaction range."

Remark 9.34. Notice that the computations above show that

$$\lim_{N\to\infty} N^2 \left\langle \underline{\psi}(V) \cdot \underline{\psi} \left(\underline{\psi}(V) \cdot \underline{\psi} \left(\frac{A(\chi)}{\chi} \right) \right) \right\rangle_{N,v}^{\mu c} = \text{const} < \infty .$$

Proof of $\sup_{N,\bar{v}\in I_{\bar{v}}} \left| \frac{\partial^4 S_N}{\partial \bar{v}^4}(\bar{v}) \right| < \infty$

The fourth derivative of $S_N(\bar{v})$ is given by the expression

$$\frac{\partial^4 S_N}{\partial \bar{v}^4}(\bar{v}) = N^3 \left\{ \frac{\Omega^{iv}(v,N)}{\Omega(v,N)} - 4\frac{\Omega'''(v,N)\,\Omega'(v,N)}{(\Omega(v,N))^2} - 3\left(\frac{\Omega''(v,N)}{\Omega(v,N)} \right)^2 \right\}$$

$$+N^3 \left\{ 12\frac{\Omega''(v,N)\,(\Omega'(v,N))^2}{(\Omega(v,N))^3} - 6\left(\frac{\Omega'(v,N)}{\Omega(v,N)} \right)^4 \right\} .$$

Again we make use of the Federer operator A to rewrite it as

$$\frac{\partial^4 S_N}{\partial \bar{v}^4}(\bar{v}) = N^3 \left\{ \left\langle \frac{A^4(\chi)}{\chi} \right\rangle_{N,v}^{\mu c} - 4\left\langle \frac{A^3(\chi)}{\chi} \right\rangle_{N,v}^{\mu c} \left\langle \frac{A(\chi)}{\chi} \right\rangle_{N,v}^{\mu c} \right\}$$

$$-N^3 \left\{ 3\left(\left\langle \frac{A^2(\chi)}{\chi} \right\rangle_{N,v}^{\mu c} \right)^2 - 12\left\langle \frac{A^2(\chi)}{\chi} \right\rangle_{N,v}^{\mu c} \left(\left\langle \frac{A(\chi)}{\chi} \right\rangle_{N,v}^{\mu c} \right)^2 \right\}$$

$$-6N^3 \left(\left\langle \frac{A(\chi)}{\chi} \right\rangle_{N,v}^{\mu c} \right)^4 ,$$

where, after trivial algebra,

$$\frac{A^4(\chi)}{\chi} = \left(\frac{A(\chi)}{\chi} \right)^4 + 6\left(\frac{A(\chi)}{\chi} \right)^2 \psi(V) \cdot \psi\left(\frac{A(\chi)}{\chi} \right)$$

$$+3\left(\psi(V) \cdot \psi\left(\frac{A(\chi)}{\chi} \right) \right)^2 + 4\frac{A(\chi)}{\chi} \psi(V) \cdot \psi\left(\psi(V) \cdot \psi\left(\frac{A(\chi)}{\chi} \right) \right)$$

$$+\psi(V) \cdot \psi\left[\psi(V) \cdot \psi\left(\psi(V) \cdot \psi\left(\frac{A(\chi)}{\chi} \right) \right) \right] . \tag{9.96}$$

To make the notation more compact we use

$$\mathcal{A} = \frac{A(\chi)}{\chi} , \qquad \mathcal{P} = \psi(V) \cdot \psi\left(\frac{A(\chi)}{\chi} \right) ,$$

$$\mathcal{W} = \psi(V) \cdot \psi\left(\psi(V) \cdot \psi\left(\frac{A(\chi)}{\chi} \right) \right) ,$$

so that, using again equations (9.91)–(9.92), we obtain

$$
\left| \frac{\partial^4 S_N}{\partial \bar{v}^4}(\bar{v}) \right| \leq N^3 \left| \langle \psi(V) \cdot \psi(\mathcal{W}) \rangle_{N,v}^{\mu c} \right|
$$

$$
+3N^3 \left| \langle \mathcal{P}^2 \rangle_{N,v}^{\mu c} - \left(\langle \mathcal{P} \rangle_{N,v}^{\mu c} \right)^2 \right|
$$

$$
+4N^3 \left| \langle \mathcal{A}\mathcal{W} \rangle_{N,v}^{\mu c} - \langle \mathcal{A} \rangle_{N,v}^{\mu c} \langle \mathcal{W} \rangle_{N,v}^{\mu c} \right| \tag{9.97}
$$

$$
+6N^3 \left| \left\langle \left(\mathcal{A} - \langle \mathcal{A} \rangle_{N,v}^{\mu c} \right)^2 \left(\mathcal{P} - \langle \mathcal{P} \rangle_{N,v}^{\mu c} \right) \right\rangle_{N,v}^{\mu c} \right|
$$

$$
+N^3 \left| \left\langle \left(\mathcal{A} - \langle \mathcal{A} \rangle_{N,v}^{\mu c} \right)^4 \right\rangle_{N,v}^{\mu c} - 3 \left(\left\langle \left(\mathcal{A} - \langle \mathcal{A} \rangle_{N,v}^{\mu c} \right)^2 \right\rangle_{N,v}^{\mu c} \right)^2 \right| .
$$

Consider the first term of equation (9.97). It is an iterative term already considered for the third derivative. This term stems from the application of the operator $\psi(V) \cdot \psi(\cdot)$ to the term $\mathcal{W}$, which in its turn stems from the application of the same operator to the term $\mathcal{P}$. The effect of this operator is to lower the N dependence of the function to which it is applied by a factor N (which is simply due to the factor $1/\|\nabla V\|^2$). Deriving with respect to $\bar{v}$ brings about a factor N in comparison to the derivation with respect to v. Therefore the first term of equation (9.97) is of the same order of $N^2 \langle \mathcal{W} \rangle_{N,v}^{\mu c}$ and consequently, according to the Remark 9.34, it has a finite upper bound independent of N in the limit $N \to \infty$.

Consider now the second term of the right hand-side of equation (9.97). Remark 9.33 ensures that $\lim_{N\to\infty} N \langle \mathcal{P} \rangle_{N,v}^{\mu c} < \infty$. Moreover, by Lemma 9.25,

$$
\lim_{N\to\infty} N^3 \left(\left\langle \mathcal{P} - \langle \mathcal{P} \rangle_{N,v}^{\mu c} \right\rangle_{N,v}^{\mu c} \right)^2 < \infty . \tag{9.98}
$$

Consider now the third term of the right hand-side of equation (9.97). Remarks 9.32 and 9.34 entail $\lim_{N\to\infty} \langle \mathcal{A} \rangle_{N,v}^{\mu c} < \infty$ and $\lim_{N\to\infty} N^2 \langle \mathcal{W} \rangle_{N,v}^{\mu c} < \infty$. Thus, after Lemma 9.25,

$$
\lim_{N\to\infty} N^{\frac{1}{2}} \left(\left\langle \mathcal{A} - \langle \mathcal{A} \rangle_{N,v}^{\mu c} \right\rangle_{N,v}^{\mu c} \right) < \infty ,
$$

$$
\lim_{N\to\infty} N^{\frac{5}{2}} \left(\left\langle \mathcal{W} - \langle \mathcal{W} \rangle_{N,v}^{\mu c} \right\rangle_{N,v}^{\mu c} \right) < \infty ,
$$

whence

$$
\lim_{N\to\infty} N^3 \left| \langle \mathcal{A}\mathcal{W} \rangle_{N,v}^{\mu c} - \langle \mathcal{A} \rangle_{N,v}^{\mu c} \langle \mathcal{W} \rangle_{N,v}^{\mu c} \right|
$$

$$
= \lim_{N\to\infty} N^3 \left| \left\langle \mathcal{A} - \langle \mathcal{A} \rangle_{N,v}^{\mu c} \right\rangle_{N,v}^{\mu c} \right| \left| \left\langle \mathcal{W} - \langle \mathcal{W} \rangle_{N,v}^{\mu c} \right\rangle_{N,v}^{\mu c} \right| < \infty . \tag{9.99}
$$

Consider now the fourth term of the right hand-side of equation (9.97). If we write

$$\mathcal{A} = \frac{1}{N} \sum_{i=1}^{N} a_i , \qquad \mathcal{P} = \frac{1}{N^2} \sum_{i=1}^{N} p_i ,$$

with a_i and p_i terms of order 1, we have

$$N^3 \left| \left\langle \left(\mathcal{A} - \langle \mathcal{A} \rangle_{N,v}^{\mu c} \right)^2 \left(\mathcal{P} - \langle \mathcal{P} \rangle_{N,v}^{\mu c} \right) \right\rangle_{N,v}^{\mu c} \right|$$

$$= \frac{1}{N} \sum_{i,j,k=1}^{N} \left\langle \left(a_i - \langle a_i \rangle_{N,v}^{\mu c} \right) \left(a_j - \langle a_j \rangle_{N,v}^{\mu c} \right) \left(p_k - \langle p_k \rangle_{N,v}^{\mu c} \right) \right\rangle_{N,v}^{\mu c}$$

$$= \frac{1}{N} \sum_{\rangle i,j,k \langle} \left\langle \left(a_i - \langle a_i \rangle_{N,v}^{\mu c} \right) \left(a_j - \langle a_j \rangle_{N,v}^{\mu c} \right) \left(p_k - \langle p_k \rangle_{N,v}^{\mu c} \right) \right\rangle_{N,v}^{\mu c}$$

$$+ \frac{1}{N} \sum_{\langle i,j,k \rangle} \left\langle \left(a_i - \langle a_i \rangle_{N,v}^{\mu c} \right) \left(a_j - \langle a_j \rangle_{N,v}^{\mu c} \right) \left(p_k - \langle p_k \rangle_{N,v}^{\mu c} \right) \right\rangle_{N,v}^{\mu c} ,$$

where $\rangle i,j,k \langle$ means that at least two of the three indexes refer to non-nearest-neighbor site, whereas $\langle i,j,k \rangle$ means that the three indexes are nearest neighbors. If i,j,k are such that $\rangle i,j,k \langle$, then at least two of the three terms a_i, a_j, and p_k have no common configurational variables. The microcanonical averages are again estimated according to Lemma 9.22 through a stochastic process on the configurational coordinates. The random processes associated with a_i, a_j, and p_k are thus completely decorrelated, and one has

$$for \ any \ i,j,k, \ such \ that \ \rangle i,j,k \langle,$$

$$\left\langle \left(a_i - \langle a_i \rangle_{N,v}^{\mu c} \right) \left(a_j - \langle a_j \rangle_{N,v}^{\mu c} \right) \left(p_k - \langle p_k \rangle_{N,v}^{\mu c} \right) \right\rangle_{N,v}^{\mu c} = 0 .$$

Now, if we consider i,j,k such that $\langle i,j,k \rangle$, the three terms a_i, a_j, and p_k are certainly correlated, but we notice that there are only $N n_p^2$ terms of this kind. Thus we have

$$\frac{1}{N} \sum_{\langle i,j,k \rangle} \left\langle \left(a_i - \langle a_i \rangle_{N,v}^{\mu c} \right) \left(a_j - \langle a_j \rangle_{N,v}^{\mu c} \right) \left(p_k - \langle p_k \rangle_{N,v}^{\mu c} \right) \right\rangle_{N,v}^{\mu c}$$

$$\leq n_c^2 \max_{\langle i,k \rangle} \left\{ \left(a_i - \langle a_i \rangle_{N,v}^{\mu c} \right) , \left(p_k - \langle p_k \rangle_{N,v}^{\mu c} \right) \right\} .$$

Since the terms a_i and p_k are of order 1, the largest term of the preceding equation is independent of N, and we have thus found the upper bound of the fourth term of the right hand-side of equation (9.97).

Finally, the last term of the right hand-side of equation (9.97) is the fourth cumulant of the stochastic variable $A(\chi)/\chi$ (multiplied by N^3). As already seen above, we write $A(\chi)/\chi = 1/N \sum_{i=1}^{N} N\partial_{ii}^2 V/\|\nabla V\|^2$ so that Lemma 9.25 applies and ensures that the distribution of $A(\chi)/\chi$ has a fourth cumulant K'_N such that $\lim_{N\to\infty} N^3 K'_N = 0$.

The ensemble of the upper bounds thus obtained yields the final desired result.   □

9.5 Main Theorems: Theorem 2

In this section we prove the second of the two main theorems resumed at beginning of the present chapter.

In view of formulating and proving Theorem 2, we have to define some neighborhoods, which we call "pseudo-cylindrical," of critical points of a potential function V_N. Before defining these pseudocylindrical neighborhoods of critical points, let us recall the following basic result in Morse theory.

Theorem 9.35. *Let f be a smooth real-valued function on a compact finite-dimensional manifold M. Let $a < b$, and suppose that the set*

$$f^{-1}([a, b]) \equiv \mathcal{M} = \{x \in M | a \le f(x) \le b\} \tag{9.100}$$

is compact and contains no critical points of f, that is, $\|\nabla f\| \ge C > 0$ with C a constant. Let $y \in (a, b)$. Then there exists a diffeomorphism

$$\sigma : (a, b) \times f^{-1}(y) \to f^{-1}(a, b) \quad by \quad (v, x) \longmapsto \sigma(v, x). \tag{9.101}$$

Corollary 9.36. *The manifolds $f^{-1}(y)$, $a < y < b$, are all diffeomorphic.*

This result is based on the existence of a one-parameter group of diffeomorphisms

$$\sigma_v : M \to \mathcal{M} \quad by \quad x \longmapsto \sigma(v, x) \tag{9.102}$$

associated with the vector field $X = -\nabla f(x)/\|\nabla f(x)\|^2$ with $v \longmapsto \sigma(v, x)$ a solution of the differential equation on M

$$\frac{d\sigma(v, x)}{dv} = -\frac{\nabla f[\sigma(v, x)]}{\|\nabla f[\sigma(v, x)]\|^2},$$

$$\sigma(0, x) = x; \tag{9.103}$$

$\sigma(v, x)$ is defined for all $v \in \mathbb{R}$ and $x \in M$. Details can be found in standard references such as [23, 210, 227].

Applied to the configuration space M, if the function f is identified with the potential V_N, then in the absence of critical points of V in the interval (v_0, v_1) the hypersurfaces $\Sigma_v = V_N^{-1}(v)$, $v \in (v_0, v_1)$, are all diffeomorphic.

Now we define what we call pseudocylindrical-neighborhoods of critical points. On a given critical level we consider a disk around a critical point, then we consider its backward and forward projections on non-critical equipotential surfaces. The projections are made by the flow of a vector field that defines the walls of the neighborhood and its caps on the mentioned equipotential surfaces. This construction guarantees that after the excision of one or more of these neighborhoods, the Σ_v and the M_v remain diffeomorphic.

As we discussed in the introduction of the present chapter, a typical situation is that of a minimum energy Δv required to pass from a critical level v_c^j to the next one v_c^{j+1}, independently of N. While in this case the length of $[\bar{v}_c^j, \bar{v}_c^{j+1}]$ shrinks with N, the length of the interval $[v_c^j, v_c^{j+1}]$ remains constant. Thus, we set the thickness of these neighborhoods equal to the minimum length of these intervals. In more generic cases, this thickness can be arbitrarily fixed to a small constant.

Definition 9.37 (Pseudocylindrical Neighborhoods). *Let Σ_{v_c} be a critical level set of V_N, that is, a level set containing at least one critical point of V_N. Around any critical point $q_c^{(i)}$, consider the set of points $\gamma(q_c^{(i)}, \rho; v_c) \subset \Sigma_{v_c}$ at a distance equal to $\rho > 0$ from $q_c^{(i)}$, that is, $q \in \gamma(q_c^{(i)}, \rho; v_c) \Rightarrow |q - q_c^{(i)}| = \rho$, and ρ is such that $\rho < \frac{1}{2} \min_{i,j} |q_c^{(i)} - q_c^{(j)}|$, and i, j label all the critical points on the given critical level set. Moreover, if in a given interval of potential energy density $[\bar{v}_0, \bar{v}_1]$ there is only one critical value $\bar{v}_c$, then set the thickness of all the pseudocylinders equal to a sufficiently small, finite, ε_0; if, otherwise, in the interval $[\bar{v}_0, \bar{v}_1]$ there is an arbitrary number of critical values v_c^j, then take the thickness ε_0 of all the pseudocylinders such that $0 < \varepsilon_0 < \min_{j \in \mathbb{N}}(v_c^{j+1} - v_c^j)$. By Sard's theorem, both ρ and ε_0 are finite because at finite dimension, there is a finite number of isolated critical points and consequently a finite number of critical values. We define a pseudocylindrical neighborhood $\Gamma(q_c^{(i)}, \varepsilon_0) \subset M$ of $q_c^{(i)}$ as the subset of M bounded by the following set of points. By mapping $\gamma(q_c^{(i)}, \rho; v_c)$ from Σ_{v_c} to $\Sigma_{(v_c + \varepsilon_0)}$ through the flow generated by the vector field $X = \nabla V_N(q)/\|\nabla V_N(q)\|^2$, and from Σ_{v_c} to $\Sigma_{(v_c - \varepsilon_0)}$ through the flow generated by the vector field $X = -\nabla V_N(q)/\|\nabla V_N(q)\|^2$, we obtain the walls of $\Gamma(q_c^{(i)}, \varepsilon_0)$, which are transverse to the Σ_v, and then we close the neighborhood with the pieces of $\Sigma_{(v_c + \varepsilon_0)}$ and $\Sigma_{(v_c - \varepsilon_0)}$ bounded by the images $\gamma(q_c^{(i)}, \rho; (v_c + \varepsilon_0))$ and $\gamma(q_c^{(i)}, \rho; (v_c - \varepsilon_0))$ of $\gamma(q_c^{(i)}, \rho; v_c)$ through $\sigma(v, x)$, respectively. Moreover, we require that the radius ρ of $\Gamma(q_c^{(i)}, \varepsilon_0)$ exceeds a minimum value $\rho_{\min}$ – which depends on ε_0 – so that $\Sigma_{(v_c - \varepsilon_0)} \backslash \Gamma(q_c^{(i)}) \cap \Sigma_{(v_c - \varepsilon_0)}$ is diffeomorphic to $\Sigma_{(v_c + \varepsilon_0)} \backslash \Gamma(q_c^{(i)}, \varepsilon_0) \cap \Sigma_{(v_c + \varepsilon_0)}$. If $\rho_{\min}$ exceeds $\frac{1}{2} \min_{i,j} |q_c^{(i)} - q_c^{(j)}|$, then ε_0 is suitably reduced.*

Lemma 9.38. *Let V_N be a standard, smooth, confining, short-range potential bounded from below (Definitions 9.5–9.9),*

$$V_N : \mathcal{B}_N \subset \mathbb{R}^N \to \mathbb{R} ,$$

with V_N given by Definition 9.5 (fluid case), or by Definition 9.6 (lattice case).

Let $\{\Sigma_v\}_{v\in\mathbb{R}}$ *be the family of* $(N-1)$-*dimensional hypersurfaces* $\Sigma_v :=$ $V_N^{-1}(v)$, $v \in \mathbb{R}$, *of* $\mathbb{R}^N$. *Let* $\{M_v\}_{v\in\mathbb{R}}$ *be the family of* N-*dimensional subsets* $M_v := V_N^{-1}((-\infty, v])$, $v \in \mathbb{R}$, *of* $\mathbb{R}^N$. *Let* $\{\overline{M}_v\}_{v\in\mathbb{R}}$ *be the family of* N-*dimensional subsets* $\overline{M}_v := M_v \setminus \bigcup_{i=1}^{\mathcal{N}(v)} \Gamma(q_c^{(i)}, \varepsilon)$, $v \in \mathbb{R}$, *of* $\mathbb{R}^N$, *where* $\Gamma(q_c^{(i)}, \varepsilon)$ *are the pseudocylindrical neighborhoods of the critical points* $q_c^{(i)}$ *of* $V_N(q)$ *contained in* M_v *and* $\mathcal{N}(v)$ *is the number of critical points in* M_v. *Let* $\{\overline{\Sigma}_v\}_{v\in\mathbb{R}}$ *be the family of* $(N-1)$-*dimensional subsets of* $\mathbb{R}^N$ *defined as* $\overline{\Sigma}_v := \Sigma_v \setminus \bigcup_{i=1}^{\mathcal{N}(v)}[\Gamma(q_c^{(i)}, \varepsilon) \cap \Sigma_v]$.
Let $\bar{v}_0 = v_0/N, \bar{v}_1 = v_1/N \in \mathbb{R}$, $\bar{v}_0 < \bar{v}_1$, *and let* $\bar{v}_c = v_c/N$ *be the only critical value of* V_N *in the interval* $I_{\bar{v}} = [\bar{v}_0, \bar{v}_1]$, *and let* $\Gamma^*(q_c^{(i)}, \varepsilon^*)$, $q_c^{(i)} \in V_N^{-1}(v_c)$ *such that* $\varepsilon^* > v_1 - v_c, v_c - v_0$. *The following two statements hold:*

(a) for any $\bar{v}, \bar{v}' \in [\bar{v}_0, \bar{v}_1]$ *it is*

$$\overline{\Sigma}_{N\bar{v}} \text{ is } \mathcal{C}^\infty-\text{diffeomorphic to } \overline{\Sigma}_{N\bar{v}'};$$

(b) putting $\overline{M}(v, N) = \text{vol}(\overline{M}_v)$, *the quantities* $[d\overline{M}(v, N)/dv]/\overline{M}(v, N)$ *and* $(d^k/dv^k)\{[d\overline{M}(v, N)/dv]/\overline{M}(v, N)\}$, $k = 1, 2, 3$, *are uniformly bounded in* N *in the interval* $[\bar{v}_0, \bar{v}_1]$.

Proof. Concerning point *(a)*, we note that the flow associated with the $\mathcal{C}^\infty$ vector field $X = -\nabla V_N(q)/\|\nabla V_N(q)\|^2$ is well defined at any point $q \in \overline{M}_{v_1} \setminus \overline{M}_{v_0}$. Thus the set $\overline{M}_{v_1} \setminus \overline{M}_{v_0}$ is diffeomorphic to the noncritical neck $\partial \overline{M}_{v_0} \times [v_0, v_1]$. Then, by the "noncritical neck theorem" [210], for any $\bar{v}, \bar{v}' \in I_{\bar{v}} = [\bar{v}_0, \bar{v}_1]$, we have $\overline{\Sigma}_{N\bar{v}} \approx \overline{\Sigma}_{N\bar{v}'}$. Incidentally, this entails also $\overline{M}_{N\bar{v}} \approx \overline{M}_{N\bar{v}'}$ for any $\bar{v}, \bar{v}' \in [\bar{v}_0, \bar{v}_1]$.

Now let us consider point *(b)*. Define $\overline{S}_N(\bar{v}) = \frac{1}{N} \log[\overline{\Omega}(N\bar{v}, N)]$, where $\overline{\Omega}(N\bar{v}, N) = \text{vol}(\overline{\Sigma}_{\bar{v}N})]$. Having proved the statement *(a)*, we can apply Lemma 9.4 of Section 9.2, which entails that

$$\sup_{N, \bar{v} \in I_{\bar{v}}} |\overline{S}_N(\bar{v})| < \infty \quad and \quad \sup_{N, \bar{v} \in I_{\bar{v}}} \left| \frac{\partial^k \overline{S}_N}{\partial \bar{v}^k}(\bar{v}) \right| < \infty, \quad k = 1, 2, 3, 4.$$

Thus by Lemma 9.17 of Section 9.2, it follows that $\overline{S}_\infty(\bar{v}) = \lim_{N\to\infty} \overline{S}_N(\bar{v}) \in \mathcal{C}^3(I_{\bar{v}})$.

The next step is to prove that also $\overline{S}_\infty^{(-)}(\bar{v}) \in \mathcal{C}^3(I_{\bar{v}})$, where

$$\overline{S}_\infty^{(-)}(\bar{v}) := \lim_{N\to\infty} \overline{S}_N^{(-)}(\bar{v}) = \lim_{N\to\infty} \frac{1}{N} \log[\text{vol}(\overline{M}_{\bar{v}N})],$$

because, by Lemmas 9.17 and 9.4 of Section 9.2, this entails the truth of statement *(b)*. Let us begin by considering the microcanonical configurational inverse temperature. From its definition $\beta_N(\bar{v}) = \partial S_N^{(-)}/\partial \bar{v}$ one obtains $\beta_N(\bar{v}) = \Omega(N\bar{v}, N)/M(N\bar{v}, N)$. The function $\beta_N(\bar{v})$ is well known to be intensive and well defined also in the thermodynamic limit, at least for extensive potential energy functions. Then we work out a representation of $\beta_N(\bar{v})$ in the form of a microcanonical average of a suitable function. To this end we derive $\Omega(N\bar{v}, N)$ with respect to v by means of Federer's derivation formula and then we integrate it back. Thus we write

$$\Omega(N\bar{v}, N) = \int_0^v d\eta \int_{\Sigma_\eta} \|\nabla V\| \, A\left(\frac{1}{\|\nabla V\|}\right) \frac{d\sigma}{\|\nabla V\|}$$

$$= \int_{M_v} d\mu \, \|\nabla V\| \, A\left(\frac{1}{\|\nabla V\|}\right) , \qquad (9.104)$$

where $d\mu = d^N q$, so that we finally obtain

$$\beta_N(\bar{v}) = \left[\int_{M_v} d\mu\right]^{-1} \int_{M_v} d\mu \, \|\nabla V\| \, A\left(\frac{1}{\|\nabla V\|}\right) = \left\langle \|\nabla V\| \, A\left(\frac{1}{\|\nabla V\|}\right) \right\rangle_{M_v} ,$$
$$(9.105)$$

which holds for $\bar{v} \in [0, \bar{v}_0)$. An important remark is in order. We have used Federer's derivation formula apparently ignoring that it applies in the absence of critical points of the potential function. However, if the potential V is a good Morse function (not a very restrictive condition at all), we know, by Sard's theorem [210], that the ensemble of critical values, here of the potential, is a point set. Therefore, any finite interval of values of the potential is the union of a finite number of open intervals where no critical value is present, and correspondingly no critical point on the $\{\Sigma_v\}$ exists. On all these open sets, free of critical points, Federer's derivation formula can be legally applied. Moreover, the results found by applying Federer's formula on each open interval free of critical values of V can be regularly glued together because of the existence of the thermodynamic limit of $\beta_N(\bar{v})$.

Let us now consider $\overline{\Omega}(N\bar{v}, N)$ for $\bar{v} \in [\bar{v}_0, \bar{v}_1]$. Since all the hypersurfaces $\Sigma_{\bar{v}}$ labeled by $\bar{v} \in [\bar{v}_0, \bar{v}_1]$ are diffeomorphic, we can use Federer's derivation formula to obtain an expression for $\overline{\Omega}(N\bar{v}, N)$ similar to that given in (9.104) for $\Omega(N\bar{v}, N)$, that is,

$$\overline{\Omega}(N\bar{v}, N) = \int_{v_0}^v d\eta \int_{\overline{\Sigma}_\eta} \|\nabla V\| \, A\left(\frac{1}{\|\nabla V\|}\right) \frac{d\sigma}{\|\nabla V\|} + \Omega(N\bar{v}_0, N)$$

$$= \int_{\overline{M}_v \setminus M_{v_0}} d\mu \, \|\nabla V\| \, A\left(\frac{1}{\|\nabla V\|}\right) + \int_{M_{v_0}} d\mu \, \|\nabla V\| \, A\left(\frac{1}{\|\nabla V\|}\right)$$

$$= \int_{M_v \setminus \Gamma^*} d\mu \, \|\nabla V\| \, A\left(\frac{1}{\|\nabla V\|}\right) , \qquad (9.106)$$

where Γ^* stands for the union of all the pseudocylindrical neighborhoods of the critical points of V in the interval $[\bar{v}_0, \bar{v}_1]$. Then we consider the restriction $\overline{\beta}_N(\bar{v})$ of the function $\beta_N(\bar{v})$ to the subset $M_v \setminus \Gamma^*$; from

$$\overline{\beta}_N(\bar{v}) = \frac{\overline{\Omega}(N\bar{v}, N)}{\overline{M}(N\bar{v}, N)} \tag{9.107}$$

we get

$$\overline{\beta}_N(\bar{v}) = \left[\int_{M_v \backslash \Gamma^*} d\mu\right]^{-1} \int_{M_v \backslash \Gamma^*} d\mu \, \|\nabla V\| \, A\left(\frac{1}{\|\nabla V\|}\right)$$

$$= \left\langle \|\nabla V\| \, A\left(\frac{1}{\|\nabla V\|}\right)\right\rangle_{M_v \backslash \Gamma^*}. \tag{9.108}$$

By comparing (9.105) with (9.108), we see that also $\overline{\beta}_N(\bar{v})$ has to be intensive up to the $N \to \infty$ limit, like $\beta_N(\bar{v})$. In fact, the excision of the set Γ^* out of M_v, no matter how the measure of Γ^* depends on N, cannot change the intensive character of $\overline{\beta}_N(\bar{v})$. The relationship among $\overline{S}_N(\bar{v})$, $\overline{S}_N^{(-)}(\bar{v})$, and $\overline{\beta}_N(\bar{v})$ is given by the logarithm of both sides of (9.107),

$$\frac{1}{N} \log \overline{\Omega}(\bar{v}N, N) = \frac{1}{N} \log \overline{M}(\bar{v}N, N) + \frac{1}{N} \log \overline{\beta}_N(\bar{v}),$$

whence, using $\lim_{N \to \infty} \frac{1}{N} \log \overline{\beta}_N(\bar{v}) = 0$, we obtain $\overline{S}_\infty^{(-)}(\bar{v}) = \overline{S}_\infty(\bar{v})$ and thus $\overline{S}_\infty^{(-)}(\bar{v}) \in \mathcal{C}^3(I_{\bar{v}})$.

Finally, $\overline{S}_\infty^{(-)}(\bar{v}) \in \mathcal{C}^3(I_{\bar{v}})$ entails

$$\sup_{N, \bar{v} \in I_{\bar{v}}} \left|\overline{S}_N^{(-)}(\bar{v})\right| < \infty \quad and \quad \sup_{N, \bar{v} \in I_{\bar{v}}} \left|\frac{\partial^k \overline{S}_N^{(-)}}{\partial \bar{v}^k}(\bar{v})\right| < \infty, \quad k = 1, 2, 3, 4,$$

so that resorting to Lemma 9.17 of Section 9.2, the truth of statement *(b)* follows.
□

Theorem 9.39 (Entropy and Topology). *Let $V_N(q_1, \ldots, q_N) : \mathbb{R}^N \to \mathbb{R}$, be a smooth, nonsingular, finite-range potential. Denote by $M_v := V_N^{-1}((-\infty, v])$, $v \in \mathbb{R}$, the generic submanifold of configuration space bounded by Σ_v. Let $\{q_c^{(i)} \in \mathbb{R}^N\}_{i \in [1, \mathcal{N}(v)]}$ be the set of critical points of the potential, that is, such that $\nabla V_N(q_c^{(i)}) = 0$, and let $\mathcal{N}(v)$ be the number of critical points up to the potential energy value v. Denote by $\bar{v} = v/N$ the potential energy density. Let the number of isolated critical points on the critical level sets $\Sigma_{N\bar{v}_c}$ be arbitrary functions of N. Let $\Gamma(q_c^{(i)}, \varepsilon_0)$ be the pseudocylindrical neighborhood of the critical point $q_c^{(i)}$, and let $\mu_i(M_v)$ be the Morse indexes of M_v. Then there exist real numbers $A(N, i, \varepsilon_0)$, g_i and real almost-everywhere-smooth functions $B(N, i, v, \varepsilon_0)$ such that the following equation for the microcanonical configurational entropy $S_N^{(-)}(v)$ holds:*

$$S_N^{(-)}(v) = \frac{1}{N} \log \left[\int_{M_v \setminus \bigcup_{i=1}^{\mathcal{N}^{(v)}} \Gamma(q_c^{(i)}, \varepsilon_0) \atop \mathcal{N}_{cp}^{\nu(v)}} d^N q + \sum_{i=0}^{N} A(N, i, \varepsilon_0) \, g_i \, \mu_i(M_{v-\varepsilon_0}) \right.$$

$$\left. + \sum_{n=1}^{N} B(N, i(n), v - v_c^{\nu(v)}, \varepsilon_0) \right],$$

(9.109)

where $\nu(v) = \max\{j | v_c^j \leq v\}$. *Moreover, assume that in some interval* $[\bar{v}_0, \bar{v}_1]$ *there exists only one critical value* $\bar{v}_c$, *then in this interval an unbounded growth with* N *of one of the derivatives* $|\partial^k S^{(-)}(v)/\partial v^k|$, *for* $k = 3, 4$, *and thus the occurrence of a first- or second-order phase transition can be entailed only by the topological term* $\sum_{i=0}^{N} A(N, i, \varepsilon_0) \, g_i \, \mu_i(M_{v-\varepsilon_0})$.

The proof of formula (9.109) is worked out constructively. This formula relates thermodynamic entropy, defined in the microcanonical configurational ensemble, with quantities of topological meaning (the Morse indexes) of the configuration-space submanifolds $M_v = V_N^{-1}((-\infty, v]) = \{q = (q_1, \ldots, q_N) \in \mathbb{R}^N | V_N(q) \leq v\}$.

By Morse theory, topological changes of the manifolds M_v can be put in one-to-one correspondence with the existence of critical points of the potential function $V_N(q_1, \ldots, q_N)$. A point q_c is a *critical point* if $\nabla V_N(q)|_{q=q_c} = 0$. The potential energy value $v_c = V_N(q_c)$ is said to be a *critical value* for the potential function. Passing a critical value v_c, the manifolds M_v change topology. Within the framework of Morse theory, if the potential V_N is a *good Morse function*, that is, a regular function bounded below and with nondegenerate Hessian (that is, the Hessian has no vanishing eigenvalue), then topological changes occur through the *attachment of handles* in the neighborhoods of the critical points (see Appendix C). Therefore, in order to establish the relationship between entropy and configuration-space topology, we have to unfold the contribution given to the volume of M_v by suitably defined neighborhoods of all the critical points contained in M_v because it is within these neighborhoods that the relevant information about topology is contained.

This result is made possible by the idea of exploiting the existence of the so-called Morse chart in the neighborhood of any nondegenerate critical point of the potential function V_N. In fact, the Morse chart allows one to represent the *local* analytic form of the equipotential hypersurfaces in a universal form independent of the potential energy value at the critical point, and dependent only on the index of the critical point (equal to the number of negative eigenvalues of the Hessian of the potential) and, obviously, on the dimension N of configuration space, hence the possibility of a formal computation of the contribution of the neighborhoods of all the critical points to the volume of M_v as a function of v.

Proof. Let us consider the definition of the configurational microcanonical entropy $S_N^{(-)}(v)$, already given in (9.3),

$$S_N^{(-)}(v) = \frac{1}{N} \log M(v, N) , \qquad (9.110)$$

with

$$M(v,N) = \int_{V_N(q)\leq v} d^N q = \int_0^v d\eta \int_{(\Lambda^d)^{\times n}} d^N q\, \delta[V_N(q)-\eta]$$
$$= \int_0^v d\eta \int_{\Sigma_\eta} \frac{d\sigma}{\|\nabla V_N\|}. \tag{9.111}$$

Let $\{\Sigma_{v_c^j}\}$ be the family of all the critical level sets (in general not differentiable manifolds) of the potential, that is, the constant-potential-energy hypersurfaces that contain at least one *critical point* $q_c^{(i)}$, where $\nabla V_N(q_c^{(i)}) = 0$. For a potential that is a good Morse function, by Sard's theorem (see Corollary 2 on p. 200 of [210]), at any finite dimension N, and below any finite upper bound of the potential energy, the number of critical points in configuration space and thus also the set of critical values $\{v_c^j\}_{j\in\mathbb{N}}$ are *finite and isolated*.

In order to split the integration on M_v into two parts, the integration on the union of the neighborhoods of all the critical points contained in M_v and the integration on its complement in M_v, we have defined for each critical point $q_c^{(i)}$ its *pseudocylindrical* neighborhood $\Gamma(q_c^{(i)}, \varepsilon_0)$; ε_0 is the thickness—in potential energy—of the neighborhood.

Let us now split the integration on M_v into the integration on $M_v \cap \bigcup_i \Gamma(q_c^{(i)}, \varepsilon_0)$ and on its complement $M_v \setminus \bigcup_i \Gamma(q_c^{(i)}, \varepsilon_0)$. We have

$$\int_{M_v} d^N q = \int_{M_v\setminus\bigcup_{i=1}^{\mathcal{N}(v)} \Gamma(q_c^{(i)},\varepsilon_0)} d^N q + \int_{M_v\cap\bigcup_{i=1}^{\mathcal{N}(v)} \Gamma(q_c^{(i)},\varepsilon_0)} d^N q, \tag{9.112}$$

where $\mathcal{N}(v)$ is the number of critical points of $V_N(q)$ up to the level v. We can equivalently write

$$\text{vol}(M_v) = \int_{M_v\setminus\bigcup_{i=1}^{\mathcal{N}(v)} \Gamma(q_c^{(i)},\varepsilon_0)} d^N q + \sum_{j=1}^{\mathcal{N}_{cl}(v)} \sum_{m=1}^{\mathcal{N}_{cp}^j} \int_{M_v\cap\Gamma_j(q_c^{(m)},\varepsilon_0)} d^N q, \tag{9.113}$$

where $\mathcal{N}_{cl}(v)$ is the number of critical levels $\Sigma_{v_c^j}$ such that $v_c^j < v$, and $\mathcal{N}_{cp}^j$ is the number of critical points on the critical hypersurface $\Sigma_{v_c^j}$ and where we have changed the notation of the pseudocylindrical neighborhoods to $\Gamma_j(q_c^{(m)}, \varepsilon_0)$, labeling with j the level set to which it belongs and numbering with m the critical points on a given level set. Notice that $\mathcal{N}(v) = \sum_{j=1}^{\mathcal{N}_{cl}(v)} \mathcal{N}_{cp}^j$.

Then we use the coarea formula in the right hand-side of (9.111) to rewrite (9.113); a distinction is necessary between two cases for $\Sigma_v = \partial M_v$, depending on whether its label v is closer than ε_0 to a critical level; thus we obtain

$$\text{vol}(M_v) = \int_{M_v\setminus\bigcup_{i=1}^{\mathcal{N}(v+\varepsilon_0)} \Gamma(q_c^{(i)},\varepsilon_0)} d^N q$$
$$+ \sum_{j=1}^{\mathcal{N}_{cl}(v+\varepsilon_0)} \sum_{m=1}^{\mathcal{N}_{cp}^j} \int_{v_c^j-\varepsilon_0}^{v_c^j+\varepsilon_0} d\eta \int_{\Gamma_j(q_c^{(m)},\varepsilon_0)} d^N q\, \delta[V_N(q)-\eta] \tag{9.114}$$

when $v > v_c^{\nu(v)} + \varepsilon_0$ and $v < v_c^{\nu(v)+1} - \varepsilon_0$, where $\nu(v)$ is such that $v_c^{\nu(v)} < v < v_c^{\nu(v)+1}$, whereas

$$
\begin{aligned}
\mathrm{vol}(M_v) = {} & \int_{M_v \setminus \bigcup_{i=1}^{\mathcal{N}(v+\varepsilon_0)} \Gamma(q_c^{(i)}, \varepsilon_0)} d^N q \\
& + \sum_{j=1}^{\mathcal{N}_{cl}(v)-1} \sum_{m=1}^{\mathcal{N}_{cp}^j} \int_{v_c^j - \varepsilon_0}^{v_c^j + \varepsilon_0} d\eta \int_{\Gamma_j(q_c^{(m)}, \varepsilon_0)} d^N q \, \delta[V_N(q) - \eta] \\
& + \sum_{m=1}^{\mathcal{N}_{cp}^{\nu(v+\varepsilon_0)}} \int_{v_c^{\nu(v)} - \varepsilon_0}^{v} d\eta \int_{\Gamma_{\nu(v)}(q_c^{(m)}, \varepsilon_0)} d^N q \, \delta[V_N(q) - \eta] \quad (9.115)
\end{aligned}
$$

when $v_c^{\nu(v)+1} - \varepsilon_0 < v$ or $v < v_c^{\nu(v)} + \varepsilon_0$.

Near any critical point, a second-order power series expansion of $V(q)$ reads

$$
V_N^{(2)}(q) = V_N(q_c) + \frac{1}{2} \sum_{i,j} \frac{\partial^2 V_N}{\partial q_i \partial q_j} (q^i - q_c^i)(q^j - q_c^j) \,.
$$

For sufficiently small ε_0, the integrals $\int_{\Gamma_j(q_c, \varepsilon_0)} d^N q \, \delta[V_N(q) - \eta]$ can be replaced with arbitrary precision by $\int_{\Gamma_j(q_c, \varepsilon_0)} d^N q \, \delta[V_N^{(2)}(q) - \eta]$. Moreover, if $V_N(q)$ is a good Morse function, then a coordinate transformation exists to the so-called *Morse chart* [23] such that

$$
\widetilde{V}_N^{(2)}(x) = V_N(q_c) - \sum_{l=1}^{k} x_l^2 + \sum_{l=k+1}^{N} x_l^2 \,,
$$

where k is the Morse index of q_c. Using the Morse chart we have

$$
\int_{\Gamma_j(q_c, \varepsilon_0)} d^N q \, \delta[V_N^{(2)}(q) - \eta] = \int_{\Gamma_j(q_c, \varepsilon_0)} d^N x \, |\det J| \, \delta[\widetilde{V}_N^{(2)}(x) - \eta] \,, \quad (9.116)
$$

where J is the Jacobian of the coordinate transformation.

Using Morse coordinates inside the pseudocylinder $\Gamma_j(q_c^{(m)}, \varepsilon_0)$ around the critical point $q_c^{(m)}$, we see that each part of a hypersurface $\Sigma_\eta \cap \Gamma_j(q_c^{(m)}, \varepsilon_0)$ is a quadric

$$
\xi = \eta - v_c^j = -\sum_{l=1}^{k_m} x_l^2 + \sum_{l=k_m+1}^{N} x_l^2 = -|X|^2 + |Y|^2 \,, \quad (9.117)
$$

where the Morse index of $q_c^{(m)}$ is denoted by k_m, so that $|X|^2 = \sum_{l=1}^{k_m} x_m^2$ and $|Y|^2 = \sum_{l=k_m+1}^{N} x_l^2$. Thus we rewrite the right hand-side of (9.116) as

$$
\begin{aligned}
& |\det J| \int_{\Gamma_j(q_c^{(m)}, \varepsilon_0)} d^N x \, \delta(-|X|^2 + |Y|^2 - \xi) \\
= {} & |\det J| \int_{\Gamma_j(q_c^{(m)}, \varepsilon_0)} d\Omega^s \, d\Omega^t d|X| d|Y| |X|^{k_m-1} |Y|^{N-k_m-1} \delta(-|X|^2 + |Y|^2 - \xi)
\end{aligned}
$$

$$
\tag{9.118}
$$

where $d\Omega^s \equiv d\Omega^{k_m-1}$ is the solid-angle element in $(k_m - 1)$ dimensions, and $d\Omega^t \equiv d\Omega^{N-k_m-1}$ is the solid-angle in $(N - k_m - 1)$ dimensions, whose integrations yield the volumes C_s and C_t of the hyperspheres of unit radius in s and t dimensions respectively. Putting $z = |X|^2$ and integrating on the angular coordinates we get

$$\frac{1}{2} \, |\det J| \, C_{N-k_m-1} C_{k_m-1}$$
$$\times \int_{\Gamma_j(q_c^{(m)}, \varepsilon_0)} d|Y| \, dz \, \frac{1}{\sqrt{z}} \, |Y|^{N-k_m-1} z^{(k_m-2)/2} \delta(-z + |Y|^2 - \xi) \, .$$
$$(9.119)$$

The domains $\Gamma_j(q_c, \varepsilon_0)$ are defined so that their boundaries are orthogonal to the potential level sets defined by (9.117). These boundaries are thus given by the equation $|X||Y| = r$. From (9.117), that is, $|Y|^2 = z + \xi$, when $\xi < 0$ we have that $z > -\xi$, and from $|X||Y| = r$ we get $z = r^2/|Y|^2$, so $-r^2/|Y|^2 + |Y|^2 = -|\xi|$, whence $|Y| = \left(-\frac{1}{2}\xi + \frac{1}{2}\sqrt{|\xi|^2 + 4r^2}\right)^{1/2}$. Moreover, when $\xi < 0$, at $|X|^2 = 0$ it must be $|Y|^2 = 0$, so that $|Y|$ ranges from 0 to $\beta(\xi, r) = \sqrt{(\sqrt{\xi^2 + 4r^2} - \xi)/2}$, and z ranges from 0 to $\alpha(\xi, r) = (\sqrt{\xi^2 + 4r^2} - \xi)/2$.

At variance, for $\xi > 0$, at $|X|^2 = 0$ it is $|Y|^2 = \xi > 0$, so that from $-r^2/|Y|^2 + |Y|^2 = \xi$ we get again that z ranges from 0 to $\alpha(\xi, r)$, and that $|Y|$ ranges from $\sqrt{\xi}$ to $\beta(\xi, r) = \sqrt{(\sqrt{\xi^2 + 4r^2} + \xi)/2}$.

Thus we obtain from (9.119)

$$\frac{1}{2} \, |\det J| \, C_{N-k_m-1} C_{k_m-1}$$
$$\times \int_0^{\alpha(\xi,r)} dz \int_{0,\sqrt{\xi}}^{\beta(\xi,r)} d|Y| |Y|^{N-k_m-1} z^{(k_m-2)/2} \delta(-z + |Y|^2 - \xi) \, . \quad (9.120)$$

Finally, by putting $y = |Y|$, when $\xi > 0$ we obtain

$$\frac{1}{2} J_{jm} \, C_{N-k_m-1} C_{k_m-1} \int_{\sqrt{\xi}}^{\beta(\xi,r)} dy \, y^{N-k_m-1} \, (y^2 - \xi)^{(k_m-2)/2} \, , \quad (9.121)$$

and when $\xi < 0$ we obtain

$$\frac{1}{2} J_{jm} \, C_{N-k_m-1} C_{k_m-1} \int_0^{\beta(\xi,r)} dy \, y^{N-k_m-1} \, (y^2 - \xi)^{(k_m-2)/2} \, , \quad (9.122)$$

where C_{N-k_m-1} and C_{k_m} are volumes of hyperspheres of unit radii, that is, $C_n = 2\pi^{n/2}/(n/2 - 1)!$ (for n even) and $C_n = 2^{(n+1)/2}\pi^{(n-1)/2}/(n - 2)!!$ (for n odd); J_{jm} stands for the numerical absolute value of the determinant of J computed at the critical level v_c^j and at the critical point $q_c^{(m)}$. By defining

$$F_+(\xi, k_m, N) = \int_{\sqrt{\xi}}^{\beta(\xi,r)} dy \, y^{N-k_m-1} \, (y^2 - \xi)^{(k_m-2)/2} \quad (9.123)$$

and

$$F_-(\xi, k_m, N) = \int_0^{\beta(\xi, r)} dy \, y^{N-k_m-1} \, (y^2 - \xi)^{(k_m-2)/2} \,, \tag{9.124}$$

we can now write[6]

$$\text{vol}(M_v) = \int_{M_v \setminus \bigcup_{i=1}^{\mathcal{N}(v+\varepsilon_0)} \Gamma(q_c^{(i)}, \varepsilon_0)} d^N q$$

$$+ \sum_{j=1}^{\mathcal{N}_{cl}(v+\varepsilon_0)} \sum_{m=1}^{\mathcal{N}_{cp}^j} \frac{1}{2} C_{N-k_m-1} C_{k_m} J_{jm} \int_{-\varepsilon_0}^{\varepsilon_0} d\xi \, F(\xi, k_m, N) \tag{9.125}$$

when $v > v_c^{\nu(v)} + \varepsilon_0$ and $v < v_c^{\nu(v)+1}$, where $\nu(v) = \max\{j | v_c^j \le v\}$, or

$$\text{vol}(M_v) = \int_{M_v \setminus \bigcup_{i=1}^{\mathcal{N}(v+\varepsilon_0)} \Gamma(q_c^{(i)}, \varepsilon_0)} d^N q$$

$$+ \sum_{j=1}^{\mathcal{N}_{cl}(v)-1} \sum_{m=1}^{\mathcal{N}_{cp}^j} \frac{1}{2} C_{N-k_m-1} C_{k_m} J_{jm} \int_{-\varepsilon_0}^{\varepsilon_0} d\xi \, F(\xi, k_m, N)$$

$$+ \sum_{m=1}^{\mathcal{N}_{cp}^{\nu(v+\varepsilon_0)}} \frac{1}{2} C_{N-k_m-1} C_{k_m} J_{\tilde{j}m} \int_{-\varepsilon_0}^{v-v_c^{\nu(v)}} d\xi \, F(\xi, k_m, N) \tag{9.126}$$

with $\tilde{j} = \mathcal{N}_{cl}(v)$ when $v_c^{\nu(v)+1} - \varepsilon_0 < v$ or $v < v_c^{\nu(v)} + \varepsilon_0$. In (9.125) and (9.126) we have put $\int_{-\varepsilon_0}^{\varepsilon_0} d\xi \, F(\xi, k_m, N) = \int_{-\varepsilon_0}^0 d\xi \, F_-(\xi, k_m, N) + \int_0^{\varepsilon_0} d\xi \, F_+(\xi, k_m, N)$.

Notice that $\mathcal{N}(v) = \sum_{i=0}^N \mu_i(M_v)$, where $\mu_i(M_v)$ are the multiplicities of the critical points of index i (there are at most $N+1$ values for the indexes of critical points at dimension N) below the energy value v.

Therefore, we can rearrange the double summation in (9.125), (9.126) by expressing it as a double summation on all the possible values of the Morse indexes and on the number of critical points for each value of the Morse index, that is,

$$\sum_{i=0}^N \sum_{k=1}^{\mu_i(M_v)} A(N, i, \varepsilon_0) J_{j(i,k)m(i,k)} \,, \tag{9.127}$$

where, since the integrals in (9.125), (9.126) are independent of the index j, we have defined a set of coefficients $A(N, i, \varepsilon_0)$ as

$$A(N, i, \varepsilon_0) = \frac{1}{2} C_{N-i-1} C_i \int_{-\varepsilon_0}^{\varepsilon_0} d\xi \, F(\xi, i, N) \,. \tag{9.128}$$

---

[6] Note that in (9.123) and (9.124) $k_m < N$, in fact, if $k_m = N$ then in (9.120) $|Y| = 0$.

From the set of positive numbers $J_{j(i,k)m(i,k)}$ we define

$$g_i = \frac{1}{\mu_i(M_v)} \sum_{k=1}^{\mu_i(M_v)} J_{j(i,k)m(i,k)} \tag{9.129}$$

and rewrite the second term of the right hand-side of (9.125) as

$$\sum_{i=0}^{N} A(N, i, \varepsilon_0)\, g_i\, \mu_i(M_v). \tag{9.130}$$

Moreover, we introduce the coefficients

$$B(N, i, v - v_c^{\nu(v)}, \varepsilon_0) = \frac{1}{2} C_{N-i-1} C_i J_{\tilde{j}m(i,k(i))} \int_{-\varepsilon_0}^{v - v_c^{\nu(v)}} d\xi\, F(\xi, i, N) \,, \tag{9.131}$$

where $k(i)$ stems from $j(i,k) = \tilde{j}$, such that for $v = v_c^{\nu(v)}$ we have $B(N, i, -\varepsilon_0, \varepsilon_0) = 0$, and for $v - v_c^{\nu(v)} = \varepsilon_0$ we have $B(N, i, \varepsilon_0, \varepsilon_0) = A(N, i, \varepsilon_0) g_i$.

For the purposes of the present proof, we are not concerned about the complication of the coefficients $A(N, i, \varepsilon_0)$ and $B(N, i, v - v_c^{\nu(v)}, \varepsilon_0)$ because all that we need in order to make the link between configuration-space topology and thermodynamics is that the second term in the volume splitting in (9.112) can be written in the form (9.130). In fact, now we can write the entropy per degree of freedom as

$$S_N^{(-)}(v) = \frac{1}{N} \log M(v, N) = \frac{1}{N} \log \int_{M_v} d^N q \tag{9.132}$$

$$= \frac{1}{N} \log \left[\int_{M_v \backslash \bigcup_{i=1}^{\mathcal{N}(v+\varepsilon_0)} \Gamma(q_c^{(i)}, \varepsilon_0)} d^N q + \int_{M_v \cap \bigcup_{i=1}^{\mathcal{N}(v+\varepsilon_0)} \Gamma(q_c^{(i)}, \varepsilon_0)} d^N q \right]$$

$$= \frac{1}{N} \log \left[\int_{M_v \backslash \bigcup_{i=1}^{\mathcal{N}(v+\varepsilon_0)} \Gamma(q_c^{(i)}, \varepsilon_0)} d^N q + \sum_{i=0}^{N} A(N, i, \varepsilon_0) g_i\, \mu_i(M_v) \right]$$

when $v < v_c^{\nu(v)+1} - \varepsilon_0$ or $v > v_c^{\nu(v)} + \varepsilon_0$, or

$$S_N^{(-)}(v) = \frac{1}{N} \log \left[\int_{M_v \backslash \bigcup_{i=1}^{\mathcal{N}(v+\varepsilon_0)} \Gamma(q_c^{(i)}, \varepsilon_0)} d^N q + \sum_{i=0}^{N} A(N, i, \varepsilon_0)\, g_i \mu_i(M_{v-\varepsilon_0}) \right.$$

$$\left. + \sum_{n=1}^{\mathcal{N}_{cp}^{\nu(v+\varepsilon_0)+1}} B(N, i(n), v - v_c^{\nu(v)}, \varepsilon_0) \right] \tag{9.133}$$

when $v > v_c^{\nu(v)+1} - \varepsilon_0$ or $v < v_c^{\nu(v)} + \varepsilon_0$.

The equation above links thermodynamic entropy with the Morse indexes of the configuration-space submanifolds M_v, that is, with their topology. In fact, according to Bott's "critical-neck theorem" [219], any change with v of any index $\mu_i(M_v)$, $i = 0, \ldots, N$, which can only be due to the crossing of a critical level, is associated with a topological change of the M_v. Conversely, any topological change, in the sense of a loss of diffeomorphicity, occurring to the M_v when v is varied, is signaled by one or more changes of the Morse indexes $\mu_i(M_v)$, because by the "noncritical neck theorem" [210], this has to be the consequence of the crossing of a critical level.

In order to show that the coefficients $B(N, i, v - v_c^{\nu(v)}, \varepsilon_0)$ are smooth functions of v, one has to note that

$$\frac{d^k B(N, i, v, \varepsilon_0)}{dv^k} = \frac{d^{(k-1)} F(\xi, k_m, N)}{d\xi^{(k-1)}}\bigg|_{\xi = v - v_c},$$

and then focus on the smoothness of $F_\pm(\xi, k_m, N)$ defined in (9.123) and (9.124).

The smoothness of $F_+(\xi, k_m, N)$, for $\xi \neq 0$, is easily verified by inspection of the following outcome of direct integration

$$\int_{\sqrt{\xi}}^{\beta(\xi, r)} dy\, y^m\, (y^2 - \xi)^n$$
$$= \big\{ \xi^{(1+m+2n)/2} n(1 + m + 2n)\, \Gamma(-(1 + m + 2n)/2)\, \Gamma(1 + n)$$
$$\quad - \xi^{(1+m+2n)/2}(1 + m)\, \Gamma((1 - m - 2n)/2)\, \Gamma(1 + n)$$
$$\quad + \beta^{(1+m+2n)}(1 + m + 2n)\, \Gamma((1 - m)/2)$$
$$\quad \times\, {}_2F_1(-(1 + m + 2n)/2, -n; (1 - m - 2n)/2; \xi/\beta^2) \big\}$$
$$/\, \big[(1 + m + 2n)^2 \Gamma((1 - m)/2)\big] \tag{9.134}$$

where the function ${}_2F_1(a, b; c; x)$ is an hypergeometric function, whose argument x, for $\xi > 0$, is smaller than 1.

Also the smoothness of $F_-(\xi, k_m, N)$, for $\xi \neq 0$, is easily verified by inspection of the following result of direct integration

$$\int_0^{\beta(\xi, r)} dy\, y^m\, (y^2 - \xi)^n = \frac{1}{1 + m} \left\{ 2^{-(1+m)/2}(-\xi)^n \left[\frac{1}{\xi + \sqrt{\xi^2 + 4r^2}} \right]^{-(1+m)/2} \right.$$
$$\left. \times\, {}_2F_1((1 + m)/2, -n; (3 + m)/2; (\xi + \sqrt{\xi^2 + 4r^2})/(2\xi)) \right\} \tag{9.135}$$

where, being $\xi < 0$ and as we can always take $r < \sqrt{2\xi}$, the argument is $\left| (\xi + \sqrt{\xi^2 + 4r^2})/(2\xi) \right| < 1$.

At $\xi = 0$ it can be verified that smoothness of $F_\pm(\xi, k_m, N)$ is replaced by a high but finite differentiability class, that is, $C^{\mathcal{O}(N/2)}$.[7]

In conclusion, the smoothness of the coefficients $B(N, i, v - v_c^{\nu(v)}, \varepsilon_0)$ is proved.[8]

---

[7] The function ${}_2F_1(a, b; c; x)$ has a regular singular point at $x = 0$, allowing a polynomial representation in its neighborhood.

[8] Of course, also the differentiability class of the coefficients $B(N, i, v - v_c^{\nu(v)}, \varepsilon_0)$ drops to $C^{\mathcal{O}(N/2)}$ at $v - v_c^{\nu(v)} = 0$.

Let us now come to the proof of the statement of Theorem 9.39, which says that the source of a phase transition can only be the second term in square brackets in (9.109), which is of topological meaning. To this end we have to resort to Theorem 9.14 and Corollary 9.19.

We work under the assumption that only one critical value $\bar{v}_c$ exists in a given interval $[\bar{v}_0, \bar{v}_1]$. By Sard's theorem [210], at any finite N there is a finite number of isolated critical points on $\Sigma_{N\bar{v}_c}$. For any arbitrarily small $\delta > 0$, Theorem 9.14 and Corollary 9.19 still apply to the two subintervals $[\bar{v}_0, \bar{v}_c - \delta]$ and $[\bar{v}_c + \delta, \bar{v}_1]$. In order to understand why a breakdown of uniform boundedness in N of $|\partial^k S_N^{(-)}/\partial v^k|$ for $k = 3$ or $k = 4$ can be originated only by the topological term in right hand-side of (9.109), we consider each critical point $q_c^{(i)}$ on $\Sigma_{N\bar{v}_c}$ enclosed in a small pseudocylindrical neighborhood $\Gamma(q_c^{(i)}, \varepsilon_0)$ of thickness ε_0 and we take $\varepsilon < \varepsilon_0$ arbitrarily close to ε_0. From Morse theory, we know that passing a critical value v_c entails that for each critical point of index k_i a k_i-handle H^{N,k_i} is attached to $M_{v<v_c}$, so that the following diffeomorphism holds:

$$M_{(v_c+\varepsilon)} \approx M_{(v_c-\varepsilon)} \underset{\phi_1}{\bigcup} H^{N,k_1} \underset{\phi_2}{\bigcup} H^{N,k_2} \cdots \underset{\phi_n}{\bigcup} H^{N,k_n} , \qquad (9.136)$$

where $\bigcup_{\phi_i}$ stands for the attachment of H^{N,k_i} to $M_{(v_c-\varepsilon)}$ through the embedding $\phi_i : \mathbb{S}^{k_i-1} \times D^{N-k_i} \to \partial M_{(v_c-\varepsilon)}$. Details can be found in [23, 210, 227], see also Appendix C.

The excision of the pseudocylindrical neighborhoods $\Gamma(q_c^i, \varepsilon_0)$ of all the critical points $q_c^{(i)} \in \Sigma_{N\bar{v}_c}$ implies that all the manifolds $\overline{M}_v := M_v \setminus \bigcup_{i=1}^{\# \ crit.pts.} \Gamma(q_c^{(i)}, \varepsilon_0)$ with $N\bar{v}_c - \varepsilon_0 < v < N\bar{v}_c + \varepsilon_0$ are free of critical points and consequently are diffeomorphic. In fact, for any $v, v' \in \mathbb{R}$ such that $N\bar{v}_c - \varepsilon_0 < v < v' < N\bar{v}_c + \varepsilon_0$, $\overline{M}_v$ is a deformation retraction of $\overline{M}_{v'}$ through the flow associated with the vector field $X = -\nabla V_N/\|\nabla V_N\|^2$ [23, 210].

Now, defining $\overline{M}(v, N) = \mathrm{vol}(\overline{M}_v)$ and $\Gamma(v, N) = \mathrm{vol}[\bigcup_{i=1}^{\# \ crit.pts.} \Gamma(q_c^{(i)}, \varepsilon_0)]$, equation (9.109) becomes

$$S_N^{(-)}(v) = \frac{1}{N} \log \left[\overline{M}(v, N) + \Gamma(v, N) \right]$$
$$= \frac{1}{N} \log \left[\overline{M}(v, N) \right] + \frac{1}{N} \log \left[1 + \frac{\Gamma(v, N)}{\overline{M}(v, N)} \right] . \qquad (9.137)$$

By applying Theorem 9.14 and its Corollary 9.19 to the first term in the right hand-side of the equation above, we know that $\frac{1}{N}|\partial^k \log \left[\overline{M}(v, N) \right] / \partial v^k|$ for $k = 1, \ldots, 4$, are uniformly bounded in N, and thus no phase transition can arise from this term.

Then, let us consider the second term of the right hand-side of the equation above. By computing its first derivative we obtain

$$\frac{d}{dv} \frac{1}{N} \log \left[1 + \frac{\Gamma(v, N)}{\overline{M}(v, N)} \right] = \frac{1}{N} \frac{\Gamma'}{\overline{M} + \Gamma} - \frac{1}{N} \frac{\Gamma}{\overline{M} + \Gamma} \left(\frac{\overline{M}'}{\overline{M}} \right) , \qquad (9.138)$$

where $(\overline{M}'/\overline{M})$ stands for $[d\overline{M}(v,N)/dv]/\overline{M}(v,N)$. By Lemma 9.38, $(\overline{M}'/\overline{M})$ is uniformly bounded in N and therefore so is the second term in the right hand-side of (9.138). Therefore, if $|\partial S_N^{(-)}/\partial v|$ were to grow with N, this could not be due to the term $\overline{M}(v,N)$.

Then we compute the second derivative

$$
\frac{d^2}{dv^2}\frac{1}{N}\log\left[1+\frac{\Gamma(v,N)}{\overline{M}(v,N)}\right] = \frac{1}{N}\frac{\Gamma''}{\overline{M}+\Gamma} + \frac{1}{N}\frac{\Gamma\overline{M}(\overline{M}'/\overline{M})+\Gamma\Gamma'}{(\overline{M}+\Gamma)^2}\left(\frac{\overline{M}'}{\overline{M}}-\frac{\Gamma'}{\Gamma}\right)
$$
$$
-\frac{1}{N}\frac{\Gamma}{\overline{M}+\Gamma}\left(\frac{\overline{M}'}{\overline{M}}\right) + \frac{1}{N}\frac{d}{dv}\left(\frac{\overline{M}'}{\overline{M}}\right). \tag{9.139}
$$

Again, we can observe that the uniform boundedness with N of both $(\overline{M}'/\overline{M})$ and $(d/dv)(\overline{M}'/\overline{M}) = (\overline{M}''/\overline{M})-(\overline{M}'/\overline{M})^2$, by Lemma 9.38, entails that if $|\partial^2 S_N^{(-)}/\partial v^2|$ were to grow with N, this could not be due to the term $\overline{M}(v,N)$.

Similarly, after a lengthy but trivial computation of the third and fourth derivatives of the second term in the right hand-side of (9.137), one finds that $\overline{M}(v,N)$ enters the various terms obtained through the ratio $\overline{M}'/\overline{M}$ and through its derivatives $\frac{d^k}{dv^k}[\overline{M}'/\overline{M}]$ with $k=1,2,3$. Thus, as a consequence of Lemma 9.38, the uniform boundedness in N of $[d\overline{M}(v,N)/dv]/\overline{M}(v,N)$ and $\frac{d^k}{dv^k}\{[d\overline{M}(v,N)/dv]/\overline{M}(v,N)\}$ with $k=1,2,3$ implies that if $|\partial^3 S_N^{(-)}/\partial v^3|$ or $|\partial^4 S_N^{(-)}/\partial v^4|$ were to grow with N, this could not be due to the term $\overline{M}(v,N)$.

In conclusion, the first term within square brackets in (9.109) cannot be at the origin of a phase transition, nor can it be the third one, which is the sum of smooth functions. Only the second term of the right hand-side of (9.109), which is in one-to-one correspondence with topological changes of the M_v, can originate an unbound growth with N of a derivative $|\partial^k S_N^{(-)}/\partial v^k|$ for some k, thus entailing a phase transition.
□

Remark 9.40. A comment about the analytic expression of the volume splitting is in order. At any finite N, the volume $M(v,N)$ is a highly regular function of v (of differentiability class $C^{O(N/2)}$), as is the entropy $S_N^{(-)}(v)$. The term $\sum_{i=0}^N A(N,i,\varepsilon_0)g_i\mu_i(M_v)$ entering (9.109) and (9.133) is a discontinuous function of v because it depends on the integer valued functions $\mu_i(v)$, thus, at first sight, this term could seem conflicting with the high differentiability class of volume and entropy. Of course, at finite N, the high regularity of volume and entropy is not lost. In fact, the term $\sum_{i=0}^N A(N,i,\varepsilon_0)g_i\mu_i(M_v)$ is constant in any open interval $(v_c^{\nu(v)}+\varepsilon_0, v_c^{\nu(v)+1}-\varepsilon_0)$ and the functions $v\longmapsto B(N,i,v)$ smoothly connect in the interval $(v_c^{\nu(v)}-\varepsilon_0, v_c^{\nu(v)}+\varepsilon_0)$ the values that the function $\sum_{i=0}^N A(N,i,\varepsilon_0)g_i\mu_i(M_v)$ takes in the intervals $(v_c^{\nu(v)-1}+\varepsilon_0, v_c^{\nu(v)}-\varepsilon_0)$ and $(v_c^{\nu(v)}+\varepsilon_0, v_c^{\nu(v)+1}-\varepsilon_0)$.

Loosely speaking, $\sum_i A(N,i,\varepsilon_0)g_i\mu_i(M_v)+\sum_n B(N,i(n),v)$ is shaped as a "staircase" with "rounded corners".

Remark 9.41 (Domain of physical applications). About the applicability domain of what has been proved in this chapter, note that V is required to be a finite range interaction potential and a good Morse function. The former is a physical assumption, the latter a mathematical property of V. Finite range potentials are typical in condensed matter systems, where the interatomic and intermolecular interaction potentials are of the type of Lennard-Jones, Morse, van der Waals potentials, and where also classical spin potentials satisfy these requirements. Also regularized Coulomb interactions in condensed matter can be included, in fact, these are effective only at a finite distance because of the Debye shielding. The mathematical property of being a good Morse function is absolutely generic, in fact it requires that a potential function is bounded from below and that the Hessian of the potential is nondegenerate (i.e. its eigenvalues never vanish). Moreover, given a real-valued function f of class $\mathcal{C}^2$ defined on an arbitrary open subset X of $\mathbb{R}^N$, the mapping $x \longmapsto f(x) - (a_1 x_1 + \cdots + a_N x_N) : X \longmapsto \mathbb{R}$ is nondegenerate for almost all $(a_1, \ldots, a_N) \in \mathbb{R}^N$ (see Chapter 6 of [234]). This means that nondegeneracy is generic whereas degeneracy is exceptional. Continuous symmetries are the only physically relevant source of degeneracy, however this kind of degeneracy can be removed by adding a generic term $(a_1 x_1 + \cdots + a_N x_N)$ to the potential with an arbitrarily small vector $(a_1, \ldots, a_N)$. This removal of degeneracy is a rephrasing, within the framework of Morse theory, of a standard procedure undertaken in statistical mechanics to explicitly break a continuous symmetry, that is the addition of an external field whose limit to zero is taken after the limit $N \to \infty$.

Remark 9.42 (Sufficiency conditions). Summarizing, let us emphasize that the converse of Theorem 9.14 is not true. In other words, there is not a one-to-one correspondence between any topological change of the energy level sets and phase transitions. That this could not be the case is made clear by the existence of many systems undergoing a large number of topological transitions in the absence of thermodynamic phase transitions.

On the other hand, an analytic relation between configurational entropy and Morse indexes of the submanifolds M_v of configuration space can be established. By proceeding in two independent ways, one ends up with two different such relations—an approximate one in (8.36), and an exact one in (9.109)—that, however, are in very good qualitative agreement. Moreover, both these analytic relations suggest that "mild" variations with v of the topology of the M_v are compatible with a regular v-dependence of the entropy, thus with the absence of phase transitions. But the same equations (9.109) and (8.36) allow one to think that sufficiently "strong" variations with v and with N of the topological contribution to the entropy could be *sufficient* to entail a phase transition. By resorting to Theorem 9.14 we can prove in Theorem 9.39 that—passing a critical level of the potential—a phase transition can stem only from the topological term, that is, topological changes of the M_v are *necessary*, though *not sufficient*.

In order to go beyond qualitative arguments, one has to figure out which classes of topological changes are actually *sufficient* to make unboundedly grow with N some of the entropy derivatives with respect to v, thus entailing a phase transition.

Qualitative hints in this perspective are given in the following chapter.

We conclude this chapter with the following:

Conjecture 1. *Let $V_N(q_1, \ldots, q_N) : \mathbb{R}^N \to \mathbb{R}$, be a smooth, nonsingular, finite-range potential. If in a given interval $[\bar{v}_0, \bar{v}_1]$ of potential energy density the number of critical values of the potential grows at most linearly with N, then the statement of Theorem 9.39 establishing that an unbounded growth with N of one of the derivatives $|\partial^k S^{(-)}(v)/\partial v^k|$, for $k = 3, 4$, and thus the occurrence of a first- or second-order phase transition can be entailed only by the topological term $\sum_{i=0}^N A(N, i, \varepsilon_0) \, g_i \, \mu_i(M_{v-\varepsilon_0})$ in (9.109), is still true.*

The assumption of a linear growth with N of the number of critical values of V – in a potential energy density interval of finite length – is suggested by what happens in lattice systems where critical levels are separated by a finite minimum energy amount, as is the case of "spin flips" in the $1d$-XY model [206], mean-field XY model [206], p-trig model [223], or of elementary configurational changes that, in lattices and fluids, correspond to the appearance of a new critical value of V at an energetic cost which is independent of N. Note that the assumption that the number of critical values $\bar{v}_c^j$ is at most linearly growing with N entails, together with Sard theorem, that ε_0 is finite and can be chosen independent of N.

This conjecture is intuitively very reasonable. In fact, once we have proved that in the presence of a single critical level in $[\bar{v}_0, \bar{v}_1]$ the topological term in (9.109) is the only possible source of the breaking of uniform convergence with N of the entropy, it is hardly conceivable that this situation could be reversed just by adding critical levels in $[\bar{v}_0, \bar{v}_1]$.

Chapter 10

Phase Transitions and Topology: Exact Results

The preceding chapter contains a major theoretical achievement: the unbounded growth with N of certain thermodynamic observables, eventually leading to singularities in the $N \to \infty$ limit, which are used to define the occurrence of an equilibrium phase transition, is *necessarily* due to appropriate topological transitions in configuration space. The relevance of topology is made especially clear by the explicit dependence of thermodynamic configurational entropy on a weighed sum of Morse indexes of configuration-space submanifolds, a relation that, loosely speaking, has some analogy with the Yang–Lee "circle theorem," which relates thermodynamic observables to a fundamental mathematical object in the Yang–Lee theory of phase transitions: the angular distribution of the zeros of the grand-partition function on a circle in the complex fugacity plane.

The big challenge is now that of understanding what "appropriate topological changes" really means for the different kinds of phase transitions: first- and second-order transitions, Kosterlitz–Thouless transitions, glassy transitions, Θ-transitions of polymers, and so on. From a mathematical point of view, the problem is to find at least some *sufficient conditions* for Theorem 9.39 of Chapter 9. Again we have to start by collecting hints from the study of particular models.

We have already seen in Chapter 7 that the numerical investigation of both thermodynamics and topology in the special case of the lattice φ^4 model revealed a clear and unambiguous signature in the way topology changes when the system is found to undergo a phase transition. However, as is always the case with numerical results, the information obtained, though of utmost conceptual importance, is mainly of a qualitative nature. On the other hand, the above-mentioned relationship between thermodynamics and topology, expressed through the functional dependence of entropy on Morse indexes and reported in Chapters 8 and 9, allows us to understand also analytically the fact that "mild variations of topology" with potential energy—which are to be considered generic—are not sufficient to entail phase transitions. More drastic changes are required to produce an unbounded growth with N of some derivative of the entropy with respect to the potential energy.

In order to make these statements more precise, in the present chapter we report on a few exactly solvable models, i.e. for which both thermodynamics and topology can be computed explicitly [206, 222, 223]. By "computing topology" we mean "computing the Euler characteristic," which is quite far from exhausting the study of topology! But at present it is "the only game in town." As we shall see throughout this chapter, *prior to* and *independently of* the definition of any statistical measure in configuration space, the relevant information about the macroscopic transitional behavior is already contained in the microscopic interaction potential and concealed in its way of shaping configuration-space submanifolds.

All the models share an important property: they are "mean-field" models. In other words, they describe idealized systems with infinite-range interactions, which makes life easier in performing all the computations, but puts these systems out of the domain of applicability of the results given in Chapter 9. In fact, both Theorems 9.14 and 9.39 have been proved therein for short-range interactions, but we believe that additivity is the truly necessary property, and a generalization in this sense should be feasible, thus encompassing also the systems considered below.

10.1 The Mean-Field XY Model

The mean-field XY model is defined by the Hamiltonian [110]

$$H(p, \varphi) = \sum_{i=1}^{N} \frac{p_i^2}{2} + \frac{J}{2N} \sum_{i,j=1}^{N} [1 - \cos(\varphi_i - \varphi_j)] - h \sum_{i=1}^{N} \cos \varphi_i . \quad (10.1)$$

Here $\varphi_i \in [0, 2\pi]$ is the rotation angle of the ith rotator and h is an external field. Defining at each site i a classical spin vector $\mathbf{m}_i = (\cos \varphi_i, \sin \varphi_i)$, the model describes a planar (XY) Heisenberg system with interactions of equal strength among all the spins. We consider only the ferromagnetic case $J > 0$; for the sake of simplicity, we set $J = 1$. The equilibrium statistical mechanics of this system is exactly described, in the thermodynamic limit, by mean-field theory. In the limit $h \to 0$, the system has a continuous phase transition, with classical critical exponents, at $T_c = 1/2$, or $E_c/N = 3/4$, the critical energy per particle [110].

10.1.1 Canonical Ensemble Thermodynamics

The exact equilibrium solution of the model (10.1) for vanishing external field ($h = 0$) stems from the computation of the canonical partition function

$$Z(\beta, N) = \int \prod_{i=1}^{N} dp_i \, d\varphi_i \, \exp(-\beta H) ,$$

where, as usual, $\beta = 1/T$ (in units $k_B = 1$). This partition function factorizes into a kinetic term and a configurational part as follows:

$$Z(\beta, N) = \left(\frac{2\pi}{\beta}\right)^{N/2} \exp\left(-\frac{\beta J N}{2}\right) Z_c ,$$

where

$$Z_c = \int_{-\pi}^{\pi} \prod_{i=1}^{N} d\varphi_i \, \exp\left[\frac{\beta J}{2N} \sum_{i,j=1}^{N} \cos(\varphi_i - \varphi_j)\right] ,$$

which can be also written as

$$Z_c = \int_{-\pi}^{\pi} \prod_{i=1}^{N} d\varphi_i \, \exp\left[\frac{\beta J}{2N} \left(\sum_{i=1}^{N} \mathbf{m}_i\right)^2\right] , \qquad (10.2)$$

where the above-defined classical spin vectors $\mathbf{m}_i$ are used. In the ferromagnetic case $J > 0$, making use of the Hubbard–Stratonovich transformation

$$\exp\left(\frac{A\mathbf{x}^2}{2}\right) = \frac{1}{\pi} \int_{-\infty}^{\infty} \int_{-\infty}^{\infty} d\mathbf{y} \, \exp(-\mathbf{y}^2 + \sqrt{2A}\mathbf{x} \cdot \mathbf{y})$$

for the $A > 0$ case, (10.2) gives

$$Z_c = \frac{1}{\pi} \int_{-\pi}^{\pi} \prod_{i=1}^{N} d\varphi_i \int_{-\infty}^{\infty} \int_{-\infty}^{\infty} d\mathbf{y} \, \exp\left(-\mathbf{y}^2 + \sqrt{2A} \, \mathbf{M} \cdot \mathbf{y}\right) ,$$

with $A = \beta J/N$ and $\mathbf{M} = \sum_{i=1}^{N} \mathbf{m}_i$. By exchanging the integrals, we observe that the integration on the φ_i can be factorized; thus performing it and putting $\mathbf{z} = \mathbf{y}\sqrt{N/2\beta J}$, one has

$$Z_c = \frac{1}{\pi} \frac{N}{2\beta J} \int_{-\infty}^{\infty} \int_{-\infty}^{\infty} d\mathbf{z} \, \exp\left[-N\left(\frac{z^2}{2\beta J} - \ln(2\pi I_0(z))\right)\right]$$

where I_0 is the modified Bessel function and z is the modulus of $\mathbf{z}$. The free energy $F = \lim_{N \to \infty} -(\beta N)^{-1} \ln Z(\beta, N)$ is then obtained by computing the above integral by means of the saddle point method

$$-\beta F = \frac{1}{2} \ln\left(\frac{2\pi}{\beta}\right) - \frac{\beta J}{2} + \max_{z}\left(\frac{z^2}{2\beta J} - \ln(2\pi I_0(z))\right) . \qquad (10.3)$$

There are infinitely many minima of the free energy for each solution of

$$\frac{z}{\beta J} - \frac{I_1(z)}{I_0(z)} = 0 .$$

For $\beta J < 2$, $z = 0$ is the solution corresponding to the minimal free energy and to a vanishing magnetization. For $\beta J > 2$, the solution $\bar{z}$ is a value of z that is a function of β, correspondingly the magnetization

$$|\mathbf{M}| = \frac{I_1(\bar{z})}{I_0(\bar{z})} ,$$

obtained by repeating the computation given above with $h \neq 0$ and then taking the limit $h \to 0$, is nonvanishing and can be seen to bifurcate with a classical critical exponent $(1/2)$ for $\beta \geq \beta_c = 2/J$. The internal energy per particle $U = E/N$ is

$$U = \frac{\partial(\beta F)}{\partial \beta} = \frac{1}{2\beta} + \frac{J}{2}\left(1 - |\mathbf{M}|^2\right) .$$

In Figure 10.1 both temperature and magnetization (modulus) are given as functions of the internal energy per degree of freedom and display patterns typical of a second order phase transition.

10.1.2 Microcanonical Ensemble Thermodynamics

As we have already seen in the preceding chapter in the case of configuration space, the microcanonical volume in phase space

$$\Omega(E, N) = \int d^N p_i \, d^N \varphi_i \, \delta[H(p, \varphi) - E]$$

is related to the canonical partition function through

$$Z(\beta, N) = \int_0^\infty dE \, e^{-\beta E} \Omega(E, N) ,$$

which can be rewritten as

$$Z(\beta, N) = N \int_0^\infty dU \, \exp\left[N\left(-\beta U + \frac{1}{N}\ln\Omega(E, N)\right)\right] . \qquad (10.4)$$

In the thermodynamic limit the entropy density is

$$S(U) = \lim_{N\to\infty} \frac{1}{N}\ln\Omega(E, N) ,$$

so that by evaluating the integral in (10.4) through the saddle-point method, we get

$$-\beta F(\beta) = \max_U[-\beta U + S(U)] , \quad \beta = \frac{\partial S}{\partial U}. \qquad (10.5)$$

In general, one first computes the free energy and then derives from it the entropy. This is possible provided that the entropy is a concave function of the energy, so that

$$S(U) = \min_{\beta>0}\{\beta[U - F(\beta)]\}\ , \quad U = \frac{\partial(\beta F)}{\partial \beta}\ .$$

The entropy is not concave for long-range interaction systems near a first-order phase transition where negative specific heat appears. But in the case of the XY mean-field model, canonical and microcanonical ensembles are equivalent [226]; there is only a technical difficulty to analytically work out the microcanonical entropy. However, in this case the numerical simulation of the dynamics provides a reliable estimate of the microcanonical observables. These are reported in Figures 10.1 and 10.2 and found in excellent agreement with the (equivalent) canonical ensemble predictions.

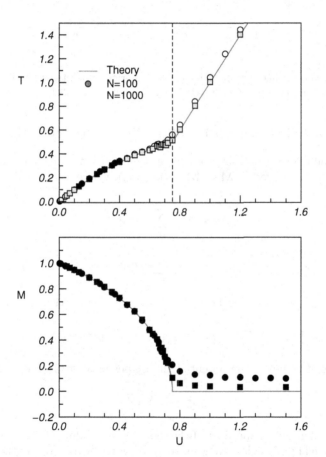

Fig. 10.1. Temperature T and magnetization M as a function of the energy per particle U. Circles and squares refer to MD computations, equivalent to microcanonical averages. Solid lines refer to the canonical predictions. See [224].

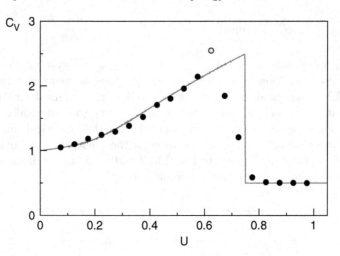

Fig. 10.2. Specific heat as a function of U. Numerical simulations (circles), obtained for $N = 500$, are compared to the theoretical result. From [224].

10.1.3 Analytic Computation of the Euler Characteristic

In what follows the analysis proceeds in configuration space only. Introducing the magnetization vector $\mathbf{M}$ as $\mathbf{M} = (m_x, m_y)$, where

$$m_x = \frac{1}{N} \sum_{i=1}^{N} \cos \varphi_i \ , \tag{10.6}$$

$$m_y = \frac{1}{N} \sum_{i=1}^{N} \sin \varphi_i \ , \tag{10.7}$$

the potential energy V can be written as a function of $\mathbf{M}$ as follows:

$$V(\varphi) = V(m_x, m_y) = \frac{N}{2}(1 - m_x^2 - m_y^2) - hN m_x \ . \tag{10.8}$$

The range of values of the potential energy per particle, $\mathcal{V} = V/N$, is then

$$-h \leq \mathcal{V} \leq \frac{1}{2} + \frac{h^2}{2} \ . \tag{10.9}$$

The configuration space M of the model is an N-dimensional torus, being parametrized by N angles. We want to investigate the topology of the following family of submanifolds of M:

$$M_v = \mathcal{V}^{-1}(-\infty, v] = \{\varphi \in M : \mathcal{V}(\varphi) \leq v\} \ , \tag{10.10}$$

i.e., each M_v is the set $\{\varphi_i\}_{i=1}^{N}$ such that the potential energy per particle does not exceed a given value v. This is the same as the $M_v = M_{E-K}$ defined above (v has been rescaled by $\frac{1}{N}$ because we choose $\mathcal{V} = V/N$ as a Morse function in order to make the comparison of systems with different N easier). As v is increased from $-\infty$ to $+\infty$, this family covers successively the whole manifold M ($M_v \equiv \emptyset$ when $v < -h$) .

As we have already seen in the preceding chapters, Morse theory [227] states that topology changes of M_v can occur only in correspondence with critical points of $\mathcal{V}$, i.e., those points where the differential of $\mathcal{V}$ vanishes. This immediately implies that no topological changes can occur when $v > 1/2 + h^2/2$, i.e., all the M_v with $v > 1/2 + h^2/2$ must be diffeomorphic to the whole M, that is, they must be N-tori. Moreover, if $\mathcal{V}$ is a Morse function (i.e., it has only nondegenerate critical points), then topological changes of M_v are actually in one-to-one correspondence with critical points of $\mathcal{V}$, and they can be characterized completely. At any critical level of $\mathcal{V}$ the topology of M_v changes in a way completely determined by the local properties of the Morse function: a k-handle $H^{N,k}$ is attached (see Appendix C), where k is the *index* of the critical point, i.e., the number of negative eigenvalues of the Hessian matrix of $\mathcal{V}$ at this point. Notice that if there are $m > 1$ critical points on the same critical level, with indices $k_1, \ldots, k_m$, then the topological change is made by attaching m disjoint handles $H^{N,k_1}, \ldots, H^{N,k_m}$. This way, by increasing v, the topology of the full configuration space M can be constructed sequentially from the M_v. Knowing the index of all the critical points below a given level v, we can obtain *exactly* the Euler characteristic of the manifolds M_v, defined by

$$\chi(M_v) = \sum_{k=0}^{N} (-1)^k \mu_k(M_v) , \tag{10.11}$$

where the *Morse number* μ_k is the number of critical points of $\mathcal{V}$ that have index k [227]. The Euler characteristic χ is a *topological invariant* (i.e. it is not affected by a diffeomorphic deformation of M_v): any change in $\chi(M_v)$ implies a topological change in the M_v. It will turn out that as long as $h > 0$, $\mathcal{V}$ is indeed a Morse function at least in the interval $-h \leq v < 1/2 + h^2/2$, while the maximum value $v = 1/2 + h^2/2$ may be pathological in that it may correspond to a critical level with degenerate critical points.

Thus, in order to detect and characterize topological changes in M_v we have to find the critical points and the critical values of $\mathcal{V}$, which means solving the equations

$$\frac{\partial \mathcal{V}(\varphi)}{\partial \varphi_i} = 0 , \qquad i = 1, \ldots, N , \tag{10.12}$$

and to compute the indices of *all* the critical points of $\mathcal{V}$, i.e., the number of negative eigenvalues of its Hessian

$$H_{ij} = \frac{\partial^2 \mathcal{V}}{\partial \varphi_i \partial \varphi_j} , \qquad i, j = 1, \ldots, N . \tag{10.13}$$

Taking advantage of (10.8), we can rewrite equations (10.12) as

$$(m_x + h) \sin \varphi_i - m_y \cos \varphi_i = 0 , \qquad i = 1, \ldots, N . \tag{10.14}$$

As long as $(m_x + h)$ and m_y are not simultaneously zero (the violation of this condition is possible only on the level $v = 1/2 + h^2/2$), the solutions of (10.14) are all configurations in which the angles are either 0 or π. In particular, the configuration

$$\varphi_i = 0 \qquad \forall i \tag{10.15}$$

is the absolute minimum of V, while all the other configurations correspond to a value of v that depends only on the number of angles that are equal to π. If we denote by n_π this number, we have that the N critical values are

$$v(n_\pi) = \frac{1}{2} \left[1 - \frac{1}{N^2} (N - 2n_\pi)^2 \right] - \frac{h}{N} (N - 2n_\pi) . \tag{10.16}$$

Inverting this relation yields n_π as a function of the level value v:

$$n_\pi(v) = \text{int} \left[\frac{1+h}{2} N \pm \frac{N}{2} \sqrt{h^2 - 2 \left(v - \frac{1}{2} \right)} \right] , \tag{10.17}$$

where int $[a]$ stands for the integer part of a. We can also compute the number $C(n_\pi)$ of critical points having a given n_π, which is the number of distinct binary strings of length N having n_π occurrences of one of the symbols, which is given by the binomial coefficient

$$C(n_\pi) = \binom{N}{n_\pi} = \frac{N!}{n_\pi! \, (N - n_\pi)!} . \tag{10.18}$$

We have thus shown that as v changes from its minimum $-h$ (corresponding to $n_\pi = 0$) to $\frac{1}{2}$ (corresponding to $n_\pi = \frac{N}{2}$), the manifolds M_v undergo a sequence of topological changes at the N critical values $v(n_\pi)$ given by (10.16). We expect that there is another topological change located at the last (maximum) critical value,

$$v_c = \frac{1}{2} + \frac{h^2}{2} . \tag{10.19}$$

However, the above argument does not prove this, since the critical points of V corresponding to this critical level may be degenerate. This is because the solutions of the two equations in N variables $m_x = m_y = 0$ need not be isolated, so that then on this level, V would not be a proper Morse function. Then a critical value v_c is still a necessary condition for the existence of a topological change, but it is no longer sufficient [210]. However, as already argued in [24, 206], it is just this topological change occurring at v_c given in (10.19), that is related to the thermodynamic phase transition of the mean-field XY model, since the temperature T, the energy per particle ε, and the average potential energy per particle $u = \langle \mathcal{V} \rangle$ obey, in the thermodynamic limit, the equation

$$2\varepsilon = T + 2u(T) ; \tag{10.20}$$

substituting in this equation the values of the critical energy per particle and the critical temperature, we get

$$u_c = u(T_c) = 1/2 \tag{10.21}$$

as $h \to 0$, $v_c \to \frac{1}{2}$, so that $v_c = u_c$. Thus a topological change in the family of manifolds M_v occurring at this v_c, where v_c is *independent* of N, is connected with the thermodynamic phase transition occurring in the limit $N \to \infty$, and $h \to 0$.

Let us now see that a topological change at v_c actually exists. One first characterizes completely all the topological changes occurring at $v < v_c$: this, together with the knowledge that at $v > v_c$ the manifold M_v must be an N-torus, will prove that a topological change at v_c must actually occur, also making clear in what sense it is different from the other topological changes occurring at $0 \le v < v_c$. Morse theory allows a complete characterization of the topological changes occurring in the M_v if the *indices* of the critical points of $\mathcal{V}$ are known. In order to determine the indices of the critical points (that is, the number of negative eigenvalues of the Hessian of $\mathcal{V}$ at the critical point) we proceed as follows. Since the diagonal elements of the Hessian are

$$H_{ii} = d_i = \frac{1}{N} \left[(m_x + h) \cos \varphi_i + m_y \sin \varphi_i \right] - \frac{1}{N^2} , \tag{10.22}$$

and the off-diagonal elements are

$$H_{ij} = -\frac{1}{N^2} \left(\sin \varphi_i \sin \varphi_j + \cos \varphi_i \cos \varphi_j \right) , \tag{10.23}$$

one can write the Hessian as the sum of a diagonal matrix D whose nonzero elements are

$$\delta_i = \frac{1}{N} \left[(m_x + h) \cos \varphi_i + m_y \sin \varphi_i \right] , \qquad i = 1, \ldots, N , \tag{10.24}$$

and of a matrix B whose elements are just the H_{ij} given in (10.23), also for $i = j$ (the diagonal elements being $-1/N^2$). Since the ratio between the elements of B and those of D is $\mathcal{O}(1/N)$, one would expect at first sight that only the diagonal elements survive when $N \gg 1$, so that the Hessian approaches a diagonal matrix equal to D. However, this is not, in principle, necessarily true: one cannot immediately say that at large N the eigenvalues of the Hessian are the δ's given in (10.24) plus a correction vanishing as $N \to \infty$, because the number of elements of B is N^2, so that the contribution of the matrix B to the eigenvalues of the Hessian does *not*, in general, vanish at large N. That nevertheless the argument for this crucial point is correct in this special case is shown in Section 10.4, and is due to the particular structure of the matrix B. The latter is of rank one, and then it can be proved that at the critical points of $\mathcal{V}$ the number of negative eigenvalues of H equals the number of negative diagonal elements δ ± 1, so that as $N \gg 1$ we can conveniently approximate the index of the critical points with the number of negative δ's at x,

$$\text{index } (x) \approx \#(\delta_i < 0) \,. \tag{10.25}$$

At a given critical point, with given n_π, where the x-component of the magnetization vector is

$$m_x = 1 - \frac{2n_\pi}{N} \,, \tag{10.26}$$

so that $m_x > 0$ (respectively < 0) if $n_\pi \leq \frac{N}{2}$ (respectively $> \frac{N}{2}$), the eigenvalues of D are

$$\delta_i = m_x + h \,, \qquad i = 1, \dots, N - n_\pi \,; \tag{10.27}$$
$$\delta_i = -(m_x + h) \,, \qquad i = N - n_\pi + 1, \dots, N \,. \tag{10.28}$$

Then, if the external field h is sufficiently small,

$$(m_x + h) > 0 \qquad \text{if } n_\pi \leq \frac{N}{2} \,,$$
$$(m_x + h) < 0 \qquad \text{if } n_\pi > \frac{N}{2} \,, \tag{10.29}$$

so that, denoting by $\text{index}(n_\pi)$ the index of a critical point with n_π angles equal to π, we can write

$$\text{index}(n_\pi) = n_\pi \qquad \text{if } n_\pi \leq \frac{N}{2} \,, \tag{10.30}$$
$$\text{index}(n_\pi) = N - n_\pi \qquad \text{if } n_\pi > \frac{N}{2} \,. \tag{10.31}$$

From these equations combined with (10.18) one can obtain for the Morse numbers μ_k, i.e., for the number of critical points of index k, as a function of the level v, as long as $-h \leq v < 1/2 + h^2/2$ (i.e., excluding the limiting value $v = 1/2 + h^2/2$) the following expression:

$$\mu_k(v) = \binom{N}{k} \left[1 - \Theta(k - n^{(-)}(v)) + \Theta(N - k - n^{(+)}(v)) \right] \,, \tag{10.32}$$

where $\Theta(x)$ is the Heaviside theta function and $n^{(\pm)}(v)$ are the limits of the allowed n_π's for a given value of v, i.e., from (10.17),

$$n^{(\pm)}(v) = \frac{N}{2} \left[1 + h \pm \sqrt{h^2 - 2\left(v - \frac{1}{2}\right)} \right] \,. \tag{10.33}$$

We note that $0 \leq n_\pi^{(-)} \leq \frac{N}{2}$ and $\frac{N}{2} + 1 \leq n_\pi^{(+)} \leq N$, so that (10.32) implies

$$\mu_k(v) = 0 \qquad \forall k > \frac{N}{2} \,, \tag{10.34}$$

i.e., no critical points with index larger than $N/2$ exist as long as $v < v_c = 1/2 + h^2/2$.

This is the crucial observation to prove that a topological change *must* occur at v_c. Since the Betti numbers of a manifold are positive (or zero) numbers, using the Morse inequalities, which state that the Morse numbers are upper bounds of the Betti numbers [210], i.e.,

$$b_k \leq \mu_k \quad \text{for} \quad k = 0, \ldots, N, \tag{10.35}$$

we can immediately conclude that as long as $v < v_c = 1/2 + h^2/2$, we have

$$b_k(M_v) = 0 \quad \forall k > \frac{N}{2} . \tag{10.36}$$

Thus, since $\frac{1}{2} \leq v < \frac{1}{2} + \frac{h^2}{2}$ the manifold is only "half" an N-torus, and since we know that for $v > \frac{1}{2} + \frac{h^2}{2}$, M_v *is* a (full) N-torus, whose Betti numbers are

$$b_k(\mathbb{T}^N) = \binom{N}{k} , \quad k = 0, 1, \ldots, N , \tag{10.37}$$

we conclude that at $v = v_c = \frac{1}{2} + \frac{h^2}{2}$ a topological change *must* occur, which involves the attaching of $\binom{N}{k}$ different k-handles for each k ranging from $\frac{N}{2} + 1$ to N, i.e., a change of $\mathcal{O}(N)$ of the number of Betti numbers.

Let us remark that such a topological change not only exists: it is surely a "big" topological change, for all of a sudden, "half" an N-torus becomes a full N-torus, via the attaching for each different k (ranging from $N/2 + 1$ to N) of $\binom{N}{k}$ k-handles. More precisely, a number of Betti numbers, which is $\mathcal{O}(N)$, changes, and changes by amounts that are of the same order as their maximum possible values. On the other hand, all the other topological changes correspond to the attaching of handles of the same type (index). In fact, as long as $v < v_c$, each critical level contains only critical points of the same index, and the index grows with v, i.e., if x_c and x'_c are critical points and $\mathcal{V}(x'_c) > \mathcal{V}(x_c)$, then index$(x'_c) > $ index(x_c). The potential energy per degree of freedom $\mathcal{V}$ is a *regular* Morse function (or a Morse–Smale function [210]) as long as $v < v_c$, but this is no longer true for $v \geq v_c$; actually, as we have already observed, $\mathcal{V}$ could even no longer be a Morse function at all, because the level v_c might contain degenerate critical points. Nevertheless, as we have shown, this does not prevent us from giving a complete analysis of the topology of the M_v's for *all* the values of v, since we can exploit our explicit knowledge of the topology of the M_v's for any $v > v_c$.

To illustrate what has been described so far, the Morse numbers μ_k are shown in Figure 10.3 as a function of k for two values of v: $v = \frac{1}{4}$, i.e., an intermediate value between the minimum and the maximum of $\mathcal{V}$, shown in Figure 10.3 (a), and $v = \frac{1}{2}$, shown in Figure 10.3 (b). We see that the μ_k with $0 \leq k \leq \frac{N}{2}$ grow regularly as v grows until $v_c = \frac{1+h^2}{2}$, while all the

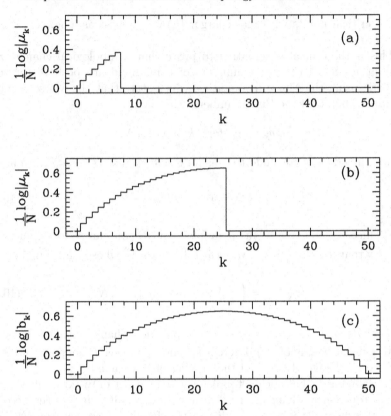

Fig. 10.3. Mean-field XY model. (a) Histogram of $\log(\mu_k(M_v))/N$ as a function of k for $v = 1/4$; (b) Histogram of $\log(\mu_k(M_v))/N$ as a function of k for $v = 1/2$. In both cases $N = 50$ and $h = 0.01$. (c) For comparison, histogram of $\log(b_k(\mathbb{T}^N))/N$ as a function of k for an N-torus $\mathbb{T}^N$, with $N = 50$, which is the lower bound of $\log(\mu_k(M_v))/N$ for any $v \geq v_c$.

μ_k with $k > \frac{N}{2}$ remain zero, so that the corresponding Betti numbers must also vanish. But at v_c a dramatic event occurs, because for all the values of $v > v_c$ the Betti numbres b_k *must* be those of an N-torus, which are reported for comparison in Figure 10.3 (c). A sudden transition from the situation depicted in Figure 10.3 (b) to that of Figure 10.3 (c) occurs at v_c, i.e., $\frac{N}{2}$ Betti numbers simultaneoulsy become nonzero.

These topological transitions can be described by the variation with v of topological invariants. The best choice would be that of computing all the $N + 1$ Betti numbers of the manifolds M_v as functions of v: unfortunately, this is infeasible. We can only set upper bounds to them, using the Morse inequalities (10.35).

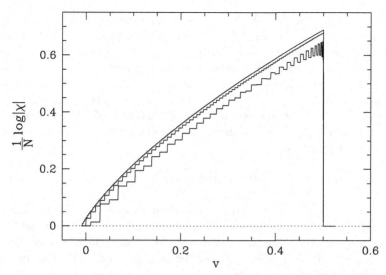

Fig. 10.4. Mean-field XY model. Plot of $\log(|\chi|(M_v))/N$ as a function of v. $N = 50,200,800$ (from bottom to top) and $h = 0.01$; $v_c = 0.5 + \mathcal{O}(h^2)$. From [206].

Nevertheless, through the knowledge of all the μ_k we can compute another (though somewhat weaker) topological invariant, the Euler characteristic of the manifolds M_v, using (10.11), (10.17), and (10.32).

It turns out then that χ jumps from positive to negative values, so that it is easier to look at $|\chi|$. In Figure 10.4, $\log(|\chi(M_v)|)/N$ is plotted as a function of v for various values of N ranging from 50 to 800. The "big" topological change occurring at the maximum value of $\mathcal{V}$, which corresponds in the thermodynamic limit to the phase transition, implies a discontinuous jump of $|\chi|$ from a big value to zero.

10.2 The One-Dimensional XY Model

We have already seen that topological changes are necessary but not sufficient to induce a phase transition. This circumstance is well illustrated by the one-dimensional XY model with nearest-neighbor interactions, whose Hamiltonian is a standard one with interaction potential

$$V(\varphi) = \frac{1}{4} \sum_{i=1}^{N} [1 - \cos(\varphi_{i+1} - \varphi_i)] - h \sum_{i=1}^{N} \cos \varphi_i . \qquad (10.38)$$

The configuration space of this system undergoes topological changes very similar to those of the mean-field XY model, but no phase transition occurs in the 1D XY model with nearest-neighbor interactions. In this case the configuration space M is still an N-torus, and using again the interaction energy per degree of freedom $\mathcal{V} = V/N$ as a Morse function, we can prove that also here there are many topological changes in the submanifolds M_v as v is varied from its minimum to its maximum value. The critical points are again those where the φ_i's are equal either to 0 or to π. However, in contrast to the mean-field XY model, the critical values are now determined by the number of domain walls, n_d, i.e., the number of boundaries between connected regions on the chain where the angles are all equal ("islands" of π's and "islands" of 0's). The number of π's leads only to a correction $\mathcal{O}(h)$ to the critical value of v, which is given by

$$v(n_d; n_\pi) = \frac{n_d}{2N} + h\, n_\pi \ . \tag{10.39}$$

Since $n_d \in [0, N-1]$ (with free boundary conditions, $n_d = 0, 1, \ldots, N-1$, while with periodic boundary conditions n_d is still bounded by 0 and $N-1$, but can only be even), the critical values lie in the same interval as in the case of the mean-field XY model. But now the maximum critical value, instead of corresponding to a huge number of critical points, which grows rapidly with N, corresponds to *only two* configurations with $N-1$ domain walls, which are $\varphi_{2k} = 0$, $\varphi_{2k+1} = \pi$, with $k = 1, \ldots, N/2$, and the reversed one.

The number of critical points with n_d domain walls is therefore (assuming free boundary conditions)

$$N(n_d) = 2\binom{N-1}{n_d} \ . \tag{10.40}$$

We can compute the index of the critical points also in this case (see Section 10.4.2 for details). It turns out that

$$\text{index}(n_d) = n_d \ , \tag{10.41}$$

so that

$$\mu_k(n_d) = 2\binom{N-1}{k}\Theta(n_d - k) \ . \tag{10.42}$$

It is evident then that any topological change here corresponds to the attaching of handles of the same type. However, no "big" change like the one at v_c in the case of the mean-field model exists, although $\mathcal{V}$ is a Morse–Smale function on the whole manifold M. To illustrate this, we plot in Figure 10.5 the values of the Morse indices μ_k as a function of k, as we have already done for the mean-field XY model in Figure 10.3. Comparing Figure 10.5 with Figure 10.3, we see that in the mean-field model there is a critical value where $\frac{N}{2}$ Betti numbers become simultaneously nonzero, i.e., there exists a topological change that corresponds to the simultaneous attaching of handles of $\frac{N}{2}$ different types, while here in the one-dimensional nearest-neighbor model nothing like that happens.

Also in this case, using (10.40) and (10.41), we can compute the Euler characteristic of the submanifolds M_v:

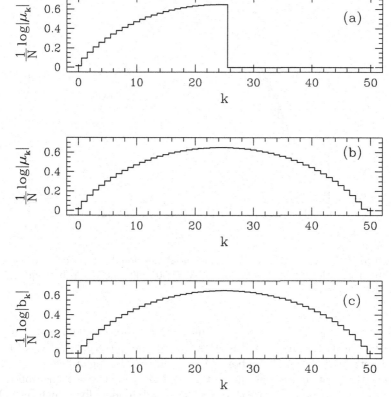

Fig. 10.5. The same as Figure 10.3 for the one-dimensional XY model with nearest-neighbor interactions. (a) Histogram of $\log(\mu_k(M_v))/N$ as a function of k for $v = 1/4$; (b) Histogram of $\log(\mu_k(M_v))/N$ as a function of k for $v = 1/2$. In both cases $N = 50$ and $h = 0.01$. (c) For comparison, histogram of $\log(b_k(\mathbb{T}^N))/N$ as a function of k for an N-torus $\mathbb{T}^N$, with $N = 50$. From [206].

$$\chi(M_v) = 2 \sum_{k=0}^{n_d(v)} (-1)^k \binom{N-1}{k} = 2\,(-1)^{n_d(v)} \binom{N-2}{n_d(v)} , \qquad (10.43)$$

where, due to (10.39),

$$n_d(v) = 2Nv + \mathcal{O}(h) . \qquad (10.44)$$

The Euler characteristic for the one-dimensional nearest-neighbor case is shown in Figure 10.6. Comparing this figure with Figure 10.4, we see that there is here no jump in the Euler characteristic.

Two remarks are in order. The first is that the topology changes with v also in the absence of phase transitions, as is the case of the one-dimensional XY model. We already found in Chapter 7 a similar result concerning the one-dimensional lattice φ^4 model. In Chapters 8 and 9, we explained that

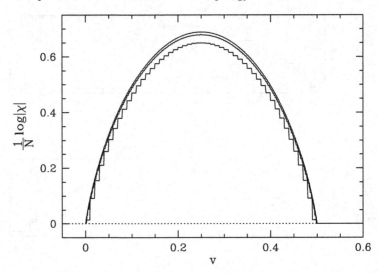

Fig. 10.6. Plot of $\log(|\chi|(M_v))/N$ for the one-dimensional XY model with nearest-neighbor interactions as a function of v. $N = 50, 200, 800$ (from bottom to top). From [206].

"mild" variations of topology, as a function of v, are not sufficient to entail a phase transition. The second remark is that the patterns of $\chi(M_v)$ for the one-dimensional nearest-neighbor XY model and the mean-field XY model are very different and clearly mark the presence of a phase transition at its proper place. In contrast to the 2D lattice ϕ^4 model, for which $\chi(\Sigma_v)$ is found to undergo a sudden variation in its rate of change with v at the phase transition point (see Chapter 7), $\chi(M_v)$ is here found to make a big sudden jump at the transition point.

Finally, after the relationship that has been worked out in Chapters 7 and 8 between entropy and topology, and in particular with the sum $\mu(v) = \mu_0(M_v) + 2\sum_{i=1}^{N-1} \mu_i(M_v) + \mu_N(M_v)$ of Morse indexes, it is interesting to compute the quantity

$$N_c(M_v) = \sum_{i=0}^{N} \mu_i(M_v) , \qquad (10.45)$$

where $N_c(M_v)$ is the total number of critical points of the function $\mathcal{V}$ in the manifold M_v. Therefore, at large N, if μ_0 and μ_N are not much larger than the other indexes (which is true in our case; see Figure 10.3), apart from an additive constant, we can write approximately

$$\tilde{S}(v) \approx \frac{1}{N} \log N_c(v) \qquad (10.46)$$

for the entropy fraction $\tilde{S}$, which is explicitly linked to the above given sum $\mu(v)$. The pattern of $\tilde{S}(v)$ is plotted as a function of v, at different N, in

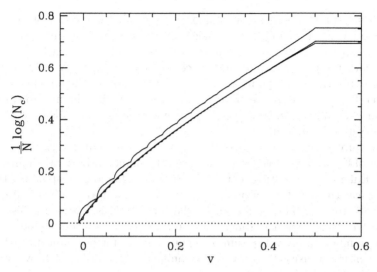

Fig. 10.7. Mean-field XY model. Plot of $\log(N_c)/N$ as a function of v. $N = 50, 200, 800$ (from top to bottom) and $h = 0.01$; $v_c = 0.5 + \mathcal{O}(h^2)$. From [206].

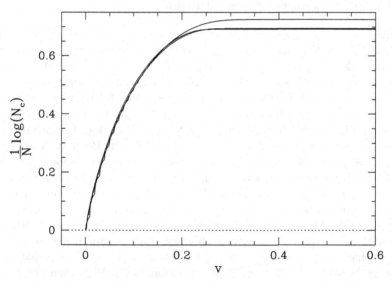

Fig. 10.8. One-dimensional XY model with nearest-neighbor interactions. Plot of $\log(N_c)/N$ as a function of v. $N = 50, 200, 800$ (from top to bottom) and $h = 0.01$. From [206].

Figure 10.7 for the mean-field XY model and in Figure 10.8 for the one-dimensional XY model. First, we note that in both cases the "topological contribution" to the entropy behaves qualitatively as expected for the configurational entropy, i.e., it grows monotonically up to the maximum value of v, after which it remains constant. Moreover, we see that in the case of the mean-field model (Figure 10.7), the topological change at v_c (the phase transition point) corresponds to a discontinuity in the slope of $\tilde{S}(v)$, which thus seems to be both the precursor and the source of the nonanalyticity of the entropy at the phase transition point as $N \to \infty$. In the case of the one-dimensional XY model, where no phase transition is present, unlike the mean-field case, the curve is smooth for any v, consistent with the fact that also the entropy of the system is smooth.

We emphasize that the discontinuity in Figure 10.4 at $v = v_c$ corresponds to that of Figure 10.7 at the same value of v. Even more, the many small jumps in the Euler characteristic occurring in Figure 10.4, and corresponding to the topological changes that occur at $v < v_c$, are smoothed out in Figure 10.7, where the topological contribution to the entropy, $\tilde{S}$, is reported. This is due to the fact that while the Euler characteristic is the alternating sum of the Morse numbers μ_i, $\tilde{S}$ is the sum of them: this is a further indication that to yield a phase transition, i.e., a discontinuity in a derivative of S, a "strong" topological change is needed.

10.2.1 The Role of the External Field h

Studying the topological changes in the configuration space of XY models, mean-field as well as one-dimensional, we have considered the presence of an external field $h \neq 0$ that explicitly breaks the $O(2)$-invariance of the potential energy, and then, discussing the connection with phase transitions, we have considered the limit in which h tends to zero. We did that for the sake of simplicity, for if we set $h = 0$ from the outset, the potential energy per degree of freedom $\mathcal{V}$ is *not*, rigorously speaking, a Morse function, because its $O(2)$-invariance entails the presence of a zero eigenvalue in its Hessian. When $h = 0$, the critical points of $\mathcal{V}$ are *not* isolated, but form one-dimensional manifolds (topologically equivalent to circles) that are left unchanged by the action of the $O(2)$ continuous symmetry group so that the critical points become in this case critical manifolds. However, in the case of the mean-field XY model, as far as the presence and the nature of topological changes are concerned, studying the case with $h = 0$ from the outset, we find *exactly* the same behavior as in the case we have discussed in this book, i.e., as long as $v < v_c$, only handles of the same type are attached, while at $v = v_c$ handles of $N/2$ different types are attached, the only difference between the two cases being that when $h = 0$ the handles are not attached at isolated points, but rather to the entire critical manifold [210, 227]. However, putting $h = 0$ from the beginning makes the computation of the Euler characteristic $\chi(M_v)$ via the Morse numbers much more difficult, because now one has to take into account the contributions to χ coming from the Betti numbers of the critical manifolds (see [228] for details).

In the case of the one-dimensional nearest-neighbor XY model, where no phase transition is present, the use of $h = 0$ from the outset implies a further complication, i.e., that the critical points consist not only of the configurations made of 0's and π's, but also of spin waves, that is, configurations

$$\varphi_j = \varphi_0 \, e^{ikj} \, , \tag{10.47}$$

with wave numbers k depending on boundary conditions.

For all these reasons it is convenient to force the potential energy to be a Morse function via the explicit breaking of the $O(2)$ symmetry using an external field $h \neq 0$. Incidentally, we notice that in the mean-field XY model, as long as $h \neq 0$, the topological changes that do not correspond to any phase transition (i.e, those occurring at $v < v_c$) occur at a number of values of v that grows with N, and these values become closer and closer as N grows, eventually filling the whole interval $[0, \frac{1}{2}]$ as $N \to \infty$. In contrast, the value v_c, which corresponds to the "big" topology change connected to the phase transition, remains separated from the others by an amount $\mathcal{O}(h^2)$ also in the thermodynamic limit, and tends to $\frac{1}{2}$ only when $h \to 0$. This is reminiscent of a similar fact occurring in statistical mechanics, where one observes a spontaneous symmetry breaking, signaled, e.g., by the onset of a finite magnetization even at zero external field, if one assumes the presence of an external field and then lets it tend to zero only after the thermodynamic limit is taken.

10.3 Two-Dimensional Toy Model of Topological Changes

Before discussing the relevance that these results may have for the general problem of the relation between topology and phase transitions, it is illustrative to consider two abstract simplified models of topological transitions that occurred in the two models we have considered so far. A two-dimensional model of the topological transition occurring in configuration space of the physical models we have discussed, which could perhaps help the intuition, can be built as follows. Let us consider a two-dimensional torus $\mathbb{T}$, and place it in a plane. Let $h_{\max}$ be the maximum height of the surface above the plane (see Figure 10.9 (a)). Then deform the torus (by means of a diffeomorphism) until the upper end of the hole is at height $h_{\max} - \varepsilon$, obtaining the surface M shown in Figure 10.9 (b). It is apparent that ε can be made as small as we want.

Let us now consider the height function $\mathcal{H}$ above the plane as a Morse function, so that the manifolds

$$\mathbb{T}_h = \mathcal{H}^{-1}(-\infty, h] = \{x \in \mathbb{T} : \mathcal{H}(x) \le h\} \, , \tag{10.48}$$
$$M_h = \mathcal{H}^{-1}(-\infty, h] = \{x \in M : \mathcal{H}(x) \le h\} \, , \tag{10.49}$$

are defined. As h varies from its minimum to its maximum values ($h = 0$ and $h = 3$, respectively, in Figure 10.9), the manifolds $\mathbb{T}_h$ and M_h cover the

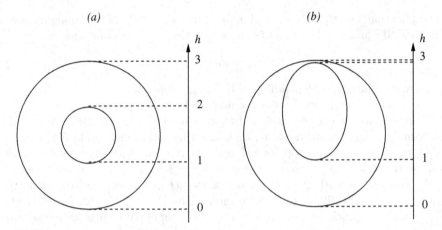

Fig. 10.9. (*a*) A torus $\mathbb{T}$ and its height function. Here $h_{\max} = 3$. (*b*) A deformation M of such a torus as explained in the text. From [206].

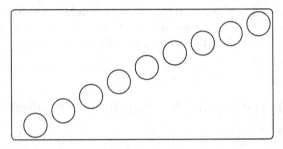

Fig. 10.10. A surface of high genus (here $g = 9$). From [206].

whole torus; as long as h is lower than the top of the hole ($h = 2$ in Figure 10.9), both $\mathbb{T}_h$ and M_h are "half-tori," but then the $\mathbb{T}_h$ become gradually a full torus, with topological changes that are equally spaced in h (by construction), while the M_h jump abruptly from a half-torus to a full torus as h is changed by ε. Identifying the height function with the potential energy, the case of the $\mathbb{T}_h$ clearly recalls the behavior of the one-dimensional XY model with no phase transition, while the case of the M_h seems close to what happens in the mean-field XY model, and the "jump" from the half-torus to the full torus is similar to the topological change that is connected to the phase transition.

The analogy with the mean-field models becomes even clearer if, instead of a torus, we consider a compact surface of genus[1] $g \gg 1$, i.e., with many holes, and deform the surface with a diffeomorphism until the upper end of all the holes is at height $h_{\max} - \varepsilon$, as shown in Figures 10.10 and 10.11. Again, ε can be made as small as we want. Let us denote by $M^{(g)}$ the deformed surface.

---

[1] The *genus* g is the number of handles of a two-dimensional surface, or equivalently, of holes without boundary.

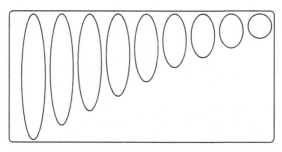

Fig. 10.11. A deformation $M^{(g)}$ of the surface of Figure 10.10 as explained in the text. From [206].

The Betti numbers of the surface $M^{(g)}$ are (see Appendix A)

$$b_0 = b_2 = 1, \qquad b_1 = 2g, \tag{10.50}$$

and the Euler characteristic is

$$\chi(M^{(g)}) = \chi(M^{(g)}) = 2 - 2g, \tag{10.51}$$

i.e., a big negative number. Let us now consider, as in the previous case of the torus, the height function $\mathcal{H}$ above the plane as a Morse function. The manifolds

$$M_h^{(g)} = \mathcal{H}^{-1}(-\infty, h] = \{x \in M^{(g)} : \mathcal{H}(x) \le h\} \tag{10.52}$$

will be topologically very different from the whole $M^{(g)}$ as long as $h < h_{\max} - \varepsilon$ but h is sufficiently large that all the critical levels corresponding to the bottoms of all the holes have already been crossed: in fact, their Betti numbers will be

$$b_0(M_h^{(g)}) = 1, \qquad b_1(M_h^{(g)}) = g, \qquad b_2(M_h^{(g)}) = 0, \tag{10.53}$$

and the Euler characteristic will be

$$\chi(M_h^{(g)}) = 1 - g. \tag{10.54}$$

Then, by changing the value of the height by an amount ε as small as one wants, one changes the Betti number b_1 from g to $2g$, the b_2 from 0 to 1, and χ from $1 - g$ to $2 - 2g$. This is a topological change that involves a change of $\mathcal{O}(d)$ Betti numbers (d is the dimension of the manifold); moreover, the size of the change is of the order of the value of the Betti numbers. This topological change also involves a change of the Euler characteristic χ which is again of the same order as its value. Identifying again the height function with the potential energy, we see that this is just what happens in the case of the M_v of the mean-field XY model, although there the dimension of the manifolds is N and very large, while in this low-dimensional analogy it is only $d = 2$.

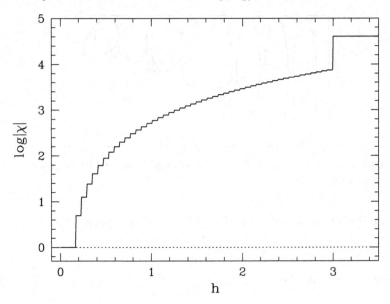

Fig. 10.12. Plot of the logarithm of the absolute value of the Euler characteristic of the submanifolds of given height of a surface $M^{(g)}$ like the one depicted in Figure 10.11, with $g = 50$ and $h_{\max} = 3$, as a function of the height h. The small jumps are the topological transitions corresponding to the crossing of the bottoms of the holes: the last big jump is the one occurring at $h_{\max}$, when $\varepsilon \to 0$. From [206].

The behavior of $|\chi(M_h^{(g)})|$ as a function of h is plotted in Figure 10.12. We see that the behavior of $|\chi|$ is indeed very similar to the case of the mean-field XY model, the only big difference being that in the latter, $|\chi|$ jumps to zero, while here it jumps to a nonzero value. However, this difference is due to the fact that the Euler characteristic of a torus is zero while that of a surface of genus g is $2 - 2g$.

10.4 Technical Remark on the Computation of the Indexes of the Critical Points

In this section we enter some details concerning the computation of the indexes of the critical points that have been used in Sections 10.1 and 10.2.

10.4.1 Mean–Field XY Model

Let us prove the crucial estimate (10.25) that is used in Section 10.1 to compute the index of the critical points for the mean-field XY model.

Here the aim is to compute the number of negative eigenvalues of the Hessian matrix of the function $\mathcal{V}$, i.e., of the matrix H whose elements H_{ij} are

$$H_{ij} = \frac{\partial^2 \mathcal{V}}{\partial \varphi_i \partial \varphi_j} \ , \qquad i,j = 1,\ldots,N \ , \tag{10.55}$$

where $\mathcal{V} = V/N$, and V is the potential energy of the mean-field XY model defined in (10.1). The diagonal elements of this matrix are

$$H_{ii} = d_i = \frac{1}{N} \left[(m_x + h) \cos \varphi_i + m_y \sin \varphi_i \right] - \frac{1}{N^2} \ , \tag{10.56}$$

and the off-diagonal ones are

$$H_{ij} = -\frac{1}{N^2} \left(\sin \varphi_i \sin \varphi_j + \cos \varphi_i \cos \varphi_j \right) \ . \tag{10.57}$$

At the critical points of $\mathcal{V}$, the angles are either 0 or π, so that the sines are all zero and the cosines are ± 1. Moreover, since we are interested only in the signs of the eigenvalues of H and not in their absolute values, we multiply H by N in order to get rid of the $1/N$ factor in front of it. We can then write the matrix H (multiplied by N) as

$$H = D + B \ , \tag{10.58}$$

where D is a diagonal matrix,

$$D = \mathrm{diag}(\delta_i) \ , \tag{10.59}$$

whose elements δ are

$$\delta_i = (m_x + h) \cos \varphi_i \ , \tag{10.60}$$

where the φ_i's $(i = 1,\ldots,N)$ are computed at the critical point, and the elements of B can be written in terms of a vector σ whose N elements are either 1 or -1:

$$b_{ij} = -\frac{1}{N} \sigma_i \sigma_j \ , \tag{10.61}$$

where

$$\sigma_i = +1(-1) \ \text{if} \ \varphi_i = 0(\pi) \ . \tag{10.62}$$

This holds because when the angles are either 0 or π, the sines in (10.23) vanish, so that then

$$N H_{ij} = -\frac{1}{N} \cos \varphi_i \cos \varphi_j \ , \tag{10.63}$$

and

$$\cos \varphi_i = \sigma_i \ . \tag{10.64}$$

Having fixed the notation, our goal is to show that, at least when N is large, the number of negative eigenvalues of the full matrix H, i.e., the index of the critical point, can be conveniently approximated by the number of negative eigenvalues of D, that is, by the number of negative δ's. To do that, we proceed in two steps: (i) we show that the matrix B is of rank one (which implies that B has $N - 1$ zero eigenvalues and only one nonzero eigenvalue), and (ii) we adapt a theorem due to Wilkinson [229] to this case, thus proving our assertion.

As to step (i), let us consider for example a case with $N = 3$, and the critical point corresponding to, say, $(\varphi_1, \varphi_2, \varphi_3) = (\pi, 0, 0)$. The vector σ is then

$$(\sigma_1, \sigma_2, \sigma_3) = (-1, 1, 1) . \tag{10.65}$$

Using (10.61), the matrix B is

$$-\frac{1}{3} \begin{pmatrix} 1 & -1 & -1 \\ -1 & 1 & 1 \\ -1 & 1 & 1 \end{pmatrix} . \tag{10.66}$$

We see that the second row is equal to the first multiplied by -1, and the same holds for the third row. This is true for any N, and is a consequence of (10.61): any row of the matrix B is equal to another row multiplied by either $+1$ or -1. This means that $N - 1$ rows are not linearly independent and that the rank of the matrix is one.

We have then proved that our Hessian matrix H is the sum of a diagonal matrix and a matrix of rank one.

Let us now pass to step (ii). First, we recall a theorem of Wilkinson found in [229]:

Theorem 10.1 (Wilkinson). *Let A and B be $N \times N$ real symmetric matrices and let*

$$C = A + B .$$

Let γ_i, α_i, and β_i ($i = 1, 2, \ldots, N$) be the (real) eigenvalues of C, A, and B, respectively, arranged in nonincreasing order, i.e.,

$$\gamma_1 \geq \gamma_2 \geq \cdots \geq \gamma_N ;$$
$$\alpha_1 \geq \alpha_2 \geq \cdots \geq \alpha_N ;$$
$$\beta_1 \geq \beta_2 \geq \cdots \geq \beta_N .$$

Then

$$\gamma_{r+s-1} \leq \alpha_r + \beta_s \qquad \forall\, r + s - 1 \leq N . \tag{10.67}$$

Notice that we can also write

$$A = C + (-B) , \tag{10.68}$$

and since the eigenvalues of $-B$ arranged in nonincreasing order are

$$-\beta_{N-i+1} ,$$

we can write

$$\alpha_{r+s-1} \leq \gamma_r - \beta_{N-s+1} . \tag{10.69}$$

Now we are interested in a special case, i.e., the case in which the matrix B is of rank one (has only one nonzero eigenvalue). What does Wilkinson's theorem say when applied to such a special case? We consider the two possible cases:

(a) the nonzero eigenvalue is negative:

$$\beta_i = 0 \text{ for } i = 1, 2, \ldots, N - 1, \beta_N = -\varrho;$$

(b) the nonzero eigenvalue is positive:

$$\beta_N = \varrho, \ \beta_i = 0 \text{ for } i = 2, 3, \ldots, N.$$

Case (a). Choosing $s = 1$ in (10.67), we get $\beta_s = 0$, so that

$$\gamma_r \le \alpha_r, \qquad r = 1, \ldots, N, \tag{10.70}$$

while choosing $s = 2$ in (10.69), we get $-\beta N - s + 1 = 0$ again, whence

$$\alpha_{r+1} \le \gamma_r, \qquad r = 1, \ldots, N - 1. \tag{10.71}$$

Combining (10.70) and (10.71), we obtain

$$\alpha_{r+1} \le \gamma_r \le \alpha_r, \qquad r = 1, \ldots, N - 1; \tag{10.72}$$

$$\gamma_N \le \alpha_N. \tag{10.73}$$

We have thus shown that all the eigenvalues of C (except for the smallest one) are bounded between two successive eigenvalues of A. As to γ_N, we can say only that it is smaller than (or equal to) the smallest eigenvalue of A.

Case (b). Choosing $s = 2$ in (10.67), we get $\beta_s = 0$, so that

$$\gamma_{r+1} \le \alpha_r, \qquad r = 1, \ldots, N - 1, \tag{10.74}$$

while choosing $s = 1$ in (10.69), we obtain

$$\alpha_r \le \gamma_r, \qquad r = 1, \ldots, N. \tag{10.75}$$

Combining (10.74) and (10.75), we obtain

$$\alpha_r \le \gamma_r \le \alpha_{r-1}, \qquad r = 2, \ldots, N; \tag{10.76}$$

$$\gamma_1 \ge \alpha_1. \tag{10.77}$$

We have thus shown again that all the eigenvalues of C (except, in this case, for the largest one) are bounded between two successive eigenvalues of A. As to γ_1, we can say only that it is larger than (or equal to) the smallest eigenvalue of A, α_1.

Let us now apply these results to our problem, i.e., to the computation of the number of negative eigenvalues of the matrix H in (10.58). Denoting by η_i its eigenvalues, by δ_i those of D, and by β_i those of B, we have

$$\beta_i = 0 \qquad \forall i \neq 1, N, \tag{10.78}$$

and either

$$\beta_1 = 0, \qquad \beta_N = -\varrho$$

or

$$\beta_1 = \varrho, \qquad \beta_N = 0.$$

At a given critical point, with n_π angles equal to π, the eigenvalues of D are (for the moment we do not order them)

$$\delta_i = m_x + h, \qquad i = 1, \ldots, N - n_\pi; \tag{10.79}$$

$$\delta_i = -(m_x + h), \qquad i = N - n_\pi + 1, \ldots, N. \tag{10.80}$$

The x-component of the magnetization vector is given by

$$m_x = 1 - \frac{2n_\pi}{N}, \tag{10.81}$$

so that

$$m_x > 0 \qquad \text{if } n_\pi \leq \frac{N}{2}, \tag{10.82}$$

$$m_x < 0 \qquad \text{if } n_\pi > \frac{N}{2}. \tag{10.83}$$

Then, if the external field h is sufficiently small, if $n_\pi \leq N/2$, then

$$\delta_i = m_x + h > 0, \qquad i = 1, \ldots, N - n_\pi, \tag{10.84}$$

$$\delta_i = -(m_x + h) < 0, \qquad i = N - n_\pi + 1, \ldots, N, \tag{10.85}$$

i.e., there are $N - n_\pi$ positive and n_π negative δ's; while if $n_\pi \leq N/2$, then

$$\delta_i = -(m_x + h) > 0, \qquad i = 1, \ldots, n_\pi, \tag{10.86}$$

$$\delta_i = m_x + h < 0, \qquad i = n_\pi + 1, \ldots, N, \tag{10.87}$$

i.e., there are n_π positive and $N - n_\pi$ negative δ's.

Now we claim that, at least as N gets large, we can estimate the number of negative η's, i.e., the index of the critical point, by saying that it is equal to the number of negative δ's. More precisely, we claim that the error of our estimate is not larger than 1, i.e.,

$$\text{index}(H) = \#(\eta < 0) = \#(\delta < 0) \pm 1, \tag{10.88}$$

and as N gets large this error becomes obviously negligible. To prove this statement, let us consider the case in which $n_\pi < N/2$. We observe that we do not know whether we are in case (a) or in case (b), i.e., we do not know whether the matrix B has a negative or a positive eigenvalue. But we can try one of the two cases, say (a). Using (10.72) and (10.73) we can then say that

$$\delta_{r+1} \le \eta_r \le \delta_r < 0 \ , \qquad r = N - n_\pi + 1, \dots, N - 1 \qquad (10.89)$$

(note that these are $n_\pi - 1$ equations), and that

$$\eta_N \le \delta_N < 0 \,. \qquad (10.90)$$

Thus we conclude that the number of negative η's is just equal to that of negative δ's, i.e., n_π. If we guessed correctly the sign of the nonzero eigenvalue of B, then our estimate is exact. But in case we guessed wrong, i.e., if we were in case (b) and not (a), then using (10.72) and (10.73), we would have overestimated the number of negative η's by 1. Conversely, if we had used the equations of case (b) in a situation that belonged to case (a) we would have underestimated the index by 1. So, we conclude that the error of our estimate is always ± 1.

10.4.2 One-Dimensional XY Model

Here we want to discuss the details of the result reported in (10.41), i.e., that in the case of the one-dimensional XY model with nearest-neighbor interactions the index of the critical points equals the number n_d of "domain walls" in the configuration.

First of all, let us notice that in the present case the Hessian matrix H is tridiagonal, i.e., it can be written as

$$H = \begin{pmatrix} \alpha_1 & \beta_1 & 0 & 0 & \cdots & 0 \\ \beta_1 & \alpha_2 & \beta_2 & 0 & \cdots & 0 \\ 0 & \ddots & \ddots & \ddots & \ddots & \vdots \\ \vdots & \ddots & \ddots & \ddots & \ddots & 0 \\ 0 & \cdots & 0 & \beta_{N-2} & \alpha_{N-1} & \beta_{N-1} \\ 0 & \cdots & 0 & 0 & \beta_{N-1} & \alpha_N \end{pmatrix} , \qquad (10.91)$$

where, assuming free boundary conditions,

$$\alpha_1 = \cos(\varphi_2 - \varphi_1) + h\cos(\varphi_1) \ ;$$
$$\alpha_i = \cos(\varphi_{i+1} - \varphi_i) + \cos(\varphi_i - \varphi_{i-1}) + h\cos(\varphi_i) \ , \qquad i = 2, \dots, N-1 \ ;$$
$$\alpha_N = \cos(\varphi_N - \varphi_{N-1}) + h\cos(\varphi_N) \ ,$$

and

$$\beta_i = -\cos(\varphi_{i+1} - \varphi_i) \ , \qquad i = 1, \dots, N-1 \ , \qquad (10.92)$$

since at critical points $\varphi_i = 0$ or π, we have that for any i and for any critical point

$$\beta_i = \pm 1 \ , \qquad (10.93)$$

while the diagonal elements α_i are

$$\alpha_1 = 1 \pm h \ ,$$
$$\alpha_i = 2 \pm h \ , \qquad i = 2, \ldots, N-1 \ ,$$
$$\alpha_N = 1 \pm h \ ,$$

if there are no domain walls, i.e., if $n_d = 0$, while they can assume also the values $\pm h$ and $-2 \pm h$ ($-1 \pm h$ if $i = 1$ or $i = N$) if $n_d \neq 0$, i.e., if there are domain walls.

Let us now prove that $n_d \neq 0$ is a *necessary* condition for the presence of negative eigenvalues of the Hessian, i.e., for a nonvanishing index of a critical point. To do that, we recall a theorem due to Gershgorin (see, e.g., [229]), which, in the simple case of a real symmetric matrix, can be stated as follows:

Theorem 10.2 (Gershgorin). *Let A be a real $n \times n$ symmetric matrix whose elements are a_{ij}, and let*

$$r_i = \sum_{j \neq i} |a_{ij}| \ , \qquad i = 1, \ldots, n \ .$$

Then the eigenvalues of A lie in the intervals

$$X_i = \{x \in \mathbb{R} \ : \ |x - a_{ii}| < r_i\} \ ,$$

and if m of the X_i form a disjoint set, then precisely m eigenvalues (counted with their multiplicity) lie in it.

In our case, due to (10.93), at any critical point we have

$$r_1 = r_N = 1 \ ;$$
$$r_i = 2 \ , \qquad i = 2, \ldots, N-1 \ ,$$

so that if $n_d = 0$ and $h \to 0$, then (10.94) and Gershgorin's theorem imply that all the eigenvalues lie in the interval $|x - 2| < 2$, so that there are no negative eigenvalues and the index is zero. On the other hand, if $n_d \neq 0$ and $h \to 0$, then the intervals X_i are either $|x| < 2$ or $|x+2| < 2$; hence the eigenvalues lie in the interval $(-4, 2)$, so that the index can be nonvanishing. However, Gershgorin's theorem is useless to compute the number of negative eigenvalues, because the intervals X_i overlap each other, and thus the eigenvalues cannot be localized more strictly.

Anyway, the fact that the Hessian is tridiagonal allows us to compute directly its characteristic polynomial $\det(H - \lambda I)$, whose roots $\lambda_1, \ldots, \lambda_N$ are the eigenvalues, by means of a recurrence formula. Let

$$p_0(\lambda) = 1 \ ;$$
$$p_1(\lambda) = \alpha_1 - \lambda \ ;$$
$$p_k(\lambda) = (\alpha_k - \lambda)p_{k-1}(\lambda) - \beta_{k-1}^2 p_{k-2}(\lambda) \ .$$

Then, since

$$
p_k(\lambda) = \det
\begin{pmatrix}
\alpha_1 - \lambda & \beta_1 & 0 & 0 & \ldots & & 0 \\
\beta_1 & \alpha_2 - \lambda & \beta_2 & 0 & \ldots & & 0 \\
0 & \ddots & \ddots & \ddots & & \ddots & \vdots \\
\vdots & & \ddots & \ddots & \ddots & & 0 \\
0 & \ldots & 0 & \beta_{k-2} & \alpha_{k-1} - \lambda & \beta_{k-1} \\
0 & \ldots & 0 & 0 & \beta_{k-1} & \alpha_k - \lambda
\end{pmatrix}
, \quad k = 2, \ldots, N,
$$

$$(10.94)$$

the characteristic polynomial of H is given by $p_N(\lambda)$. Since at the critical points all the β's are ± 1 (see (10.93)), we have that

$$p_0(\lambda) = 1 \ , \tag{10.95}$$
$$p_1(\lambda) = \alpha_1 - \lambda \ , \tag{10.96}$$
$$p_k(\lambda) = (\alpha_k - \lambda)p_{k-1}(\lambda) - p_{k-2}(\lambda) \ , \tag{10.97}$$

so that the characteristic polynomial $p_N(\lambda)$ depends only on the α's. Moreover, the following theorem holds (see, e.g., [230]):

Theorem 10.3. *Let H be a tridiagonal symmetric matrix defined as in (10.91). Define the sequence*

$$\{p_0(\lambda), p_1(\lambda), \ldots, p_N(\lambda)\} \tag{10.98}$$

as in (10.94); then the number of sign changes in the sequence (with the rule that if $p_i(\lambda) = 0$ then it has the opposite sign of $p_{i-1}(\lambda)$) equals the number of eigenvalues of H that are less than or equal to λ.

Then the number n_c of sign changes in the sequence

$$\{p_0(0), p_1(0), \ldots, p_N(0)\} \tag{10.99}$$

equals the number of negative eigenvalues, i.e., the index of the critical point because no eigenvalues are zero. If one puts $h = 0$, then there is one eigenvalue that becomes zero at any critical point, so that the index equals $n_c - 1$, but in this case one easily sees by direct computation (which can be performed exactly on a computer at any N because in this case the α's are integers) that $n_d = n_c - 1$, so that one finds the result reported in (10.41).

10.5 The k-Trigonometric Model

In this section we present a study of the thermodynamic properties of the mean-field k-trigonometric model (kTM), as well as of the topological properties of its configuration space. It is worth mentioning that this model—because of long-range interactions—may also undergo first-order phase transitions. As a consequence we a priori expect canonical and microcanonical thermodynamic functions to be different, at least close to first-order transitions [225]. The kTM is defined by the Hamiltonian:

$$H_k = \sum_{j=1}^{N} \frac{1}{2}p_j^2 + V_k(\varphi_1, \ldots, \varphi_N) \, , \qquad (10.100)$$

where $\{\varphi_j\}$ are angular variables: $\varphi_j \in [0, 2\pi)$, $\{p_j\}$ are the conjugated momenta, and the potential energy V is given by

$$V_k = \frac{\Delta}{N^{k-1}} \sum_{j_1,\ldots,j_k} [1 - \cos(\varphi_{j_1} + \ldots + \varphi_{j_k})] \, , \qquad (10.101)$$

where Δ is the coupling constant. Similarly to what we have done with the other models, only the potential energy part is considered. This interaction energy is apparently of a mean-field nature, in that each degree of freedom interacts with all the others; moreover, the interactions are k-body ones.

The kTM is a generalization of the trigonometric model (TM) introduced by Madan and Keyes [231] as a simple model for the potential energy surface of simple liquids. The TM is a model for N independent degrees of freedom with Hamiltonian (10.100) with $k = 1$: $H_{k=1}$. It shares with Lennard-Jones-like systems [232] the existence of a regular organization of the critical points of the potential energy above a given minimum (the elevation in energy of the critical points is proportional to their index) and a regular distribution of the minima in the configuration space (nearest-neighbor minima lie at a well-defined Euclidean distance). The potential energy surface of the kTM maintains the main features of the TM [233], introducing, however, a more realistic feature, namely the interaction among the degrees of freedom (in the form of a k-body interaction).

Using the relation

$$\cos(\varphi_{j_1} + \cdots + \varphi_{j_k}) = \mathrm{Re}(e^{i\varphi_{j_1}} \cdots e^{i\varphi_{j_k}}) \, , \qquad (10.102)$$

the configurational part of the Hamiltonian can be written as

$$V_k = N\Delta \left[1 - \mathrm{Re}(c + is)^k\right]$$
$$= N\Delta \left[1 - \sum_{n=0}^{[k/2]} \binom{k}{2n} (-1)^n \, c^{k-2n} \, s^{2n}\right] \, , \qquad (10.103)$$

where c and s are collective variables (components of the function whose statistical average is the order parameter, i.e., the components of the "magnetization vector"), that are functions of $\{\varphi_j\}$:

$$c = \frac{1}{N} \sum_j \cos \varphi_j ,$$

$$s = \frac{1}{N} \sum_j \sin \varphi_j . \tag{10.104}$$

We observe also that the model has a symmetry group obtained by the transformations

$$\varphi_j \rightarrow \varphi_j + \ell \frac{2\pi}{k} ,$$
$$\varphi_j \rightarrow -\varphi_j .$$

If we think of φ_j as the angle between a unitary vector in a plane and the horizontal axis of this plane, we find that the first transformations are rotations in this plane of an angle $\ell \frac{2\pi}{k}$, and the second is the reflection with respect to the horizontal axis. This group is also called C_{kv}.

Let us now derive the thermodynamic properties of the kTM.

10.5.1 Canonical Ensemble Thermodynamics

The partition function is

$$Z_k(\beta, N) = \int_0^{2\pi} \prod_{j=1}^N d\varphi_j \, \exp(-\beta H_k)$$

$$= \int_0^{2\pi} \prod_{j=1}^N d\varphi_j \, \exp\{-\beta N \Delta [1 - \text{Re}(c + is)^k]\} . \tag{10.105}$$

Introducing δ-functions for the variables c and s,

$$Z_k(\beta, N) = \int_0^{2\pi} \prod_{j=1}^N d\varphi_j \int_{-\infty}^{\infty} dx \, dy \, \delta(x - c) \, \delta(y - s) \, e^{-\beta N \Delta [1 - \text{Re}(x + iy)^k]} ,$$

and using the integral representation of the δ-function, we obtain for Z_k

$$Z_k(\beta, N)$$

$$= \int_0^{2\pi} d^N \varphi_j \int_{-\infty}^{\infty} dx dy \int_{-\infty}^{\infty} N^2 \frac{d\lambda \, d\mu}{2\pi \, 2\pi} e^{iN\lambda(x-c)} e^{iN\mu(y-s)} e^{-\beta \Delta N[1 - \text{Re}(x+iy)^k]}$$

$$= \int_{-\infty}^{\infty} dx \, dy \, e^{-\beta \Delta N[1 - \text{Re}(x+iy)^k]}$$

$$\times \int_{-\infty}^{\infty} \frac{d\lambda \, d\mu}{2\pi \, 2\pi} N^2 \, e^{iN(\lambda x + \mu y)} \int_0^{2\pi} d^N \varphi_j \, e^{-iN(\lambda c + \mu s)} .$$

The saddle-point evaluation of this multiple integral requires us to look for the minima of the exponent in the complex λ, μ, x, y plane. These minima have to lie on the imaginary axes of the λ, μ planes; otherwise, the free energy of the model would be imaginary. Thus the integration path is rotated in the λ, μ planes by substituting $i\lambda \to \lambda$ and $i\mu \to \mu$. Then $i\Lambda \to \Lambda$ and $2\pi J_0(\Lambda) = \int_0^{2\pi} d\varphi \, e^{-i\Lambda \cos(\varphi-\psi)} \to 2\pi I_0(\Lambda)$, where I_0 is the modified Bessel function:

$$I_0(\Lambda) = \frac{1}{2\pi} \int_0^{2\pi} d\varphi \, e^{\Lambda \cos \varphi} . \tag{10.106}$$

In conclusion, one obtains

$$Z_k = N^2 (2\pi)^{N-2} \int_{-\infty}^{\infty} dx \, dy \, d\lambda \, d\mu \, e^{-N g_k(x,y,\lambda,\mu;\beta)} , \tag{10.107}$$

where g_k is the real function

$$g_k(x, y, \lambda, \mu; \beta) = \beta\Delta - \lambda x - \mu y - \beta\Delta \, \mathrm{Re}(x + iy)^k - \log(I_0(\Lambda)) . \tag{10.108}$$

In order to find the stationary points, we first determine the subspace defined by the equations

$$\frac{\partial g_k}{\partial x} = 0 , \tag{10.109}$$

$$\frac{\partial g_k}{\partial y} = 0 , \tag{10.110}$$

obtaining the relations

$$\lambda = -\beta \, \Delta \, k \, \mathrm{Re}(x + iy)^{k-1} , \tag{10.111}$$

$$\mu = \beta \, \Delta \, k \, \mathrm{Im}(x + iy)^{k-1} . \tag{10.112}$$

Thus we get

$$\Lambda = \beta\Delta k \, |(x + iy)^{k-1}|. \tag{10.113}$$

Now, using (10.111) and (10.112), we can substitute λ and μ with x and y in (10.108), obtaining, in terms of the complex number $z = x + iy$,

$$g_k(z; \beta) = \beta \, \Delta + \beta \, \Delta \, (k-1) \, \mathrm{Re} \, z^k - \log I_0(\beta\Delta p \, |k^{p-1}|) , \tag{10.114}$$

and using the polar representation $z = \rho e^{i\psi}$,

$$g_k(\rho, \psi; \beta) = \beta \, \Delta + \beta \, \Delta \, (k-1) \, \rho^k \, \cos(k\psi) - \log I_0(\beta\Delta k\rho^{k-1}) . \tag{10.115}$$

The derivative with respect to ψ leads to

$$-\beta\Delta(k-1)k\rho^k \sin(k\psi) = 0 , \tag{10.116}$$

so that there are $2k$ solutions

$$\psi_n = \frac{n\pi}{k} \quad (n = 1, \ldots, 2k) . \tag{10.117}$$

Observing that $\cos(k\psi_n) = (-1)^n$, we obtain

$$g_k(\rho, n; \beta) = \beta \Delta + (-1)^n \beta \Delta (k-1) \rho^k - \log I_0(\beta \Delta k \rho^{k-1}) , \tag{10.118}$$

and we can restrict ourselves to $n = 0, 1$. Finally, the derivative with respect to ρ leads to the stationary-points equation

$$(-1)^n \rho = \frac{I_1(\beta \Delta k \rho^{k-1})}{I_0(\beta \Delta k \rho^{k-1})} , \tag{10.119}$$

where the modified Bessel function I_1 is defined by

$$I_1(\Lambda) = \frac{1}{2\pi} \int_0^{2\pi} d\varphi \, \cos\varphi \, e^{\Lambda \cos\varphi} = I_0'(\Lambda) . \tag{10.120}$$

For $n = 1$ we have only the trivial solution $\rho = 0$, because the I functions are always positive. Using an expansion for small ρ one can show that this solution is a maximum for g. So we can study only the case $n = 0$. We note that if there is a nontrivial solution (i.e., $\tilde{\rho}(\beta) \neq 0$) of (10.119), then calling $\tilde{g}_k(\beta)$ the value of $g_k(\beta, \tilde{\rho}(\beta))$, we have

$$Z_k \approx N^2 \, (2\pi)^{N-2} \, e^{-N\tilde{g}_k(\beta)} , \tag{10.121}$$

and the free energy and internal energy are, respectively,

$$f_k(\beta) = \beta^{-1} \tilde{g}_k(\beta) - \beta^{-1} \log(2\pi) , \tag{10.122}$$
$$e_k(\beta) = \Delta(1 - \tilde{\rho}^k) . \tag{10.123}$$

Let us now analyze the case $k = 1$. In this case the solutions $\rho = 0$ are not present, so that we have only the solution

$$\tilde{\rho} = \frac{I_1(\beta\Delta)}{I_0(\beta\Delta)} . \tag{10.124}$$

There is no phase transition, and using (10.123) we have

$$e_1(\beta) = \Delta \left(1 - \frac{I_1(\beta\Delta)}{I_0(\beta\Delta)} \right) . \tag{10.125}$$

This is the free energy of trigonometric model that has been mentioned before.

For $k = 2$ the solution $\rho = 0$ is stable for high temperatures, but a non-trivial solution of (10.119) appears at $\beta\Delta = 1$. The transition temperature is given by the condition

$$\left.\frac{d^2 g_2(\rho; \beta_c)}{d\rho^2}\right|_{\rho=0} = 2\beta_c \Delta(1 - \beta_c \Delta) = 0 , \qquad (10.126)$$

so that we obtain $\beta_c \Delta = 1$; the transition is continuous, and the order parameter is $\tilde{\rho}$. It is easy to show that $\tilde{x} = \langle c \rangle$ and $\tilde{y} = \langle s \rangle$ (e.g., by adding an external field of the form $-N(hc + ks)$ to the Hamiltonian and taking the limit $h, k \to 0$); then the vector $(\tilde{x}, \tilde{y})$ is the mean magnetization of the spins represented by the φ_i. Since $\tilde{\rho}$ is the modulus of the magnetization, for $\beta\Delta > 1$, when $\tilde{\rho} \neq 0$, the C_{2v} symmetry is broken.

When $k > 2$, the nontrivial solution of (10.119) appears at a given β' but becomes stable only at $\beta'' > \beta'$, so that $\tilde{\rho}(\beta)$ and $e(\beta)$ are discontinuous at β''; instead of the instability region $\beta' < \beta < \beta''$, in the microcanonical ensemble a region where the specific heat is negative appears, as we shall see below. The C_{kv} symmetry is broken in the low-temperature phase, so that $\tilde{\rho}$ can be used as an order parameter in revealing the symmetry-breaking, even if it is not continuous at β''. The transition is then of first order, but keeps the symmetry structure of a second-order transition, i.e., in the low-temperature phase there are k pure states related by the symmetry group also in the case of the first-order transition.

In Figure 10.13 we report the caloric curve, i.e., the temperature $T = \beta^{-1}$ as a function of the average energy (per degree of freedom) e, for three values of k, $k = 1, 2$ and 3. As previously discussed, the temperature is an analytic function of e for $k = 1$; for $k = 2$ the system undergoes a second-order phase transition at a critical temperature $T_c = \Delta$, which changes to first order for $k > 2$.

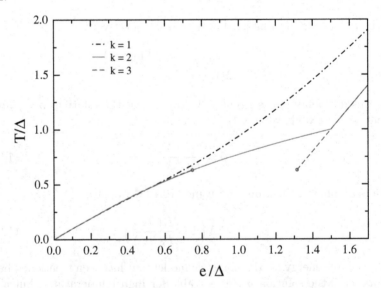

Fig. 10.13. Temperature T as a function of canonical average energy e for three different values of k; for $k=1$ there is no phase transition, while for $k=2$ there is a second order transition and for $k > 2$ a first order one. From [223].

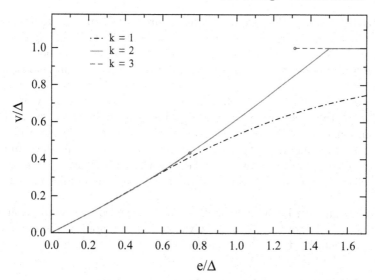

Fig. 10.14. Canonical average potential energy v as a function of canonical average energy e for k=1, 2, and 3. The upper phase transition point is, for $\forall k \geq 2$, $v_c = \Delta$. From [223].

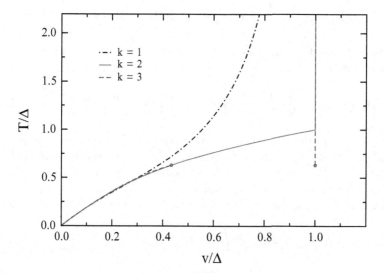

Fig. 10.15. Temperature T as a function of canonical average potential energy v for three different values of k. From [223].

In Figures 10.14 and 10.15 we report the average potential energy v as a function of the average energy e and the temperature T as a function of v, respectively. It is apparent that for $k \geq 2$, the phase transition point always corresponds to $v_c = \Delta$.

Another feature that shows up in Figures 10.14 and 10.15 is that the average potential energy v never exceeds the value Δ, i.e., although the maximum of V/N is equal to 2Δ, the region $v > \Delta$ is not thermodynamically accessible to the system. The reason for this is in the mean-field nature of the system and in the fact that we are working in the thermodynamic limit $N \to \infty$. According to (10.103), the potential energy can be written as a function of the collective variables c and s defined in (10.104), which are the components of the function whose statistical average is the order parameter, i.e., the "magnetization." In the thermodynamic limit these functions become constants, whose value coincides with their statistical average, and since $\langle c \rangle = \langle s \rangle = 0$ for $T > T_c$, then from (10.103) this implies $v = \Delta$ for all $T > T_c$.

As we shall see below, this fact remains true also in the microcanonical ensemble, which, however, is *not* equivalent to the canonical ensemble for the present model, due to the long-range nature of the interactions.

10.5.2 Microcanonical Thermodynamics

The microcanonical phase space volume for the kTM,

$$\Omega_k(E, N) = \frac{1}{N!} \int d^N \pi_j \, d^N \varphi_j \, \delta(H_k(\pi, \varphi) - E) \,, \tag{10.127}$$

can be computed by introducing the standard integral representation of the delta function, that is,

$$\Omega_k(E, N) = \frac{1}{N!} \int_{-\infty}^{\infty} \frac{d\beta}{2\pi} \int d^N \pi_j \, d^N \varphi_j \, e^{-i\beta[H_k(\pi, \varphi) - E]} \,. \tag{10.128}$$

Now, after rotating the integration path on the imaginary axis in the complex-β plane, since, as in the canonical case, the saddle-point is located on this axis, the integral on β is evaluated by means of the saddle-point method. Integrating over the momenta, and using $V_k(\varphi) = V_k(c(\varphi), s(\varphi))$ with $\varphi = (\varphi_1, \ldots, \varphi_N)$, see (10.103), one obtains

$$\Omega_k(E, N) = C_N \rho^N \int_{-\infty}^{\infty} d\beta \, d\xi \, d\eta \, \beta^{-\frac{N}{2}} e^{\beta(E - V_k(\xi, \eta))}$$
$$\times \int d^N \varphi_j \, \delta[N(\xi - c(\varphi))] \, \delta[N(\eta - s(\varphi))] \,, \tag{10.129}$$

where $\rho = N/L$ and the constant C_N gives only a constant contribution to the entropy per particle, i.e., it is at most of order e^N. The last integral is evaluated using again the integral representation of the delta function, and then rotating the integration path as before, whence

$$\int_{-\infty}^{\infty} \frac{d\mu\,d\nu}{(2\pi)^2}\, e^{-N(\mu\xi+\nu\eta)} \int_0^{2\pi} d^N\varphi_j\, e^{\sum_j(\mu\cos\varphi_j+\nu\sin\varphi_j)}$$

$$= \int_{-\infty}^{\infty} \frac{d\mu\,d\nu}{(2\pi)^2}\, e^{-N(\mu\xi+\nu\eta)}[2\pi I_0(\Lambda)]^N\ ,$$

where $\Lambda = \sqrt{\mu^2+\nu^2}$, and I_0 is the modified Bessel function. Thus, the microcanonical volume reads as

$$\Omega_k(e,N) = C_N\,\rho^N \int d\mathbf{u}\, e^{Nf_k(\mathbf{u},e)}\ , \tag{10.130}$$

where $\mathbf{u} \equiv (\beta,\xi,\eta,\mu,\nu)$, $e = E/N$, and

$$f_k(\mathbf{u},e) = \beta e - \beta\Delta[1 - \mathrm{Re}(\xi+i\eta)^k] - \frac{1}{2}\log\beta - \mu\xi - \nu\eta + \log I_0(\Lambda)\ .$$

Then, using the saddle-point theorem, the entropy per particle, $s = S/N$, is given by ($k_B = 1$)

$$s_k(e) = \lim_{N\to\infty} \frac{1}{N}\log\Omega_k(e,N) = \max_{\mathbf{u}} f_k(\mathbf{u},e)\ . \tag{10.131}$$

To find the maximum of $f_k(\mathbf{u},\mathbf{e})$ one can calculate analytically some derivatives of f to obtain a one-dimensional problem that can be easily solved

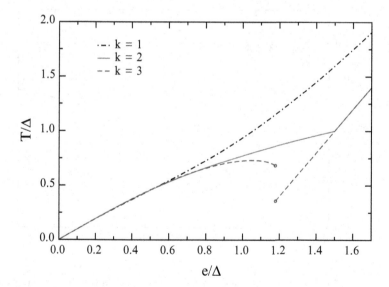

Fig. 10.16. Microcanonical temperature T as a function of energy e for three different values of k; for $k=1$ there is no phase transition, while for $k=2$ there is a second-order transition and for $k > 2$ a first-order one. From [223].

numerically with standard methods. As already done in the case of the canonical ensemble, in Figure 10.16 we report the microcanonical caloric curve, i.e., the temperature T as a function of the energy (per degree of freedom) e, $T(e) = [\partial s/\partial e]^{-1}$ for three values of k, $k = 1, 2$ and 3. As in the canonical case, the temperature is an analytic function of e for $k = 1$, while for $k = 2$ the system undergoes a second-order phase transition at a certain energy value e_c, that changes to first order for $k > 2$.

We note that for $k > 2$, in a region of energies smaller than the critical energy e_c of the first-order phase transition, the curve $T(e)$ has a negative slope, i.e., the system has a negative specific heat. This is not surprising since we are considering the *microcanonical* thermodynamics of a system with long-range interactions (see, e.g., [225] for other examples and a general discussion); such a region is *not* present when we consider the canonical ensemble, as shown above; there, the region of negative specific heat corresponds to the region of instability of the nontrivial solution of the saddle-point equations.

In Figures 10.17 and 10.18 we report the average microcanonical potential energy v as a function of e and the microcanonical temperature T as a function of v, respectively. It is apparent that for $k \geq 2$, the phase transition point always corresponds to $v_c = \Delta$.

As in the canonical case, the average potential energy v never exceeds the value Δ, i.e., the region $v > \Delta$ is not thermodynamically accessible to the system also in the microcanonical ensemble.

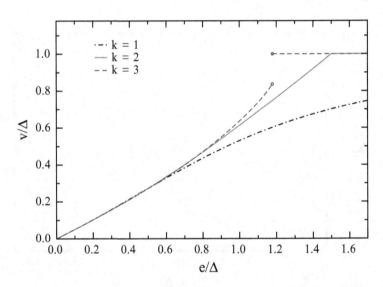

Fig. 10.17. Microcanonical average potential energy v as a function of energy e for k=1, 2 and 3. The phase transition point is, for $\forall k \geq 2$, $v_c = \Delta$. From [223].

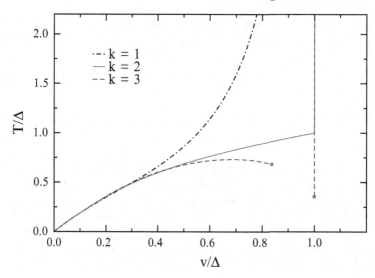

Fig. 10.18. Microcanonical temperature T as a function of microcanonical average potential energy energy v for three different values of k. From [223].

10.5.3 Topology of Configuration Space

Let us now consider the relation between the phase transitions occurring in the kTM and the topological changes of the submanifolds M_v of its configuration space.

Also for this model we probe these topological changes through the potential-energy dependence of the Euler characteristic, which is computed, as in the case of the mean-field XY model, through the knowledge of all the critical points $\tilde{\varphi}$ of the potential and of the corresponding indices. Again it is necessary to locate the points where $dV_k(\tilde{\varphi}) = 0$ and to compute at these points the number of negative eigenvalues of the Hessian matrix

$$H_{ij}^{(k)}(\tilde{\varphi}) = \left(\frac{\partial^2 V_k}{\partial \varphi_i \partial \varphi_j} \right)\Big|_{\tilde{\varphi}} . \tag{10.132}$$

To determine the critical points we have then to solve the system

$$\frac{\partial V_k}{\partial \varphi_j} = 0 , \qquad \forall j = 1, \ldots, N , \tag{10.133}$$

that is, inserting (10.103) in the equations above,

$$-\Delta k \operatorname{Re}[i(c+is)^{k-1}e^{i\varphi_j}] = \Delta k \, \zeta^{k-1} \sin[(k-1)\psi + \varphi_j] = 0 \quad \forall j = 1, \ldots, N , \tag{10.134}$$

where we have defined $c + is = \zeta e^{i\psi}$. From (10.103) we have

$$V_k(\varphi) = N\Delta[1 - \zeta^k \cos(k\psi)] ; \tag{10.135}$$

then all the critical points with $\zeta(\tilde{\varphi}) = 0$ have energy $v = V(\tilde{\varphi})/N = \Delta$. We note that they correspond to vanishing magnetization. Let us now consider all the critical points with $\zeta(\tilde{\varphi}) \neq 0$. Then (10.134) becomes

$$\sin[(k-1)\psi + \varphi_j] = 0 \qquad \forall j = 1, \ldots, N , \qquad (10.136)$$

and its solutions are

$$\tilde{\varphi}_j^{\mathbf{m}} = [m_j\pi - (k-1)\psi]_{\mathrm{mod}\ 2\pi} , \qquad (10.137)$$

where $m_j \in \{0,1\}$. Since in (10.134), ζ appears to the $(k-1)$th power, in the case $k = 1$ (10.134) and (10.136) coincide. This means that the solutions given in (10.137) are *all* the critical points, regardless of their energy, in the case $k = 1$ and all the critical points *but* those with energy $v = \Delta$ in the case $k > 1$. The critical point $\tilde{\varphi}^{\mathbf{m}}$ is then characterized by the set $\mathbf{m} \equiv \{m_j\}$. To determine the unknown constant ψ we have to substitute (10.137) in the self-consistency equation

$$\zeta e^{i\psi} = c + is = N^{-1} \sum_j e^{i\varphi_j} = N^{-1} e^{-i\psi(k-1)} \sum_j (-1)^{n_j} . \qquad (10.138)$$

If we introduce the quantity $n(\tilde{\varphi})$ defined by

$$n = N^{-1} \sum_j m_j , \qquad (10.139)$$

which means

$$1 - 2n = N^{-1} \sum_j (-1)^{n_j} , \qquad (10.140)$$

we have from (10.138)

$$\zeta = |1 - 2n| , \qquad (10.141)$$

$$\psi_l = 2l\pi/k \qquad\qquad \text{for} \quad n < 1/2 , \qquad (10.142)$$

$$\psi_l = (2l+1)\pi/k \qquad \text{for} \quad n > 1/2 , \qquad (10.143)$$

where $l \in \mathbb{Z}$. Then the choice of the set $\{m_j\}$ is not sufficient to specify the set $\{\tilde{\varphi}_j\}$, because the constant ψ can assume some different values. This fact is connected with the symmetry structure of the potential-energy surface: the different values of ψ_l correspond to the symmetry-related critical points under the group C_{kv}.

We can then state that all the critical points with $\zeta \neq 0$, whose energy v is not equal to Δ, have the form

$$\tilde{\varphi}_j^{\mathbf{m},l} = [m_j\pi - (k-1)\psi_l]_{\mathrm{mod}\ 2\pi} . \qquad (10.144)$$

The Hessian matrix is given by

$$H_{ij}^{(k)} = \Delta \, k \, \mathrm{Re}[N^{-1}(k-1)(c+is)^{k-2}e^{i(\varphi_i+\varphi_j)} + \delta_{ij}(c+is)^{k-1}e^{i\varphi_i}] \, . \quad (10.145)$$

In the thermodynamic limit it becomes diagonal:

$$H_{ij}^{(k)} = \delta_{ij} \, \Delta \, k \, \zeta^{k-1} \, \cos\left(\psi(k-1)+\varphi_i\right) \, . \quad (10.146)$$

One cannot a priori neglect the contribution of the off-diagonal terms to the eigenvalues of $H^{(k)}$, but accurate numerical checks show that they change at most the sign of only one eigenvalue out of N. In the case of the mean-field XY model we have explicitly proved this fact. Neglecting the off-diagonal contributions, the eigenvalues of the Hessian are calculated at any critical point $\tilde{\varphi}$ by substituting (10.144) in (10.146),

$$\lambda_j = (-1)^{m_j}\Delta \, k \, \zeta^{k-1} \, , \quad (10.147)$$

so the index of the critical point is simply the number of $m_j = 1$ in the set $\mathbf{m}$; we can identify the quantity $n(\tilde{\varphi})$ given by (10.139) with the *fractional index* n/N of the critical point $\tilde{\varphi}$. Then, from (10.103), (10.141), (10.142) and (10.143) we get a relation between the fractional index $n(\tilde{\varphi})$ and the potential energy $v(\tilde{\varphi}) = V(\tilde{\varphi})/N$ at each critical point $\tilde{\varphi}$:

$$n(v) = \frac{1}{2}\left[1 - \mathrm{sgn}\left(1-\frac{v}{\Delta}\right)\left|1-\frac{v}{\Delta}\right|^{1/k}\right] \, . \quad (10.148)$$

Moreover, the number of critical points of given index ν is simply the number of ways in which one can choose ν times 1 among the $\{m_j\}$, see (10.144), multiplied by a constant $\mathcal{A}_k$ that accounts for the degeneracy introduced by (10.143).

Hitherto, the critical points with $\zeta \neq 0$ have been completely characterized. The knowledge of the critical points considered so far is sufficient to compute the Euler characteristic of the manifolds M_v, because the critical points with $\zeta = 0$ can be neglected. Let us discuss why. The critical points with $\zeta = 0$ are degenerate: the Hessian determinant vanishes at these points. This means that the potential energy is no longer a proper Morse function when $v \geq \Delta$, and therefore we can use its critical points to compute the Euler characteristic of the manifolds M_v only when $v < \Delta$. To overcome this difficulty we proceed as in the case of the mean-field XY model, thus we consider as our Morse function the function $\tilde{V}_k$ obtained by adding to the potential energy a linear term that can be made arbitrarily small:

$$\tilde{V}_k = V_k + \sum_{i=1}^{N} h_i\varphi_i \, , \quad (10.149)$$

where $h \in \mathbb{R}^N$. The perturbation changes only slightly the critical points with $\zeta \neq 0$, but completely removes the points with $\zeta = 0$ for any $h \neq 0$, no matter how small. All the critical points of this function are given by the solutions of the equations

$$\sin[(k-1)\psi + \varphi_j] = h_j \qquad \forall j = 1, \ldots, N, \tag{10.150}$$

which are only a slight deformation of (10.136), so that provided all the h's are very small, the numerical values of critical points and critical levels will essentially coincide with those computed so far, in the case $h = 0$ but assuming $\zeta \neq 0$.

The fractional index $n = \nu/N$ of the critical points is a well-defined monotonic function of their potential energy v, given by (10.148), and the number of critical points of a given index n is $\mathcal{A}_k\binom{N}{n}$. Then the Morse indexes $\mu_n(M_v)$ of the manifold M_v are given by $\mathcal{A}_k\binom{N}{n}$ if $n/N \le x(v)$ and 0 otherwise, and the Euler characteristic is

$$\chi(M_v) = \mathcal{A}_k \sum_{n=0}^{Nx(v)} (-1)^n \binom{N}{n} = \mathcal{A}_k(-1)^{Nx(v)} \binom{N-1}{Nx(v)}, \tag{10.151}$$

where the relation

$$\sum_{n=0}^{m} (-1)^n \binom{N}{n} = (-1)^m \binom{N-1}{m}$$

has been used. In Figure 10.19 we plot $\sigma(v) = \lim_{N\to\infty} \frac{1}{N} \log |\chi(v)|$, which, from (10.151), is given by

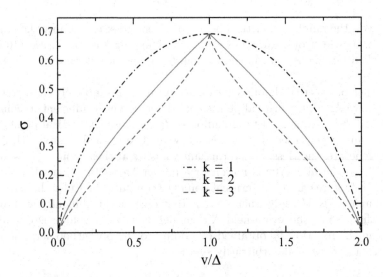

Fig. 10.19. Logarithmic Euler characteristic of the M_v manifolds $\sigma(v)$ (see text) as a function of the potential energy v. The phase transition is signaled as a singularity of the first derivative at $v_c = \Delta$; the sign of the second derivative around the singular point allows to predict the order of the transition. The region $v > \Delta$, in which $\sigma'(v) < 0$, in not reached by the system (see text). From [222].

$$\sigma(v) = -x(v)\log x(v) - (1 - x(v))\log(1 - x(v)) \ . \qquad (10.152)$$

Of course $\sigma(v)$ is a purely *topological* quantity, being related only to the properties of the configuration-space submanifolds M_v, defined through $V_k(\varphi)$, and, in particular, through the energy distribution of the critical points of $V_k(\varphi)$. From Figure 10.19 we can see that there is an evident signature of the phase transition in the v-shape of $\sigma(v)$. First, we observe that the region $v > \Delta$ is never reached by the system, as discussed before and shown in Figures 10.17 and 10.18 as to the microcanonical case, and in Figures 10.14 and 10.15 as to the canonical case; this region is characterized by $\sigma'(v) < 0$. The main observations are that: (i) for $k=1$, where there is no phase transition, the function $\sigma(v)$ is analytic; (ii) for $k=2$, when we observe a second-order phase transition, the first derivative of $\sigma(v)$ is discontinuous at $v_c = v(e_c) = \Delta$, and its second derivative is *negative* around the singular point, (iii) for $k \geq 3$ the first derivative of $\sigma(v)$ is also discontinuous at the transition point $v_c = \Delta$, but its second derivative is *positive* around v_c. In this case *a first-order transition takes place*. The interesting consequence is that through the pattern of $\sigma(v)$ we can establish both the critical value v_c where the phase transition occurs, and the *order* of the transition. Again we see that everything is "read" just in the configuration-space topology induced by the potential function, without resorting to any effect due to the properties of statistical measures.

The above-listed items concerning the pattern of $\sigma(v)$ confirm, at least qualitatively, the already discussed relation between topology and thermodynamic entropy. For example, a first-order transition with a discontinuity in the energy is generally accompanied [37] by a region of negative specific heat, i.e., of positive second-order derivative of the entropy (compare with the third item given above), and it seems that the jump in the second derivative of the entropy stems from the jump in the second derivative of $\sigma(v(e))$.

In Chapters 8 and 9 we have seen that an explicit analytic link between thermodynamics and topology actually exists. However, thermodynamic entropy is found to be related either to the sum of Morse indexes or to a suitably weighted sum of them. In Figure 10.20 the sum of the logarithms of the Morse indexes, $\mu(v)$, is displayed for the three cases of $k = 1, 2, 3$. Also in this case, a sharp difference of the shape of $\mu(v) = \frac{1}{N}\log\left[\mu_0 + \sum_{i=1}^{N-1} 2\mu_i(M_v) + \mu_N\right]$ is found for each version of the model, that is, when no phase transition takes place, when a second-order transition is present, or when a first-order transition is present, respectively. The close resemblance of $\sigma(v)$ and $\mu(v)$ below the transition point can be explained as a consequence of the growth with i of the $\mu_i(M_v)$, as shown by (10.151), and of the sum formula in (10.152). In fact, $\mu_{N-1}(M_v)$ is the dominant term in the sum defining $\mu(v)$, whence the close similarity with $\chi(M_v)$ and $\sigma(v)$. Note that this is a model-dependent circumstance.

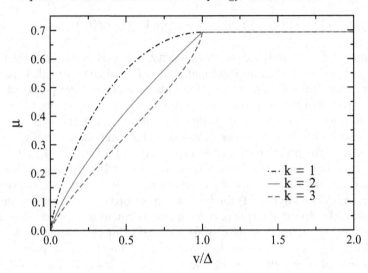

Fig. 10.20. Sum of the logarithms of the Morse indexes $\mu = \frac{1}{N}\log\left[\mu_0 + \sum_{i=1}^{N-1} 2\mu_i(M_v) + \mu_N\right]$ of the manifolds M_v versus the energy density v, scaled with Δ, for $k = 1, 2, 3$. From [223].

10.5.4 Topology of the Order Parameter Space

A feature of many mean-field models (although not of all of them) is that the potential energy can be written as a function of a collective variable, whose statistical average is the order parameter. In the case of the k-trigonometric model this variable is the two-dimensional "magnetization" vector defined as $\mathbf{m} = (c, s)$, where (see 10.104)

$$
\begin{cases}
c = \dfrac{1}{N}\sum_{i=1}^{N}\cos(\varphi_i)\,, \\[4mm]
s = \dfrac{1}{N}\sum_{i=1}^{N}\sin(\varphi_i)\,.
\end{cases}
\tag{10.153}
$$

Written in terms of (c, s), the potential energy is a function defined on the unit disk in the real plane, which is given by (see (10.103))

$$
V_k(c, s) = N\Delta \left[1 - \sum_{n=0}^{[k/2]} \binom{k}{2n} (-1)^n\, c^{k-2n}\, s^{2n}\right].
\tag{10.154}
$$

In the particular cases $k = 1, 2, 3$ the potential energy V_k reads as

$$
V_1(c, s) = N\Delta(1 - c)\,,
\tag{10.155}
$$

$$
V_2(c, s) = N\Delta(1 - c^2 + s^2)\,,
\tag{10.156}
$$

$$
V_3(c, s) = N\Delta(1 - c^3 + 3cs^2)\,,
\tag{10.157}
$$

and it is then natural to investigate the topology of the M_v's seen as submanifolds of the unit disk in the plane, i.e., we now consider the submanifolds

$$\mathcal{M}_v \equiv \{(c, s) \in D^2 \mid V_k(c, s) \leq Nv\} \,, \tag{10.158}$$

where $D^2 \equiv \{(c, s) \in \mathbb{R}^2 \mid c^2 + s^2 \leq 1\}$. The $\mathcal{M}_v$'s are nothing but the M_v's projected onto the "magnetization" plane.

The topology of these manifolds can be studied directly, by simply drawing them. In the case $k = 1$, where no phase transition is present, no topological changes occur in the $\mathcal{M}_v$'s, i.e., all of them are topologically equivalent to a single disk D^2 (Figure 10.21). When $k = 2, 3$, and a phase transition is present, there is a topological change precisely at $v_c = \Delta$, where k disks merge into a single disk (see Figures 10.22 and 10.23). The detail of the transition, i.e., the number of disks that merge into one, clearly reflects the nature of the

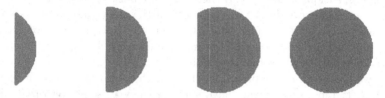

Fig. 10.21. The submanifolds $\mathcal{M}_v$ in the case $k = 1$ for $v = 0.5\Delta, \Delta, 1.5\Delta, 2\Delta$ (from left to right). All the submanifolds are topologically equivalent to a single disk. From [223].

Fig. 10.22. The submanifolds $\mathcal{M}_v$ in the case $k = 2$ for $v = 0.5\Delta, \Delta, 1.5\Delta, 2\Delta$ (from left to right). For $v < v_c = \Delta$ the submanifolds are topologically equivalent to two disconnected disks, while for $v > v_c$ they are equivalent to a single disk. From [223].

Fig. 10.23. The submanifolds $\mathcal{M}_v$ in the case $k = 3$ for $v = 0.5\Delta, \Delta, 1.5\Delta, 2\Delta$ (from left to right). As $v < v_c = \Delta$ the submanifolds are topologically equivalent to three disconnected disks, while as $v > v_c$ they are equivalent to a single disk. From [223].

symmetry-breaking for the particular value of k considered (similar pictures are obtained for $k > 3$). Thus, when projected onto the order-parameter space, the correspondence between topological changes and phase transitions becomes one-to-one (this was already found in Section 7.4 for the mean-field XY model); however, in contrast to the study of the topology of the "full" M_v's, no direct way to discriminate between first- and second-order transitions seems available in this picture.

10.6 Comments on Other Exact Results

The proposal to look for a topological origin of phase transitions has been recently given increasing attention. A-priori, the best candidates to allow analytical computations of both thermodynamical and topological features are one-dimensional models and mean-field models. Of course, in order to deepen our understanding of the subject, all the computable models among these ones are worth consideration. However, the interpretation of the outcomes is a delicate point on which some confusion has been made.

Let us try to clarify some basic points.

As we have seen in the preceding chapter, there is a wide class of systems for which the loss of uniform convergence with N in some low-order differentiablity class of basic thermodynamic functions (thus the appearance of a phase transition) is *necessarily* driven by a topological transition in configuration space. Some questions immediately arise. The first concerns the weakening of the conditions which the microscopic potential has to satisfy. We considered short-range potentials only, but, as the results given in the present chapter witness, some extension also to long-range potentials of the theorems given in Chapter 9 should be possible. At least for potentials bounded from above, as is the case of the XY mean-field model and of the k-trigonometric model. And what about unbound potentials?

The second–very challenging—question concerns *sufficiency* conditions. Here the problem is to find out which kind of topological transitions can entail a thermodynamic phase transition. Precious hints can be obtained by studying particular models, provided that the models considered satisfy the hypotheses of the theorems given in Chapter 9.

The third question concerns the way of investigating topological transitions. Apart from resorting to the Gauss–Bonnet–Hopf theorem, as we did in Chapter 7, the main constructive, computational way of studying topology is based on the analysis of the critical points of the potential. However, the study of topology cannot be considered exhausted at all by considering this kind of critical points. Moreover, in some cases, a good Morse function on the relevant configuration space submanifolds could be very different from the physical potential.

A fourth more general question. Do we have to think that any phase transition has a topological origin? A cautious answer is negative. However, we simply do not know whether and how much the domain of validity of the theorems given in Chapter 9 can be broadened.

Now, as far as one dimensional models are concerned, in [178] a simple model for DNA denaturation, known as the Peyrard–Bishop model, was considered. The Hamiltonian describing this model is

$$H = \sum_n \left[\frac{m}{2} \dot{y}_n^2 + \frac{K}{2} (y_n - y_{n+1})^2 + D(e^{-a y_n} - 1)^2 \right] . \qquad (10.159)$$

This model is considered in [235] too, where the authors consider also the modified model

$$H = \sum_n \left[\frac{m}{2} \dot{y}_n^2 + \frac{K}{2} (y_n - y_{n+1})^2 + D(e^{-a|y_n|} - 1)^2 \right] . \qquad (10.160)$$

The claim in [235] is that—for both models—the topological transition occurs at a critical value of the potential energy which is lower than the thermodynamic transition energy. The first obvious problem with these models is that the associated configuration space submanifolds are *noncompact*, and the critical manifolds are infinitely large. There is only one critical point (apart from the trivial one) whose coordinates are all infinite! Somewhat pathological indeed. Even though we can claim that—since the above models live on noncompact manifolds—no contradiction exists with the theorems of the preceding chapter, in [177] it has been shown that the energy-pattern of the largest Lyapunov exponent clearly marks the unbinding phase transition of model (10.159), whence it is evident that a microcanonical computation of the total energy per degree of freedom—*on any arbitrarily large but finite (!)* region of phase space—will give $u = E/N \approx k_B T + D$, the first term, after energy equipartition, is the average kinetic plus harmonic potential energy per degree of freedom, the second constant term comes from the Morse potential for large deformations. The vanishing of harmonic energy on the infinitely large critical equipotential energy surface tells us that the limit is singular (using equations (5) and (6) of [177] draw the caloric curve $T = T(U/N)$ and consider the value U/N for $T = T_c$). We believe that a more careful analysis[2] of the results in [178] should encompass also this model in the family of "good" models.

In [236] another one-dimensional model, known as the Burkhardt model for localization-delocalization transition of interfaces, has been investigated. It is described by the Hamiltonian

$$H = \sum_n \left[\frac{1}{2} p_n^2 + |q_{n+1} - q_n| + U(q_n) \right] \qquad (10.161)$$

---

[2] One should compute the caloric curve, that is $T = T(E) = (\partial S/\partial E)^{-1}$, through the microcanonical entropy $S(E)$, compute the average potential energy as a function of the total energy, hence obtain the transition value of the latter from the critical value of the former, and finally check that approaching the topological transition from below the good transition temperature is retrieved.

where $U(q_n)$, called pinning potential, is a square-well. This model has been studied also in [235] with a somewhat different version of the pinning potential. Both [236] and [235] report the non-coincidence of topological and thermodynamical transitions. However, with respect to the theorems in Chapter 9, this model has two bad features: the configuration-space submanifolds are noncompact and the pinning potentials considered are singular (this is evident after the definition given in [235], and implicit in [236] where infinitely steep barriers of potential are needed in order to constrain the coordinates on a semi-infinite positive line).

At first sight, another puzzling result concerns the mean-field φ^4 model described by the Hamiltonian

$$H = \sum_i \left[\frac{1}{2} p_i^2 - \frac{1}{2} \phi_i^2 + \frac{1}{4} \phi_i^4 \right] - \frac{J}{2N} \left(\sum_i \phi_i \right)^2 . \tag{10.162}$$

In this case, depending on the choice of the parameters of the model, the phase transition point can lack a counterpart in the Euler characteristic $\chi(v)$; worse, there is not a single critical point of the potential at the transition point [237–239].

In [238] the authors simply conclude that the claim that the topological and thermodynamical transitions points coincide is not valid in general. In [237] the authors propose a weakening of the topological hypothesis and introduce a mapping to recover some correspondence between the topological and thermodynamical transitions. In [240, 241] the authors surmise that for non-confining potentials, as in the case of Burkhardt model, as well as for long-range interactions, rather than from a topological transition the thermodynamic singularity stems from the maximization over one variable of a non-concave entropy function of two variables. All these contributions are certainly useful, provided that we keep in mind that the general theory proposed in the preceding chapter has no pretence to encompass *all* the existing phase transitions. Long-range-interaction models with unbound potentials, as is the mean-field φ^4 model, might well be outside the domain of validity of the topological theory. But are we sure that no topological transition exists in this model? The mean-field φ^4 model undergoes a $\mathbb{Z}_2$-symmetry-breaking phase transition. Therefore, a major change of topology at the phase transition point exists: in the $N \to \infty$ limit, in the broken symmetry phase, the configuration space splits into two disjoint submanifolds, a major topological change indeed. When the number of connected components changes, so does the zeroth cohomology group H^0 and, correspondingly, the Betti number b_0. We can think that the larger N, the "thinner" the bridging between these two components of configuration space. However, this kind of topological transition is not detectable through the critical points of a Morse function. Of course, we can wonder whether other cohomology groups besides H^0 are involved despite the above mentioned results. The problem, overlooked in the above quoted papers, is that for the mean-field φ^4 model the good geometrical objects whose topological changes have to be investigated are *not* the Σ_v, or the M_v, but other submanifolds of configuration space. In fact, in this model there is not a one-to-one correspondence between the order parameter and the potential function. The configurational entropy now depends on two independent variables, that is, $S = S(v, m)$ with v the potential energy and m the magnetization, whereas in the case of the mean-field XY and k-trigonometric models it is $S = S(v, m(v))$. Thus the question is whether for any given value of the magnetization the corresponding subset of configuration space qualifies as a good manifold, if this is the case, then we have to find out a good Morse function on these constant-magnetization submanifolds of configuration space. Yet a lot of work remains to be done on this model to clarify whether the phase transition it undergoes is driven also by higher cohomology groups, in any case, since at least H^0 is involved, the mean-field φ^4 model does not seem outside the domain of the topological theory.

Another interesting element enters the game with the mean-field Berlin–Kac spherical model: statistical ensemble nonequivalence. This model is described by the Hamiltonian

$$H = \sum_{i=1}^{N} \left[\frac{1}{2} p_i^2 - \frac{J}{2N} \sum_{\substack{j \neq i, j=1}}^{N} s_i s_j - h s_i \right] . \tag{10.163}$$

where $s_i \in \mathbb{R}$, and h is an external field; the spin variables are subject to the spherical constraint

$$\sum_{i=1}^{N} s_i^2 = N . \tag{10.164}$$

In [242] it has been found that there is not so much difference between the two cases of no phase transition (nonzero external field) and continuous phase transition (zero external field). Moreover, in the latter case there is not a big change in the topology of configuration space. Also this model, at first sight, could seem to disprove the topological theory. However, the second-order phase transition—in the limit of vanishing external field—is predicted by the canonical ensemble, whereas in the microcanonical ensemble no phase transition exists at all [243, 244], that is, there is nonequivalence of statistical ensembles for this model. Thus, since the microcanonical ensemble is the natural framework of the topological theory of phase transitions, there is no contradiction.[3]

In [245] the following modified version of the Berlin-Kac model has been considered

$$H = \sum_{i=1}^{N} \left[\frac{1}{2} p_i^2 - \frac{1}{2} \lambda_i x_i^2 - h \sqrt{N} x_1 \right] . \tag{10.165}$$

---

[3] Moreover, the microcanonical ensemble is the fundamental statistical ensemble because the microcanonical measure is the natural ergodic invariant measure for the microscopic Hamiltonian dynamics, and because there is no extra parameter, such as temperature, aside from the microscopic Hamiltonian.

where $x_i \in \mathbb{R}$, λ_i are certain real coefficients, h is an external field and again the system is constrained by

$$\sum_{i=1}^{N} x_i^2 = N \ .$$

The authors claim that this is the first example of a short-range, confining potential for which there is not a topological transition originating the thermodynamic phase transition of the model. This claim is simply *wrong*. This is again a *long-range* interaction model because of the spherical constraint that limits the freedom of all the degrees of freedom to vary independently from one another. Of course, one could argue that this is a weak constraint, however, it plays a fundamental role because without this spherical constraint the model is obviously trivial. The long-range interaction introduced by the constraint is weaker than a mean-field long-range interaction, and in fact this model has no phase transitions in less than three dimensions (whereas mean-field models undergo a phase transition at any dimension), nevertheless, the spherical model shares with mean-field models a discontinuous but non-diverging pattern of the specific heat. The spherical model (10.165) is *not* a short-range interaction one.

Chapter 11

Future Developments

The theoretical scenario depicted in this monograph is *not* a rephrasing of already known facts in an unusual mathematical language.

In fact, the Riemannian theory of Hamiltonian chaos, though still formulated at a somewhat primitive level (in that it does not yet include the role of nontrivial topology of the mechanical manifolds), provides a natural explanation of the origin of the chaotic instability of classical dynamics, substantially in the absence of competing theories.[1]

As far as the topological theory of phase transitions is concerned, which applies to a sufficiently broad class of physically relevant systems described by continuous variables, it shows that the conventional mathematical explanation of the origin of phase transitions (as due to the loss of analyticity of macroscopic observables in the $N \to \infty$ limit) is *not* the primitive, fundamental source of the phenomenon. In other words, there is a deeper phenomenon that drives the appearance of nonanalytic behaviors of statistical-mechanical averages. This deeper phenomenon is due only to the microscopic interaction potential and to the energy variation of the topology of its level sets in configuration space. The topological properties of the leaves of the foliation of configuration space, which are the level sets of the potential function, are what they are *independent* of the definition of statistical measures in configuration space. This is a theoretical step forward. As such, its value is both conceptual and

---

[1] The standard explanation of the origin of Hamiltonian chaos, based on homoclinic intersections and sketched in Chapter 2, has a very limited validity because it requires the explicit analytic knowledge of the separatrices; the theorem that tells that resonant tori break into an even number of fixed points, half of them elliptic and the other half hyperbolic, works for two-degrees-of-freedom systems but has no general validity at arbitrary N; the system must be given in action-angle coordinates and it has to be quasi-integrable; no relation between the explanation of the origin of chaos and the standard way of measuring its strength, through Lyapunov exponents, is given. At $N > 2$, using the natural coordinates of a standard Hamiltonian system and at any energy, the theory of homoclinic intersections cannot be considered a satisfactory theory of Hamiltonian chaos.

practical. From the conceptual point of view, it sheds a completely new light on the relationship between microscopic dynamics and macroscopic thermodynamics, including emergent phenomena such as phase transitions. This is due to the fact that microscopic dynamics and macroscopic statistical properties of a system are rooted in the same common ground: configuration-space topology. From the practical point of view, in one of the following sections, we give emphasis to some preliminary results concerning the protein folding problem, because these results paradigmatically illustrate the innovative potentialities of the topological theory of phase transitions.

11.1 Theoretical Developments

The many open points requiring further investigation have been put in evidence throughout the book. However, let us summarize some of them.

For what concerns the Riemannian theory of Hamiltonian chaos, we have seen that the quasi-isotropy assumption works strikingly well in the case of the FPU β model, whereas it requires a reasonable, but somewhat ad hoc, correction for the XY chain. We have also highlighted a major difference between these models. The configuration space of the FPU model is *topologically trivial*, whereas it is nontrivial for the XY chain. We have explained, thus, the necessity of including topological information in an improved version of the theory hitherto proposed. Parametric instability of the trajectories, due to the variability of curvature along them, plus hyperbolic scattering near critical points of the potential, and possibly the interplay between these two mechanisms, should account for the origin of Hamiltonian chaos in most of the systems described by standard Hamiltonians. Pursuing such an improvement directly in the form of a general theory would perhaps be like attempting to climb a mountain along the steepest path: some "diagonal paths" are advisable. For example, an interesting key study could be the XY chain for which, as we have seen in Chapter 10, the energy distribution of critical points is exactly known.

Let us remark that the aim of the Riemannian theory of chaos is not to provide recipes to compute Lyapunov exponents; rather, it aims at understanding why Newtonian dynamics is essentially unpredictable, with a few well-known exceptions. However, any version or improvement of the theory must be tested against some experimental results. These "experimental" results are provided by the outcomes of numerical computations of Lyapunov exponents for specific models.

For what concerns the topological theory of phase transitions, it is worth remarking that it does *not* claim that *any possible phase transition* must have a topological origin. It is perhaps because of the misunderstanding of this point that some papers recently appearing in the literature claim to give counterexamples to the topological theory (see Section 10.6). Actually, none of these papers goes against what we have presented in this book.

In our opinion, relevant leaps forward could reward future efforts addressed to the following points:

(a) Finding sufficiency conditions for the theorems given in Chapter 9. This is perhaps the most challenging problem. The question is, which kinds of distributions—of the number of critical points of the potential and of their Morse indexes—as a function of v and of the dimension N, are capable of inducing an unbounded growth with N of the upper bound of the third- or the fourth-order derivatives of the configurational entropy? This is the same as wondering which kinds of topological changes of the Σ_v, or of the M_v, as a function of v and N, are capable of inducing a phase transition by breaking uniform convergence with N of configurational entropy in the appropriate differentiability class. Needless to say, one would like to discriminate between first-, second-, and infinite-order transitions, and, perhaps, glassy transitions.

Can localization theorems for integrals—such as the Duistermaat–Heckman theorem[2] and its generalizations—be of some technical help?

(b) Topology throughout the present monograph is intended from the point of view of cohomology theory. We have hardly touched homotopy theory. The connection between phase transitions and topology seen from the point of view of homotopy theory is still an uncharted territory, worth exploring.

---

[2] The Duistermaat–Heckman theorem establishes some conditions that allow the exact evaluation of oscillatory integrals of the form

$$\int_M e^{itH(x)}\eta \; ,$$

where M is a symplectic manifold, H a Hamiltonian function on M, and η is the Liouville form. The integration measure localizes at the critical points of H, and this can be seen as a general consequence of equivariant cohomology [246]. To give an example of this kind of localization results, consider an n-dimensional compact manifold M, with volume form $dVol(x)$, and let $f : M \to \mathbb{R}$ be a Morse function, then for large t one has [247]

$$\int_M e^{itf(x)}dVol(x) = \sum_{p \in Crit(f)} \left(\frac{2\pi}{t}\right)^{n/2} \frac{e^{i\pi \, \mathrm{sgn}(\mathrm{Hess}_f(p))/4}}{\sqrt{|\det \mathrm{Hess}_f(p)|}} e^{itf(p)} + O(t^{-n/2-1})$$

where $Crit(f)$ is the set of critical points of f, and $\mathrm{sgn}(\mathrm{Hess}_f(p))$ is the signature of the Hessian at p. The signature of a real symmetric matrix is the number by which positive eigenvalues outnumber negative ones, thus $\mathrm{sgn}(\mathrm{Hess}_f(p)) = n - 2k_p$, with k_p the index of the critical point p. If the integration on the left-hand-side is carried over M_v, then the right-hand-side—being a sum restricted to the set $\{p \in M_v| \ (df)_p = 0\}$—is a function of v. Moreover, the right-hand-side, being a sum over the set of critical points of f where the phase of each contribution depends on the index of the critical point, is tightly related with the topology of M_v.

(c) Developing computational methods to probe topology changes. Luckily, for some mean-field models, such as those studied in Chapter 10, one can analytically find all the critical points of the potential. For generic systems, a priori only numerical methods would be left to find the critical points of the potential. However, this is practically undoable. In fact, the number of critical points of a generic potential function can easily grow exponentially or more with the configuration-space dimension. Thus, when numerical computations are the only possible option, one has to resort to typical theorems of differential topology, like the Gauss–Bonnet–Hopf theorem relating the total Gaussian curvature of a hypersurface to its Euler characteristic. Are there other useful tools that can be used, or adapted, or developed, that probe topology through the computation of analytic mathematical objects? Is it conceivable to define vector bundles based on (microscopic) configuration space and then try to investigate their characteristic classes?

(d) Are there smart dimensional reductions that make feasible a direct analytic or numeric investigation of the configuration-space topology? One obviously would think of defining some renormalization scheme. But perhaps already a "Bethe–Peierls"-like approximation [33] could provide an efficient, nontrivial, and drastic dimensional reduction.

(e) Extension to quantum systems of the methods developed for classical systems. This point is briefly discussed in the last section of this chapter.

(f) The topological theory presented in this book has been developed for potentials that depend on continuous variables, which is realistic from a physical point of view. Nevertheless, beginning with the Ising model, many discrete-variable models have been traditionally investigated in the statistical-mechanical literature. Thus it would be very interesting to make some connections between topology and phase transitions in discrete-variable systems. But topology of which manifolds? A bridging seems possible through the Hubbard–Stratonovich transform, which maps a discrete-variable system such as the Ising Hamiltonian to a continuous-variable potential.

(g) Extend the domain of applicability of the theorems proved in Chapter 9 to encompass long-range systems, at least in the case of potentials bounded also from above. What we have seen in Chapter 10 witnesses that such an extension must be possible. In fact, the models therein considered are described by long-range forces and display unequivocal topological signatures of the phase transitions they undergo. Moreover, a clear discrimination seems possible between first- and second-order transitions. Relaxing the additivity requirement seems at present very hard.

(h) Numerical computations. We cannot forget that—as we have seen in Chapter 7—precious informations can be obtained through direct numerical computations of the energy dependence of a topological invariant, as the Euler–Poincaré characteristic, for non-pathological models. With respect to the computational effort required to work out the results reported in [198] some improvements are possible. For example, to compute the Gauss–Kronecker curvature of the equipotential hypersurfaces Σ_v, the use of (8.9) in place of (7.38) should be more efficient. A more refined way of generating a Monte Carlo Markov chain to compute $\int_{\Sigma_v} d\sigma K_G / \int_{\Sigma_v} d\sigma$ should be that of numerically solving the stochastic differential equation [248]

$$dR = P(R)dB + M_1(R)n(R)dt$$

where R_t, $t \geq 0$, is a random sequence of points on $\Sigma_v \subset \mathbb{R}^N$, $P(R) = \mathbb{I} - n(R) \otimes n^T(R)$ is the orthogonal projection of a point R on the plane tangent to Σ_v at R, provided that $n \equiv \nabla V / \|\nabla V\|$ (normal at R); B is a Brownian motion in $\mathbb{R}^N$ and $M_1(R)$ is the mean curvature at R, that is, $M_1 = -\frac{1}{N-1} \text{div } n(R)$. Then one should combine this algorithm of random generation of points on Σ_v with a standard Metropolis importance sampling adapted to the desired measure on Σ_v.

Finally, the speed of computers has considerably grown during the last years.

As far as numerical computations are concerned, it is worth keeping in mind that "singular" patterns of the energy dependence of the largest Lyapunov exponent are indirect but reliable probes of non-trivial topological transitions in configuration space.

11.2 Transitional Phenomena in Finite Systems

Needless to say, in nature, phase transitions occur also in very small systems, with N much smaller than the Avogadro number. Phase transitions are qualitative changes of physical systems, such as, for example, a small crystal that melts. The intrinsic physical phenomenon, it is worth repeating, is independent of our mathematical description of phase transitions, and also of the quantitative way of detecting them through thermodynamic observables. In Chapter 7, tackling the lattice φ^4 model, we have seen that with a lattice as small as 7×7 sites, the Euler characteristic of the potential level sets sharply marks the phase transition point.

A future understanding of which kinds of topological changes in configuration space can induce the growth with N of the upper bound of some of the derivatives of the entropy (sufficiency conditions) will give, as a by-product, a natural definition of phase transitions for *finite* systems. In fact, since topological changes of manifolds exist at any dimension N, one would properly speak of "phase transitions at finite N" in the presence of those topological changes of the submanifolds of configuration space that satisfy sufficiency conditions, thus recovering, in the arbitrary large-N limit, the standard statistical-mechanical definition. In principle, any thermodynamic observable could do the same job at a phenomenological level. However, not only do the topological invariants give sharper signals already with very small numbers of degrees of freedom, but, from a conceptual point of view, there is a big difference: in the standard framework based on analyticity of observables, at finite N, no mathematical difference can be made between systems with and without phase transitions, because analyticity is unavoidable at finite N. In contrast, within the topological framework, a sharp difference can be made also at any finite N, as we have seen in Chapters 7 and 10. This requires nontrivial mathematical work to be properly formalized, but the intuitive evidence is already there.

This topic is of prospective interest for all those systems that intrinsically do not have a thermodynamic limit, yet undergo transitional phenomena. This is the case of the filament-globule transition in homopolymers, or of the protein folding from the primary conformation to the native one (which are given some emphasis in Section 11.4), or of many physical systems of biological interest, listed in the section devoted to the classification of phase transitions in Chapter 2, or of nano- and mesoscopic systems. For example, phase transition phenomena at the nanoscopic level are of increasing interest in modern atomic physics with Bose–Einstein condensation and in general transition phenomena in radiation–matter interaction (e.g., microlasers).

11.3 Complex Systems

In this book we have dealt, on the one hand, with *geometric* and *topological complexity* of phase space, and on the other hand, with the relation that such a complexity has with microscopic dynamics and macroscopic thermodynamics of many-degrees-of-freedom systems. We introduce here the word *complexity* on purpose, to suggest that the so-called *complex systems* and their transitional phenomenology provide a natural arena for the application of topological concepts and methods. A system is generically defined as *complex* if it can assume a large number of states or conformations. This concept made its first appearance in statistical mechanics, where it was mainly related to *disordered* and *amorphous* materials, like spin-glasses and glasses. In contrast to "standard" phase transitions, for which a unique stable equilibrium state exists in each thermodynamic phase and is unambiguously defined by the unique minimum of the Helmholtz free energy, the phase transitions of glassy type are associated with the appearance of a huge number of minima of the Helmholtz free energy and by their degeneracy. This feature led to the so-called *energy landscape paradigm* to tackle complexity in the framework of physics and to bridge with complexity as it is tackled in other contexts, mainly biology. It is now quite clear that many of the puzzling properties of glasses are encoded in their "energy landscape" [249], i.e., in the structure of valleys and saddles of the potential-energy function; but this is directly connected to the structure of the submanifolds M_v and Σ_v of configuration space,[3] and in fact, topological concepts have begun to emerge in some recent papers on glasses [250].

11.4 Polymers and Proteins

Among the many interesting transitional phenomena that one encounters in the field of soft-matter physics, we highlight the Θ-transition between filamentary and globular configurations in homopolymeric chains, and the *protein folding* transition from the primary sequence of amino acids to the native (or biologically active) state in heteropolymeric chains. These phenomena, which are commonly tackled within the framework of statistical mechanics, provide clear examples of the prospective relevance of the methods proposed in the present book.

In [88], the study of the off-lattice dynamics of simple models of homopolymeric chains has shown that the Θ-transition between the swollen and globular phases is marked very clearly by a change in the energy density dependence of the largest Lyapunov exponent. In real homopolymers, this transition occurs in systems with a number of degrees of freedom that is finite and small, much smaller than the Avogadro number. Therefore, the close resemblance of this dynamical characterization of the Θ-transition to the other dynamical characterizations of phase transitions, encountered in Chapter 6, has both conceptual and practical consequences. In fact, the existence of a natural "dynamical order parameter" (the largest Lyapunov exponent) to detect the transition between the filamentary and globular phases of a chain of given length N allows one to make a link with the geometry and topology of the mechanical manifolds, thus defining the transition independently of thermodynamic observables (which, in this case, are of little use to mark the transition), and independently of the thermodynamic limit (which, in this case, is somewhat unphysical).

Let us now come to the *protein folding* problem. This is a very challenging subject at the forefront of modern research in statistical mechanics applied to biophysics. This is not only a hard theoretical problem, but a fundamental issue of a new discipline named *proteomics*. Proteins are the working machines of every living organism, performing almost all biological functions at a microscopic level. Proteins are sequences of covalently bonded amino acids forming polypeptide chains of a few hundred building blocks. Among a huge number of possible amino-acid sequences of finite length, only a very few of them—when put in an aqueous environment—are capable of folding into precise configurations that are working proteins. Starting from a "good" sequence of amino

---

[3] In current literature on glasses Σ_v are called Potential Energy Surfaces (PES), and the configurational entropy, computed as the logarithm of the number of stationary points of the potential whose energy lies in an interval $[v, v + \delta v]$, is referred to as the *complexity* of the system.

acids, in a surprisingly short time the same spatial compact conformation of a protein is always reached. These sequences are called *good folders*. In contrast, a randomly sorted sequence yields a random heteropolymer undergoing a glassy transition.

The crucial question is, how can we read in a given sequence of amino acids whether it corresponds to a *good folder*, that is, capable of folding into a protein (if put in a suitable environment), or a *bad folder*, that is, a random heteropolymer and not a protein? This is a fundamental theoretical problem, still wide open even though the field has considerably advanced in recent times, which is also of a major relevance to the so-called protein design problem.

We take this problem as a *paradigmatic example* to illustrate why the topological theory of phase transitions promises a real methodological advancement in the study of transitional phenomena in strongly inhomogeneous systems at meso- or nanoscale.

To this end, we need to enter into some detail of a recent and very interesting paper [251], where the Hamiltonian dynamics—and its underlying configuration-space geometry—have been numerically investigated for the so-called *minimal model*, originally introduced by Thirumalai and coworkers [252]. The authors consider a three-dimensional off-lattice model of a polypeptide that has only three different kinds of amino acids: polar (P), hydrophobic (H), and neutral (N). The potential energy of a chain is given by [252]

$$V = V_B + V_A + V_D + V_{NB} , \tag{11.1}$$

where

$$V_B = \sum_{i=1}^{N-1} \frac{k_r}{2}(|\mathbf{r}_i - \mathbf{r}_{i-1}| - a)^2 ; \tag{11.2}$$

$$V_A = \sum_{i=1}^{N-2} \frac{k_\vartheta}{2}(|\vartheta_i - \vartheta_{i-1}| - \vartheta_0)^2 ; \tag{11.3}$$

$$V_D = \sum_{i=1}^{N-3} \{A_i[1 + \cos\psi_i] + B_i[1 + \cos(3\psi_i)]\} ; \tag{11.4}$$

$$V_{NB} = \sum_{i=1}^{N-3} \sum_{j=i+3}^{N} V_{ij}(|\mathbf{r}_{i,j}|), \tag{11.5}$$

where $\mathbf{r}_i$ is the position vector of the ith monomer; $\mathbf{r}_{i,j} = \mathbf{r}_i - \mathbf{r}_j$; ϑ_i is the ith bond angle, i.e., the angle between $\mathbf{r}_{i+1}$ and $\mathbf{r}_i$; ψ_i the ith dihedral angle, that is, the angle between the vectors $\hat{n}_i = \mathbf{r}_{i+1,i} \times \mathbf{r}_{i+1,i+2}$ and $\hat{n}_{i+1} = \mathbf{r}_{i+2,i+1} \times \mathbf{r}_{i+2,i+3}$; $k_r = 100$; $a = 1$; $k_\vartheta = 20$; $\vartheta_0 = 105°$; $A_i = 0$; and $B_i = 0.2$ if at least two among the residues $i, i+1, i+2, i+3$ are N, and $A_i = B_i = 1.2$ otherwise. As to V_{ij}, it is

$$V_{ij} = \frac{8}{3}\left[\left(\frac{a}{r}\right)^{12} + \left(\frac{a}{r}\right)^{6}\right]$$

if $i,j = \mathrm{P},\mathrm{P}$ or $i,j = \mathrm{P},\mathrm{H}$;

$$V_{ij} = 4\left[\left(\frac{a}{r}\right)^{12} - \left(\frac{a}{r}\right)^{6}\right]$$

if $i,j = \mathrm{H},\mathrm{H}$; and

$$V_{ij} = 4\left(\frac{a}{r}\right)^{6}$$

if either i or j is N.

Five different sequences are considered, four of 22 monomers and one of 44 monomers. The sequences are as follows:

$$S_{\mathrm{g}}^{22} = \mathrm{PH}_9(\mathrm{NP})_2\mathrm{NHPH}_3\mathrm{PH},$$
$$S_{\mathrm{b}}^{22} = \mathrm{PHNPH}_3\mathrm{NHNH}_4(\mathrm{PH}_2)_2\mathrm{PH},$$
$$S_{\mathrm{i}}^{22} = \mathrm{P}_4\mathrm{H}_5\mathrm{NHN}_2\mathrm{H}_6\mathrm{P}_3,$$
$$S_{\mathrm{h}}^{22} = \mathrm{H}_{22}$$
$$S_{\mathrm{h}}^{44} = \mathrm{H}_{44}.$$

The sequence S_{g}^{22} has already been identified as a good folder [252], which is confirmed by the dynamical simulations: below a certain temperature the system always folds into the same β-sheet-like structure.

In contrast, homopolymers S_{h}^{22} and S_{h}^{44} undergo a hydrophobic collapse but do not display any tendency to attain a particular conformation in the collapsed phase. The sequence S_{b}^{22} (which consists of a rearrangement of the sequence S_{g}^{22}) behaves as a bad folder and does not reach a unique native state, while S_{i}^{22} stands in between good and bad folders: the middle of the sequence always forms the same structure, while the more external parts of the chain do not fold into a stable configuration and still fluctuate even at low temperature.

A comparison between standard thermodynamic observables, like the specific heat, and geometric observables, namely the variance of the Ricci curvature fluctuations of configuration space, both sampled by the dynamical trajectories, is very instructive.

As far as the specific heat is concerned, all the sequences show a very similar pattern; in particular, both the specific heat C_V of the homopolymer S_{h}^{22} and of the good folder S_{g}^{22} exhibit a peak at the transition point (see Figures 11.1 and 11.2 respectively), and on the sole basis of this quantity it would be hard to discriminate between a simple hydrophobic collapse and a protein-like folding. The slight asymmetry of the specific heat pattern obtained for the homopolymer cannot be given any special meaning.

On the other hand, the temperature patterns of the Ricci curvature fluctuations of configuration space (computed in the Eisenhart metric g_E) display a major qualitative difference between the homopolymer and the good folder.

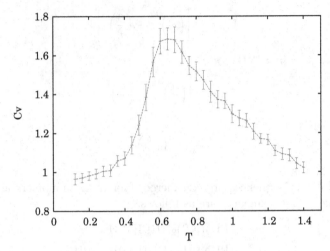

Fig. 11.1. Specific heat C_V vs. temperature T for the homopolymeric chain S_h^{22}. From [251].

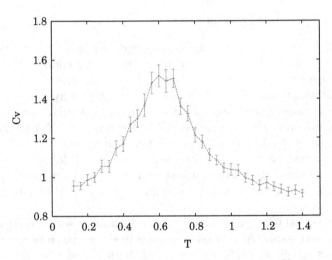

Fig. 11.2. C_V vs. T for the good heteropolymeric folder S_g^{22}. From [251].

The normalized variance of the Ricci curvature fluctuations σ measured along the numerically computed trajectories is defined as

$$\sigma^2 = \frac{N\left(\langle K_R^2 \rangle_t - \langle K_R \rangle_t^2\right)}{\langle K_R \rangle_t^2} , \tag{11.6}$$

where $\langle \cdot \rangle_t$ stands for a time average, and K_R for the Ricci curvature. In Figure 11.3, σ is reported, as a function of the temperature T, for the homopolymer S_h^{22} and for the good folder S_g^{22}. In the case of the good folder, a peak of

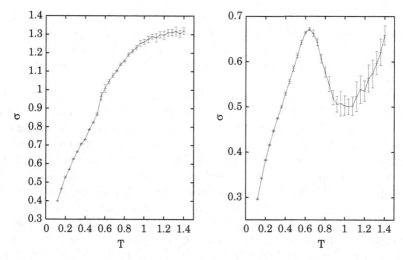

Fig. 11.3. Relative curvature fluctuation σ vs. temperature T for the homopolymer $S_{\rm h}^{22}$ (left) and for the good folder $S_{\rm g}^{22}$ (right). The solid curves are a guide to the eye. From [251].

σ shows up at a temperature value that is practically coincident with the folding temperature T_f (estimated as $T_f = 0.6 \pm 0.05$). At $T < T_f$, the system is mostly in the native state, folded into a β-sheet.

No mark of the hydrophobic collapse of the homopolymer $S_{\rm h}^{22}$ is found in the pattern of $\sigma(T)$. As to the other sequences, $\sigma(T)$ for the longer homopolymer $S_{\rm h}^{44}$ is even smoother than for $S_{\rm h}^{22}$, in contrast to the specific heat, which develops a sharper peak consistently with the presence of a thermodynamic Θ-transition as $N \to \infty$. For the bad folder $S_{\rm b}^{22}$, $\sigma(T)$ is not as smooth as for the homopolymers, but only a very weak signal is found at a lower temperature than that of the peak in C_V, i.e., at the temperature where the system starts to behave as a glass; for the "intermediate" sequence $S_{\rm i}^{22}$ a peak is present at the "quasi-folding" temperature, although considerably broader than in the case of $S_{\rm g}^{22}$.

The above-reported shapes of $\sigma(T)$ bring us back to Chapters 6 and 7, where we have discussed the early developments of the study of geometry of dynamics in the presence of phase transitions. In particular, the pattern of $\sigma(T)$ obtained for the good folder $S_{\rm g}^{22}$ is remarkably similar to that exhibited by other systems undergoing a symmetry-breaking phase transition and, in particular, to the pattern of $\sigma(v)$ found for the $\mathbb{Z}_2$-breaking phase transition in a two-dimensional lattice φ^4 model, reported in Figure 7.4 (notice that $\sigma(T)$ and $\sigma(v)$ are parametrically related through the function $v = v(T)$; thus the presence of the peak of $\sigma(T)$ at the folding transition must be kept in the shape of $\sigma(v)$ at some $v_f = v(T_f)$). This suggests that the folding of a protein-like heteropolymer does share some features of the symmetry-breaking phase transitions, at least when these are seen from a different viewpoint with respect to the standard thermodynamic one. In fact, proteins are intrinsically finite and aperiodic objects for which thermodynamic-limit singularities can hardly be defined. The topological framework, in contrast, naturally allows one to tackle finite systems, as we have already discussed throughout this monograph.

At present the results reported here *strongly* suggest a possible link between the folding transition and a topological change in configuration space. However, this still remains to be directly checked, similarly to what has been done for the lattice φ^4 model in Chapter 7 with the computation of the Euler characteristic of the potential level sets. That topology must have something to do with the protein folding problem is witnessed by a widely accepted "metaphor" that associates with a good folder a funnel-like pattern of the free energy as a function of a suitable parameter labeling different conformations; this funnel is assumed to have many local minima but one absolute minimum for the native conformation of the protein. A bad folder, in this picture, has a huge number of degenerate minima of the free energy, typical of a glassy phase. This qualitative picture can be replaced by a direct analysis of how the topology of configuration-space submanifolds changes as a function of the potential energy density, and by a comparison of the outcomes for good and bad folders. In conclusion, we have very good reasons to think that the theoretical framework discussed in the present monograph can give a relevant contribution to the development of a new strategy to distinguish between good and bad folders.

11.5 A Glance at Quantum Systems

To conclude, let us just mention that some possibilities seem to exist of exporting to quantum systems the geometric and topological methods developed throughout this book for classical systems. The most natural way stems from the following remark. The two fundamental operators in quantum physics and statistical physics, respectively, are the unitary evolution operator $e^{-i\widehat{H}t}$ and the density matrix $e^{-\beta\widehat{H}}$. These are formally related by the analytic continuation known as the Wick rotation, that is, $t \mapsto -it_E$. This operation is at the basis of a formal link that can be established between classical and quantum systems.

Let us first consider the path integral for a scalar field theory

$$\mathcal{Z} = \int \mathcal{D}\phi \, \exp\left\{\frac{i}{\hbar}\int d^d x [\frac{1}{2}\partial_\mu\phi\partial^\mu\phi - V(\phi)]\right\} . \tag{11.7}$$

By performing the Wick rotation we have

$$\mathcal{Z} = \int \mathcal{D}\phi \, \exp\left\{-\frac{1}{\hbar}\int d^d_E x [\frac{1}{2}\partial_\mu\phi\partial^\mu\phi + V(\phi)]\right\}$$

$$= \int \mathcal{D}\phi \, \exp\left\{ -\frac{1}{\hbar} \int d_E^d x \, \mathcal{L}_E(\phi) \right\} , \tag{11.8}$$

where $d^d x = -i d_E^d x$ and $d_E^d x = d^{(d-1)} x \, dt_E$.

In (11.7) we have $\partial_\mu \phi \partial^\mu \phi - V(\phi) = (\frac{\partial \phi}{\partial t})^2 - (\nabla\phi)^2 - V(\phi)$, while in (11.8) we have $\partial_\mu \phi \partial^\mu \phi - V(\phi) = (\frac{\partial \phi}{\partial t_E})^2 + (\nabla\phi)^2 + V(\phi)$. In the last term of (11.8) the so-called Euclidean Lagrangian $\mathcal{L}_E(\phi) = \frac{1}{2}(\frac{\partial \phi}{\partial t_E})^2 + \frac{1}{2}(\nabla\phi)^2 + V(\phi)$ is actually an energy density of the field ϕ, so that

$$E(\phi) = \int d_E^d x \left[\frac{1}{2}\left(\frac{\partial \phi}{\partial t_E} \right)^2 + \frac{1}{2}(\nabla\phi)^2 + V(\phi) \right]$$

is just the total energy, and the Euclidean path integral

$$\mathcal{Z} = \int \mathcal{D}\phi \, \exp\left\{ -\frac{1}{\hbar} E(\phi) \right\} \tag{11.9}$$

now looks very similar to the canonical partition function of classical statistical mechanics, and if one makes an ultraviolet regularization, by considering the field on a lattice, then (11.9) *is* the classical canonical partition function.

On the other hand, starting from the classical canonical configurational partition function with a potential $V = V(q_1, \ldots, q_N)$, by replacing i with x, q_i with $\phi(x)$, and by paying attention to the way of taking the continuum limit, one recovers (11.9) by identifying $\hbar$ with the Boltzmann factor $k_B T$. Notice that in this case the functional $E(\phi)$ corresponds to the classical configurational part only.[4]

Therefore, we see that *a Euclidean quantum field theory in a d-dimensional space-time is equivalent to classical statistical mechanics in a d-dimensional space.*

In view of this equivalence, it seems worth investigating whether the generalizations of the Duistermaat–Heckman integration formula to path-integrals [253, 254] could be useful—in both quantum and classical frameworks—for further developments of the topological theory of phase transitions.

A further interesting connection is obtained as follows. Consider the partition function of quantum statistical mechanics

$$Z = \text{Tr} \, e^{-\beta \widehat{H}} = \sum_n \langle \psi_n | e^{-\beta \widehat{H}} | \psi_n \rangle \tag{11.10}$$

and work out an integral representation of Z by means of the standard integral representation of $\langle \psi_{\text{final}} | e^{-i\widehat{H}t} | \psi_{\text{initial}} \rangle$. After having performed the analytic

---

[4] If we start from $Z = \int \prod_i dp_i dq_i e^{-\beta H(p,q)}$ we have to integrate over the p_i and then take the continuum limit of $Z_C = \int \prod_i dq_i e^{-\beta V(q)}$; otherwise, we should consider a double functional integration $\int \mathcal{D}\pi \mathcal{D}\phi(\ldots)$, which would not be the Euclidean path integral (11.9).

continuation $t \to -i\beta$, having put $|\psi_{\text{final}}\rangle = |\psi_{\text{initial}}\rangle = |\psi_n\rangle$, and then by summing over $|\psi_n\rangle$, one finally obtains [255]

$$Z = \text{Tr } e^{-\beta \widehat{H}} = \int \mathcal{D}q \, \exp\left\{ -\int_0^\beta d\tau \, \mathcal{L}_E(q) \right\} , \qquad (11.11)$$

where $\int \mathcal{D}q$ is constrained to periodic boundary conditions, that is, $q(0) = q(\beta)$; $\mathcal{L}_E(q)$ is the Euclidean Lagrangian, which, in the Euclidean time τ, is just the Hamiltonian $H = \sum_i \frac{1}{2}(dq_i/d\tau)^2 + V(q)$; the new time is bounded as follows: $0 \leq \tau < \beta$.

The extension to field theory is made as follows: Consider the Hamiltonian $\widehat{H}$ of a quantum field theory in D-dimensional space, so that the d-dimensional space-time has $d = D + 1$, and (11.11) becomes

$$Z = \text{Tr } e^{-\beta \widehat{H}} = \int_{\text{pbc}} \mathcal{D}\phi \, \exp\left\{ -\int_0^\beta d\tau \int d^D x \, \mathcal{L}_E(\phi) \right\} , \qquad (11.12)$$

where the subscript "pbc" reminds us that the integral is computed on all the paths $\phi(x, \tau)$ satisfying periodic boundary conditions, that is, $\phi(x, \tau = 0) = \phi(x, \tau = \beta)$.

We see that *a Euclidean quantum field theory in $(D + 1)$-dimensional spacetime, with $0 \leq \tau < \beta$, is equivalent to quantum statistical mechanics in D-dimensional space.*

These connections between quantum field theory and statistical mechanics have been widely exploited in the recent past. In view of the application to quantum systems of the topological theory of phase transitions, these sketchily presented connections seem to open the most natural way of approaching the problem. Actually, within this framework, in recent years much work has been done to develop *effective potential methods* that have been devised to tackle quantum statistical systems by means of *formally* classical statistical-mechanical methods [256].

Nevertheless, there can be other possibilities of exporting to quantum systems the geometric and topological methods here developed for classical systems. For example, let us mention only that the use of generalized coherent states or of Bohm's theory of quantum motions leads to formally classical descriptions of quantum systems [257, 258] that can be used to tackle chaos in quantum systems by resorting to standard definitions and methods used in classical dynamics [259].

It is worth mentioning that also purely quantum-mechanical treatments of phase transitions have revealed intriguing connections with topological changes of certain state manifolds [260].

Appendix A

Elements of Geometry and Topology of Differentiable Manifolds

The present appendix, and the following ones, are for physicists' use and contain a concise summary of some elementary concepts in differential geometry and differential topology. Despite what it might perhaps seem by simply browsing these final pages of the book, the style of the presentation is informal and should accommodate to the standard way that we, physicists, have to approach mathematics. The aim of these appendices is twofold: first, to provide those physicists who are not familiar with these subjects with the conceptual mathematical background necessary to follow and appreciate the meaning of the book content; second, to ease further mathematical readings for those who, starting from scratch in this field, might find it interesting to go more deeply into the subject.

The interested reader will benefit from the reading the books listed at the end of this appendix.

In what follows, the Einstein summation convention over repeated indices is always understood unless explicitly stated to the contrary.

A.1 Tensors

Let us consider an n-dimensional vector space V_n on a field $\mathbb{K}$, and its dual V_n^*. Any element ω of V_n^* is a linear function $\omega : V_n \mapsto \mathbb{K}$ such that, given $X \in V_n$, $\omega(X) = \langle \omega, X \rangle$; the field $\mathbb{K}$ is here assumed to be $\mathbb{R}$ or $\mathbb{C}$. Of course any $X \in V_n$ can be regarded as a linear function on V_n^*, that is $X(\omega) = \langle \omega, X \rangle$, so that $(V_n^*)^*$ can be identified with V_n. Consider $X, Y \in V_n$ and a function $f : V_n \times V_n \mapsto \mathbb{K}$ defined on the Cartesian product $V_n \times V_n$. The function f is called *bilinear* if it is linear in X, for fixed Y, and is linear in Y, for fixed X.

Then consider $\omega, \pi \in V_n^*$. The *tensor product* $\omega \otimes \pi$ is a $\mathbb{K}$-valued map, $\omega \otimes \pi : V_n \times V_n \mapsto \mathbb{K}$, such that $\omega \otimes \pi(X, Y) = \omega(X)\pi(Y)$, which is obviously bilinear. In general, a function $f : V_n \times \cdots \times V_n(q \ copies) \mapsto \mathbb{K}$ is q-linear if it is linear in each argument $X_1, \ldots, X_q \in V_n$. Considering, then, p elements of

V_n^*, that is, $\omega^1, \ldots, \omega^p \in V_n^*$, their tensor product $\omega^1 \otimes \cdots \otimes \omega^p : V_n \times \cdots \times V_n(p \; copies) \mapsto \mathbb{K}$ is

$$\omega^1 \otimes \cdots \otimes \omega^p(X_1, \ldots, X_p) = \omega^1(X_1) \cdots \omega^p(X_p) \; .$$

Let $W^{(p,q)}$ be the product space of p times the dual space V_n^* of a vector space V_n, and q times the vector space V_n itself: $W^{(p,q)} = V_n^* \times \cdots \times V_n^* \times V_n \times \cdots \times V_n$. If $f : V_n \times \cdots \times V_n(q \; copies) \mapsto \mathbb{K}$ is a q-linear map, and $g : V_n^* \times \cdots \times V_n^*(p \; copies) \mapsto \mathbb{K}$ is p-linear, their tensor product is a $(p+q)$-linear map $g \otimes f : W^{(p,q)} \mapsto \mathbb{K}$ such that

$$g \otimes f(\omega^1, \ldots, \omega^p, X_1, \ldots, X_q) = g(\omega^1, \ldots, \omega^p)f(X_1, \ldots, X_q) \; .$$

The set of all the $(p+q)$-linear mappings from $W^{(p,q)}$ to the field $\mathbb{K}$, is the space, denoted by

$$\mathcal{T}^{(p,q)} = \underbrace{V_n^* \otimes \cdots \otimes V_n^*}_{p \; \text{times}} \otimes \underbrace{V_n \otimes \cdots \otimes V_n}_{q \; \text{times}} \; ,$$

of (p,q)-type *tensors*; $r = p + q$ is the rank of the tensor, p the index of *covariance*, and q the index of *contravariance*.

So a real (p,q)-*tensor* **t** over a vector space V_n is

$$\mathbf{t} : \mathcal{T}^{(p,q)} \mapsto \mathbb{R} \; , \tag{A.1}$$

i.e., acting on p dual vectors and q vectors, **t** yields a number, and it does so in such a manner that if we fix all but one of the vectors or dual vectors, it is a linear map in the remaining variable. A $(0,0)$ tensor is a scalar, a $(0,1)$ tensor is a vector, and a $(1,0)$ tensor is a dual vector (covector).

The space $\mathcal{T}^{(p,q)}$ of the tensors of type (p,q) is a linear space; a (p,q)-tensor is defined once its action on p vectors of the dual basis and on q vectors of the basis is known, and since there are $n^p n^q$ independent ways of choosing these basis vectors, $\mathcal{T}^{(p,q)}$ is an n^{p+q}-dimensional linear space. The vector space structure of $\mathcal{T}^{(p,q)}$ is defined in the standard way: if $a \in \mathbb{K}$, and $\mathbf{t}, \mathbf{s} \in \mathcal{T}^{(p,q)}$, then

$$(a\mathbf{t})(\omega^1, \ldots, \omega^p, X_1, \ldots, X_q) = a\mathbf{t}(\omega^1, \ldots, \omega^p, X_1, \ldots, X_q) \; ,$$

$$(\mathbf{t} + \mathbf{s}) = \mathbf{t}(\omega^1, \ldots, \omega^p, X_1, \ldots, X_q) + \mathbf{s}(\omega^1, \ldots, \omega^p, X_1, \ldots, X_q) \; .$$

The introduction of a product operation between tensors (defined below) leads to the definition of a non-commutative *tensor algebra* associated to V_n. Two other natural operations can be defined on tensors. The first one is called *contraction* with respect to the ith (dual vector) and the jth (vector) arguments and is a map

$$C : \mathbf{t} \in \mathcal{T}^{(p,q)} \mapsto C\mathbf{t} \in \mathcal{T}^{(p-1,q-1)} \tag{A.2}$$

defined by

$$Ct = \sum_{\mu=1}^{n} t(\ldots, \underbrace{v^{\mu*}}_{i}, \ldots; \ldots, \underbrace{v_{\mu}}_{j}, \ldots) \, . \tag{A.3}$$

The contracted tensor Ct is independent of the choice of the basis, so that the contraction is a well-defined, invariant operation.

The second operation is the *tensor product*, which maps an element $\mathcal{T}^{(p,q)} \times \mathcal{T}^{(p',q')}$ into an element of $\mathcal{T}^{(p+p',q+q')}$, i.e., two tensors t and t' into a new tensor, denoted by $t \otimes t'$, defined as follows: given $p + p'$ dual vectors $v^{1*}, \ldots, v^{p+p'*}$ and $q + q'$ vectors $w_1, \ldots, w_{q+q'}$, then

$$t \otimes t'(v^{1*}, \ldots, v^{p+p'*}; w_1, \ldots, w_{q+q'})$$
$$= t(v^{1*}, \ldots, v^{p*}; w_1, \ldots, w_q) \, t'(v^{p+1*}, \ldots, v^{p+p'*}; w_{q+1}, \ldots, w_{q+q'}). \tag{A.4}$$

The tensor product allows one to construct a basis for $\mathcal{T}^{(p,q)}$ starting from a basis $\{v_\mu\}$ of V and its dual basis $\{v^{\nu*}\}$: such a basis is given by the n^{p+q} tensors $\{v_{\mu_1} \otimes \cdots \otimes v_{\mu_p} \otimes v^{\nu_1*} \otimes \cdots \otimes v^{\nu_q*}\}$. Thus, every tensor $t \in \mathcal{T}^{(p,q)}$ allows a decomposition

$$t = \sum_{\mu_1,\ldots,\nu_q=1}^{n} t^{\mu_1 \cdots \mu_p}{}_{\nu_1 \cdots \nu_q} v_{\mu_1} \otimes \cdots \otimes v^{\nu_q*} \, ; \tag{A.5}$$

the numbers $t^{\mu_1 \cdots \mu_p}{}_{\nu_1 \cdots \nu_q}$ are called the *components* of t in the basis $\{v_\mu\}$. The components of the contracted tensor Ct are

$$(Ct)^{\mu_1 \cdots \mu_{p-1}}{}_{\nu_1 \cdots \nu_{q-1}} = t^{\mu_1 \cdots \rho \cdots \mu_p}{}_{\nu_1 \cdots \rho \cdots \nu_q} \tag{A.6}$$

(let us keep in mind the summation convention), and the components of the tensor product $t \otimes t'$ are

$$(t \otimes t')^{\mu_1 \cdots \mu_{p+p'}}{}_{\nu_1 \cdots \nu_{q+q'}} = t^{\mu_1 \cdots \mu_p}{}_{\nu_1 \cdots \nu_q} \, t'^{\mu_{p+1} \cdots \mu_{p+p'}}{}_{\nu_{q+1} \cdots \nu_{q+q'}} \, . \tag{A.7}$$

All these results are valid for a generic vector space V_n. As we will see below, they apply in particular to the vector spaces of the tangent bundle TM of a manifold M, over which tensors (and, analogously to vector fields, *tensor fields*) can be defined as above.

Now let G_q be the index permutation group acting on $1, 2, \ldots, q$ and denote by $\sigma \in G_q$ any of its elements. Denote again by σ the mapping $\sigma : W^{(0,q)} \to W^{(0,q)}$ defined by

$$\sigma(u_{(1)}, \ldots, u_{(q)}) = (u_{\sigma(1)}, \ldots, u_{\sigma(q)}), \quad u_{(k)} \in V_n \, ;$$

if $t \in \mathcal{T}^{(0,q)}$ is a $(0,q)$ tensor, i.e., $t : W^{(0,q)} \to K$, then also $t\sigma \in \mathcal{T}^{(0,q)}$; moreover, if $a, b \in K$, $t_1, t_2 \in \mathcal{T}^{(0,q)}$, it is true that $(at_1 + bt_2)\sigma = at_1\sigma + bt_2\sigma$.

A *symmetric* tensor $\mathbf{t}_2 \in \mathcal{T}^{(0,q)}$ is such that $\mathbf{t} = \mathbf{t}\sigma$ for any $\sigma \in G_q$, whereas a tensor $\mathbf{t}_2 \in \mathcal{T}^{(0,q)}$ is called *antisymmetric* if

$$\text{for any } \sigma \in G_q , \qquad \mathbf{t} = \epsilon(\sigma)\mathbf{t}\sigma ,$$

where $\epsilon(\sigma) = \pm 1$ according to the parity of σ, $\epsilon(\sigma) = -1$ for even permutations.

Hence, for any q-tuple of vectors $(u_{(1)}, \ldots, u_{(q)}) \in W^{(0,q)}$,

$$\forall \sigma \in G_q , \qquad \mathbf{t}(u_{(1)}, \ldots, u_{(q)}) = \mathbf{t}(u_{\sigma(1)}, \ldots, u_{\sigma(q)})$$

for symmetric tensors and

$$\forall \sigma \in G_q , \qquad \mathbf{t}(u_{(1)}, \ldots, u_{(q)}) = \epsilon(\sigma)\mathbf{t}(u_{\sigma(1)}, \ldots, u_{\sigma(q)})$$

for antisymmetric tensors.

By replacing $\{u_{(k)}\}$ with $\{e_{(j_k)}\}$, $k = 1, \ldots, q$ ($\{e_{(j_k)}\}$ is any basis of V_n), we get the following relations among the tensor components:

$$t_{j_1 \ldots j_q} = t_{j_{\sigma(1)} \ldots j_{\sigma(q)}} \quad \forall \sigma \in G_q ,$$
$$t_{j_1 \ldots j_q} = \epsilon(\sigma)t_{j_{\sigma(1)} \ldots j_{\sigma(q)}} \quad \forall \sigma \in G_q .$$

The two sets of symmetric and antisymmetric tensors of $\mathcal{T}^{(0,q)}$ form two subspaces Λ_S^q and Λ_A^q of $\mathcal{T}^{(0,q)}$ such that $\Lambda_S^q \bigcap \Lambda_A^q = \{0\}$.

The elements of Λ_A^q are also called *exterior forms* of degree q or *q-forms*.

Proposition. *Any q-form* $\alpha \in \Lambda_A^q$ *vanishes on any q-tuple of linearly dependent vectors. In particular, if $q > n$ then $\Lambda_A^q = \{0\}$.*

A.1.1 Symmetrizer and Antisymmetrizer

An endomorphism $\varphi : \mathcal{T}^{(0,q)} \to \mathcal{T}^{(0,q)}$ permutable with G_q, i.e., such that $\forall \mathbf{t} \in \mathcal{T}^{(0,q)}$, $\forall \sigma \in G_q$ gives $\varphi(\mathbf{t}\sigma) = \varphi(\mathbf{t})\sigma$, is called a *symmetrization operator*, or symmetrizer, if it transforms any tensor into a symmetric tensor and leaves a symmetric tensor unchanged An endomorphism, permutable with G_q, that transforms any tensor into an antisymmetric tensor and leaves an antisymmetric tensor unchanged is called an *antisymmetrization operator*, or antisymmetrizer.

Proposition. *On $\mathcal{T}^{(0,q)}$ there exist, and are unique, a symmetrizer $\mathcal{S}$ and antisymmetrizer $\mathcal{A}$ defined by*

$$\mathcal{S}(\mathbf{t}) = \frac{1}{q!} \sum_{\tau \in G_q} \mathbf{t}\tau , \quad \forall \mathbf{t} \in \mathcal{T}^{(0,q)} ,$$

$$\mathcal{A}(\mathbf{t}) = \frac{1}{q!} \sum_{\tau \in G_q} \epsilon(\tau)\mathbf{t}\tau , \quad \forall \mathbf{t} \in \mathcal{T}^{(0,q)} . \tag{A.8}$$

Proposition. $\forall \mathbf{t} \in T^{(0,q)}$ it is true that $\mathcal{S}\mathcal{A}(\mathbf{t}) = 0$ and $\mathcal{A}\mathcal{S}(\mathbf{t}) = 0$. Thus symmetrization of an antisymmetric tensor or antisymmetrization of a symmetric tensor yields a null tensor. Moreover, $\mathbf{t} \in \Lambda_A^q \oplus \Lambda_S^q$ iff $\mathbf{t} = \mathcal{S}(\mathbf{t}) + \mathcal{A}(\mathbf{t})$, and $\mathcal{A}(\mathbf{t}) = 0$ $\forall \mathbf{t} \in T^{(0,q)}$ with $q > n$.

Everything can be also repeated for contravariant tensors and, more generally, for tensors of $T^{(p,q)}$. Symmetrization and antisymmetrization can be defined in this case as operators acting on a given number k of indexes (all of them either covariant or contravariant).

A.2 Grassmann Algebra

Let us now simply denote by Λ^q the space of q-forms associated with V_n. We define the *exterior product* of two forms $\alpha \in \Lambda^p$ and $\beta \in \Lambda^q$, denoting it by $\alpha \wedge \beta$, by antisymmetrizing the tensor $\alpha \otimes \beta$, i.e.,

$$\alpha \wedge \beta = \mathcal{A}(\alpha \otimes \beta) \ .$$

Remember that if $\alpha \in T^{(0,p)}$ and $\beta \in T^{(0,q)}$, then $\alpha \otimes \beta : W^{(0,p+q)} \to K$ is defined by

$$\alpha \otimes \beta(u_{(1)}, \ldots, u_{(p+q)}) = \alpha(u_{(1)}, \ldots, u_{(p)})\beta(u_{(p+1)}, \ldots, u_{(p+q)}) \ .$$

The exterior product, or wedge product, thus gives a $(p+q)$-form.

From the general property of the antisymmetrizer $\mathcal{A}$, i.e., $\forall \mathbf{t}_1 \in T^{(0,q_1)}$, $\mathbf{t}_2 \in T^{(0,q_2)}, \mathbf{t}_3 \in T^{(0,q_3)}$, it is true that

$$\mathcal{A}(\mathcal{A}(\mathbf{t}_1 \otimes \mathbf{t}_2) \otimes \mathbf{t}_3) = \mathcal{A}(\mathbf{t}_1 \otimes \mathbf{t}_2 \otimes \mathbf{t}_3) = \mathcal{A}(\mathbf{t}_1 \otimes \mathcal{A}(\mathbf{t}_2 \otimes \mathbf{t}_3)) \ ,$$

and the *associative* property of the exterior product follows:

$$(\alpha_1 \wedge \alpha_2) \wedge \alpha_3 = \mathcal{A}(\alpha_1 \otimes \alpha_2 \otimes \alpha_3) = \alpha_1 \wedge (\alpha_2 \wedge \alpha_3) \ .$$

More generally, if $\alpha_i \in \Lambda^{p_i}$ we have

$$(\alpha_1 \wedge \alpha_2 \wedge \cdots \wedge \alpha_k) = \mathcal{A}(\alpha_1 \otimes \alpha_2 \otimes \cdots \otimes \alpha_k) \ .$$

Let us give an example. Consider $v^{(1)}, \ldots, v^{(p)} \in V_n^* = \Lambda^1 = T^{(0,1)}$, i.e., p covariant vectors $(p \leq n)$. Their exterior product

$$v^{(1)} \wedge v^{(2)} \wedge \cdots \wedge v^{(p)}$$

belongs to $T^{(0,p)}$. Now we want to compute its components with respect to a basis of $T^{(0,p)}$. Denote by $V_i^{(r)} = \langle v^{(r)}, e_{(i)} \rangle$ the components with respect to $\{e_{(i)}\}$, a basis of V_n. Then

$$(v^{(1)} \wedge v^{(2)} \wedge \cdots \wedge v^{(p)})_{i_1 \ldots i_p} = \mathcal{A}(v^{(1)} \otimes v^{(2)} \otimes \cdots \otimes v^{(p)})(e_{(i_1)} \ldots e_{(i_p)}) \ ,$$

$$\frac{1}{p!} \sum_{\tau \in G_p} \epsilon(\tau) v^{(1)} \otimes v^{(2)} \otimes \cdots \otimes v^{(p)} (e_{i_{\tau(1)}} \cdots e_{i_{\tau(p)}})$$

$$= \frac{1}{p!} \sum_{\tau \in G_p} \epsilon(\tau) v^{(1)}_{i_{\tau(1)}} \cdots v^{(1)}_{i_{\tau(p)}} \; ;$$

hence, using $\det(A) = \det(A^1, \ldots, A^n) = \sum_\sigma \epsilon(\sigma) a_{\sigma(1),1} \cdots a_{\sigma(n),n}$, we obtain

$$(v^{(1)} \wedge \cdots \wedge v^{(p)})_{i_1 \ldots i_p} = \frac{1}{p!} \det \begin{pmatrix} v^{(1)}_{i_1} & v^{(2)}_{i_1} & \cdots & v^{(p)}_{i_1} \\ \cdots & \cdots & \cdots & \cdots \\ v^{(1)}_{i_1} & v^{(2)}_{i_1} & \cdots & v^{(p)}_{i_1} \end{pmatrix} ,$$

whence it follows that p covariant vectors are dependent iff their exterior product vanishes. Moreover

$$v^{(1)} \wedge \cdots \wedge v^{(p)} = \epsilon(\sigma) v^{\sigma(1)} \wedge \cdots \wedge v^{\sigma(p)} . \tag{A.9}$$

Now consider the p-forms of the type

$$e^{(h_1)} \wedge e^{(h_2)} \wedge \cdots \wedge e^{(h_p)} , \qquad 1 \leq h_1 < h_2 < \cdots < h_p \leq n .$$

There are $\binom{n}{p}$ such forms obtained from all the possible combinations of $h_1, \ldots, h_p$. It can easily be shown that these p-forms are *linearly independent*, thus providing a basis for Λ^p. In fact consider any $\alpha \in \Lambda^p \subseteq T^{(0,p)}$, $p \leq n$. We can write

$$\alpha = \frac{1}{p!} \alpha_{i_1 i_2 \cdots i_p} e^{(i_1)} \otimes e^{(i_2)} \otimes \cdots \otimes e^{(i_p)}$$

(here, as usual, repeated indexes mean summation on them), $\frac{1}{p!} \alpha_{i_1 i_2 \cdots i_p}$ are the components of α with respect to the basis $\{e^{(i_1)} \otimes \cdots \otimes e^{(i_p)}\}$ of $T^{(0,p)}$. Now apply to both members the antisymmetrizer $\mathcal{A}$ (remember that $\mathcal{A}(\alpha) = \alpha$). Since $\mathcal{A}(e^{(i_1)} \otimes \cdots \otimes e^{(i_p)}) = e^{(i_1)} \wedge \cdots \wedge e^{(i_p)}$,

$$\alpha = \frac{1}{p!} \alpha_{i_1 i_2 \ldots i_p} e^{(i_1)} \otimes e^{(i_2)} \otimes \cdots \otimes e^{(i_p)}$$

$$= \sum_{h_1 < \cdots < h_p} \alpha_{h_1 h_2 \ldots h_p} e^{(h_1)} \wedge e^{(h_2)} \wedge \cdots \wedge e^{(h_p)} , \tag{A.10}$$

which now gives the development of α with respect to the basis $\{e^{(h_1)} \wedge e^{(h_2)} \wedge \cdots \wedge e^{(h_p)}\}$ of Λ^p; such a basis is called a *basis associated* with the basis $\{e_{(i)}\}$ of V_n. Clearly the dimension of Λ^p is $\binom{n}{p}$. The components $\alpha_{h_1 h_2 \cdots h_p}$ are called *restricted components* of α. From (A.9) and (A.10) we have

$$\alpha^q \wedge \beta^p = (-1)^{pq} \beta^p \wedge \alpha^q , \qquad \alpha \in \Lambda^q, \quad \beta \in \Lambda^p .$$

Thus $\alpha^p \wedge \alpha^p = 0$ if p is odd.

Let us now define

$$\Lambda = \bigoplus_{p=0}^{n} \Lambda^p$$

with $\Lambda^0 = K$ and $\Lambda^1 = V_n^*$. The elements of Λ are the $(n+1)$-tuples $(\alpha^0, \alpha^1, \ldots, \alpha^n)$ with $\alpha^p \in \Lambda^p$.

The sum $(\alpha^0, \alpha^1, \ldots, \alpha^n) + (\beta^0, \beta^1, \ldots, \beta^n) = (\alpha^0 + \beta^0, \alpha^1 + \beta^1, \ldots, \alpha^n + \beta^n)$ and the product by a coefficient c: $c(\alpha^0, \alpha^1, \ldots, \alpha^n) = (c\alpha^0, c\alpha^1, \ldots, c\alpha^n)$ provide Λ with the structure of a vector space. Its dimension is $\dim \Lambda = \sum_{k=0}^{n} \binom{n}{k} = 2^n$. Now the following product, à la Cauchy, can be defined over Λ:

$$(\alpha^0, \alpha^1, \ldots, \alpha^n) \wedge (\beta^0, \beta^1, \ldots, \beta^n)$$
$$= (\alpha^0 \wedge \beta^0, \alpha^0 \wedge \beta^1 + \alpha^1 \wedge \beta^0, \alpha^0 \wedge \beta^2 + \alpha^1 \wedge \beta^1 + \alpha^2 \wedge \beta^0, \ldots)$$

with respect to this product operation Λ is an algebra. It is the *Grassmann algebra* associated with V_n.

A.3 Differentiable Manifolds

A.3.1 Topological Spaces

A basic concept, preliminary to the definition of manifolds, is that of topological space. A topological space is a set $\mathcal{X}$ of arbitrary elements, the "points" of the space, where a concept of continuity is defined. To do this, one defines a collection $\mathcal{T} = \{U(x) | x \in \mathcal{X}\}$ of certain subsets $U(x)$ of the space that are associated with the points of the space as their neighborhoods. Depending on the set of axioms that these neighborhoods are required to obey, one gets different kinds of topological spaces $(\mathcal{X}, \mathcal{T})$. The most important of them are *Hausdorff topological spaces*, for which the neighborhoods must satisfy the following axioms:

1. To each point x there corresponds at least one neighborhood $U(x)$; each neighborhood $U(x)$ contains the point x.
2. If $U(x)$ and $V(x)$ are two neighborhoods of the same point x, then there exists a neighborhood $W(x)$ that is a subset of both.
3. If the point y lies in $U(x)$, there exists a neighborhood $U(y)$ that is a subset of $U(x)$.
4. For two distinct points x, y there exist two neighborhoods $U(x)$, $U(y)$ without common points.

Using neighborhoods we can introduce the concept of continuity: a mapping f of a topological space $\mathcal{X}$ onto a subset of a topological space $\mathcal{Y}$ is called *continuous* at the point x if for every neighborhood $U(y)$ of the point $y = f(x)$ one can find a neighborhood $U(x)$ of x such that all points of $U(x)$ are mapped into points of $U(y)$ by means of f. If f is continuous at every point of $\mathcal{X}$, it is called continuous in $\mathcal{X}$. A neighborhood is said to be *open* if every point $y \in U(x)$ has a neighborhood $V(y)$ entirely contained in $U(x)$, that is, $V(y) \subset U(x)$.

The main purpose of topology is to classify spaces through some equivalence relation. Two spaces $\mathcal{X}$ and $\mathcal{Y}$ are said to be topologically equivalent if they can be continuously deformed one into the other, i.e., if there exists a *homeomorphism* (a continuous invertible map with continuous inverse) φ that maps $\mathcal{X}$ into $\mathcal{Y} = \varphi(\mathcal{X})$. If there exists a homeomorphism between $\mathcal{X}$ and $\mathcal{Y}$, then $\mathcal{X}$ is said to be homeomorphic to ($\sim$) $\mathcal{Y}$. A homeomorphism establishes an equivalence relation, so that classifying topological spaces means to devise a method of finding the equivalence classes of homeomorphisms.

This can be achieved, at least in principle, by working out all the *topological invariants* of given classes of spaces, that is, finding all the quantities conserved under the action of homeomorphisms.

If the topological spaces are differentiable manifolds, one will require equivalence under diffeomorphisms instead of homeomorphisms. A diffeomorphism is a differentiable map with differentiable inverse.

A cube (or a polyhedron) in Euclidean 3-space is topologically equivalent to a sphere under a homeomorphism but not under any diffeomorphism. A cube (or a polyhedron) in Euclidean 3-space is not topologically equivalent to a torus, neither under homeomorphisms nor under diffeomorphisms. This can be seen by the fact that a cube (or a polyhedron) has no holes, while a torus has one hole.

A.3.2 Manifolds

An n-dimensional *topological manifold* M is a topological space such that to every point $x \in M$ is associated an open neighborhood $U(x)$ homeomorphic to $\mathbb{R}^n$. In other words, M *locally* resembles Euclidean space. Open neighborhoods are required in the definition to avoid restrictions on the global topology of M.

The local homeomorphism is realized by a one-to-one map φ that associates to every point $P \in U \subset M$ of an open neighborhood an n-tuple $(x^1(P), \dots, x^n(P))$, the *coordinates* of P under φ; thus φ associates to U an open set $\varphi(U) \subset \mathbb{R}^n$. The pair $\{(U, \varphi)\}$ is called a *chart*. It is easily realized that open neighborhoods must overlap if all the points of M are to belong at least to one neighborhood. If two neighborhoods U and V of the charts (U, φ) and (V, ψ) have in common a point P such that $\varphi(P) = (x^1(P), \dots, x^n(P))$ and $\psi(P) = (y^1(P), \dots, y^n(P))$ are the coordinates of P in the two charts, then there will be a functional relationship among these coordinates, that is, the composite map $\psi \circ \varphi^{-1}$ is a coordinate transformation

$$y^1 = y^1(x^1, \dots, x^n) \,,$$
$$y^2 = y^2(x^1, \dots, x^n) \,,$$
$$\cdots \quad \cdots$$
$$y^n = y^n(x^1, \dots, x^n) \,,$$

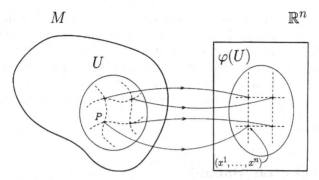

Fig. A.1. A neighborhood U in the manifold M is mapped onto $\varphi(U) \subset \mathbb{R}^n$ by the one-to-one mapping φ. This allows one to associate with any point $P \in U$ a unique set of numbers $(x^1, \ldots, x^n) \in \mathbb{R}^n$. Thus U is given a system of coordinates, as is pictorially shown by the dashed curves within U that are obtained through the action of φ^{-1} on the straight coordinate lines of $\varphi(U) \subset \mathbb{R}^n$.

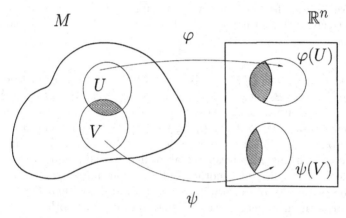

Fig. A.2. The partially overlapping neighborhoods U and V in the manifold M are mapped onto $\mathbb{R}^n$ by the maps φ and ψ, respectively. The overlap region is thus given two distinct coordinate systems by these two maps. The differentiability class of M depends of the relation between these coordinates.

which is required to be continuous and invertible on $U \cap V$. If the partial derivatives of all the functions $\{y^i\}$ with respect to the $\{x^i\}$ exist and are continuous up to the order k, then the charts (U, φ) and (V, ψ) are said to be $\mathcal{C}^k$-related. If one can construct a whole family of charts $\{(U_i, \varphi_i)\}$ such that each point of M belongs to at least one neighborhood of M, that is, $\bigcup_i U_i = M$, such that $\varphi_i(U_i) \subset \mathbb{R}^n$ is an open set, and every chart is $\mathcal{C}^k$-related with the charts with which it overlaps, then this family is called an *atlas* and the manifold is said to be a $\mathcal{C}^k$-manifold. A manifold of class $\mathcal{C}^1$ is called a *differentiable manifold* and a manifold of class $\mathcal{C}^\infty$ is called a *smooth manifold*.

If the union of two atlases is again an atlas, then these atlases are said to be compatible. This relation of compatibility defines an equivalence relation; the corresponding equivalence class is the *differentiable structure* of the manifold.

Endowing a topological manifold with these differentiable structures produces an enormous amount of structure, allowing one to define calculus on manifolds, as well as to define vector and tensor fields, differential forms, flows, Lie derivatives, and so on.

A.4 Calculus on Manifolds

Since every manifold is locally the same as some $\mathbb{R}^n$, we can extend to manifolds the usual calculus developed in $\mathbb{R}^n$. As a consequence of smoothness of the coordinate transformations, calculus on manifolds is independent of the choice of local coordinates.

Consider a map $f : M \to N$ from an m-dimensional manifold M to an n-dimensional manifold N, and let (U, φ) and (V, ψ) be two charts on M and N respectively, such that if $P \in U \subset M$ then $f(P) \in V \subset N$. If the coordinate expression

$$\psi \circ f \circ \varphi^{-1} : \mathbb{R}^m \to \mathbb{R}^n$$

is smooth, that is, $y = \psi f \varphi^{-1}(x)$ is a smooth vector-valued function defined on an open set of $\mathbb{R}^m$, then f is said to be *differentiable* at $P \in M$. This property is independent of the chosen coordinate systems (charts) in M and N. A *diffeomorphism* is a smooth mapping $f : M \to N$ having an inverse mapping that is also smooth. In this case we write $M \approx N$ and dim $M =$ dim N.

Homeomorphisms classify spaces according to the possibility of *continuously* deforming a space into another. Diffeomorphisms classify spaces according to the possibility of performing such a deformation *smoothly*. A smooth homeomorphism need not be a diffeomorphism; a smooth inverse is also

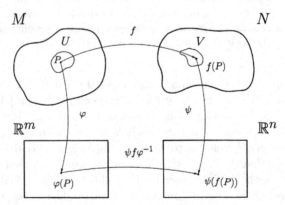

Fig. A.3. Constructing differentiable mappings between manifolds.

required. The set of diffeomorphisms $f : M \rightarrow M$, denoted by $\mathrm{Diff}(M)$, is a group.

A smooth mapping $f : M \rightarrow \mathbb{R}$ is a *function* on M. In a chart (U, φ), the coordinate presentation of f is given by $f \circ \varphi^{-1} : \mathbb{R}^m \rightarrow \mathbb{R}$, a real-valued function of m variables. The set of smooth functions on M is denoted by $\mathcal{F}(M)$.

A.4.1 Vectors

Vectors on a manifold M are defined through *tangent vectors* to a curve in M. Recall first that the tangent line to a plane curve in the xy plane

$$y - y(x_0) = \frac{dy}{dx}\bigg|_{x=x_0} (x - x_0) \tag{A.11}$$

approximates the curve, assumed differentiable, close to x_0. The tangent vectors on a manifold generalize the tangent line (A.11). So we define tangent vectors on M with the aid of a curve $c : (a, b) \rightarrow M$, with (a, b) a real open interval containing $t = 0$, and of a function $f : M \rightarrow \mathbb{R}$. The tangent vector at $c(0)$ is defined as the directional derivative of a function f along the curve $c(t)$ at $t = 0$, i.e., $f(c(t))$. This is given by

$$\frac{df(c(t))}{dt}\bigg|_{t=0} , \tag{A.12}$$

and in local coordinates by

$$\frac{\partial f}{\partial x^i} \frac{dx^i(c(t))}{dt}\bigg|_{t=0} , \tag{A.13}$$

which is equivalent to the application of the differential operator $X = X^i \frac{\partial}{\partial x^i}$, where $X^i = dx^i(c(t))/dt|_{t=0}$, to the function f:

$$\frac{df(c(t))}{dt}\bigg|_{t=0} = Xf \equiv X^i \frac{\partial f}{\partial x^i} .$$

We define as tangent vector at $p \in M$, $p = c(0)$, along the direction $c(t)$ by

$$X = X^i \frac{\partial}{\partial x^i} .$$

Let now p be in the chart (U, φ). Applying X to the coordinate functions $\varphi(c(t)) = x^i(t)$ we obtain

$$X[x^i] = \frac{dx^k}{dt} \frac{\partial x^i}{\partial x^k} = \frac{dx^i}{dt}\bigg|_{t=0} .$$

Since any two curves $c_1(t)$ and $c_2(t)$ such that $c_1(0) = c_2(0) = p \in M$ and $dx^i(c_1(t))/dt|_{t=0} = dx^i(c_2(t))/dt|_{t=0}$, yield the same tangent vector at p, we identify a tangent vector X with the equivalence class of curves at $p \in M$ with the just-remarked equivalence relation. All the tangent vectors at $p \in M$ form the *tangent space* T_pM. It is evident that $e_{(i)} = \partial/\partial x^i$, $i = 1, \ldots, n$, ($n = \dim M$), are the basis vectors of T_pM and $\{e_{(i)}\}$ the coordinate basis, so that a vector $V \in T_pM$ is written as $V = V^i e_{(i)}$. Now, T_pM is a vector space of dim $T_pM = \dim M$; thus there exists a dual vector space T_p^*M whose elements are linear functions from T_pM to $\mathbb{R}$. An element of the cotangent space T_p^*M, i.e., $\omega : T_pM \to \mathbb{R}$, is called a covector, or cotangent vector, or one-form. The differential df of a function $f \in \mathcal{F}(M)$ provides us with the simplest example of a one-form.

We know that a vector $V \in T_pM$ acts on f as $V[f] = V^i(\partial f/\partial x^i) \in \mathbb{R}$, whence the action of $df \in T_p^*M$ on $V \in T_pM$ can be defined as follows: notice that with respect to a chart (U, φ), df is expressed as $df = (\partial f/\partial x^i)dx^i$, in terms of $x = \varphi(p)$, and that an element of T_p^*M must act on an element of T_pM, yielding a real number. Thus put

$$\langle df, V \rangle := V[f] = V^i \frac{\partial f}{\partial x^i} \in \mathbb{R} \, ,$$

where $V[f]$ is the real number in question, and this definition is natural if we regard $\{dx^i\}$ as a basis for T_p^*M (in fact, from $df = (\partial f/\partial x^i)dx^i$ and $\langle df, V \rangle = V^1(\partial f/\partial x^1) + \cdots + V^n(\partial f/\partial x^n) \equiv V^i(\partial f/\partial x^i) = V[f]$). The basis $\{dx^i\}$ of T_p^*M is a dual basis; in fact,

$$\left\langle dx^i, \frac{\partial}{\partial x^k} \right\rangle = \frac{\partial x^i}{\partial x^k} = \delta_k^i \, .$$

Now, any arbitrary one-form ω will be expressed as $\omega = a_i dx^i$, where the coefficients a_i need not be partial derivatives of some function f.

The union of all the tangent spaces of the manifold M,

$$TM = \bigcup_{p \in M} T_pM \, , \tag{A.14}$$

is a manifold, called the *tangent bundle* of M; we have $\dim TM = 2 \dim M$. The dual space of TM, T^*M, is called the *cotangent bundle* of M.

Notice that both vectors and one-forms exist independently of any coordinate system; thus their coordinate independence must be expressed by suitable transformation properties of their components. Given two charts, let $p \in U_a \cap U_b$ and $x = \varphi_a(p)$, $y = \varphi_b(p)$. For a vector $V \in T_pM$ we have two equivalent representations

$$V = V^i \left(\partial/\partial x^i \right) = \tilde{V}^i \left(\partial/\partial y^i \right) \, ,$$

so that the components V^i and $\widetilde{V}^i$ are related by

$$\widetilde{V}^i = V^j \left(\partial y^i / \partial x^j \right) . \tag{A.15}$$

Similarly, for one-forms, from $dy^i = (\partial y^i / \partial x^j) dx^j$ we get

$$\widetilde{\omega}_i = \omega_j \left(\partial x^j / \partial y^i \right) . \tag{A.16}$$

A.4.2 Flows and Lie Derivatives

Consider a vector field X on an m-dimensional manifold M. A curve $c : I \subset \mathbb{R} \to M$ is an *integral curve* of X in M if it is tangent to X at all its points, that is, if $c'(t) = X_{c(t)}$, $\forall t \in I$. Using a chart (U, φ), with $X = X^\mu (\partial/\partial x^\mu)$ and with $\{x^\mu(t)\}$ for the components of $\varphi(c(t))$, this means that

$$\frac{dx^\mu}{dt} = X^\mu(x^1, \dots, x^m) , \tag{A.17}$$

which is a system of ordinary differential equations. Since the right hand side of (A.17) does not explicitly depend on the parameter t, we can intuitively think of X as giving the velocity of a steady state flow of a fluid through M. The existence and uniqueness theorem for ODE ensures that, at least locally, there is a unique solution of (A.17) with initial data $x_0^\mu = x^\mu(0)$.

A vector field is said to be *complete* if each of its integral curves is defined on the whole real line. Then we can proceed to assemble all the integral curves of a given complete vector field X into a single mapping.

We define the *flow* of a complete vector field X on M as the mapping $\sigma : \mathbb{R} \times M \to M$ given by

$$\sigma(p, t) = c_p(t)$$

for one-forms, where $c_p(t)$ is an integral curve starting at the point $p \in M$. In local coordinates we have

$$\frac{d\sigma^\mu(t, x_0)}{dt} = X^\mu(\sigma(t, x_0)) \tag{A.18}$$

with initial data $\sigma^\mu(t, x_0) = x_0^\mu$.

If $\sigma(t, x_0)$ is the flow of a complete vector field, that is, for any $x \in M$ $\sigma : \mathbb{R} \times M \to M$ is a differentiable mapping, then:

(i) $\sigma(t, 0)$ is the identity map of M.
(ii) $t \mapsto \sigma(t, x)$ is a solution of (A.18).
(iii) $\sigma(t, \sigma(s, x)) = \sigma(t + s, x)$ for any $t, s \in \mathbb{R}$.
(iv) the function $\sigma_t : M \to M$, which makes any point $x \in M$ flow for exactly time t, is a diffeomorphism of M with $\sigma_t^{-1} = \sigma_{-t}$.

Using the function σ_t above, condition (i) means that $\sigma_0 \equiv \mathrm{Id}(M)$, condition (iii) means $\sigma_t \circ \sigma_s = \sigma_{t+s}$, and these, together with (iv), make σ_t into a commutative group that is a *one-parameter group of diffeomorphisms*.

The infinitesimal version σ_ε of the transformation σ_t reads

$$\sigma_\varepsilon^\mu(x) = \sigma^\mu(\varepsilon, x) = x^\mu + \varepsilon X^\mu(x) \, ,$$

so that the vector field X is also called the *infinitesimal generator* of σ_t.

Given a complete vector field X, the corresponding flow σ is also denoted by

$$\sigma^\mu(t, x) = e^{tX} x^\mu \equiv e^{t(d/ds)} \sigma^\mu(s, x)\Big|_{s=0} \, .$$

Consider now two complete vector fields X and Y, and the flows generated by them, $\sigma(t, x)$ and $\tau(t, x)$ respectively. In order to evaluate how the vector field Y changes as a result of the dragging along the flow generated by the vector field X, we define the *Lie derivative* $\mathcal{L}_X Y$ of Y with respect to the flow of X as the vector field

$$\mathcal{L}_X Y = \lim_{\varepsilon \to 0} \frac{1}{\varepsilon} \Big[(\sigma_{-\varepsilon})_\star Y\big|_{\sigma_\varepsilon(x)} - Y\big|_x \Big] \, , \tag{A.19}$$

where the variation of Y passing from x to $x' = \sigma_\varepsilon(x)$ cannot be naively computed as $[Y\big|_{\sigma_\varepsilon(x)} - Y\big|_x]$ because $Y\big|_{\sigma_\varepsilon(x)} \in T_{\sigma_\varepsilon(x)} M$ and $Y\big|_x \in T_x M$, so that this difference is meaningless because the two vectors belong to different spaces. Thus we have to map $Y\big|_{\sigma_\varepsilon(x)}$ back to $T_x M$ by means of $(\sigma_{-\varepsilon})_\star :$ $T_{\sigma_\varepsilon(x)} M \to T_x M$. In loal coordinates, with $X = X^\mu(\partial/\partial x^\mu)$, $Y = Y^\mu(\partial/\partial x^\mu)$, and $\{e_\nu\} = \partial/\partial x^\nu$, one gets

$$\mathcal{L}_X Y = (X^\mu \partial_\mu Y^\nu - Y^\mu \partial_\mu X^\nu) e_\nu \, . \tag{A.20}$$

By defining the *Lie bracket* $[X, Y]$ as

$$[X, Y]f = X[Y[f]] - Y[X[f]]$$

for $f \in \mathcal{F}(M)$, we can see that $\mathcal{L}_X Y = [X, Y]$.

The Lie derivative of a function $f \in \mathcal{F}(M)$ along a flow generated by a vector field X is easily found to be $\mathcal{L}_X f = X[f] \equiv X^\mu(\partial f/\partial x^\mu)$, the usual directional derivative of f along X.

The Lie derivative of a one-form $\omega \in \Lambda^1(M)$ along X is found to be $\mathcal{L}_X \omega = (X^\nu \partial_\nu \omega_\mu + \omega_\nu \partial_\mu X^\nu) dx^\mu \in \Lambda^1(M)$, then, using the Leibniz rule

$$\mathcal{L}_X(Y \otimes \omega) = Y \otimes (\mathcal{L}_X \omega) + (\mathcal{L}_X Y) \otimes \omega$$

we can extend the Lie derivative to more general cases.

A.4.3 Tensors and Forms on Manifolds

Similarly to what has been defined for vector spaces, we can now construct a tensor of (r, q)-type at the point p of the manifold M by taking r factors of $T_p M$ and q factors of $T_p^* M$. Hence $T_p^{(r,q)}$, the space of (r, q)-type tensors, is the space whose elements $\mathbf{t}$ are

$$\mathbf{t} : \overset{r}{\bigotimes} T_p M \overset{q}{\bigotimes} T_p^* M \to \mathbb{R} \, ,$$

written, in the bases given above, as

$$\mathbf{t} = t^{i_1 \cdots i_r}_{j_1 \cdots j_q} \frac{\partial}{\partial x^{i_1}} \cdots \frac{\partial}{\partial x^{i_r}} dx^{j_1} \cdots dx^{j_q} \ .$$

The action of this tensor on r one-forms (covectors) and q vectors results in the real number

$$\mathbf{t}(\omega_1, \ldots, \omega_r; V_1, \ldots, V_q) = t^{i_1 \cdots i_r}_{j_1 \cdots j_q} \omega_{1 i_1} \cdots \omega_{r i_r} V_1^{j_1} \cdots V_q^{j_q} \ .$$

A smooth map that associates a tensor to each point $p \in M$ defines a *tensor field* on M.

The subspace Λ_p^q of $T_p^{(0,q)}$ of completely antisymmetric tensors is the space of *differential q-forms* tangent at p to M. Thus if $\omega \in \Lambda_p^q$, we have

$$\omega = \frac{1}{q!} \, \omega_{i_1 \cdots i_q} dx^{i_1} \wedge \cdots \wedge dx^{i_q}. \tag{A.21}$$

Let us give some useful definitions:

Given a vector $V \in T_p M$ and a one-form $\omega \in T_p^* M$, their *inner product* $\langle \cdot, \cdot \rangle : T_p^* M \otimes T_p M \to \mathbb{R}$ is defined by

$$\langle \omega, V \rangle = \omega_i V^k \langle dx^i, \frac{\partial}{\partial x^k} \rangle = \omega_i V^k \delta_k^i = \omega_i V^i \ .$$

Notice that the inner product is not defined between two vectors or two one-forms.

A smooth map $f : M \to N$ naturally induces the differential map f_* according to the definition

$$f_* : T_p M \to T_{f(p)} N \ ,$$

whose explicit form uses the definition of a tangent vector as follows. If $gf \in \mathcal{F}(M)$, with $g \in \mathcal{F}(N)$ ($\mathcal{F}(N)$ is the set of smooth real functions on N), then $V \in T_p M$ acts on gf to give a real number $V[gf]$, so we define $f_* V \in T_{f(p)} N$ by

$$(f_* V)[g] = V[gf]$$

in local charts (U, φ) on M and $(\mathcal{U}, \psi)$ on N:

$$(f_* V)[g\psi^{-1}(y)] \equiv V[gf\varphi^{-1}(x)]$$

with $x = \varphi(p)$ and $y = \psi(f(p))$. Now, put $V = V^i \frac{\partial}{\partial x^i}$ and $f_* V = W^j \frac{\partial}{\partial y^j}$. From the equation above we obtain

$$W^j \frac{\partial}{\partial y^j} [g\psi^{-1}(y)] = V^i \frac{\partial}{\partial x^i} [gf\varphi^{-1}(x)] \ .$$

Putting $g = y^j$ we get

$$W^j = V^i \frac{\partial y^j(x)}{\partial x^i} \ .$$

Obviously the matrix $(\partial y^j / \partial x^i)$ is the Jacobian of $f : M \to N$. The differential map f_* is naturally extended to contravariant tensors

$$f_* : T_p^{(r,0)}(M) \to T_{f(p)}^{(r,0)}(N) \ . \tag{A.22}$$

A map $f : M \to N$ also induces a map f^* such that

$$f^* : T_{f(p)}^* N \to T_p M \ . \tag{A.23}$$

Since f^* goes "backward" with respect to f, it is called a *pullback*. If we consider $\omega \in T_{f(p)}^* N$ and $V \in T_p M$, the pullback of ω by f^* is defined by

$$\langle f^* \omega, V \rangle = \langle \omega, f_* V \rangle \ .$$

Then the pullback f^* is naturally extended to covariant tensors

$$f^* : T_{f(p)}^{(0,q)}(N) \to T_p^{(0,q)}(M) \ . \tag{A.24}$$

If $\omega = \omega_i dy^i \in T_{f(p)}^* N$, the induced form $f^* \omega = \eta_k dx^k \in T_p M$ is

$$\eta_k = \omega_i \frac{\partial y^i}{\partial x^k} \ , \tag{A.25}$$

where again, $(\partial y^j / \partial x^i)$ is the Jacobian matrix J.

Remark 1: $(gf)_* = g_* f_*$ and $(gf)^* = g^* f^*$.

Remark 2: The extension of $f : M \to N$ to $T^{(r,q)}$ is possible only if f is a diffeomorphism, so that J (the Jacobian) is defined also for f^{-1}.

In this case—consider for example $T^{(1,1)}$—we have

$$f_* \left(t_k^i \frac{\partial}{\partial x^i} \otimes dx^k \right) = t_k^i \left(\frac{\partial y^j}{\partial x^i} \right) \left(\frac{\partial x^k}{\partial y^l} \right) \frac{\partial}{\partial y^j} \otimes dy^l \ . \tag{A.26}$$

Let us now come back to (A.21) and to the vector space $\Lambda_p^q(M)$ of differential q-forms at $p \in M$. By strictly following what we already did in the previous section, we can define the *Grassmann algebra* of M. Let us take $\Lambda_p^0(M) = \mathcal{F}(M)$, $\Lambda_p^1(M) = T_p^* M$, $\Lambda_p^2(M), \ldots, \Lambda_p^n(M)$, $\Lambda_p^{n+1}(M) = \{0\}$, $\Lambda_p^r(M) = \{0\}$ for $r \geq n+1$; $n = \dim M$, and form the space of all differential forms at p,

$$\Lambda_p^*(M) = \Lambda_p^0(M) \oplus \Lambda_p^1(M) \oplus \cdots \oplus \Lambda_p^n(M) = \bigoplus_{i=0}^n \Lambda_p^i(M) \ , \tag{A.27}$$

to yield the Grassmann algebra of M once it is equipped with the *exterior product* $\wedge$ of differential forms, that is, $\wedge : \Lambda_p^q(M) \times \Lambda_p^r(M) \to \Lambda_p^{q+r}(M)$.

The space $\Lambda_p^*(M)$ is closed under the exterior product which acts, let us recall, "à la Cauchy":

$$\omega, \xi \in \Lambda_p^*(M), \quad \omega = (\omega^0, \omega^1, \omega^2, \ldots), \quad \xi = (\xi^0, \xi^1, \xi^2, \ldots) ,$$

$$(\omega^0, \omega^1, \omega^2, \ldots) \wedge (\xi^0, \xi^1, \xi^2, \ldots)$$
$$= (\omega^0 \wedge \xi^0, \omega^0 \wedge \xi^1 + \omega^1 \wedge \xi^0, \omega^0 \wedge \xi^2 + \omega^1 \wedge \xi^1 + \omega^2 \wedge \xi^0, \ldots). \quad (A.28)$$

Then we can smoothly assign to each point of a manifold M an r-form. Hence, from each vector space $\Lambda_p^r(M)$ we generalize to the space $\Lambda^r(M)$ of r-differential forms on M. We have

space	basis	dimension
$\Lambda^0(M) = \mathcal{F}(M)$	$\{1\}$	$\dim \Lambda^0(M) = 1$
$\Lambda^1(M) = T^*M$	$\{dx^k\}$	$\dim \Lambda^1(M) = n$
$\Lambda^2(M)$	$\{dx^{i_1} \wedge dx^{i_2}\}$	$\dim \Lambda^2(M) = \frac{1}{2}n(n-1)$
$\Lambda^3(M)$	$\{dx^{i_1} \wedge dx^{i_2} \wedge dx^{i_3}\}$	$\dim \Lambda^3(M) = \frac{1}{6}n(n-1)(n-2)$
$\vdots$	$\vdots$	$\vdots$
$\Lambda^n(M)$	$\{dx^1 \wedge dx^2 \wedge \ldots \wedge dx^n\}$	$\dim \Lambda^n(M) = 1$

For example, we have

$$dx^i \wedge dx^k = dx^i \otimes dx^k - dx^k \otimes dx^i ,$$
$$dx^i \wedge dx^k \wedge dx^l = \frac{1}{2}(dx^i \otimes dx^k \otimes dx^l + dx^l \otimes dx^i \otimes dx^k)$$
$$+ \frac{1}{6}(dx^k \otimes dx^l \otimes dx^i - dx^i \otimes dx^l \otimes dx^k$$
$$- dx^l \otimes dx^k \otimes dx^i - dx^k \otimes dx^i \otimes dx^l) . \quad (A.29)$$

Obviously $\Lambda_p^*(M)$ generalizes to $\Lambda^*(M)$ on the whole manifold. Each space $\Lambda^p(M)$, $p = 1, \ldots, n$, is also called a homogeneous component of degree p of $\Lambda^*(M)$. Its elements are $\omega^p = (0, 0, \ldots, 0, \omega^p, 0, \ldots)$, and any differential form $\omega \in \Lambda^*(M)$ can be written as sum of homogeneous forms as $\omega = \sum_{p=0}^n \omega^p$. If $\xi = \sum_{q=0}^n \xi^q$ we have (see (A.28))

$$\omega \wedge \xi = \sum_{p=0}^n \sum_{q=0}^n \omega^p \wedge \xi^q .$$

The map $\tilde{\ } : \omega \in \Lambda^*(M) \to \tilde{\omega} \in \Lambda^*(M)$, such that $\tilde{\omega} = \sum_{p=0}^n (-1)^p \omega^p$, is a bijective and involutory (i.e., $\tilde{\tilde{\omega}} = \omega$, $\forall \omega \in \Lambda^*$) endomorphism of $\Lambda^*(M)$.

We have already seen that to any differentiable map between two manifolds, $f : M \to N$, we can associate its pullback f^* that acts between the spaces $T_p^{(0,q)}(M)$ and $T_{f(p)}^{(0,q)}(N)$. Thus f^* associates to a q-form at $f(P) \in N$, $\omega^q \in \Lambda_{f(P)}^q(N)$, a q-form $f^*\omega^q$ tangent at P to M. One can easily verify that

$$f^*(a\omega^r + b\xi^q) = af^*\omega^r + bf^*\xi^q , \qquad a, b \in \mathbb{R} ,$$
$$f^*(\omega^r \wedge \xi^q) = f^*\omega^r \wedge f^*\xi^q . \tag{A.30}$$

These "pointwise" properties can be extended to the whole manifold, so that $f^* : \Lambda^q(N) \subseteq T^{(0,q)}(N) \to \Lambda^q(M) \subseteq T^{(0,q)}(M)$ has the properties (A.30). This mapping also defines a *homomorphism* between the two subalgebras $\Lambda^q(N)$ and $\Lambda^q(M)$ of $\Lambda^*(N)$ and $\Lambda^*(M)$ respectively; this homomorphism maps q-forms to forms of the same degree; hence f^* associates to any $\omega = \sum_{q=0}^{N} \omega^q \in \Lambda^*(N)$ the form $f^*\omega = \sum_{q=0}^{N} f^*\omega^q \in \Lambda^*(M)$. Summarizing, to any differentiable manifold M we can associate its Grassmann algebra $\Lambda^*(M)$, and to any differentiable map $f : M \to N$ the homomorphism $f^* : \Lambda^*(N) \to \Lambda^*(M)$. If f is a *diffeomorphism* between M and N, then their corresponding Grassmann algebras are *isomorphic*.

A.4.4 Exterior Derivatives

The *exterior derivative* d_r of an r-form is a mapping $d_r : \Lambda^r(M) \to \Lambda^{r+1}(M)$ that is defined by

$$\omega = \frac{1}{r!}\omega_{\mu_1\cdots\mu_r}dx^{\mu_1}\wedge\cdots\wedge dx^{\mu_r} \to d_r\omega = \frac{1}{r!}\left(\frac{\partial\omega_{\mu_1\cdots\mu_r}}{\partial x^\nu}\right)dx^\nu\wedge dx^{\mu_1}\wedge\cdots\wedge dx^{\mu_r} . \tag{A.31}$$

The exterior derivative has the following fundamental property:

$$d_{r+1}d_r = 0 \tag{A.32}$$

or, dropping the suffix, $d^2 = 0$.

In fact, considering $\omega = \frac{1}{r!}\omega_{\mu_1\cdots\mu_r}dx^{\mu_1} \wedge \cdots \wedge dx^{\mu_r} , \in \Lambda^r(M)$, we have

$$d^2\omega = \frac{1}{r!}\left(\frac{\partial^2\omega_{\mu_1\cdots\mu_r}}{\partial x^\lambda\partial x^\nu}\right)dx^\lambda \wedge dx^\nu \wedge dx^{\mu_1} \wedge \cdots \wedge dx^{\mu_r} , \tag{A.33}$$

which is identically zero because $(\partial^2\omega_{\mu_1\cdots\mu_r}/\partial x^\lambda\partial x^\nu)$ is symmetric with respect to λ and ν, whereas $dx^\lambda \wedge dx^\nu$ is antisymmetric.

An example can be useful. In three dimensional-space all the r-forms are of the following kind:

$$\omega^0 = f(x, y, z) ,$$
$$\omega^1 = \omega_x(x, y, z)dx + \omega_y(x, y, z)dy + \omega_z(x, y, z)dz ,$$
$$\omega^2 = \omega_{xy}(x, y, z)dx \wedge dy + \omega_{yz}(x, y, z)dy \wedge dz + \omega_{zx}(x, y, z)dz \wedge dx ,$$
$$\omega^3 = \omega_{xyz}(x, y, z)dx \wedge dy \wedge dz , \tag{A.34}$$

and the action of the exterior derivative turns out to be quite familiar, in fact

$$d\omega^0 = (\partial_x f)dx + (\partial_y f)dy + (\partial_z f)dz \ ,$$
$$d\omega^1 = (\partial_x \omega_y - \partial_y \omega_x)dx \wedge dy + (\partial_y \omega_z - \partial_z \omega_y)dy \wedge dz + (\partial_z \omega_x - \partial_x \omega_z)dz \wedge dx$$
$$d\omega^2 = (\partial_x \omega_{yz} + \partial_y \omega_{zx} + \partial_z \omega_{xy})dx \wedge dy \wedge dz$$
$$d\omega^3 = 0 \ , \tag{A.35}$$

where we can recognize the actions of the basic operators of vector calculus: $d\omega^0 \equiv \mathrm{grad}(f)$, $d\omega^1 \equiv \mathrm{rot}(\omega^1)$ and $d\omega^2 \equiv \mathrm{div}(\omega^2)$; moreover, the elementary properties that $\mathrm{rot}[\mathrm{grad}(f)] \equiv 0$ and $\mathrm{div}[\mathrm{rot}(\mathbf{v})] \equiv 0$ exemplify the meaning of $d^2\omega = 0$ in this particular case.

de Rham Cohomology Groups

An r-form $\omega \in \Lambda^r(M)$ is said to be *closed* if its exterior derivative vanishes, $d_r\omega = 0$, and it is said to be *exact* if it is the derivative of an $(r-1)$-form $\psi \in \Lambda^{r-1}(M)$, i.e., $\omega = d\psi$. A closed r-form $\omega \in \Lambda^r(M)$ belongs to $\mathrm{Ker}(d_r)$, while an exact r-form $\omega \in \Lambda^r(M)$ belongs to $\mathrm{Im}(d_{r-1})$.

Definition. The rth (real) de Rham cohomology group of a manifold M, $H^r(M)$, is the set of the equivalence classes of the closed r-differential forms on M that differ only by exact r-forms, that is, the quotient group

$$H^r(M) = \mathrm{Ker}(d_r)/\mathrm{Im}(d_{r-1}) \tag{A.36}$$

and

$$[\omega] \in H^r(M) \Leftrightarrow [\omega] = \{\omega' \in \mathrm{Ker}(d_r)|\omega' = \omega + d\psi, \ \psi \in \Lambda^{r-1}(M)\} \ .$$

Given an n-dimensional manifold, one can thus define n such groups $H^r(M)$ for $r = 0, \ldots, n$.

A.4.5 Interior Product

Another important operation with forms is the *interior product* of a form ω and a vector X, denoted by $i_X\omega$, or $X \lrcorner \ \omega$, or also $i(X)\omega$. Let $\omega \in \Lambda^r(M)$ be an r-form and $X \in \mathcal{X}(M)$ a vector field on M. The interior product is a mapping $i_X : \Lambda^r(M) \to \Lambda^{r-1}(M)$ such that, using $X = X^\mu \partial/\partial x^\mu$ and $\omega = \frac{1}{r!}\omega_{\mu_1 \cdots \mu_r}dx^{\mu_1} \wedge \cdots \wedge dx^{\mu_r}$,

$$i_X(\omega) = \frac{1}{(r-1)!}X^\nu \omega_{\nu\mu_2 \cdots \mu_r}dx^{\mu_2} \wedge \cdots \wedge dx^{\mu_r}$$
$$= \frac{1}{r!}\sum_{s=1}^{r} X^{\mu_s}\omega_{\mu_1 \cdots \mu_s \cdots \mu_r}(-1)^{s-1}dx^{\mu_1} \wedge \cdots \wedge \widehat{dx}^{\mu_s} \wedge \cdots \wedge dx^{\mu_r} \ ,$$

where $\widehat{dx}^{\mu_s}$ means that this term is omitted. The interior product has, among others, the following properties: $i_X(\omega \wedge \eta) = i_X\omega \wedge \eta + (-1)^r\omega \wedge i_X\eta$ (antiderivation); $i_X f = 0$; $i_X^2 = 0$ (nilpotency); $di_X + i_X d = \mathcal{L}_X$ the Lie derivative with respect to X (this important formula is due to Cartan).

To give again an example in $\mathbb{R}^3$, with coordinates (x, y, z), one easily checks that

$$i_{e_x}(dx \wedge dy) = dy, \qquad i_{e_x}(dy \wedge dz) = 0, \qquad i_{e_x}(dz \wedge dx) = -dz .$$

A.4.6 Integration of Forms on Manifolds

Differential forms can be integrated only on *orientable* manifolds. Thus, let us define orientability. Consider a connected m-dimensional manifold M and a point $p \in M$. Then consider two charts (U_i, φ_i) and (U_j, φ_j) such that $p \in U_i \cap U_j$. Let x^μ be the local coordinates on U_i and y^α the local coordinates on U_j. Then let $\{e_\mu\} = \{\partial/\partial x^\mu\}$ and $\{\bar{e}_\alpha\} = \{\partial/\partial y^\alpha\}$ be the corresponding bases on T_pM. We have the relation

$$\bar{e}_\alpha = \frac{\partial x^\mu}{\partial y^\alpha}e_\mu .$$

If $J = \det(\partial x^\mu/\partial y^\alpha e_\mu) > 0$ on the set $U_i \cap U_j$, then the two bases are said to define the same *orientation* on $U_i \cap U_j$; if $J < 0$ then they define an opposite orientation. Now, consider a family of sets $\{U_i\}$ that covers M, i.e., $\bigcup_i U_i = M$. If for any two sets U_i, U_j such that $U_i \cap U_j \neq \emptyset$ the local coordinates $\{x^\mu\}$ and $\{y^\alpha\}$ respectively are such that $J > 0$, then the manifold M is said to be *orientable*.

Suppose we want to compute the integral $\int_M f\omega$, where $f : M \to \mathbb{R}$ is a function, and ω is a volume form, that is, an m-form $\omega = h(p)dx^1 \wedge \cdots \wedge dx^m$. By resorting to an open covering $\{U_i\}$ of M such that any point $p \in M$ is covered by a finite number of U_i, we define

$$\int_{U_i} f\omega \equiv \int_{\varphi_i(U_i)} f(\varphi_i^{-1}(x))h(\varphi_i^{-1}(x))dx^1 \cdots dx^m \tag{A.37}$$

and then we define a *partition of unity*, subordinate to the covering $\{U_i\}$, as a family of differentiable functions $\{\varepsilon_i(p)\}$ such that (i) $0 \leq \varepsilon_i(p) \leq 1$; (ii) $\varepsilon_i(p) = 0$ if $p \notin U_i$; (iii) $\sum_k \varepsilon_k(p) = 1$, $\forall p \in M$. We readily see that

$$f(p) = \sum_i f(p)\varepsilon_i(p) := \sum_i f_i(p) , \qquad f_i(p) \neq 0 \text{ iff } p \in U_i .$$

Using (A.37) to define the integral of f_i on U_i, we have

$$\int_M f\omega = \sum_i \int_{U_i} f_i\omega . \tag{A.38}$$

This integral is independent of the atlas chosen on M.

A.5 The Fundamental Group

Let us now sketchily review some basic facts about topology.

The idea of classifying all topological spaces is hopeless. This notwithstanding, some strategy to decide whether two spaces are homeomorphic can be developed. The next section will be devoted to the study of the conditions that two spaces have to satisfy to be diffeomorphic.

An obvious way to prove that, for example, two given spaces are homeomorphic is to construct a homeomorphism between them. The converse, that is proving that two spaces are not homeomorphic, is much worse from a constructive viewpoint because we cannot check all the possible functions acting between two spaces and then prove that none of them is a homeomorphism. A successful way out of this difficulty is to look for *topological invariants*, which can be very different things, such as a geometrical property of the space, a number such as the Euler characteristic of the space, an algebraic stucture, such as a group or a vector space constructed from the topological space under study. The requirement is that a given invariant has to be preserved by homeomorphisms. Then, by computing the same topological invariant for two given spaces, if one gets different results, this means that the two spaces are not homeomorphic. Sometimes it may happen that the chosen topological invariant is not able to discriminate between topologically equivalent spaces, so that some refinement is required.

An algebraic way of investigating homeomorphicity between spaces is due to Poincaré, who devised a method to associate with each topological space an invariant group such that *homeomorphic spaces have isomorphic groups*. This amounts to constructing the so-called *fundamental* or *first homotopy group* associated with a topological space $\mathcal{X}$. This has to do with multiple connectedness of a topological space.[1,2]

To begin with an intuitive example, consider two spaces, $\mathcal{X}$ and $\mathcal{Y}$, as shown in Figure A.4. The space $\mathcal{X}$ has a hole; thus it is not simply connected. The presence of the hole makes us think that the two spaces are not homeomorphic. A simple way of detecting the presence of the hole is to fix a point in the space and then to consider all the loops starting and ending at this point. If a loop encircles the hole it cannot be continuously shrunk to a point.

---

[1] If, given a topological space $\mathcal{X}$, there exist two nonempty, open, and disjoint subsets and $B \subset \mathcal{X}$ such that $A \cup B = \mathcal{X}$, then $\mathcal{X}$ is *disconnected*. $\mathcal{X}$ is locally connected if any neighborhood of any point $x \in \mathcal{X}$ contains a connected neighborhood. The space $\mathcal{X}$ is *simply connected* if it is connected and locally connected and if every covering space $(\widetilde{\mathcal{X}}, f)$ is isomorphic to $(\mathcal{X}, \mathrm{Id})$, where Id is the identity mapping. The covering space $(\widetilde{\mathcal{X}}, f)$ is a *universal covering space* for $\mathcal{X}$ if it is simply connected.

[2] A covering space of a topological space $\mathcal{X}$ is a pair $(\widetilde{\mathcal{X}}, f)$ with $f : \widetilde{X} \to \mathcal{X}$ continuous onto, and such that for any $x \in \mathcal{X}$ there is a neighborhood $I(x)$ such that $f_{|C_\alpha} : C_\alpha \to I(x)$ is a homeomorphism onto, C_α being a connected component of the preimage $f^{-1}(I(x))$.

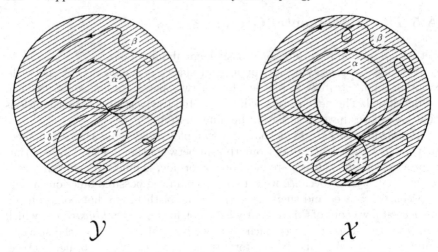

$\mathcal{Y}$ $\mathcal{X}$

Fig. A.4. In a disc (left) any loop can be continuously shrunk to a point. In presence of a hole (right) the loops embracing it cannot be shrunk to a point.

Thus in the space $\mathcal{X}$ we have two different kinds of loops: those that can be shrunk to a point and those that cannot. In contrast, in the space $\mathcal{Y}$ all the possible loops can be shrunk to a point. Poincaré's construction makes use of loops to build the fundamental group to be used as a detector of the presence of holes in a space. More precisely, since different loops are equivalently able to detect the presence of the same hole in the space, one is led to build the fundamental group using equivalence classes of loops. This requires an equivalence relation.

Two loops are equivalent if they are *homotopic*, that is, if they can be continuously deformed one into the other. Homotopic loops form a homotopy class. The space $\mathcal{Y}$ has only one such class. However, the space $\mathcal{X}$ has a homotopy class for each integer n equal to the number of encirclements of the hole by the loops. This makes a difference between the two spaces. Hence we understand that working with these homotopy classes should allow to study some fundamental topological properties of spaces. These homotopy classes can be given a group structure, and this is the first homotopy group $\pi_1(\mathcal{X})$.

Let us now give a few basic definitions.

By a *path* in a space $\mathcal{X}$ going from x_0 to x_1 we mean a continuous map $\alpha : [0, 1] \to \mathcal{X}$ such that $\alpha(0) = x_0$, $\alpha(1) = x_1$.

A topological space $\mathcal{X}$ is said to be *arcwise connected* or *path connected* if for all $x_0, x_1 \in \mathcal{X}$ there is a path α joining them.

By a *loop* in the space $\mathcal{X}$ we mean a path α such that $\alpha(0) = \alpha(1) = x_0$. Such a loop is said to be *based* at the point x_0.

If α and β are two loops based at x_0, one defines the *product* $\alpha \star \beta$ to be the loop given by the formula

$$(\alpha \star \beta)(s) = \begin{cases} \alpha(2s), & 0 \le s \le 1/2 \,, \\ \beta(2s - 1), & 1/2 \le s \le 1 \,; \end{cases} \tag{A.39}$$

$\alpha \star \beta$ is continuous and maps $[0, \frac{1}{2}]$ onto the image of α in $\mathcal{X}$, and maps $[\frac{1}{2}, 1]$ onto the image of β in $\mathcal{X}$.

The *inverse* loop α^{-1} based at $x_0 \in \mathcal{X}$ is defined as

$$\alpha^{-1}(s) = \alpha(1 - s), \qquad 0 \le s \le 1 \,.$$

The *constant* loop c based at $x_0 \in \mathcal{X}$ is a map $c : [0, 1] \to \mathcal{X}$ such that $c(s) = x_0, 0 \le s \le 1$.

However, this multiplication does not give a group structure on the set of loops based at a given point. To overcome this difficulty and to obtain a group, we identify two loops if they can be continuously deformed one into the other, keeping the base point fixed.

If $f, g : \mathcal{X} \to \mathcal{Y}$ are maps, we say that f is *homotopic* to g if there exists a map $F : \mathcal{X} \times I \to \mathcal{Y}$, $I = [0, 1]$, such that $F(x, 0) = f(x)$ and $F(x, 1) = g(x)$ for all points $x \in \mathcal{X}$. The map F is called a *homotopy* from f to g.

Two loops α and β based at x_0 are homotopic, $\alpha \simeq \beta$, if there exists a continuous map

$$H : [0, 1] \times [0, 1] \to \mathcal{X}$$

that satisfies the following conditions:

$$\begin{aligned} H(s, 0) &= \alpha(s), & 0 \le s \le 1 \,, \\ H(s, 1) &= \beta(s), & 0 \le s \le 1 \,, \\ H(0, t) &= H(1, t) = x_0, & 0 \le t \le 1 \,. \end{aligned}$$

The map H is called a homotopy between α and β.

If $\alpha_0, \beta_0, \gamma_0, \alpha_1, \beta_1, \gamma_1, \dots$ are loops based at $x_0 \in \mathcal{X}$ then the following relations hold:

(a) $\alpha_0 \simeq \alpha_0$;
(b) $\alpha_0 \simeq \beta_0 \Rightarrow \beta_0 \simeq \alpha_0$;
(c) $\alpha_0 \simeq \beta_0, \beta_0 \simeq \gamma_0 \Rightarrow \alpha_0 \simeq \gamma_0$;
(d) $\alpha_0 \simeq \alpha_1 \Rightarrow \alpha_0^{-1} \simeq \alpha_1^{-1}$;
(e) $\alpha_0 \simeq \beta_0, \beta_0 \simeq \beta_1 \Rightarrow \alpha_0 \star \beta_0 \simeq \alpha_1 \star \beta_1$.

These properties imply that homotopy is an equivalence relation. Let us denote by $[\alpha]$ an equivalence class of loops (all the loops in $[\alpha]$ are mutually homotopic), so that the set of loops can be partitioned into disjoint equivalent classes. We denote by $\pi_1(\mathcal{X}, x_0)$ the set of homotopy classes of loops based at x_0.

Given any two equivalence classes $[\alpha], [\beta] \in \pi_1(\mathcal{X}, x_0)$, their product is defined as

$$[\alpha] \circ [\beta] = [\alpha \star \beta] \ .$$

In view of the properties $(a) - (e)$ listed above, this product is well defined and provides $\pi_1(\mathcal{X}, x_0)$ with a group structure, whose unit element is the homotopy class $[c]$ of the constant loop based at x_0.

$\pi_1(\mathcal{X}, x_0)$ is the *first homotopy group*, or fundamental group, of the arcwise connected topological space $\mathcal{X}$.

An important property of $\pi_1(\mathcal{X}, x_0)$ so defined is that it does not depend on the base point x_0, in the sense that if $x_0, x_1 \in \mathcal{X}$ and $\mathcal{X}$ is path-connected, then $\pi_1(\mathcal{X}, x_0)$ can be proved to be isomorphic to $\pi_1(\mathcal{X}, x_1)$.

Moreover, the fundamental group can be proved to be a *topological invariant* of a path-connected space $\mathcal{X}$. In fact, the following theorems can be proved:

Theorem. *Two path-connected spaces $\mathcal{X}$ and $\mathcal{Y}$ are said to be of the same homotopy type if there are continuous maps f and g such that $f : \mathcal{Y} \to \mathcal{X}$, $g : \mathcal{X} \to \mathcal{Y}$, and $f \circ g$ is not necessarily $\mathrm{Id}_{\mathcal{Y}}$ but just homotopic to it. If $\mathcal{X}$ and $\mathcal{Y}$ are of the same homotopy type and $x_0 \in \mathcal{X}$, $y_0 \in \mathcal{Y}$, then $\pi_1(\mathcal{X}, x_0)$ is isomorphic to $\pi_1(\mathcal{Y}, y_0)$.*

Theorem. *Given $\mathcal{X}$ and $\mathcal{Y}$ path-connected topological spaces, if $\mathcal{X}$ is homeomorphic to $\mathcal{Y}$, then $\pi_1(\mathcal{X}, x_0)$ is isomorphic to $\pi_1(\mathcal{Y}, y_0)$.*

A subset $R \subset \mathcal{X}$ of a topological space is said to be a *retract* of $\mathcal{X}$ if there exists a continuous map, which is called a *retraction*,

$$f : \mathcal{X} \to R$$

such that $f(r) = r$ for any $r \in R$. A subset R of a topological space $\mathcal{X}$ is said to be a *deformation retract* of $\mathcal{X}$ if a retraction exists together with a homotopy $H : \mathcal{X} \times [0, 1] \to \mathcal{X}$ for which

$$
\begin{aligned}
H(x, 0) &= x \ , \\
H(s, 1) &= f(x) \ , \\
H(r, t) &= r, \qquad r \in R, \quad 0 \le t \le 1 \ .
\end{aligned}
$$

The relevance of deformation retracts is due to the fact that if R is a deformation retract of $\mathcal{X}$, then $\pi_1(\mathcal{X}, r)$ is isomorphic to $\pi_1(R, r)$, $r \in R$.

We just mention an interesting result that provides a tool to calculate fundamental groups: *if $\mathcal{X}$ and $\mathcal{Y}$ are path-connected spaces, then*

$$\pi_1(\mathcal{X} \times \mathcal{Y}, (x_0, y_0)) \quad \text{is isomorphic to} \quad \pi_1(\mathcal{X}, x_0) \oplus \pi_1(\mathcal{Y}, y_0) \ ,$$

where $\oplus$ denotes as the product of the two groups. For example, having proved that $\pi_1(\mathbb{S}^1) \cong \mathbb{Z}$, for the torus $\mathbb{T}^2 = \mathbb{S}^1 \times \mathbb{S}^1$ we immediately find that $\pi_1(\mathbb{T}^2) \cong \mathbb{Z} \oplus \mathbb{Z}$.

Finally, we just mention that in many cases $\pi_1(|K|)$, where $|K|$ is a polyhedron, is the same as the first homology group $H_1(|K|)$ of the corresponding simplicial complex K (see the next section). In particular, for a connected topological space $\mathcal{X}$, $\pi_1(\mathcal{X})$ is isomorphic to $H_1(\mathcal{X})$ iff $\pi_1(\mathcal{X})$ is commutative. Moreover, if two spaces $\mathcal{X}$ and $\mathcal{Y}$ are of the same homotopy type, then their first homology groups are the same: $H_1(\mathcal{X}) = H_1(\mathcal{Y})$.

A.6 Homology and Cohomology

In this section we deal with topology of differentiable manifolds. To this end we provide a concise account of homology groups associated with a topological space $\mathcal{X}$ that is triangulable, that is, homeomorphic to a polyhedron. Then we define cohomology groups that are the duals of homology groups. We anticipate that homology is constructed by means of the so-called boundary operator, which is a *global* operator on a manifold and thus requires some global information about the manifold. In contrast, cohomology is constructed by means of the exterior derivative, a first-order *local* differential operator that does not require global information about the manifold.

Through Stokes's theorem, a duality between homology and cohomology groups is established that allows one to relate *global* topological properties of a manifold with *local* differentiable properties of the same manifold. This is typical of differential topology.

A.6.1 Homology Groups

As we have already observed in Chapter 7, a basic way to analyze a manifold (more generally a topological space) is to fragment it into other more familiar objects and then to examine how these pieces fit together. The classical example is that of making a triangulation of a surface in $\mathbb{R}^3$, that is, to slice it into curved triangles, and then to count the number F of faces of the triangles, the number E of edges, and the number V of vertices on the tesselated surface. If the surface is compact, independently of the way we choose to triangulate it, the quantity $\chi = F - E + V$ is a characteristic number of the surface, which is invariant under diffeomorphisms of the surface. Thus this is a topological invariant, known as the Euler characteristic of the surface. At generic dimension, this procedure can be generalized by defining high-dimensional versions of triangles: simplices.

An r-simplex in $\mathbb{R}^n$, with $n \geq r$, is defined as the following set of points of $\mathbb{R}^n$:

$$\sigma_r = \left\{ x \in \mathbb{R}^n \;\middle|\; x = \sum_{i=0}^{r} c_i p_i; c_i \geq 0, \sum_{i=0}^{r} c_i = 1 \right\} \qquad (A.40)$$

where $p_0, \ldots, p_r$ are geometrically independent points of $\mathbb{R}^n$, that is, no $(r-1)$-dimensional hyperplane contains all the $r + 1$ points. The set σ_r is a compact subset of $\mathbb{R}^n$, being bounded and closed. A 0-simplex $[p_0]$ is a vertex (point), a 1-simplex $[p_0 p_1]$ is an edge (line), a 2-simplex $[p_0 p_1 p_2]$ is a face (triangle), a 3-simplex $[p_0 p_1 p_2 p_3]$ is a solid tetrahedron. An oriented r-simplex $(p_0 \ldots p_r)$ is an r-simplex $[p_0 \ldots p_r]$ with an assigned ordering for its vertices so that, for example, $(p_0 p_1) = -(p_1 p_0)$, where the minus sign means that $(p_0 p_1)$ is the inverse of $(p_1 p_0)$. Positive or negative signs are assigned according to the parity of the permutation with respect to a reference ordering.

A *simplicial complex* K is defined as a finite collection of simplices of $\mathbb{R}^n$ such that *(i)* if $\sigma_r \in K$ then also all its faces belong to K; *(ii)* if $\sigma_r \in K$ and $\sigma_s \in K$, then either $\sigma_r \cap \sigma_s = \emptyset$ or $\sigma_r \cap \sigma_s \leq \sigma_r$ and $\sigma_r \cap \sigma_s \leq \sigma_s$, that is, their intersection is a common face. The dimension of a simplicial complex K is given by the maximum of the dimensions of the simplices in the complex K.

Let us now give the definition of a *free abelian group*. If every element g of an abelian (commutative) group G can be represented as $g = \sum_{i=1}^m n_i g_i$, with $n_i \in \mathbb{Z}$, and $\{g_i\}$ a set of elements of G, called generators of G, and if this representation is unique, that is the $\{g_i\}$ are linearly independent over $\mathbb{Z}$, then G is a free abelian group and the $\{g_i\}$ form a basis.

Then we define the *r-chain group* $C_r(K)$ of an n-dimensional simplicial complex K, assumed to contain I_r r-simplices, as the free abelian group generated by the oriented r-simplices of K. The elements $c_r \in C_r(K)$ are called *r-chains* and are formally given by

$$c_r = \sum_{i=1}^{I_r} c_i \, \sigma_{r,i} \, , \qquad c_i \in \mathbb{Z} \, .$$

The group structure is given by the sum of two r-chains, $c_r + c_r' = \sum_i (c_i + c_i') \sigma_{r,i}$, by the existence of an inverse element $-c_r$ of c_r, that is, $-c_r = \sum_i (-c_i) \sigma_{r,i}$, and by the existence of the neutral element, $0 = \sum_i 0 \sigma_{r,i}$. If $r > \dim K$ then $C_r(K) = 0$ by definition.

Now we define the *boundary operator* ∂_r as the mapping

$$\partial_r : C_r(K) \to C_{r-1}(K)$$

having the following properties: *(i)* $\partial_r [\sum_i c_i \sigma_{r,i}] = \sum_i c_i \partial_r \sigma_{r,i}$ that is, ∂_r is linear; *(ii)* to an oriented r-simplex $\sigma_r = (p_0 p_1 \ldots p_r)$ the operator ∂_r associates its boundary as follows:

$$\partial_r \sigma_r \equiv \sum_{i=0}^{r} (-1)^i (p_0 p_1 \ldots \widehat{p_i} \ldots p_r) \, ,$$

where $\widehat{p_i}$ means that the point p_i has been omitted. The operator ∂_r is a homomorphism.

An r-chain $z_r \in C_r(K)$ is called an *r-cycle* if $\partial_r z_r = 0$. The set $Z_r(K)$ of r-cycles is the *r-cycle group* and $Z_r(K) = \text{Ker}(\partial_r) \subset C_r(K)$. If $r = 0$ then $Z_0(K) = C_0(K)$.

An r-chain $\beta_r \in C_r(K)$ is called an r-boundary if there exists an $(r+1)$-chain $c_{r+1} \in C_{r+1}(K)$ such that $\beta_r = \partial_{r+1} c_{r+1}$. The set $B_r(K)$ of r-boundaries is a group, the r-boundary group, and $B_r(K) = \text{Im}(\partial_{r+1}) \subset C_r(K)$. If $r = n$ then one defines $B_n(K) = 0$.

A fundamental property of the boundary operator ∂_r is that the mapping

$$\partial_r \cdot \partial_{r+1} : C_{r+1}(K) \rightarrow C_{r-1}(K)$$

is a zero mapping, that is, $\partial_r \cdot \partial_{r+1} = 0$ for any $c_{r+1} \in C_{r+1}(K)$. As a consequence, any element $\beta_r \in B_r(K)$ has the property that $\partial_r \beta_r = 0$. Hence $B_r(K)$ is a subgroup of $Z_r(K)$; in fact, the former is the set of those chains that are boundaries of other chains, and the latter is the set of chains without boundary, and the boundary of a boundary is always the empty set.

If we regard each simplex of a simplicial complex K as a subset of $\mathbb{R}^n$, the union of all the simplices is a still a subset of $\mathbb{R}^n$, which is called the polyhedron $|K|$ of the complex K. If, given a topological space $\mathcal{X}$, there exist a simplicial complex K and a homeomorphism $f : |K| \rightarrow \mathcal{X}$, then (K, f) is called a triangulation (not unique) of $\mathcal{X}$. Now we can wonder whether and how we can use the above-defined groups, $C_r(K)$, $Z_r(K)$, and $B_r(K)$, to catch some topological property of the space $\mathcal{X}$ whose triangulation is K, that is, some property that is invariant under homeomorphisms of $\mathcal{X}$. It turns out that none of them qualifies as a topological invariant. In contrast, for an n-dimensional complex K the rth simplicial homology group $H_r(K; \mathbb{Z})$ is defined as[3]

$$H_r(K; \mathbb{Z}) \equiv Z_r(K)/B_r(K) .$$

This is well defined since $B_r(K)$ is a subgroup of $Z_r(K)$. The homology group $H_r(K; \mathbb{Z})$ is the set of equivalence classes of r-cycles $z_r \in Z_r(K)$ that differ by a boundary, that is, $z_r \sim z_r'$ if there exists an $(r+1)$-chain such that $z_r = z_r' + \partial_{r+1} w_{r+1}$.

There are two important theorems about homology groups:

Theorem 1. If $\mathcal{X}$ and $\mathcal{Y}$ are two homeomorphic topological spaces, then $H_r(\mathcal{X})$ is isomorphic to $H_r(\mathcal{Y})$ for all r. We write $H_r(\mathcal{X}) \cong H_r(\mathcal{Y})$.

The consequence of this fact is that homology groups are topological invariants.

Theorem 2. If (K_1, f) and (K_2, g) are two triangulations of the same topological space then

$$H_r(K_1) \cong H_r(K_2) \qquad \forall r .$$

This means that homology groups are meaningful not only for simplicial complexes, but also for a topological space that is not a simplex but is triangulable.

---

[3] The notation $H_r(K; \mathbb{Z})$ puts in evidence that the group structure is defined with integer coefficients; for most of these groups we have $H_r(K; G)$ with $G = \mathbb{Z}_2$, $\mathbb{R}$, $\mathbb{C}$.

Among other properties of homology groups we have that for a connected complex K,

$$H_0(K) \cong \mathbb{Z} ,$$

and if K is the disjoint union of N connected components $K_1, K_2, \ldots, K_N$ then

$$H_0(K) \cong \underbrace{\mathbb{Z} \oplus \cdots \oplus \mathbb{Z}}_{N}$$

and also

$$H_r(K) = H_r(K_1) \oplus H_r(K_2) \oplus \cdots \oplus H_r(K_N) .$$

Even if $Z_r(K)$ and $B_r(K)$ are free abelian groups (because they are subgroups of the free abelian group $C_r(K)$) this is not necessarily true for $H_r(K) = Z_r(K)/B_r(K)$. By a theorem on finite cyclic groups[4] [261], the most general form of $H_r(K)$ is then

$$H_r(K) \cong \underbrace{\mathbb{Z} \oplus \cdots \oplus \mathbb{Z}}_{m} \oplus \mathbb{Z}_{k_1} \oplus \cdots \oplus \mathbb{Z}_{k_p} \tag{A.41}$$

with $r = m + p$, where the first m factors form a free abelian group and the remaining p factors form a subgroup of $H_r(K)$ called its *torsion group*.

The number of generators of $H_r(K)$ counts the number of $(n + 1)$-dimensional holes in the polyhedron $|K|$. For $c_r \in C_r(K; \mathbb{Z})$ we write $c_r = \sum_{i=1}^{I_r} g_i \sigma_{r,i}$ with $g_i \in \mathbb{Z}$, but considering $g_i \in \mathbb{R}$, we have real-coefficients homology groups, and (A.41) becomes

$$H_r(K; \mathbb{R}) \cong \underbrace{\mathbb{R} \oplus \cdots \oplus \mathbb{R}}_{m} ,$$

that is, we lose the torsion subgroup, which detects the "twisting" of $|K|$. The curious feature of homology theory is that the integer-coefficients homology contains more information than rational-, real-, or complex-coefficients homology, because the torsion subgroup is lost with other than integer coefficients.

Given a simplicial complex K, the rth Betti number $b_r(K)$ is defined as

$$b_r(K) = \dim H_r(K; \mathbb{R}) ,$$

that is, $b_r(K)$ is the rank of the free abelian part of $H_r(K)$.

If now I_r is the number of r-simplices in K, then

$$\chi(K) = \sum_{r=0}^{n} (-1)^r I_r = \sum_{r=0}^{n} (-1)^r b_r(K) \tag{A.42}$$

is the Euler characteristic of K, which generalizes the classic formula $\chi = V - E + F$. Since Betti numbers are topological invariants, that is, they do

---

[4] A group G generated by one element a, that is $G = \{0, \pm a, \pm 2a, \ldots\}$, is called a cyclic group.

not change under the action of homeomorphisms, $\chi(K)$ too is a topological invariant.

If $f : |K| \to \mathcal{X}$ and $g : |K'| \to \mathcal{X}$ are two triangulations of a topological space $\mathcal{X}$, we have $\chi(K) = \chi(K')$. Hence one defines $\chi(\mathcal{X})$ by means of $\chi(K)$ for any triangulation (K, f) of $\mathcal{X}$.

A.6.2 Cohomology Groups

If the topological space $\mathcal{X}$ for which we have defined the homology group is a compact differentiable manifold M, we can construct the dual of homology groups by means of differential forms defined on M. The dual groups are the so-called de Rham cohomology groups. The essential tool to construct the cohomology of M is Stokes's theorem, given below.

Stokes' Theorem

Let M be an n-dimensional manifold with a nonempty boundary ∂M. For any differentiable $(r - 1)$-form ω^{r-1} one has the Stokes's formula

$$\int_M d\omega^{r-1} = \int_{\partial M} \omega^{r-1} .$$

This is of fundamental importance in the study of de Rham's cohomology groups.

In the special case $M = \mathbb{R}^3$, if $\omega = A dx + B dy + C dz$ and $\boldsymbol{\omega} = (A, B, C)$, we have, using standard notation,

$$\int_\Sigma \text{rot}(\boldsymbol{\omega}) \cdot \mathbf{n} d\sigma = \int_\gamma \boldsymbol{\omega} \cdot \boldsymbol{\tau} d\ell ,$$

where γ is the boundary of the surface Σ. Moreover, with $\psi = \frac{1}{2}\psi_{\mu\nu}dx^\mu \wedge dx^\nu$ we have

$$\int_V \text{div}(\boldsymbol{\psi})dV = \int_\Sigma \boldsymbol{\psi} \cdot \mathbf{n} d\sigma ,$$

where V is the volume enclosed by the surface Σ. This is commonly known as Gauss's theorem.

Stokes's theorem also applies to lower-dimensional subsets of a manifold M, and homology theory appears to naturally provide a class of them. However, we have first to extend to compact differentiable manifolds what we have introduced above for arbitrary topological spaces. Thus we define r-chains, r-cycles, and r-boundaries for an n-dimensional manifold.

Consider an r-simplex σ_r in $\mathbb{R}^n$ and a continuous mapping $f : \sigma_r \to M$. The image $s_r = f(\sigma_r) \in M$ is called a *singular r-simplex* in M. This simplex is called singular because it does not provide a triangulation of M. Denoting by $\{s_{r,i}\}$ the set of r-simplices of M, we can define an r-chain in M as the formal finite sum with real coefficients

$$c_r = \sum_{i=1}^{I_r} a_i s_{r,i} \,, \qquad a_i \in \mathbb{R} \,,$$

where $s_{r,i}$ are r-dimensional oriented submanifolds of M. The ensemble of r-chains is the *chain group* $C_r(M)$ of M. Moreover, the images in M $\partial s_r \equiv f(\partial \sigma_r)$ of the boundaries of the simplices σ_r form the ensemble of $(r-1)$-simplices in M that are the geometrical boundaries of the s_r. Hence we have a mapping that associates to a manifold its oriented boundary

$$\partial : C_r(M) \to C_{r-1}(M)$$

with $\partial^2 = 0$.

We define the *cycle group* $Z_r(M)$ as the set of r-cycles, that is, of r-chains without boundary, and we define the *boundary group* $B_r(M)$ as the set of r-chains that are the boundary of an $(r+1)$-dimensional chain, i.e., $c_r = \partial c_{r+1}$. Since the boundary of a boundary is always the empty set, $B_r(M) \subset Z_r(M)$, and we can define the *singular homology group* of the manifold M

$$H_r(M; \mathbb{R}) = Z_r(M)/B_r(M) \,.$$

If M is orientable, the following property, called *Poincaré duality*, holds:

$$H_p(M; \mathbb{R}) = H_{n-p}(M; \mathbb{R}) \,.$$

Now we can proceed to define the integration of an r-form over an r-chain in M, with $0 \le r \le n$. First put, for an r-simplex of M,

$$\int_{s_r} \omega = \int_{\overline{\sigma}_r} f^* \omega \,,$$

where $\overline{\sigma}_r$ is a standard r-simplex of $\mathbb{R}^r$, $f : \overline{\sigma}_r \to M$ is a smooth mapping, $s_r = f(\overline{\sigma}_r)$, $f^* \omega$ is an r-form in $\mathbb{R}^r$, so that the right hand side is a standard r-dimensional integral. Then the integral for an r-chain $c \in C_r(M)$ is defined as

$$\int_c \omega = \sum_{i=1}^{I_r} a_i \int_{s_{r,i}} \omega \,. \tag{A.43}$$

A more general version of Stokes's theorem is the following. Let $\omega \in \Lambda^{r-1}(M)$ and $c \in C_r(M)$. Then

$$\int_c d\omega = \int_{\partial c} \omega \,. \tag{A.44}$$

The form $\omega \in \Lambda^r(M)$ associates to $c \in C_r(M)$ a real number through the integral (A.43), so we can write

$$\omega : C_r(M) \to \mathbb{R} \,,$$

$$c \mapsto \langle \omega, c \rangle := \int_c \omega \,.$$

In other words, ω is an element of the dual of $C_r(M)$, i.e., it is a *cochain*. Stokes's theorem in (A.44), with the notation above, reads

$$\langle d\omega, c \rangle = \langle \omega, \partial c \rangle ,$$

showing that the boundary operator ∂ and the exterior derivative d are adjoints of one another, so that we can define the dual of the space $H_r(M; \mathbb{R})$ as

$$H^r(M; \mathbb{R}) = Z^r(M)/B^r(M) , \qquad (A.45)$$

that is, the quotient space of the *cocycle* group and the *coboundary group* which is called the rth *cohomology group*. These cohomology groups, as is shown below, are just the de Rham cohomology groups defined in (A.36).

In fact, by Stokes's theorem, for any $c_r \in Z_r(M)$ and $\omega_r \in Z^r(M)$ the following identities hold:

$$\int_{c_r} \omega_r + d\psi_{r-1} = \int_{c_r} \omega_r + \int_{\partial c_r} \psi_{r-1} = \int_{c_r} \omega_r$$

and

$$\int_{c_r + \partial a_{r+1}} \omega_r = \int_{c_r} \omega_r + \int_{a_{r+1}} d\omega_r = \int_{c_r} \omega_r .$$

Hence, the product is independent of the choices made in the respective equivalence classes, so that the following mapping is well defined:

$$\langle \omega, c \rangle : H_r(M; \mathbb{R}) \otimes H^r(M; \mathbb{R}) \to \mathbb{R} .$$

Then, for compact manifolds M without boundary, let $\{c_i : i = 1, \ldots, \dim H_r(M; \mathbb{R})\}$ be a set of independent r-cycles forming a basis for $H_r(M; \mathbb{R})$. Given an r-form ω, and having defined the *periods* ν_i as $\nu_i = \langle c_i, \omega \rangle$, de Rham proved two important theorems:

Theorem 1. *For any given set of real numbers* ν_i, $i = 1, \ldots, \dim H_r(M; \mathbb{R})$, *there exists a closed form* ω *for which*

$$\nu_i = \langle c_i, \omega \rangle = \int_{c_i} \omega , \qquad i = 1, \ldots, \dim H_r(M; \mathbb{R}) .$$

Theorem 2. *If all the periods* ν_i, $i = 1, \ldots, \dim H_r(M; \mathbb{R})$, *for an* r-form ω *are vanishing,*

$$0 = \langle c_i, \omega \rangle = \int_{c_i} \omega \qquad i = 1, \ldots, \dim H_r(M; \mathbb{R}) ,$$

then ω *is exact.*

The meaning of these theorems is that for a compact manifold M without boundary, if $\{\omega_i\}$ is a basis for $H^r(M; \mathbb{R})$ and $\{c_i\}$ is a basis for $H_r(M; \mathbb{R})$, then the matrix of periods

$$\nu_{ij} = \langle c_i, \omega_j \rangle$$

is invertible; hence $H^r(M;\mathbb{R})$ is the dual space of $H_r(M;\mathbb{R})$ with respect to the product $\langle c,\omega \rangle$ and the two spaces are isomorphic:

$$H^r(M;\mathbb{R}) \cong H_r^\star(M;\mathbb{R}) \ .$$

A standard way of summarizing the content of the construction of $H^r(M;\mathbb{R})$ and $H_r(M;\mathbb{R})$ is through the following topological diagrams:

$$0 \xleftarrow{\partial_0} C_0(M) \xleftarrow{\partial_1} C_1(M) \xleftarrow{\partial_2} \cdots \xleftarrow{\partial_{n-2}} C_{n-1}(M) \xleftarrow{\partial_n} C_n(M) \longleftarrow 0$$

$$H_r(M;\mathbb{R}) = Z_r(M)/B_r(M) = \mathrm{Ker}\ \partial_r/\mathrm{Im}\ \partial_{r+1}$$

$$0 \xrightarrow{\ \hookrightarrow\ } \Lambda^0(M) \xrightarrow{d_0} \Lambda^1(M) \xrightarrow{d_1} \cdots \xrightarrow{d_{m-2}} \Lambda^{m-1}(M) \xrightarrow{d_{m-1}} \Lambda^m(M) \xrightarrow{d_m} 0$$

$$H^r(M;\mathbb{R}) = Z^r(M)/B^r(M) = \mathrm{Ker}\ d_{r+1}/\mathrm{Im}\ d_r \ .$$

The former is called the *chain complex* $C(M)$ and the latter is known as the *de Rham complex* $\Lambda^*(M)$. These are *finite* sequences telling that the only nontrivial cohomology groups are those with $0 \le r \le \dim M$. The inverted directions of the arrows recall the duality between the operators d and ∂.

Let us remark that $H^0(M;\mathbb{R})$ is a special case because (-1)-forms do not exist and thus 0-forms (functions) cannot be expressed as the exterior derivative of a form. Since closed 0-forms are functions f such that $df = 0$, we have $H^0(M;\mathbb{R}) = \{space\ of\ constant\ functions\}$, that is, $H^0(M;\mathbb{R}) = \mathbb{R}$; if M has n connected components, then $H^0(M;\mathbb{R}) = \oplus^n\mathbb{R}$, and thus in general,

$$\dim H^0(M;\mathbb{R}) = \{\#\ of\ connected\ components\ of\ M\} \ .$$

We add, without proof, that for a simply connected manifold, $H^1(M;\mathbb{R}) = 0$, and that in general,[5]

$$\dim H^k(M;\mathbb{R}) = \{\#\ of\ (k+1) - dimensional\ \text{``holes''}\ in\ M\} \ .$$

Note that the cohomology of the Euclidean space $\mathbb{R}^n$ is always trivial. In fact, since $\mathbb{R}^n$ is contractible, that is, $\mathbb{R}^n$ is contracted to the point 0 by the map $\alpha : \mathbb{R}^n \times [0,1] \to \mathbb{R}^n$ defined by $(x,t) \mapsto (1-t)x$, it can be shown that every closed form in $\mathbb{R}^n$ is also exact. Thus

$$H^r(\mathbb{R}^n;\mathbb{R}) = 0 \ , \qquad r = 1,\ldots,n \ ,$$
$$H^0(\mathbb{R}^n;\mathbb{R}) = \mathbb{R} \ .$$

We have seen that a differentiable manifold is locally homeomorphic to a Euclidean $\mathbb{R}^n$ space, so that the consequence of the homological triviality of

---

[5] A hole is well defined for two-dimensional surfaces; for example, a torus has a three-dimensional hole in the sense that we can put a ball inside it. In higher dimensions one has to resort to the fundamental group π_1 to define "higher-dimensional holes."

$\mathbb{R}^n$ is the Poincaré lemma: *If a coordinate neighborhood U of a manifold M is contractible to a point, then any closed r-form on U is also exact.*

Remark. Since any closed form on a manifold is locally exact, what prevents global exactness is just a nontrivial topology, that is, loosely speaking, when local charts are necessarily glued together in a nontrivial way. Nonvanishing de Rham cohomology groups detect topological obstructions to global exactness of closed forms.

To give a classic example, consider $M = \mathbb{R}^2 - \{0\}$ (punctured plane where the closed paths encircling the origin cannot be contracted to the point 0) and the 1-form

$$\omega = \frac{-y}{x^2 + y^2} dx + \frac{x}{x^2 + y^2} dy \ .$$

The point $x = y = 0$ is singular, whence the need to exclude the origin $\{0\}$. Computing

$$d\omega = \frac{\partial}{\partial x} \frac{-y}{x^2 + y^2} dy \wedge dx + \frac{\partial}{\partial y} \frac{x}{x^2 + y^2} dx \wedge dy \ ,$$

we immediately see that $d\omega = 0$; the form is closed. Introducing the 0-form $\eta(x,y) = \arctan(y/x)$, so that $\partial\eta/\partial x = -y/(x^2 + y^2)$ and $\partial\eta/\partial y = x/(x^2 + y^2)$, we might think that $\omega = d\eta$, i.e., that ω is also exact, but since η is the angle in the polar representation of the plane, single-valuedness imposes that $\eta(x,y)$ is defined only on $\mathbb{R}^2 - \mathbb{R}_+$. No function exists such that $\omega = df$ on the whole of $M = \mathbb{R}^2 - \{0\}$. However, $\omega = d\eta$ *locally* holds in neighborhoods excluding the origin.

A.6.3 Betti Numbers

The direct sum $H^*(M) = \oplus_{r=1}^n H^r(M)$ is the de Rham *cohomology algebra*, which coincides with the quotient between the closed-forms subalgebra of the Grassmann algebra of M and its subset (actually an ideal) of exact forms. Just below (A.30) we have seen that to any differentiable map $f : M \to N$ between two differentiable manifolds we can associate a homomorphism f^* between their associated Grassmann algebras, that is, $f^* : \Lambda^*(N) \to \Lambda^*(M)$. This homomorphism maps each homogeneous component of $H^*(N)$ into the homogeneous component of $H^*(M)$ of the same degree. If $f : M \to N$ is a *diffeomorphism*, then all the cohomology spaces $H^r(M)$ and $H^r(N)$ are *isomorphic* for $r = 0, 1, \ldots, n$. As a consequence, the dimension of each cohomology space of a manifold is invariant under diffeomorphisms of this manifold, whence the set of integers

$$b_r = \dim H_r(M; \mathbb{R}) = \dim H^r(M; \mathbb{R}) \ , \qquad 0 \le r \le \dim M \ ,$$

known as the Betti numbers of M, is a set of *topological invariants* of M.

Poincaré Duality

Consider a compact n-dimensional manifold M and two forms $\omega \in H^r(M)$ and $\eta \in H^{n-r}(M)$. Define a product $\langle \cdot, \cdot \rangle : H^r(M) \times H^{n-r}(M) \to \mathbb{R}$ as

$$\langle \omega, \eta \rangle = \int_M \omega \wedge \eta .$$

Since $\omega \wedge \eta$ is a volume form, it does not vanish unless one or both forms vanish, so we have a nonsingular bilinear product that defines a duality between $H^r(M)$ and $H^{n-r}(M)$, that is,

$$H^r(M) \cong (H^{n-r}(M))^\star ,$$

which is called *Poincaré duality* and entails the relation

$$b_r = b_{n-r} .$$

Through Betti numbers we can also define the Euler characteristic for differentiable manifolds, as

$$\chi(M) = \sum_{r=0}^{n} (-1)^r b_r(M) . \tag{A.46}$$

This remarkable formula relates a purely topological quantity (left hand side) with an analytic property of a manifold (right hand side) stemming from the condition of closeness for differential forms. Such a situation is common in differential topology. We have already encountered it with the duality between homology and cohomology groups and we will find it again with Morse theory.

Notice that by Poincaré duality, the Euler characteristic for odd-dimensional manifolds is identically vanishing.

Cup Product

Let us just mention the possibility that different manifolds can have the same cohomology groups yet be topologically different. An example is given by the two manifolds $M = \mathbb{S}^2 \times \mathbb{S}^4$ and $N = \mathbb{C}P^3$, complex-projective space, which are neither diffeomorphic nor homeomorphic yet have the same cohomology groups. To distinguish these spaces from one another, and, in general, to make cohomology theory somewhat finer in detecting manifolds' topologies, we provide the de Rham cohomology algebra with the so-called *cup product*

$$\cup : H^*(M; \mathbb{R}) \times H^*(M; \mathbb{R}) \to H^*(M; \mathbb{R}) ,$$

which acts as follows: given $[\omega] \in H^p(M; \mathbb{R})$ and $[\nu] \in H^q(M; \mathbb{R})$, where $[\omega]$ stands for the cohomology class of the form ω, the product

$$[\omega] \cup [\nu] = [\omega \wedge \nu] ,$$

with $[\omega] \cup [\nu] \in H^{p+q}(M; \mathbb{R})$, gives $H^*(M; \mathbb{R})$ a (graded commutative) ring structure.[6] Different such ring structures are found for the above manifolds M and N.

Kunneth Formula

A particularly important case is the case of *product manifolds*, for which the following *Kunneth formula* holds: Let Q be a product manifold, i.e., $Q = M \times N$. Then, the kth real de Rham cohomology group of Q, $H^k(Q)$, is given by

$$H^k(Q) = H^k(M \times N) = \bigoplus_{p+q=k} [H^p(M) \otimes H^q(N)] , \qquad (A.47)$$

and its Betti numbers are

$$b_k(Q) = b_k(M \times N) = \sum_{p+q=k} b_p(M) \, b_q(N) . \qquad (A.48)$$

As we have seen in Chapters 8 and 9, this is a very useful formula.

Bibliography

1. C. Nash and S. Sen, *Topology and Geometry for Physicists* (Academic Press, London 1989).
2. M. Nakahara, *Geometry, Topology and Physics* (Adam Hilger, Bristol 1991).
3. M.P. do Carmo, *Riemannian Geometry* (Birkhäuser, Boston 1992).
4. Y. Choquet-Bruhat, C. De Witt-Morette, and M. Dillard-Bleick, *Analysis, Manifolds and Physics* (North-Holland, Amsterdam 1985).
5. M.W. Hirsch, *Differential Topology* (Springer, New York 1976).
6. V. Guillemin and A. Pollack, *Differential Topology* (Prentice Hall, Englewood Cliffs 1974).
7. R.S. Palais and C. Terng, *Critical Point Theory and Submanifold Geometry* (Springer, New York 1988).
8. C.W. Misner, K.S. Thorne, and J.A. Wheeler, *Gravitation* (Freeman, San Francisco 1973).

---

[6] Let us recall that a ring is a set with two binary operators (addition and multiplication) with the properties of additive associativity and commutativity, existence of additive identity and inverse, left and right distributivity, and multiplicative associativity.

9. M. Spivak, *A Comprehensive Introduction to Differential Geometry* (Publish or Perish, Boston 1970).

10. R. Abraham, and J.E. Marsden, *Foundations of Mechanics*, Second Edition (Addison–Wesley, Redwood City, 1987).

The books quoted in items 5 and 7 are the more advanced. Several chapters of the book in item 8 are an excellent introduction to differential geometry for physicists.

Appendix B

Elements of Riemannian Geometry

The present appendix provides a concise and informal treatment of some essential concepts of Riemannian differential geometry. Here we make use of concepts and definitions given in appendix A.

For a more elaborate discussion, we refer the reader to the books listed at the end of the preceding appendix.

B.1 Riemannian Manifolds

Riemannian geometry is the natural extension of the theory of surfaces in $\mathbb{R}^3$. For surfaces $S \subset \mathbb{R}^3$, Gauss proved that all their intrinsic geometry is completely determined by the inner product applied to tangent vectors at S. In order to generalize this result to the study of the geometry of arbitrary n-dimensional manifolds, Riemann introduced an inner product on each tangent space of a manifold. In particular, this inner product provides a way of measuring infinitesimal distances. Roughly speaking, if P and $P + dP$ are two nearby points, their distance is measured by the norm of the "tangent vector" dP.

To introduce the subject at an intuitive level, let x, y, z be the coordinates in an ordinary Euclidean $\mathbb{R}^3$ space. Consider

$$x = R\cos\theta \ ,$$

$$y = R\sin\theta \ ,$$

$$z = \zeta \ ,$$

with constant R. By varying θ and ζ, the point $P(x, y, z)$ describes a cylindrical surface. The (squared) distance between two close points $P(x, y, z)$ and $P(x + dx, y + dy, z + dz)$ is given by

$$ds^2 = dx^2 + dy^2 + dz^2 = R^2 d\theta^2 + d\zeta^2 \ . \tag{B.1}$$

Then consider

$$x = R \sin \theta \cos \varphi \; ,$$
$$y = R \sin \theta \sin \varphi \; ,$$
$$z = R \cos \theta \; ,$$

again with R =const. In this case, by varying θ and φ the point $P(x, y, z)$ describes a spherical surface of radius R. We have

$$ds^2 = dx^2 + dy^2 + dz^2 = R^2 d\theta^2 + R^2 \sin^2 \theta d\varphi^2 \; . \qquad (B.2)$$

The quadratic differential forms (B.1) and (B.2) represent the metrics of the cylinder and the sphere respectively. The former has a Pythagorean form, whereas the second does not because the coefficient of $d\varphi^2$ depends on the coordinate θ. There is no coordinate system on the sphere that casts (B.2) into a Pythagorean form. Thus, the cylinder metric (B.1) describes a *Euclidean two-dimensional manifold*, whereas the sphere metric (B.2) describes a *non-Euclidean two-dimensional manifold*.

More generally, for an m-dimensional surface $S_{(m)} \subseteq \mathbb{R}^n$, we have

$$ds^2 = dP \cdot dP = \sum_{\alpha=1}^{n} (dx^\alpha)^2 \equiv \sum_{i,k=1}^{m} a_{ik} dy^i dy^k$$

with

$$a_{ik} = \frac{\partial P}{\partial y^i} \cdot \frac{\partial P}{\partial y^k} = \sum_{\alpha=1}^{n} \frac{\partial x^\alpha}{\partial y^i} \frac{\partial x^\alpha}{\partial y^k} \; ,$$

where the $\{y^i\}$ are curvilinear coordinates on $S_{(m)}$ and the $\{x^\alpha\}$ are coordinates in $\mathbb{R}^n$. (For the sake of clarity, here we have not yet adopted the Einstein summation convention for repeated indexes.) The knowledge of the metric tensor a_{ik} allows us to work out all the intrinsic geometry of these kinds of spaces.

B.1.1 Riemannian Metrics on Differentiable Manifolds

Given an n-dimensional differentiable manifold M and a point $p \in M$, a $(0, 2)$-type tensor $g_p \in \mathcal{T}_p^{(0,2)}$ is said to be a *Riemannian metric tensor* provided it is *symmetric*, i.e., $g_p(X, Y) = g_p(Y, X)$, and *nondegenerate*, i.e., $g_p(X, Y) = 0 \; \forall X \in T_p M$ if and only if $Y = 0$.

Then a symmetric *tensor field* $g \in \mathcal{T}^{(0,2)}(M)$ on M that satisfies these conditions is said to be a *Riemannian metric tensor* of M. The pair (M, g) is a *Riemannian manifold*. Let (M, g) be a Riemannian manifold, $p \in M$, and $X, Y \in T_p M$. The inner (or scalar) product $\langle X, Y \rangle$ between these vectors is the real number

$$\langle X, Y \rangle = g(X, Y) \; .$$

This induces on the tangent bundle TM a nondegenerate quadratic mapping

$$g(\cdot,\cdot) : TM \otimes TM \to \mathbb{R} . \tag{B.3}$$

If the quadratic form (B.3) is positive definite, then one speaks of a (proper) *Riemannian metric*, which makes it possible to measure lengths on a differentiable manifold where the squared length element ds^2 is always positive. If the quadratic form (B.3) is not positive definite, then the manifold (M, g) is called a *pseudo-Riemannian manifold*, and the scalar product is referred to as a *pseudo-Riemannian structure* on M.

In a chart (U, φ) on $U \subset M$, we have

$$g|_U = g_{ij}\, dx^i \otimes dx^j , \tag{B.4}$$

where the components $g_{ij}(x^1, \ldots, x^n) = \langle \partial_i, \partial_j \rangle$, with $\partial_i \equiv \partial/\partial x^i$, are differentiable functions on U.

The scalar product of two vectors $X = X^i \partial_i$ and $Y = Y^j \partial_j$ is given, in terms of g, by

$$\langle X, Y \rangle = \langle X^i \partial_i, Y^j \partial_j \rangle = X^i Y^j \langle \partial_i, \partial_j \rangle = g_{ij} X^i Y^j \equiv X_j Y^j = X^i Y_i . \tag{B.5}$$

The last two equalities in the right hand side of the above equation follow from the fact that g estabilishes a one-to-one correspondence between vectors and dual vectors, that is, an isomorphism between $T_p M$ and $T_p^* M$, which, in local coordinates, reads

$$g_{ij} X^j = X_i . \tag{B.6}$$

For this reason, the components of the inverse metric g^{-1} (which exist everywhere because of nondegeneracy of the metric tensor) are simply denoted by g^{ij}, instead of $(g^{-1})^{ij}$, and allow one to pass from dual vector (covariant) components to vector (contravariant) components:

$$g^{ij} X_j = X^i . \tag{B.7}$$

This operation of raising and lowering the indices can be applied not only to vector, but also to tensor components. This allows us to pass from (k, l) tensor components to the corresponding $(k + 1, l - 1)$ or $(k - 1, l + 1)$ tensor components and vice versa. This operation will be tacitly assumed in the following. What does not change in the operation is the sum $k + l$, which is called the *rank* (or the *order*) of the tensor.

By considering an infinitesimal displacement $ds = (ds)^i X_i = dx^i (\partial/\partial x^i) \in T_p M$, we can compute

$$ds^2 = g\left(dx^i \partial_i, dx^j \partial_j\right) = dx^i dx^j g(\partial_i, \partial_j) = g_{ij} dx^i dx^j . \tag{B.8}$$

Sometimes, mainly in the physics literature, this invariant (squared) length element of the manifold, ds^2, is called the metric.

Let M and N be two Riemannian manifolds. A diffeomorphism $f : M \to N$ is said to be an *isometry* if for all $p \in M$ and $u, v \in T_pM$, we have

$$\langle u, v \rangle_p = \langle df_p(u), df_p(v) \rangle_{f(p)} .$$

A differentiable mapping $f : M \to N$ is called a *local isometry* at p if there is a neighborhood U of p such that $f : U \subset M \to f(U) \subset N$ is a diffeomorphism satisfying the above conditions for isometries.

It is a remarkable fact that *any differentiable manifold can be endowed with a Riemannian metric*. This follows from the use of a differentiable partition of unity $\{\varepsilon_\alpha\}$ on M subordinate to a covering $\{U_\alpha\}$ of M by coordinate neighborhoods. $\{U_\alpha\}$ is a locally finite covering. The functions $\{\varepsilon_\alpha\}$ satisfy the conditions $\varepsilon_\alpha \geq 0$, $\varepsilon_\alpha = 0$ on the complement of the closed set $\overline{U}_\alpha$, and $\sum_\alpha \varepsilon_\alpha(p) = 1$, for all $p \in M$. One defines a Riemannian metric $\langle \cdot, \cdot \rangle^\alpha$ on each U_α using the metric induced by the system of local coordinates. Then one sets $\langle X, Y \rangle_p = \sum_\alpha \varepsilon_\alpha(p) \langle X, Y \rangle_p^\alpha$ for all $p \in M$ and $X, Y \in T_pM$. Thus a Riemannian metric can be always defined on a differentiable manifold M.

We just mention another remarkable result:

Theorem (Nash). *Any simply connected, compact, n-dimensional Riemannian manifold (M, g) of class C^k, with $3 \leq k \leq \infty$, can be C^k-isometrically embedded in a Euclidean $\mathbb{R}^N$ space with $N = \frac{1}{2}n(3n + 11)$. Every noncompact n-dimensional Riemannian manifold (M, g) of class C^k, with $3 \leq k \leq \infty$, can be C^k-isometrically embedded in a Euclidean $\mathbb{R}^N$ space with $N = \frac{1}{2}n(n + 1)(3n + 11)$.*

A Riemannian metric can be used to compute the lengths of curves. Consider a curve on M defined by the differentiable mapping $c : I \subset \mathbb{R} \to M$. The velocity field of the curve c is dc/dt, and one defines the length of a segment (that is, a restriction of $c(t)$ to the interval $[A, B] \subset I$) by

$$\ell_{AB}(c) = \int_A^B \left\langle \frac{dc}{dt}, \frac{dc}{dt} \right\rangle^{1/2} dt , \tag{B.9}$$

where $\langle \dot{c}, \dot{c} \rangle = g(\dot{c}, \dot{c})$.

B.2 Linear Connections and Covariant Differentiation

The introduction of differential calculus on a non-Euclidean manifold is complicated by the fact that ordinary derivatives do not map vectors into vectors. A vector field along c is a differentiable mapping that associates the tangent vector $V(t) \in T_{c(t)}M$ at each $t \in I$. The function $t \mapsto V(t)f$ is a differentiable function on I for any differentiable function f on M.

For example, consider a surface $S \subset \mathbb{R}^3$, a curve $c : I \subset \mathbb{R} \to \mathbb{R}$ in S, and a vector field $V : I \to \mathbb{R}$ along c tangent to S. In general, the ordinary derivatives of the components of V, dV^i/dt, are *not* the components of a vector in $T_{c(t)}S$, because they do not transform according to the rule in (A.15). Thus the differentiation of a vector field is not an intrinsic geometric concept on the surface S. To overcome this difficulty, one considers the orthogonal projection of $\frac{dV}{dt}(t)$ on $T_{c(t)}S$. This we call the *covariant derivative* of $V(t)$ and we denote it by $\frac{\nabla V}{dt}(t)$. The geometric origin of this fact is that the *parallel transport* of a vector from a point p to a point q on a non-Euclidean manifold *does* depend on the path chosen to join p and q. Since in order to define the derivative of a vector at p, we have to move that vector from p to a neighboring point along a curve and then parallel-transport it back to the original point in order to measure the difference, we need a definition of parallel transport to define a derivative; conversely, given a (consistent) derivative, i.e., a derivative that maps vectors into vectors, one can define the parallel transport by imposing that a vector is parallel-transported along a curve if its derivative along the curve is zero. The two ways are conceptually equivalent: we follow the first way, by introducing the notion of a *connection*, and then we use it to define the covariant derivative.

Let us denote by $\mathcal{X}(M)$ the set of all smooth vector fields on M. A (linear) *connection* on the manifold M is a map $\nabla : \mathcal{X}(M) \times \mathcal{X}(M) \to \mathcal{X}(M)$ such that given two vector fields X and Y, it yields a third field $\nabla_X Y$, and satisfies the following properties:

(i) $\nabla_X Y$ is bilinear in X and Y, i.e., $\nabla_X(\alpha Y + \beta Z) = \alpha \nabla_X Y + \beta \nabla_X Z$ and $\nabla_{\alpha X + \beta Y} Z = \alpha \nabla_X Z + \beta \nabla_Y Z$;

(ii) $\nabla_{fX} Y = f(\nabla_X Y)$;

(iii) $\nabla_X f Y = X(f)Y + f(\nabla_X Y)$ (Leibniz rule).

Note that ∇ is *local*, that is, if $X_1, X_2, Y_1\, Y_2$ are vector fields such that there exists an open set U of M where $X_1 = X_2$ and $Y_1 = Y_2$, then $\forall p \in U$ $(\nabla_{X_1} Y_1)(p) = (\nabla_{X_2} Y_2)(p)$.

Thus ∇ naturally restricts to open sets of M.

If (U, φ) is a chart for $U \subset M$, with $x = \varphi(p)$, and $\{e_{(i)}\} = \{\partial/\partial x^i\}$ is a coordinate basis for $T_p M$, then

$$\nabla_{\partial_i} \partial_j = \Gamma_{ij}^k \partial_k \,, \tag{B.10}$$

where the n^3 numbers $\Gamma_{ij}^k = \langle dx^k, \nabla_{\partial_i} \partial_j \rangle$ are called the *Christoffel coefficients* of the connection ∇ relative to the chart (U, φ). If $(\widetilde{U}, \widetilde{\varphi})$ is another chart on M in the open set $\widetilde{U}$ with $U \cap \widetilde{U} \neq \emptyset$ and $y = \widetilde{\varphi}(p)$, then we have

$$\widetilde{\partial}_i \equiv \frac{\partial}{\partial y^i} = \left(\frac{\partial x^j}{\partial y^i} \right) \partial_j \equiv \left(\widetilde{\partial}_i x^j \right) \partial_j$$

and thus

$$\widetilde{\Gamma}_{ij}^k \equiv dy^k \left(\nabla_{\widetilde{\partial}_i} \widetilde{\partial}_j \right) = \frac{\partial x^m}{\partial y^i} \frac{\partial x^l}{\partial y^j} \frac{\partial y^k}{\partial x^r} \Gamma_{ml}^r + \frac{\partial y^k}{\partial x^m} \left(\frac{\partial^2 x^m}{\partial y^i \partial y^j} \right) \,.$$

Because of the last term, with second derivatives, in the right hand side of the equation above, Γ_{ij}^k are *not* the components of a tensor on M.

Consider $X, Y \in \mathcal{X}(M)$, expressed as $X = X^i \partial_i$ and $Y = Y^i \partial_i$ in the chart (U, φ). By making use of the above-listed properties of the linear connection ∇, one obtains

$$\nabla_X Y = X^j \left(\nabla_{\partial_j} Y^i \partial_i \right) = X^j \left(\partial_j [Y^i] \partial_i + Y^i \nabla_{\partial_j} \partial_i \right)$$
$$= \left(X^j \frac{\partial Y^k}{\partial x^j} + \Gamma_{ji}^k X^j Y^i \right) \partial_k . \tag{B.11}$$

It is evident that $\nabla_X Y(p)$ depends only, for any $p \in M$, on $X(p)$ and not on its derivatives; hence it is sensible to define the covariant derivative of vector and tensor fields properly generalizing the concept of directional derivative of functions.

If ∇ is a linear connection on M, one defines the *torsion tensor* of ∇ as the $(1, 2)$-type tensor on M as

$$\mathrm{Tor}_\nabla(X, Y) = \nabla_X Y - \nabla_Y X - [X, Y]$$

for all $X, Y \in \mathcal{X}(M)$. Then $\mathrm{Tor}_\nabla(X, Y) = -\mathrm{Tor}_\nabla(Y, X)$. If $\mathrm{Tor}_\nabla = 0$ then the linear connection ∇ is said to be *symmetric*. In local coordinates on an open set $U \subset M$, we have

$$\mathrm{Tor}_\nabla (\partial_i, \partial_j) = T_{ij}^k \partial_k,$$

with $T_{ij}^k = \Gamma_{ij}^k - \Gamma_{ji}^k$. Thus ∇ is symmetric if and only if in any chart, $\Gamma_{ij}^k = \Gamma_{ji}^k$.

A remarkable property of Riemannian geometry is the existence of a *unique* connection satisfying the following properties:

(i) $[X, Y] = \nabla_X Y - \nabla_Y X$;

(ii) $X \langle Y, Z \rangle = \langle \nabla_X Y, Z \rangle + \langle Y, \nabla_X Z \rangle$;

for all $X, Y, Z \in \mathcal{X}(M)$. Such a connection is called *metric connection*, or *Levi-Civita connection*, and, by (i) and (ii), it can be proved to satisfy the following formula:

$$2 \langle \nabla_Y Z, X \rangle = Y \langle Z, X \rangle + Z \langle X, Y \rangle - X \langle Y, Z \rangle$$
$$- \langle Y, [Z, X] \rangle + \langle Z, [X, Y] \rangle + \langle X, [Y, Z] \rangle . \tag{B.12}$$

By applying this formula to the special case $\langle \nabla_{\partial_i} \partial_j, \partial_k \rangle$, since $[\partial_i, \partial_j] = 0$ and $g_{ij} = \langle \partial_i, \partial_j \rangle$, and using the definition of the Christoffel coefficients $\nabla_{\partial_i} \partial_j = \Gamma_{ij}^k \partial_k$, one easily obtains

$$\Gamma_{ij}^k = \frac{1}{2} g^{km} \left(\frac{\partial g_{jm}}{\partial x^i} + \frac{\partial g_{mi}}{\partial x^j} - \frac{\partial g_{ij}}{\partial x^m} \right) . \tag{B.13}$$

The notion of *covariant derivative* now can be given. Let $c : I \to M$ be a curve on M and $X \in \mathcal{X}(M)$. There is a unique map which associates to the vector field X along c another vector field $\nabla X/dt$ along c, that is called the covariant derivative of X along c, satisfying the following conditions:

(i) $\frac{\nabla}{dt}(X + Y) = \frac{\nabla X}{dt} + \frac{\nabla Y}{dt}$, where $Y \in \mathcal{X}(M)$;

(ii) $\frac{\nabla}{dt}(fX) = \frac{df}{dt}X + f\frac{\nabla X}{dt}$, where f is a differentiable function on I;

(iii) if $Y \in \mathcal{X}(M)$ and $X(t) = Y(c(t))$, then $\frac{\nabla X}{dt} = \nabla_{dc/dt}Y$.

Using a chart, $c(t)$ is represented in local coordinates by $(x^1(t), \ldots, x^n(t))$, $t \in I$, and the vector field is locally represented as $X = X^i\partial_i$, where $X^i = X^i(t)$ and $\partial_i = \partial_i(c(t))$. Thus, according to the above-listed properties, we have

$$\frac{\nabla X}{dt} = \frac{dX^i}{dt}\partial_i + X^i\frac{\nabla \partial_i}{dt} ,$$

and from the general properties of linear connections listed above, we have

$$\frac{\nabla \partial_i}{dt} = \nabla_{dc/dt}\partial_i = \nabla_{\dot{x}^j\partial_j}\partial_i = \frac{dx^j}{dt}\nabla_{\partial_j}\partial_i , \qquad (B.14)$$

whence

$$\frac{\nabla X}{dt} = \frac{dX^i}{dt}\partial_i + \frac{dx^j}{dt}\nabla_{\partial_j}\partial_i ,$$

and using again $\nabla_{\partial_i}\partial_j = \Gamma_{ij}^k\partial_k$, we obtain

$$\frac{\nabla X}{dt} = \left[\frac{dX^k}{dt} + \Gamma_{ij}^k\frac{dx^i}{dt}X^j\right]\partial_k . \qquad (B.15)$$

Now the notion of *parallel transport* can be naturally given. If we consider a curve $c : I \to M$ and $v_0 \in T_{c(0)}M$, the equation

$$\nabla_{\dot{c}}X \equiv \frac{\nabla X}{dt} = 0$$

defines for all $t \in I$ a unique vector field $X(t) \in T_{c(t)}M$ such that $X(0) \equiv v_0$. Thus if $(s, t) \in I \times I$, the following isomorphism between vector spaces exists,

$$\mathcal{P}_{s,t}^c : T_{c(s)}M \to T_{c(t)}M ,$$

which associates with every $v_s \in T_{c(s)}M$ the unique vector $v_t \in T_{c(t)}M$ given by $X(c(t))$, where X satisfies the conditions $\nabla_{\dot{c}}X = 0$ and $X(c(s)) = v_s$.

The isomorphism $\mathcal{P}_{s,t}^c$ is called *parallel translation* along c from $c(s)$ to $c(t)$, and v_t is said to be obtained from v_s by parallel transport along the curve c.

Parallel translation is a linear isometry.

The covariant derivative can be extended linearly to tensors, that is, by requiring that the directional covariant derivative satisfies

(i) $\nabla_X f = X(f)$;

(ii) $\nabla_X(T_1 + T_2) = \nabla_X T_1 + \nabla_X T_2$, where T_1 and T_2 ; are tensors of any rank

(iii) $\nabla_X(T_1 \otimes T_2) = (\nabla_X T_1) \otimes T_2 + T_1 \otimes \nabla_X T_2$.

moreover ∇_X is required to commute with tensor contraction. Putting $f = \langle \omega, Y \rangle \in \mathcal{F}(M)$, with $\omega \in \Lambda^1(M)$ and $Y \in \mathcal{X}(M)$, with the aid of the Leibniz rule one obtains

$$(\nabla_X \omega)_i = X^j \partial_j \omega_i - X^j \Gamma_{ji}^k \omega_k \ ,$$

and with $\omega = dx^i$ we have $\nabla_{\partial_j} dx^k = -\Gamma_{ji}^k dx^i$. By resorting to this result and to the above-listed requirements, the covariant derivative of a (q,p)-type tensor T on M relative to X is defined as

$$(\nabla_X T)(X_1, \ldots, X_q, \omega_1, \ldots, \omega_p) = \nabla_X(T(X_1, \ldots, X_q, \omega_1, \ldots, \omega_p))$$

$$- \sum_{i=1}^{q} T(X_1, \ldots, \nabla_X X_i, \ldots, X_q, \omega_1, \ldots, \omega_p)$$

$$- \sum_{i=1}^{p} T(X_1, \ldots, X_q, \omega_1, \ldots, \nabla_X \omega_i, \ldots, \omega_p),$$

which in components reads

$$\nabla_{\partial_k} T_{j_1 \cdots j_p}^{i_1 \cdots i_q} = \frac{\partial T_{j_1 \cdots j_p}^{i_1 \cdots i_q}}{\partial x^k} - T_{j j_2 \cdots j_p}^{i_1 \cdots i_q} \Gamma_{j_1 k}^j - \cdots - T_{j_1 \cdots j_{p-1} j}^{i_1 \cdots i_q} \Gamma_{j_p k}^j$$

$$+ T_{j_1 \cdots j_p}^{i i_2 \cdots i_q} \Gamma_{ik}^{i_1} + \cdots + T_{j_1 \cdots j_p}^{i_1 \cdots i_{q-1} i} \Gamma_{ik}^{i_q} \ . \qquad \text{(B.16)}$$

B.2.1 Geodesics

A curve $\gamma : I \to M$ is said to be a *geodesic* with respect to the connection ∇ on M if $\dot{\gamma}$ is parallel along γ, that is, if

$$\nabla_{\dot{\gamma}} \dot{\gamma} = 0 \ .$$

According to this definition, *geodesics* are *autoparallel curves*, i.e., curves such that the tangent vector $\dot{\gamma}$ is always parallel-transported. With a chart (U, φ) such that $x = \varphi(\gamma(t_0))$, in U we have $\gamma(t) = (x^1(t), \ldots, x^n(t))$ and

$$\dot{\gamma}(t) = \frac{dx^i}{dt}(t) \partial_{i_{|\gamma(t)}} \ ,$$

so that

$$0 = \nabla_{\dot{\gamma}} \dot{\gamma} = \frac{\nabla \dot{\gamma}}{dt} = \left[\frac{d}{dt} \frac{dx^k}{dt}(t) + \Gamma_{ij}^k(\gamma(t)) \frac{dx^i}{dt} \frac{dx^j}{dt}(t) \right] \partial_{k_{|\gamma(t)}} \ , \qquad \text{(B.17)}$$

whence the geodesic lines are solutions of the equations

$$\frac{d^2x^k}{dt^2} + \Gamma_{ij}^k \frac{dx^i}{dt}\frac{dx^j}{dt} = 0 \ , \tag{B.18}$$

which is a system of second-order differential equations (all the indexes run from 1 to dim M).

From the theorem of existence and uniqueness of the solutions of ordinary differential equations we can infer that for every point $x \in M$ and for every $v \in T_x M$, there exists exactly one geodesic with initial conditions (x, v), that is, a unique geodesic $\gamma(t)$ defined in an interval $(-\varepsilon, \varepsilon)$, with $\varepsilon > 0$, such that $\gamma(0) = x$ and $\dot\gamma(0) = v$.

Since the norm of the tangent vector $\dot\gamma$ of a geodesic is constant, $|d\gamma/dt| = C$, the arc length of a geodesic is proportional to the parameter:

$$s(t) = \int_{t_1}^{t_2} \left|\frac{d\gamma}{dt}\right| dt = C(t_2 - t_1) \ . \tag{B.19}$$

When the parameter is actually the arc length, i.e., $C = 1$, we say that the geodesic is *normalized*. Whenever we consider a geodesic, we assume that it is normalized, if not explicitly stated otherwise. It can be shown that the equations (B.18) are the Euler–Lagrange equations for the length functional defined on the set of curves $\gamma(s)$ joining two fixed endpoints A and B and parametrized by the arc length. In other words, geodesics stem from the variational condition

$$\delta\ell_{AB}(\gamma) = \delta \int_A^B ds = 0, \tag{B.20}$$

and thus are curves of extremal length on a manifold, or, loosely speaking, the straightest possible lines in a non-Euclidean space.

Given a geodesic $\gamma(s)$ on M, there exists a unique vector field G on TM such that its trajectories are $s \mapsto (\gamma(s), \dot\gamma(s))$. Such a vector field is called the *geodesic field* on TM; its flow satisfies the system of equations

$$\frac{dx^k}{ds} = y^k \ ,$$

$$\frac{dy^k}{ds} = -\Gamma_{ij}^k y^i y^j \ , \qquad i, j, k = 1, \ldots, \dim M \ , \tag{B.21}$$

and is called *geodesic flow* on TM.

B.2.2 The Exponential Map

In the literature on Riemannian geometry, a commonly used tool is the so-called *exponential map*, which allows one to define the *normal neighborhood* of any point of a manifold.

Consider an n-dimensional Riemannian manifold (M, g). For each $p \in M$ and $X \in T_p M$ there is a unique geodesic issuing from p with velocity X at the instant $t = 0$, that is, $\gamma(0) = p$ and $\dot{\gamma}(0) = X$. With a suitable $c \in \mathbb{R}$, the geodesic equations entail $\gamma_{cX}(t) = \gamma_X(ct)$; thus the endpoint $\gamma_{cX}(1)$ is defined if $\gamma_X(c)$ is defined. As a consequence, $\gamma_X(1)$ is a well-defined point in M if X is sufficiently small.

Now, for each $X \in T_p M$ we define $q = \exp_p X$ as the point $q \in M$ given by $\gamma_X(1)$. The mapping $\exp_p$ is said to be the *exponential map* at p.

Given a Riemannian manifold M, for all points $p \in M$ there exists a number $\varepsilon > 0$ and a corresponding open neighborhood $B_\varepsilon(0) \subset T_p M$ of the origin of $T_p M$ such that if $X \in B_\varepsilon(0)$, then $\|X\| < \varepsilon$, and there exists an open neighborhood $U \subset M$ of p such that the mapping $\exp_p : B_\varepsilon(0) \subset T_p M \to U \subset M$ is a diffeomorphism of $B_\varepsilon(0)$ onto U.

Given any $p \in M$, a neighborhood $B_\varepsilon(0)$ of the null vector in $T_p M$ is said to be *normal* if:

(i) it is *starlike*, that is, if for an arbitrary vector $Y \in B_\varepsilon(0)$ and an arbitrary number $0 \le c \le 1$, the vector cY still belongs to $B_\varepsilon(0)$;
(ii) $\exp_p$ is defined on $B_\varepsilon(0)$;
(iii) $\exp_p$ diffeomorphically maps $B_\varepsilon(0)$ onto an open neighborhood $U(p) \in M$.

The neighborhood $U(p) = \exp_p B_\varepsilon(0)$ is called a *normal neighborhood* of p.

Let $\{e_{(1)}, \dots, e_{(n)}\}$ be an orthonormal basis for $T_p M$. Considering $X_q \in T_p M$ such that $\gamma_{X_q}(1) = q$, we can associate to each point $q \in M$ a tangent vector $X_q = X_q^i e_{(i)}$, at least if p and q are sufficiently close, and the mapping $\varphi : q \to X_q^i$ provides a local coordinate system for $U(p)$. These are called *normal coordinates* of the point q based at p. Their interesting property is that they locally give $g_{ik}(p) = \delta_{ik}$ and $\Gamma_{jk}^i = 0$, so that the covariant derivatives coincide with standard derivatives and the geodesics are locally represented as linear functions of t, that is, $\varphi(\gamma(t)) = \{X_q^i t\}$. Details can be found, for example, in [126].

B.3 Curvature

The concept of curvature plays a central role in Riemannian geometry. Roughly speaking, curvature tells us how much a manifold fails to be locally Euclidean. For example, Lie derivatives satisfy the identity $\mathcal{L}_{[X,Y]} = [\mathcal{L}_X, \mathcal{L}_Y] = \mathcal{L}_X \mathcal{L}_Y - \mathcal{L}_Y \mathcal{L}_X$. For coordinate vector fields, since $[\partial_i, \partial_j] = 0$, we see that $\mathcal{L}_X$ and $\mathcal{L}_Y$ commute. However, this is not the case for the covariant derivative ∇_X. Thus, if ∇ is the Levi-Civita connection on a Riemannian manifold (M, g), the function $R : \mathcal{X}(M) \times \mathcal{X}(M) \times \mathcal{X}(M) \to \mathcal{X}(M)$ defined as

$$R(X, Y)Z = \nabla_X \nabla_Y Z - \nabla_Y \nabla_X Z - \nabla_{[X,Y]} Z , \tag{B.22}$$

is a $(1, 3)$-type tensor field on M known as the *Riemann curvature tensor*. Observe that R measures the noncommutativity of the covariant derivative.

Moreover, if $M = \mathbb{R}^n$, then $R(X, Y)Z = 0$ for all the tangent vectors X, Y, Z, because of the commutativity of the ordinary derivatives. In a local coordinate system $(x^1, \ldots, x^n)$, we have $R(\partial_i, \partial_j)\partial_k = R^\ell_{ijk}\partial_\ell$ and since $\left[\frac{\partial}{\partial x^i}, \frac{\partial}{\partial x^j}\right] = 0$, we obtain

$$R\left(\frac{\partial}{\partial x^i}, \frac{\partial}{\partial x^j}\right)\frac{\partial}{\partial x^k} = \left[\nabla_{\partial/\partial x^i}\nabla_{\partial/\partial x^j} - \nabla_{\partial/\partial x^j}\nabla_{\partial/\partial x^i}\right]\frac{\partial}{\partial x^k} . \tag{B.23}$$

The components R^ℓ_{ijk} of the Riemann tensor are found by computing

$$
\begin{aligned}
R^\ell_{ijk} &= \langle dx^\ell, R(\partial_i, \partial_j)\partial_k\rangle = \langle dx^\ell, \nabla_{\partial_i}\nabla_{\partial_j}\partial_k - \nabla_{\partial_j}\nabla_{\partial_i}\partial_k\rangle \\
&= \langle dx^\ell, \nabla_{\partial_i}(\Gamma^m_{jk}\partial_m) - \nabla_{\partial_j}(\Gamma^m_{ik}\partial_m)\rangle \\
&= \langle dx^\ell, (\partial_i\Gamma^m_{jk})\partial_m + \Gamma^m_{jk}\Gamma^h_{im}\partial_h - (\partial_j\Gamma^m_{ik})\partial_m - \Gamma^m_{ik}\Gamma^h_{jm}\partial_h\rangle ,
\end{aligned}
$$

where we have used (B.10) and (B.11), and hence we finally get

$$R^\ell_{ijk} = \left(\frac{\partial\Gamma^\ell_{jk}}{\partial x^i} - \frac{\partial\Gamma^\ell_{ik}}{\partial x^j}\right) + \left(\Gamma^r_{jk}\Gamma^\ell_{ir} - \Gamma^r_{ik}\Gamma^\ell_{jr}\right) . \tag{B.24}$$

Thus, given a metric g, the curvature R is uniquely defined. A manifold (M, g) is called *flat* when the curvature tensor vanishes.

Let us list some important relations that are satisfied by the Riemann curvature tensor.

These are, for all $X, Y, Z, W \in \mathcal{X}(M)$, the following:

$$R(X, Y)Z + R(Y, X)Z = 0 , \tag{B.25}$$

$$R(X, Y)Z + R(Y, Z)X + R(Z, X)Y = 0 , \tag{B.26}$$

$$\langle R(X, Y)Z, W\rangle + \langle R(Y, X)W, Z\rangle = 0 , \tag{B.27}$$

$$\langle R(X, Y)Z, W\rangle - \langle R(Z, W)X, Y\rangle = 0 , \tag{B.28}$$

which, apart from (B.25), which trivially follows from the definition of R, in local coordinates give

$$R_{hijk} + R_{hjki} + R_{hkij} = 0 , \tag{B.29}$$

known as the first Bianchi identity and which is the same as (B.26);

$$R_{kjhi} = -R_{jkhi} , \tag{B.30}$$

which is the same as (B.27); and

$$R_{kjhi} = R_{ihjk} = R_{hikj} , \tag{B.31}$$

which is the same as (B.28). Moreover, there is a second Bianchi identity, a very useful one, which in coordinates reads as (setting $\nabla_i = \nabla_{\partial/\partial x^i}$)

$$\nabla_l R_{hijk} + \nabla_j R_{hikl} + \nabla_k R_{hilj} = 0 . \tag{B.32}$$

B.3.1 Sectional Curvature

Given a Riemannian manifold M and $p \in M$, let $\pi(p) \subset T_pM$ be a two-dimensional plane spanned by an orthonormal basis X, Y. The *sectional curvature* $K(p, \pi)$ at p with respect to the plane π is defined as

$$K(p, \pi) = R(X, Y, X, Y) . \tag{B.33}$$

Let us verify that this definition is sensible, that is, independent of the choice of X, Y. In fact, a change of basis from X, Y to X', Y' has to satisfy the conditions

$$X' = aX + bY , \qquad Y' = -bX + aY ,$$

with $a^2 + b^2 = 1$. Now

$$
\begin{aligned}
R(X', Y', X', Y') &= R(aX + bY, -bX + aY, aX + bY, -bX + aY) \\
&= R(aX, aY, aX, aY) + R(aX, aY, bY, -bX) \\
&\quad + R(bY, -bX, aX, aY) + R(bY, -bX, bY, -bX) \\
&= (a^2 + b^2)^2 R(X, Y, X, Y) , \tag{B.34}
\end{aligned}
$$

where we have used (B.25) and (B.26). We have found that $R(X', Y', X', Y') = R(X, Y, X, Y)$, that is, the definition of sectional curvature is independent of the choice of X and Y and thus it is an intrinsic geometric property of M at p.

The remarkable fact is that the knowledge of the sectional curvatures at $p \in M$ for all 2-planes $\pi(p) \subset T_pM$ completely determines the curvature tensor R at p. To prove this, it is enough simply to show that $R(X, Y, X, Y) = 0$ implies $R(X, Y, Z, W) = 0$ for all $X, Y, Z, W \in \mathcal{X}(M)$. Thus let $R(X, Y, X, Y) = 0$. Using (B.28) we have

$$
\begin{aligned}
0 = R(X, Y + W, X, Y + W) &= R(X, Y, X, W) + R(X, W, X, Y) \\
&= 2R(X, Y, X, W);
\end{aligned}
$$

hence $R(X, Y, X, Y) = 0 \Rightarrow R(X, Y, X, W) = 0$ for all X, Y, W. Then, replacing X in $R(X, Y, X, W) = 0$ by $X + Z$, consider

$$
\begin{aligned}
0 = R(X + Z, Y, X + Z, W) &= R(X, Y, Z, W) + R(Z, Y, X, W) \\
&= R(X, Y, Z, W) + R(X, W, Z, Y) \\
&= R(X, Y, Z, W) - R(X, W, Y, Z), \tag{B.35}
\end{aligned}
$$

where we have used (B.28) and (B.26).

The last equality means that $R(X, Y, Z, W) = R(X, W, Y, Z)$. Then one replaces Y, Z, W by Z, W, Y respectively, so that $R(X, Y, Z, W) = R(X, Z, W, Y)$. Using this together with the last equality in (B.35) and using the Bianchi identity (B.26), we finally obtain

$$3R(X, Y, Z, W) = R(X, Y, Z, W) + R(X, W, Y, Z) + R(X, Z, W, Y) = 0 .$$

Summarizing, we have seen that $R(X,Y,X,Y) = 0$ implies $R(X,Y,Z,W) = 0$ for all $X,Y,Z,W \in \mathcal{X}(M)$. Thus the sectional curvatures determine the curvature tensor.

Consider then the case of two generic vectors, v_1 and v_2, that span $\pi(p) \subset T_pM$. They can be orthonormalized by putting

$$X = \frac{v_1}{\langle v_1, v_1 \rangle^{1/2}},$$

$$Y = \frac{\langle v_1, v_1 \rangle v_2 - \langle v_1, v_2 \rangle v_1}{\langle v_1, v_1 \rangle^{1/2}[\langle v_1, v_1 \rangle \langle v_2, v_2 \rangle - \langle v_1, v_2 \rangle^2]^{1/2}}.$$

Now that X, Y are orthonormal, we have

$$K(p, \pi) = R(X,Y,X,Y) = \frac{R(v_1, v_2, v_1, v_2)}{\langle v_1, v_1 \rangle \langle v_2, v_2 \rangle - \langle v_1, v_2 \rangle^2}, \qquad (B.36)$$

and with $v_1 = \lambda^i \partial/\partial x^i$, $v_2 = \mu^j \partial/\partial x^j$, in components this is

$$K(p, \pi) = \frac{R_{ijkl}\lambda^i \mu^j \lambda^k \mu^l}{(g_{ab}\lambda^a \lambda^b)(g_{mn}\mu^m \mu^n) - (g_{rs}\lambda^r \mu^s)^2}. \qquad (B.37)$$

Let $n = \dim M$, and let $(z_{(1)}, \ldots, z_{(n)})$ be an orthonormal set of vectors in T_pM. Put $x = z_{(n)}$ and let $(z_{(1)}, \ldots, z_{(n-1)})$ be a basis of the hyperplane in T_pM orthogonal to x. The quantity

$$K_{\mathrm{R}}(x) = \sum_{i=1}^{n-1} R(x, z_{(i)}, x, z_{(i)}) = \sum_{i=1}^{n-1} \langle R(x, z_{(i)})x, z_{(i)} \rangle \qquad (B.38)$$

is called the *Ricci curvature* at p in the direction x. This scalar quantity can be also derived from the contraction of a $(0, 2)$-type tensor, called the Ricci curvature tensor, obtained from the Riemann tensor as

$$\mathrm{Ric}(X, Y) = \langle dx^i, R(e_{(i)}, Y)X \rangle \qquad (B.39)$$

which in components is

$$R_{ik} = \mathrm{Ric}(e_{(i)}, e_{(k)}) \equiv R^j_{ijk}. \qquad (B.40)$$

The Ricci curvature $K_{\mathrm{R}}(x)$ is thus

$$K_{\mathrm{R}}(x) = R_{ik}\, x^i\, x^k. \qquad (B.41)$$

For a generic vector $V = v^i \partial/\partial x^i$, the Ricci curvature in the direction of the tangent vector V is

$$K_{\mathrm{R}}(V) = \frac{R_{ik}\, v^i\, v^k}{g_{ik}\, v^i\, v^k}.$$

Contracting the Ricci tensor gives

$$\mathcal{R} \equiv g^{ik} \, \mathrm{Ric}(e_{(i)}, e_{(k)}) = g^{ik} \, R_{ik} \, , \tag{B.42}$$

which is called the *scalar curvature*. From (B.38), (B.39), and (B.42) we see that the scalar curvature is the sum of all the $n(n-1)$ independent sectional curvatures, that is,

$$\mathcal{R} = \sum_{j=1}^{n} K_R(z_{(j)}) = \sum_{j=1}^{n} \sum_{i=1}^{n-1} R(z_{(i)}, z_{(j)}, z_{(i)}, z_{(j)})$$

$$= \sum_{j=1}^{n} \sum_{i=1}^{n-1} R_{hkml} \, z_{(i)}^{h} \, z_{(j)}^{k} \, z_{(i)}^{m} \, z_{(j)}^{l} \, .$$

B.3.2 Isotropic Manifolds

In view of the use which is made of Schur's theorem in Chapter 5, we give some emphasis to it.

If the sectional curvatures $K(p, \pi)$ at a point $p \in M$ happen to be independent of the choice of the 2-plane π, then we say that M has *isotropic curvature* at p. The remarkable—and somewhat surprising—fact is that if M is isotropic at some point p, then it is a *constant-curvature manifold*, that is, $K(p, \pi)$ is also independent of the point p, for it is the same everywhere on M.

This can be seen by considering the tensor field R' defined as

$$R'(W, Z, X, Y) = \langle W, X \rangle \langle Z, Y \rangle - \langle Z, X \rangle \langle Y, W \rangle \, , \tag{B.43}$$

where $X, Y, Z, W \in \mathcal{X}(M)$, which satisfies the properties (B.25–B.28). From (B.43) we have

$$R'(X, Y, X, Y) = \langle X, X \rangle \langle Y, Y \rangle - \langle X, Y \rangle^2 \, ,$$

so that $K'(\pi) = 1$. The assumption of isotropy at p means that $R(X, Y, X, Y) = K(p)R'(X, Y, X, Y)$, where $K(p)$ is some scalar function on M. As we have seen above, if two tensors have the same sectional curvatures, then they are the same, that is, $R(X, Y, Z, W) = K(p)R'(X, Y, Z, W)$.

For any vector field $V \in \mathcal{X}(M)$,

$$(\nabla_V R)(W, Z, X, Y) = (VK(p))R'(W, Z, X, Y) \, , \tag{B.44}$$

since $\nabla_V R' = 0$.[1] By the second Bianchi identity

$$\nabla_V R(W, Z, X, Y) + \nabla_X R(W, Z, Y, V) + \nabla_Y R(W, Z, V, X) = 0 \, ,$$

---

[1] This immediately follows from the fact that for all $X, Y, Z \in \mathcal{X}(M)$ we have $\nabla_Z g(X, Y) = 0$.

for all $X, Y, V, W, Z \in \mathcal{X}(M)$, one has

$$
\begin{aligned}
0 = \quad & (VK(p))[\langle W, X\rangle\langle Z, Y\rangle - \langle Z, X\rangle\langle Y, W\rangle] \\
+ \; & (XK(p))[\langle W, Y\rangle\langle Z, V\rangle - \langle Z, Y\rangle\langle W, V\rangle] \\
+ \; & (YK(p))[\langle W, V\rangle\langle Z, X\rangle - \langle Z, V\rangle\langle W, X\rangle] \; .
\end{aligned}
\tag{B.45}
$$

For $n \geq 3$ it is always possible to fix X and choose Y, Z, V so that X, Y, Z are mutually orthogonal, that is, $\langle X, Y\rangle = \langle X, Z\rangle = \langle Y, Z\rangle = 0$ and $\langle Z, Z\rangle = 1$. Then, by putting also $V = Z$ at p, (B.45) simplifies to

$$
(XK(p))Y - (YK(p))X = 0 \; ,
$$

whence, since X and Y are linearly independent, $(XK(p)) = (YK(p)) = 0$, so that $K(p) =$const.

It is instructive to give an alternative proof as follows. In components, $R = KR'$ and the definition of R' in (B.43) give

$$
R_{ijhl} = K(g_{ih}g_{jl} - g_{il}g_{jh}) \; ,
\tag{B.46}
$$

and contracting it,

$$
\begin{aligned}
R_{il} = R^j_{ijl} &= K(g_{ih}g_{jl} - g_{il}g_{jh})g^{jh} \\
&= K(g_{ih}g^h_l - ng_{il}) = -(n-1)Kg_{il} \; ,
\end{aligned}
\tag{B.47}
$$

and further contraction gives the scalar curvature

$$
\mathcal{R} = -n(n-1)K \; .
$$

Now form the Einstein tensor

$$
\begin{aligned}
G^i_l = R^i_l - \frac{1}{2}\mathcal{R}g^i_l &= -n(n-1)Kg^i_l + \frac{1}{2}n(n-1)Kg^i_l \\
&= \frac{1}{2}(n-2)(n-1)Kg^i_l \; ,
\end{aligned}
\tag{B.48}
$$

and since we can prove that $\nabla_i G^i_l = 0$, it immediately follows that $\nabla_i K = 0$ and thus $K =$const on the manifold.

To prove that $\nabla_i G^i_l = 0$ we use the second Bianchi identity, which for mixed components reads $\nabla_r R^{ij}_{hl} + \nabla_h R^{ij}_{lr} + \nabla_l R^{ij}_{rh} = 0$. By contracting the indexes j and h we have

$$
\nabla_r R^i_l + \nabla_j R^{ij}_{lr} - \nabla_l R^i_r = 0 \; .
$$

Then a further contraction on i and l gives

$$
\nabla_r \mathcal{R} - \nabla_j R^i_r - \nabla_i R^i_r = 0 \; ,
$$

that is,

$$\nabla_i R_r^i - \frac{1}{2}\nabla_r \mathcal{R} = 0 \ ,$$

which can be also written

$$\nabla_i \left(R_r^i - \frac{1}{2}g_r^i \mathcal{R} \right) = 0 \ .$$

Finally, putting $G_l^i = R_r^i - \frac{1}{2}g_r^i\mathcal{R}$, we have $\nabla_i G_l^i = 0$.

A *constant-curvature* manifold is also called an *isotropic manifold*, and the components of its Riemann curvature tensor have the remarkably simple form given by (B.46). Then the components of the Ricci tensor are

$$R_{ik} = K \, g_{ik} \ , \tag{B.49}$$

and all the above-defined curvatures are constants, and are related by

$$K = \frac{1}{n-1}K_{\mathrm{R}} = \frac{1}{n(n-1)}\mathcal{R} \ . \tag{B.50}$$

Given a *positive* function f^2, a *conformal transformation* is the transformation

$$(M, g) \mapsto (M, \tilde{g}) \ ; \qquad \tilde{g} = f^2 g \ . \tag{B.51}$$

Two Riemannian manifolds are said to be *conformally related* if they are linked by a conformal transformation. In particular, a manifold is (M, g) *conformally flat* if it is possible to find a conformal transformation that sends g into a flat metric. Conformally, flat manifolds exhibit some remarkable simplifications for the calculation of the curvature tensor components (see, e.g., [163]).

B.4 The Jacobi–Levi-Civita Equation for Geodesic Spread

An important geometric aspect of the Riemann tensor concerns *geodesic deviation*, that is, the fact that nearby geodesics that start parallel do not stay parallel. In order to measure the geodesic deviation, we need to consider a *congruence* of geodesics and a vector field, called the *Jacobi* field or *geodesic separation* vector field, and to describe the way of changing this vector along a geodesic. Let us proceed to define the geodesic separation field and then derive its evolution equation along a reference geodesic.

Let M be a *complete*[2] Riemannian manifold. Let $\gamma : \mathbb{R} \to M$ be a maximal geodesic. A mapping

---

[2] A connected Riemannian manifold is complete if every maximal geodesic in it is defined on the entire real axis. As a consequence, any two points in a complete Riemannian manifold can be joined at least by one geodesic. A geodesic is said to be *maximal* if it is not the restriction of any other geodesic defined on a larger interval of $\mathbb{R}$.

$$\varphi : \mathbb{R} \times I \to M \ , \qquad\qquad (B.52)$$

$I = (-\tau_0, \tau_0)$, $\mathbb{R} \times I \subset \mathbb{R}^2$, is said to be a *variation* of the geodesic γ and, in particular, is said to be a *Jacobi variation* if for any $\tau \in I$ the curve

$$\varphi_\tau(s) = \varphi(s, \tau) \ , \qquad s \in \mathbb{R} \ ,$$

is a geodesic. For every fixed $s_0 \in \mathbb{R}$, the Jacobi variation defines a smooth curve

$$\alpha(\tau) = \varphi(s_0, \tau) \ , \qquad \tau \in \mathbb{R} \ ,$$

and along $\alpha(\tau)$, a vector field

$$A(\tau) = \frac{\partial \varphi}{\partial s}(s_0, \tau) \ , \qquad \tau \in \mathbb{R}.$$

Since $A(\tau) = \dot{\varphi}_\tau(s_0)$, the vector $A(\tau)$ unambiguously determines the geodesic $\varphi_\tau(s)$ for an arbitrary τ. Together, the curve $\alpha(\tau)$ and the vector field $A(\tau)$ unambiguously define the Jacobi variation φ. These are related to the geodesic $\gamma(s)$ by

$$\alpha(0) = \gamma(s_0) \ , \qquad A(0) = \dot{\gamma}(s_0) \equiv \frac{\partial \varphi}{\partial s}(s_0, 0) \ .$$

This Jacobi variation is defined by the formula

$$\varphi(s, \tau) = \gamma_{\alpha(\tau), A(\tau)}(s) \ , \qquad\qquad (B.53)$$

where $\gamma_{\alpha, A}(s)$ is the maximal geodesic passing though the point $\alpha(\tau)$ for $s = s_0$ with $A(\tau)$ as tangent vector at that point. The vector field $J(s)$ on a geodesic $\gamma(s)$,

$$J(s) = \frac{\partial \varphi}{\partial \tau}(s, 0) \ , \qquad s \in \mathbb{R} \ , \qquad\qquad (B.54)$$

is called a Jacobi field.

For any two vectors $X, Y \in M_{\gamma(s_0)}$, there exists on the geodesic $\gamma(s)$ a Jacobi field $J(s)$ such that

$$J(s_0) = X, \qquad \frac{\nabla J}{ds}(s_0) = Y \ .$$

This field can be constructed by considering a curve $\alpha(\tau)$ such that

$$\alpha(0) = \gamma(s_0), \qquad \dot{\alpha}(0) = X \ ,$$

and by constructing on $\alpha(\tau)$ a vector field $A(\tau)$ such that

$$A(0) = \dot{\gamma}(s_0), \qquad \frac{\nabla A}{ds}(0) = Y \ .$$

Let us show that an arbitrary Jacobi field $J(s)$ on a geodesic $\gamma(s)$ satisfies the Jacobi–Levi-Civita equation

$$\frac{\nabla}{ds}\frac{\nabla}{ds}J(s) + R(J(s),\dot{\gamma}(s))\dot{\gamma}(s) = 0 \; . \tag{B.55}$$

To prove this, let us begin by defining a *parametrized surface* in M as a differentiable mapping $\phi : A \subset \mathbb{R}^2 \to M$, where A is a connected set with piecewise differentiable boundary. A *vector field V along ϕ* is a mapping that associates to each $x \in A$ a differentiable vector $V(x) \in T_{\phi(x)}M$. Denote by (s,τ) the coordinates in $\mathbb{R}^2$. At fixed τ_0, the mapping $s \mapsto \phi(s,\tau_0)$ is a curve in M, and $(\partial\phi/\partial s)$ is a vector field along this curve. Likewise, for fixed s_0 the mapping $\tau \mapsto \phi(s_0,\tau)$ defines another curve in M and the corresponding vector field $(\partial\phi/\partial\tau)$. Given a vector field V along $\phi : A \to M$, we can define the covariant derivatives $\nabla V/\partial s$ and $\nabla V/\partial\tau$ along the curves $s \mapsto \phi(s,\tau_0)$ and $\tau \mapsto \phi(s_0,\tau)$ respectively, by considering the restrictions of V to these curves. For a Levi-Civita connection, a direct computation shows that

$$\frac{\nabla}{\partial\tau}\frac{\partial\phi}{\partial s} = \frac{\nabla}{\partial s}\frac{\partial\phi}{\partial\tau} \; . \tag{B.56}$$

Now, identifying ϕ with the variation mapping φ given above, this equation implies

$$\frac{\nabla\dot{\gamma}}{\partial\tau} = \frac{\nabla J}{\partial s} \; ,$$

and since $(\partial\phi/\partial s) = (\partial x^i/\partial s)\partial_i$ and $(\partial\phi/\partial\tau) = (\partial x^j/\partial\tau)\partial_j$, we have

$$\nabla_J\dot{\gamma} = \nabla_{\dot{\gamma}}J \; . \tag{B.57}$$

Now, using this result, and the fact that $\nabla_{\dot{\gamma}}\dot{\gamma} = 0$ because γ is a geodesic, we can write

$$\nabla_{\dot{\gamma}}^2 J = \nabla_{\dot{\gamma}}\nabla_{\dot{\gamma}}J = \nabla_{\dot{\gamma}}\nabla_J\dot{\gamma} = [\nabla_{\dot{\gamma}},\nabla_J]\dot{\gamma} \; , \tag{B.58}$$

from which, using the definition of the curvature tensor (B.22) and since (B.57) entails the vanishing of $\nabla_{[\dot{\gamma},J]}$, we get

$$\nabla_{\dot{\gamma}}^2 J = R(\dot{\gamma}, J)\dot{\gamma} \; , \tag{B.59}$$

which is (B.55) written in compact notation (recall that $R(\dot{\gamma}, J) = -R(J,\dot{\gamma})$). In local coordinates, (B.55) reads

$$\frac{\nabla^2 J^k}{ds^2} + R^k_{\;ijr}\frac{dq^i}{ds}J^j\frac{dq^r}{ds} = 0 \; . \tag{B.60}$$

It is worth noticing that the normal component $J_\perp$ of J, i.e., the component of J orthogonal to $\dot{\gamma}$ along the geodesic γ, is again a Jacobi field, since we can always write $J = J_\perp + \lambda\dot{\gamma}$. One immediately finds, then, that the velocity $\dot{\gamma}$ satisfies the Jacobi equation, so that $J_\perp$ must obey the same equation. This tells us that the relevant information about geodesic separation is conveyed by normal Jacobi fields.

Conjugate points. Let $\gamma : I \to M$, $I \subset \mathbb{R}$, be a geodesic on a Riemannian manifold M. The point $\gamma(t_0) \in M$, with $t_0 \in I$, is said to be the *conjugate point* of $\gamma(0)$ along γ if there exists a nonidentically vanishing Jacobi field J along γ such that $J(0) = 0$ and $J(t_0) = 0$.

An important property of Jacobi fields is the following. Given a geodesic $\gamma : I \to M$ and an arbitrary piecewise differentiable vector field V along γ, for all $s \in I$, provided that $\gamma(0)$ has no conjugate points in I, that $\nabla_{\dot\gamma} V(0) = 0$, $V(0) = 0$, and $\langle V, \dot\gamma \rangle = 0$, the quantity

$$I_{s_0}(V, V) = \int_0^{s_0} ds \ \{\langle \nabla_{\dot\gamma} V, \nabla_{\dot\gamma} V \rangle - \langle R(\dot\gamma, V)\dot\gamma, V \rangle\} \tag{B.61}$$

attains its minimum value when $V = J$ on I, that is, if and only if V is a Jacobi field.

Equation (B.55) can also be written using a set of orthonormal, parallel fields $(e_{(1)}, \ldots, e_{(n)})$ along γ. One has

$$J(s) = \sum_{i=1}^{n} \xi_i(s) e_{(i)} \ ,$$

$$F_{ij} = \langle R(\dot\gamma(s), e_{(i)})\dot\gamma(s), e_{(j)} \rangle \ ,$$

and (B.55) becomes

$$\frac{d^2\xi_j(s)}{ds^2} + \sum_{i=1}^{n} F_{ij}\xi_i(s) = 0 \ .$$

An interesting result that helps one to understand intuitively the relation between geodesics and curvature is the following. Let $p \in M$, $\gamma : I \to M$ be a geodesic parametrized by the arc length, so that $\|\dot\gamma\| = 1$, with $\gamma(0) = p$, $\dot\gamma(0) = v$. Consider also $J(0) = 0$ and $\nabla J/ds(0) = w$ such that $\langle v, w \rangle = 0$, and also $\|w\| = 1$. Let $K(p, \pi) = \langle R(v, w)v, w \rangle$ be the sectional curvature at p with respect to the plane π generated by $\dot\gamma(0)$ and $\nabla_{\dot\gamma} J(0)$. The "short-time" Taylor expansion of $\|J(s)\|^2$ gives

$$\|J(s)\|^2 = s^2 - \frac{1}{3}K(p, \pi)s^4 + R(s), \qquad \lim_{s \to 0} R(s)/s^4 = 0 \ . \tag{B.62}$$

This roughly tells us that locally, geodesics spread more if $K(p, \pi) < 0$ than if $K(p, \pi) > 0$.

Equation (B.62) also tells that the smaller the value of $K(p, \pi)$, the larger is $\|J(s)\|^2$. In fact, consider another Riemannian manifold $\widetilde{M}$ with a geodesic $\widetilde\gamma : I \to \widetilde{M}$ and a Jacobi field $\widetilde{J}$ along $\widetilde\gamma$ such that $\widetilde{J}(0) = 0$, $\|\nabla_{\widetilde\gamma} \widetilde{J}(0)\| = 1$, and $\langle \dot{\widetilde\gamma}(0), \nabla_{\widetilde\gamma} \widetilde{J}(0) \rangle = 0$. If

$$\widetilde{K}(\dot{\widetilde\gamma}(0), \nabla_{\widetilde\gamma} \widetilde{J}(0)) \geq K(\dot\gamma(0), \nabla_{\dot\gamma} J(0)) \ ,$$

it follows from (B.62) that for short s the following inequality holds: $\|J(s)\|^2 \geq \|\widetilde{J}(s)\|^2$.

Rauch's comparison theorem extends this result to arbitrary s.

Theorem (Rauch). *Consider two Riemannian manifolds M^n and $\widetilde{M}^{n+k}$, with $k \geq 0$, and two geodesics $\gamma : I \to M$ $\widetilde{\gamma} : I \to \widetilde{M}$ such that $\|\dot{\gamma}(s)\| = \|\dot{\widetilde{\gamma}}(s)\|$, and given two Jacobi fields J and $\widetilde{J}$ along γ and $\widetilde{\gamma}$, respectively, with the conditions*

$$J(0) = \widetilde{J}(0) = 0 \ ,$$

$$\langle \dot{\gamma}(0), \nabla_{\dot{\gamma}} J(0) \rangle = \langle \dot{\widetilde{\gamma}}(0), \nabla_{\dot{\widetilde{\gamma}}} \widetilde{J}(0) \ , \rangle$$

$$\|\nabla_{\dot{\gamma}} J(0)\| = \|\nabla_{\dot{\widetilde{\gamma}}} \widetilde{J}(0)\| \ .$$

Let $\widetilde{\gamma}$ have no conjugate points in I, and for all s and all $x \in T_{\gamma(s)}M$, $\widetilde{x} \in T_{\widetilde{\gamma}(s)}M$, let

$$\widetilde{K}(\widetilde{x}, \dot{\widetilde{\gamma}}(s)) \geq K(x, \dot{\gamma}(s)) \ ,$$

with $K(x,y)$ and $\widetilde{K}(\widetilde{x}, \widetilde{y})$ the sectional curvatures relative to the planes generated by (x,y) and $(\widetilde{x}, \widetilde{y})$ respectively. Then

$$\|J(s)\| \geq \|\widetilde{J}(s)\| \ .$$

Rauch's comparison theorem allows one to prove the following useful proposition:

Given a Riemannnian manifold M and a geodesic γ on it, if the sectional curvature K of the manifold is bounded by two positive constants L and H, that is, $0 < L \leq K \leq H$, then the distance d between two consecutive conjugate points of γ satisfies the condition

$$\frac{\pi}{\sqrt{H}} \leq d \leq \frac{\pi}{\sqrt{L}} \ .$$

B.5 Topology and Curvature

In the preceding appendix, we have seen that certain properties of the differentable structures defined on manifolds can give information about the topology of the same manifolds. Besides analysis on manifolds, geometric methods can be used to "probe" topology. In particular, on Riemannian manifolds different kinds of relationships can be established between their curvature properties and topology. This is a vast subject, and as far as the needs of the present monograph are concerned, stating the Gauss–Bonnet–Hopf result for hypersurfaces of $\mathbb{R}^n$—as we did in Chapters 7 and 8—would be enough. However, in view of future possible developments of the subjects treated in the main text of this book, it is not out of place to give a few more elements about this topic. We list some classical results without entering into details.

B.5.1 The Gauss–Bonnet Theorem

In dimension 2, the Gauss–Bonnet formula tells everything about the relationship between curvature and topology. If (M, g) is a compact Riemannian manifold, its Euler characteristic $\chi(M)$ is given by

$$\chi(M) = \frac{1}{2\pi} \int_M K(u, v)\, d\sigma , \qquad (B.63)$$

where u and v are local coordinates on M, $d\sigma = \sqrt{\det g}\, du\, dv$ is the invariant volume element, and $K(u, v)$ is the Gaussian curvature of the manifold. In the compact two-dimensional case the relation between χ and the Betti numbers is

$$\chi = 2 - 2g = b_0 - b_1 + b_2 , \qquad (B.64)$$

where g is the *genus*, which equals the number of holes of the surface. In this case, the following equality holds: $b_1 = 2g$.

The relation in (B.63) can be generalized (including manifolds with boundary) to $2n$ dimensions:

Theorem (Gauss–Bonnet–Chern–Avez). *For a $2n$-dimensional compact oriented Riemannian manifold M of Euler class γ, its Euler characteristic $\chi(M)$ is given by*

$$\chi(M) = \int_M \gamma .$$

The Euler class of M (even dimensional) is the Euler characteristic class of TM,

$$\gamma = \frac{(-1)^n}{(4\pi)^n n!} \epsilon^{1\cdots 2n}_{i_1 \cdots i_p} \Omega^{i_1}_{i_2} \wedge \cdots \wedge \Omega^{i_{2n-1}}_{i_{2n}} ,$$

where $\Omega^j_i = \frac{1}{2} R^j_{ikl} e^{(k)} \wedge e^{(l)}$ are the so-called curvature 2-forms, R^j_{ikl} are the components of the Riemann curvature tensor, and $\epsilon^{1\cdots 2n}_{i_1 \cdots i_p}$ is the Kronecker tensor with $\epsilon^{1\cdots 2n}_{i_1 \cdots i_p} = 0$ if the string $(i_1 \ldots i_p)$ is not a permutation of the string $(1 \ldots 2n)$, $\epsilon^{1\cdots 2n}_{i_1 \cdots i_p} = +1$ if the string $(i_1 \ldots i_p)$ is an even permutation of the string $(1 \ldots 2n)$, and $\epsilon^{1\cdots 2n}_{i_1 \cdots i_p} = -1$ if the string $(i_1 \ldots i_p)$ is an odd permutation of the string $(1 \ldots 2n)$.

For $\dim M = 2$ we have $\gamma = (1/2\pi)\Omega^2_1 = (1/2\pi)K d\sigma$, where K is the Gaussian curvature and $d\sigma$ the volume form, thus recovering (B.63).

A much more computationally simple version of this theorem, known as the Gauss–Bonnet–Hopf theorem, holds for $2n$-dimensional hypersurfaces of $\mathbb{R}^{2n+1}$, for example defined as level sets $\Sigma_a = f^{-1}(a)$ of a function $f : B \subset \mathbb{R}^{2n+1} \to \mathbb{R}$, which gives

$$\int_{\Sigma_a} d\sigma \, K_G = \frac{1}{2} \mathrm{vol}(\mathbb{S}^{2n-1}_1)\chi(\Sigma_a) , \qquad (B.65)$$

where $\mathbb{S}_1^{2n-1}$ is a $(2n-1)$-dimensional hypersphere of unit radius, and $\chi(\Sigma_a)$ is the Euler characteristic of the level set Σ_a and K_G its Gauss–Kronecker curvature. We have encountered and used this remarkable formula in Chapters 7 and 8.

B.5.2 Hopf-Rinow Theorem

Let (M, g) be a Riemannian manifold and $\gamma : [a, b] \to M$ a smooth curve on it. The length $L(\gamma)$ of the curve γ is obtained by means of (B.9), a definition that can be easily generalized to piecewise smooth curves. Consider a connected Riemannian manifold M. If $x, y \in M$, define $C_{x,y} = \{\gamma : [a, b] \to M \mid \gamma \text{ piecewise smooth}, \gamma(a) = x, \gamma(b) = y, \ a, b \in \mathbb{R}\}$ and

$$d(x, y) := \inf L(\gamma) , \qquad \gamma \in C_{x,y} .$$

One can easily check that this definition of $d(x, y)$ is a *distance on M*; that is, $d(x, y) \geq 0$, $d(x, y) = d(y, x)$, $d(x, y) \leq d(x, z) + d(y, z)$, $d(x, y) = 0$ if and only if $x = y$. Moreover, the topology induced by d on M (which thus becomes a metric space) is the same as the manifold topology of M.

Let us now enunciate the Hopf–Rinow theorem, which topologically characterizes the connected and geodesically complete Riemannian manifolds.

Theorem. *Let M be a simply connected Riemannian manifold. Then the following affirmations are equivalent:*

(i)   *M is geodesically complete;*

(ii)   *(M, d) is a complete metric space (i.e. all the Cauchy sequences are convergent);*

(iii) *Every closed and bounded subset of M (bounded with respect to d) is compact;*

(iv)   *The exponential mapping $\exp_p$ is defined on the whole T_pM.*

Corollary. *If (M, g) is a compact and connected Riemannian manifold, then for all $x, y \in M$ there exists a geodesic $\gamma : [a, b] \to M$ such that $\gamma(a) = x$, $\gamma(b) = y$, and $L(\gamma) = d(x, y)$.*

For example, the removal of a point from a manifold makes it noncomplete. Another example: the half-plane $\pi_+ = \{(x, y) \in \mathbb{R}^2 \mid y > 0\}$ with the Euclidean metric of $\mathbb{R}^2$ is not complete, whereas with the Poincaré hyperbolic metric $(dx^2 + dy^2)/y^2$ it is geodesically complete.

The Hopf–Rinow theorem has an interesting application in the proof of a global fact, linking curvature and topology, established by the following theorem:

Theorem (Hadamard). *If the sectional curvature $K(p, \pi)$ of a complete, simply connected n-dimensional Riemannian manifold M is such that $K(p, \pi) \leq 0$ for all $p \in M$ and for all $\pi \in T_pM$, then M is diffeomorphic to $\mathbb{R}^n$, and $\exp_p : T_pM \to M$ is a diffeomorphism.*

Other results about negative-curvature manifolds are of the following kind: *Any compact orientable surface of genus $g > 1$ carries a metric with constant negative curvature.* More generally, if (M, g_0) is a compact Riemannian manifold with negative Euler characteristic, that is, $\chi(M) < 0$, and K is such that sup $K < 0$, there exists a unique smooth conformal metric $g = f^2 g_0$ on M with curvature K.

Links between the curvature properties of a manifold M and its topology are established also by other kinds of theorems providing information about the *fundamental group* $\pi_1(M)$. We mention two results of this kind.

Theorem (Milnor). *If (M, g) is a compact manifold with strictly negative curvature, then $\pi_1(M)$ has exponential growth.*

Let us explain what exponential growth means. Given a group G of finite type, if $\mathcal{G} = \{a_1, \ldots, a_k\}$ is a set of generators, any element of $h \in G$ can be written as

$$h = \prod_i a_{k_i}^{r_i} \, , \qquad r_i \in \mathbb{Z} \, ,$$

where repetitions of the generators a_{k_i} are allowed. This representation is a *word* with respect to the generators and $\sum_i |r_i|$ is the *length* of the word. For any positive integer m, denote by $\mathcal{N}_\mathcal{G}(m)$ the number of elements of G that can be represented by words of length not exceeding m. A group G of finite type is said to have *exponential growth* if for any system $\mathcal{G}$ of generators, there is a constant $a > 0$ such that $\mathcal{N}_\mathcal{G}(m) \geq \exp(am)$. The growth is said to be *polynomial* of degree $\leq n$ if $\mathcal{N}_\mathcal{G}(m) \leq (am^n)$.

The second theorem, due to Preissman, is as follows:

Theorem . *If M is a compact Riemannian manifold with strictly negative curvature, any abelian subgroup of $\pi_1(M)$ (different from the identity) is infinite cyclic.*

To understand the meaning of this theorem, consider, for example, the torus $\mathbb{T}^3 = \mathbb{S}^1 \times \mathbb{S}^1 \times \mathbb{S}^1$. Since its fundamental group is $\mathbb{Z} \oplus \mathbb{Z} \oplus \mathbb{Z}$, it cannot be endowed with a metric of strictly negative curvature.

Conversely, by a theorem due to Synge, any compact, even-dimensional orientable Riemannian manifold with $K(p, \pi) > 0$, that is, with strictly positive curvature, is *simply connected*. Moreover, any noncompact complete Riemannian manifold with $K(p, \pi) > 0$ is *contractible*.

A theorem due to Myers states that if for a complete Riemannian manifold (M, g) the Ricci tensor is such that

$$\mathrm{Ric} \geq \frac{(n-1)}{r^2} g \, , \qquad r \in \mathbb{R}_+ \, ,$$

then[3] $\mathrm{diam}(M, g) \leq \mathrm{diam}[\mathbb{S}^n(r)]$, where $\mathbb{S}^n$ is a sphere in $\mathbb{R}^{n+1}$ of radius r, and M is *compact* with *finite fundamental group* $\pi_1(M)$.

---

[3] The diameter of a metric space $\mathcal{M}$ is the supremum of the distance function on $\mathcal{M} \times \mathcal{M}$.

Another kind of link between curvature and topology is given by the so-called "pinching" theorems. These have their most famous representative in the sphere theorem:

Theorem . *If the sectional curvature K of an n-dimensional compact and simply connected Riemannian manifold M satisfies the condition*

$$0 < \delta K_{\max} < K \leq K_{\max} ,$$

and if $\delta = 1/4$, then M is homeomorphic to a sphere.

Pinching theorems have been mainly worked out for manifolds with positive curvature, though some results exist also for δ-pinched negative curvature manifolds that carry no metric of constant curvature. An interesting fact related to these works, is that there seem to exist many more manifolds with negative curvature than with positive curvature.

Appendix C

Summary of Elementary Morse Theory

The purpose of this appendix is to recall the main ideas and concepts of Morse theory that are relevant for the main text of the book. For a more elaborate discussion we refer the reader to [210, 227, 262].

Morse theory, also referred to as critical-point theory, links the *topology* of a given manifold M with the properties of the *critical points* of smooth (i.e., with infinitely many derivatives) functions defined on it. Morse theory links *local* properties (what happens at a particular point of a manifold) with *global* properties (the topology, i.e., the shape, of the manifold as a whole). Two manifolds M and M' are topologically equivalent if they can be smoothly deformed one into the other: a teacup is topologically equivalent to a doughnut, but it is *not* topologically equivalent to a ball. In fact, a ball has no holes, while both a teacup and a doughnut have one hole. To define precisely what a "smooth deformation" is, one has to resort to the notion of a *diffeomorphism*, as we have already seen in Appendixes A and B. Thus the notion of "topological equivalence" between M and M' has a precise meaning: it requires that there exists a diffeomorphism ψ that maps M into $M' = \psi(M)$. For the sake of simplicity, we consider only compact, finite-dimensional manifolds. Most of the results can be extended not only to noncompact manifolds, but also to infinite-dimensional manifolds modeled on Hilbert spaces (see [210]).

The key ingredient of Morse theory is to look at the manifold M as decomposed into the *level sets* of a function f. Let us recall that the a-level set of a function $f : M \mapsto \mathbf{R}$ is the set

$$f^{-1}(a) = \{x \in M : f(x) = a\} , \tag{C.1}$$

i.e., the set of all the points $x \in M$ such that $f(x) = a$. Now, M being compact, any function f has a minimum, $f_{\min}$, and a maximum, $f_{\max}$, so that

$$f_{\min} \leq f(x) \leq f_{\max} \qquad \forall x \in M . \tag{C.2}$$

This means that the whole manifold M can be decomposed into the level sets of f. In fact, one can build M starting from $f^{-1}(f_{\min})$ and then adding continuously to it all the other level surfaces up to $f^{-1}(f_{\max})$. To be more precise, one defines the "part of M below a" as

$$M_a = f^{-1}(-\infty, a] = \{x \in M : f(x) \le a\}, \qquad (C.3)$$

i.e., each M_a is the set of the points $x \in M$ such that the function $f(x)$ does not exceed a given value a; as a is varied between $f_{\min}$ and $f_{\max}$, M_a describes the whole manifold M.

For our purposes, we need to restrict the class of functions we are interested in to the class of *Morse functions*, which are defined as follows. Given a manifold M of dimension n and a smooth function $f : M \mapsto \mathbf{R}$, a point $x_c \in M$ is called a *critical point* of f if $df(x_c) = 0$, while the value $f(x_c)$ is called a *critical value*. The function f is called a *Morse function* on M if its critical points are nondegenerate, i.e., if the Hessian matrix of f at x_c, whose elements in local coordinates are

$$H_{ij} = \frac{\partial^2 f}{\partial x^i \partial x^j}, \qquad (C.4)$$

has rank n, i.e., has only nonzero eigenvalues. This means that there are no directions along which one could move the critical point, so that there are no lines (or surfaces, or hypersurfaces) made of critical points. As a consequence, one can prove that the critical points x_c of a Morse function, and also its critical values, are isolated. It can be proved also that Morse functions are generic: the space of the Morse functions is a *dense* subset of the space of the smooth functions from M to $\mathbf{R}$. A level set $f^{-1}(a)$ of f is called a *critical level* if a is a critical value of f, i.e., if there is at least one critical point $x_c \in f^{-1}(a)$.

The main results of Morse theory are the following:

1. If the interval $[a, b]$ contains no critical values of f, then the topology of $f^{-1}[a, v]$ does not change for any $v \in (a, b]$. This result[1] is sometimes called the *noncritical neck theorem*. The reason for this terminology will be made clear in the following.

2. If the interval $[a, b]$ contains critical values, the topology of $f^{-1}[a, v]$ changes in correspondence with the critical values themselves, in a way that is completely determined by the properties of the Hessian of f at the critical points.

3. (*Sard's theorem*) If $f : M \to \mathbb{R}$ is a Morse function, the set of all the critical points of f is a *discrete* subset of M (the critical points are *isolated*).

4. If $f : M \to \mathbb{R}$ is a Morse function, with compact M, then on a finite interval $[a, b] \subset \mathbb{R}$ there is only a finite number of critical points p of f such that $f(p) \in [a, b]$. Moreover, the set of *critical values* of f is a *discrete* set of $\mathbb{R}$.

---

[1] We note that this result is valid even if f is not a Morse function; it is sufficient that it be a smooth function.

5. For any differentiable manifold M, the set of Morse functions on M is an open dense set in $C^r(M, \mathbb{R})$ (real functions on M of differentiability class r) for $0 \leq r \leq \infty$.[2]

6. Some topological invariants of M, i.e., quantities that are the same for all the manifolds that have the same topology as M, so that they characterize unambiguously the topology itself, can be estimated and sometimes computed exactly once all the critical points of f are known.

Without giving explicit proofs, which can be found in [210], we now discuss the items given above in more detail.

C.1 The Non-Critical Neck Theorem

If there are no critical values in the interval $[a, b]$, there exists a diffeomorphism that sends $f^{-1}[a, b]$ into the Cartesian product $f^{-1}(a) \times [a, b]$. This means that the shape of $f^{-1}[a, b]$ is that of a multidimensional cylinder, or a neck (from which the name "noncritical neck"), if $f^{-1}(a)$ is simply connected, because the Cartesian product of a circle and an interval is a cylinder. This might be better understood with the aid of a two-dimensional example. Suppose that M is two-dimensional, and that the level set $f^{-1}(a)$ is topologically equivalent to a circle (see Figure C.1). Then one can construct a diffeomorphism explicitly

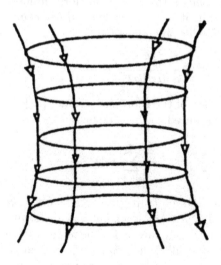

Fig. C.1. A noncritical neck. The lines with the arrows are the flow lines of ∇f, and the ellipses are the level sets of f.

---

[2] For example, if y_0 is, in a Morse chart, a degenerate critical point of $f : M \to \mathbb{R}$, an arbitrarily small perturbation $\delta f = \sum_{i=0}^{n} a_i y^i$, with $a \in \mathbb{R}^n$, removes the degeneracy without altering the indices of the critical points.

as the flow of the gradient vector field of f, ∇f, whose flow lines are orthogonal to the level surfaces of f and are depicted as the lines with the arrows in Figure C.1. This flow has no singularities if there are no critical values of f, so that the level set $f^{-1}(a)$ is transported up to $f^{-1}(b)$ along the flow lines of ∇f without changing its topology. As a consequence,

$$f^{-1}[a,b] \approx f^{-1}(a) \times [a,b] \approx f^{-1}(b) \times [a,b] \,, \qquad \text{(C.5)}$$

where "$x \approx y$" must be read as "x is diffeomorphic to y."

C.1.1 Critical Points and Topological Changes

In the neighborhood of a regular point P, $N(P)$, there always exists a coordinate system such that f can be written as its first-order Taylor expansion,[3] setting the origin of such coordinates in P, in the form

$$f(x) = f(0) + \frac{\partial f}{\partial x^i} x^i + \cdots \qquad \forall x \in N(P) \,. \qquad \text{(C.6)}$$

Geometrically, this means that in the neighborhood of a regular point the level sets of f look like hyperplanes in $\mathbf{R}^n$, because they are the level sets of a *linear* function.

But what if P is a critical point of f? A fundamental result by M. Morse, called the *Morse lemma*, is that if f is a Morse function then there always exists in $N(P)$ a coordinate system (called a *Morse chart*) such that f is given by its *second-order* Taylor polynomial:

$$f(x) = f(0) + \frac{\partial^2 f}{\partial x^i \partial x^j} x^i x^j + \cdots \qquad \forall x \in N(P) \,. \qquad \text{(C.7)}$$

With a suitable rotation of the coordinate frame, $\{x^i\} \mapsto \{y^i\}$, the expansion (C.7) can always be reduced to the canonical diagonal form

$$f(y) = f(0) - \sum_{i=1}^{k} (y^i)^2 + \sum_{i=k+1}^{n} (y^i)^2 + \cdots \qquad \forall y \in N(P) \,. \qquad \text{(C.8)}$$

Close to P, the level sets of f are the level sets of a *quadratic* function, so that geometrically, they are nondegenerate quadrics, such as hyperboloids or ellipsoids, that become degenerate at P. The number of negative eigenvalues of the Hessian matrix, k, is called the *index* of the critical point. Passing through the critical level, the shape of the level sets of f changes dramatically, in a way that is completely determined by the index k. Some examples in two and three dimensions are given in Figure C.2.

The change undergone by the submanifolds M_a as a critical level is passed is described using the concept of "attaching handles." A k-handle $H^{(k)}$ in n

---

[3] This follows from the implicit function theorem.

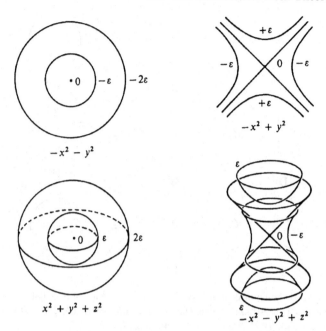

Fig. C.2. Some examples of ε-level sets near a critical point (the critical value of the function is set to 0). Upper left: $n = 2$, critical point of index $k = 2$; upper right: $n = 2$, critical point of index $k = 1$; lower left: $n = 3$, critical point of index $k = 0$; lower right: $n = 3$, critical point of index $k = 2$.

dimensions $(0 \le k \le n)$ is a product of two disks, one k-dimensional (D^k) and the other $(n - k)$-dimensional (D^{n-k}):

$$H^{(k)} = D^k \times D^{n-k} . \tag{C.9}$$

In two dimensions, we can have either 0-handles, which are 2-dimensional disks, or 1-handles, which are the product of two 1-dimensional disks, i.e., of two intervals, so that they are stripes, or 2-handles, which are again 2-dimensional disks (Figure C.3). In three dimensions, we have 0-handles, which are solid spheres; 1-handles, which are the product of a disk and an interval, so that they are solid cylinders; 2-handles, which are the same as 1-handles; and 3-handles, which are the same as 0-handles (Figure C.3). In more than three dimensions it is difficult to visualize handles: however, there is still the duality of the $n = 2$ and $n = 3$ cases, i.e., k and $n - k$ handles are topologically equivalent.

Having defined handles, we can state the main result of Morse theory.

Let M be an n-dimensional manifold with boundary ∂M, and let ϕ be a smooth embedding $\phi : \mathbb{S}^{k_i-1} \times D^{n-k_i} \to \partial M$ (where $\mathbb{S}$ is a hypersphere). Then one can build the topological space $M \bigcup_\phi H^{n,k}$, that is, M with a k-handle attached by ϕ, by considering the topological sum of M and $H^{n,k}$ (i.e. the

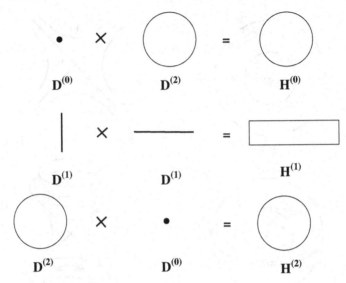

Fig. C.3. Two-dimensional handles: $H^{(0)}$ is the product of a 0-disk (a point) and a 2-disk, so that it is a 2-disk; $H^{(1)}$ is the product of two 1-disks, i.e., of two intervals, so that it is a strip; $H^{(2)}$ is again a 2-disk, as is $H^{(0)}$.

disjoint union $M \bigcup H^{n,k}$), and then identifying $x \in H^{n,k}$ with $\phi(x) \in \partial M$. It can be shown that "rounding the corners" of this topological space is possible, that is, $M \bigcup_{\phi} H^{n,k}$ can be given an n-dimensional differentiable structure that makes it a differentiable manifold with boundary. This procedure admits a generalization to the simultaneous attachment of m n-dimensional handles $H_1^{n,k_1} H_2^{n,k_2} \ldots H_m^{n,k_m}$ of indexes $k_1 \ldots k_m$. The fundamental result of Morse theory, relating critical points to the attachment of handles, is expressed as follows.

Theorem C.1. *Let* $f : M \to \mathbb{R}$, *with* $\partial M = \emptyset$. *Let* $Crit(f)$ *be the set of critical points of* f, $c \in \mathbb{R}$, *and* $Crit(f) \cap f^{-1}(c) = \{x_1, \ldots, x_m\}$, *where each* x_i, $i = 1, \ldots, m$, *is a critical point of index* k_i. *Assume that for* $\varepsilon_0 > 0$ $f^{-1}([c-\varepsilon_0, c+\varepsilon_0])$ *is a compact set with no other critical point but* $x_1, \ldots, x_m$. *Then for any* ε *such that* $0 < \varepsilon < \varepsilon_0$, *the manifold* $M_{c+\varepsilon} = f^{-1}((-\infty, c+\varepsilon])$ *is diffeomorphic to* $M_{c-\varepsilon} = f^{-1}((-\infty, c - \varepsilon])$ *with the handles* $H_1^{n,k_1} \ldots H_m^{n,k_m}$ *attached, that is,*

$$M_{v_c+\varepsilon} \approx M_{v_c-\varepsilon} \bigcup_{\phi_1} H^{n,k_1} \bigcup_{\phi_2} H^{n,k_2} \ldots \bigcup_{\phi_n} H^{n,k_n},$$

for some embeddings $\phi_i : \mathbb{S}^{k_i-1} \times D^{n-k_i} \to \partial M_{v_c-\varepsilon}$, *with* $\partial M_{v_c-\varepsilon} = f^{-1}(c-\varepsilon)$.

Let us see how this works in a simple example. Consider as our manifold M a two-dimensional torus standing on a plane (think of a tire in a ready-to-roll

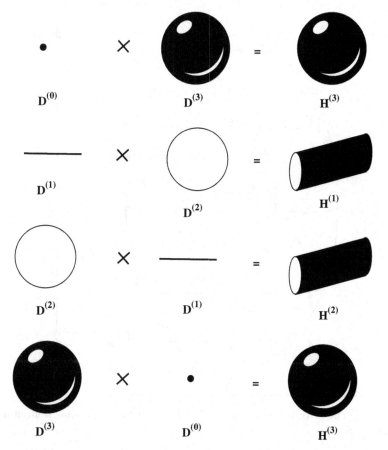

Fig. C.4. Three-dimensional handles: $H^{(0)}$ is the product of a 0-disk (a point) and a 3-disk (a ball), so that it is a ball; $H^{(1)}$ is the product of a 1-disk (an interval) and a 2-disk, so that it is a tube; $H^{(2)}$ is as $H^{(1)}$, and $H^{(3)}$ is as $H^{(0)}$.

position), and define a function f on it as the height of a point of M above the floor level. If the z-axis is vertical, f is the orthogonal projection of M onto the z-axis. Such a function has four critical points, and the corresponding four critical levels of f, which will be denoted p_0, p_1, p_2, p_3, are depicted in Figure C.5. We can build our torus in separate steps: each step will correspond to the crossing of a critical level of f. As long as $a < 0$, the manifold M_a is empty. At $a = p_0 = 0$ we cross the first critical value, corresponding to a critical point of index 0. This means that we have to attach a 0-handle (a disk) to the empty set. Any M_a with $0 < a < p_1$ is diffeomorphic to a disk, as we can see by cutting a torus at any height between 0 and p_1 and throwing away the upper part. At p_2 we meet the second critical point, which now has index 1, so that we have to attach a 1-handle (a stripe) to the previous disk, obtaining a sort of a basket. Such a basket can be smoothly deformed into a U-shaped

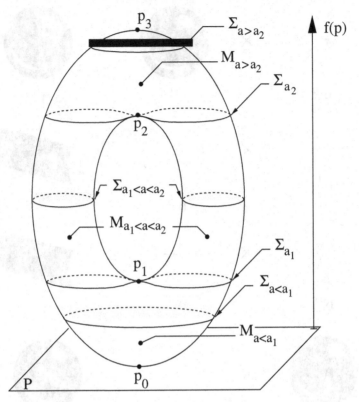

Fig. C.5. The critical points and critical levels of the height function on a two-dimensional torus.

tube: in fact, if we cut a torus at any height between p_1 and p_2 and we throw away the upper part, we get a U-shaped tube. The third critical point p_2 is again a point of index 1, so we have to glue another stripe to the tube. What we obtain can be smoothly deformed into a full torus with only the polar cap cut away from it. The last critical point has index 2, so that the crossing of it corresponds to the gluing of a 2-handle (a disk), which is just the polar cap we needed to complete the torus.

C.1.2 Morse Inequalities

There is a relationship between Morse theory and de Rham cohomology. This link is made through the *Morse inequalities*. We have the following theorem:

Theorem C.2. *Let $a < b$ be two regular values of the Morse function $f : M \to \mathbb{R}$, with M compact.*

Let $M_{a,b} = f^{-1}([a,b]) = \{x \in M | f(x) \in [a,b]\}$. Denoting by μ_k the number of critical points of index k that the function f has in $M_{a,b}$, and denoting by b_k the kth Betti number of $M_{a,b}$, one has

(Morse inequalities)

$$b_0 \leq \mu_0 \ ,$$
$$b_1 - b_0 \leq \mu_1 - m_0 \ ,$$
$$\vdots$$
$$b_k - b_{k-1} + \cdots \pm b_0 \leq \mu_k - \mu_{k-1} + \cdots \pm \mu_0$$

(weak Morse inequalities),

$$\forall k, \quad b_k \leq \mu_k \ ,$$

(Euler characteristic),

$$\chi(M_{a,b}) \equiv \sum_{k=1}^{n}(-1)^k \, b_k = \sum_{k=1}^{n}(-1)^k \, \mu_k \ .$$

The last formula provides a way of computing exactly the Euler characteristic of a manifold once all the critical points of a Morse function are known.

Among all the Morse functions on a manifold M, there is a special class (called *perfect* Morse functions) for which the Morse inequalities (10.35) hold as equalities. Perfect Morse functions characterize completely the topology of a manifold. It is possible to prove that the height function on the torus we considered above is a perfect Morse function [210]. However, there are no simple general recipes to construct perfect Morse functions.

For a (two-dimensional) surface, the number of holes of the surface is a topological invariant called its *genus* g. In this case we have $\chi = 2 - 2g$.

References

1. J. Hadamard, *J. Math. Pur. Appl.* **4**, 27 (1898); G. A. Hedlund, *Bull. Amer. Math. Soc.* **45**, 241 (1939); E. Hopf, *Proc. Nat. Acad. Sci.* **18**, 263 (1932).
2. N. S. Krylov, *Works on the Foundations of Statistical Physics* (Princeton University Press, Princeton, 1979).
3. Ya. G. Sinai (ed.), *Dynamical Systems II*, Encyclopædia of Mathematical Sciences **2** (Springer, Berlin, 1989).
4. V. I. Anosov, *Proc. Steklov Math. Inst.* **90**, 1 (1967); also reprinted in [5].
5. R. S. MacKay, and J. D. Meiss (eds.), *Hamiltonian Systems: a Reprint Selection* (Adam Hilger, Bristol, 1990).
6. Y. Aizawa, *J. Phys. Soc. Jpn.* **33**, 1693 (1972).
7. C. P. Ong, *Adv. Math.* **15**, 269 (1975).
8. J. F. C. van Velsen, *J. Phys. A: Math. Gen.* **13**, 833 (1980).
9. V. G. Gurzadyan, and G. K. Savvidy, *Astron. & Astrophys.* **160**, 203 (1986).
10. M. Szydlowski, *J. Math. Phys.* **35**, 1850 (1994).
11. Y. Aizawa, *J. Korean Phys. Soc.* **28**, S310 (1995).
12. B. Nobbe, *J. Stat. Phys.* **78**, 1591 (1995).
13. M. Szydlowski, M. Heller, and W. Sasin, *J. Math. Phys.* **37**, 346 (1996).
14. H. E. Kandrup, *Astrophys. J.* **364**, 420 (1990); *Physica A* **169**, 73 (1990); *Phys. Rev. E* **56**, 2722 (1997).
15. M. C. Gutzwiller, *J. Math. Phys.* **18**, 806 (1977).
16. A. Knauf, *Comm. Math. Phys.* **110**, 89 (1987).
17. R. Thom, in *Statistical Mechanics*, eds. S. A. Rice, K. F. Freed, and J. C. Light (University of Chicago Press, 1972), p. 93.
18. T. Poston, and I. Stewart, *Catastrophe Theory and Its Applications* (Pitman Press, London, 1978), and references therein quoted.
19. G. Ruppeiner, *Phys. Rev. A* **44**, 3583 (1991); *Riemannian Geometry in Thermodynamic Fluctuation Theory*, *Rev. Mod. Phys.* **67**, 605–659 (1995).
20. D. H. E. Gross, *Phys. Rep.* **279**, 119–202 (1997).
21. M. Rasetti, *Topological Concepts in the Theory of Phase Transitions*, in *Differential Geometric Methods in Mathematical Physics*, ed. H. D. Döbner (Springer-Verlag, New York, 1979).
22. M. Rasetti, *Structural Stability in Statistical Mechanics*, in *Springer Tracts in Math.*, ed. W. Güttinger (Springer-Verlag, New York, 1979).
23. M. W. Hirsch, *Differential Topology* (Springer, New York 1976).

24. L. Casetti, M. Pettini, and E. G. D. Cohen, *Phys. Rep.* **337**, 237–341 (2000).
25. F. H. Stillinger, *Science* **267**, 1935 (1995).
26. S. Sastry, P. G. Debenedetti and F. H. Stillinger, *Nature* **393**, 554 (1998).
27. A. Bellemans, and J. Orban, in *Topics in Nonlinear Physics*, ed. N. J. Zabusky (Springer, Berlin, 1968), p. 236.
28. L. Szilard, *Z. Phys.* **32**, 753 (1925).
29. J. Bockris, and S. Khan, *Surface Electrochemistry* (Plenum Press, New York, 1993), Chapter 7.
30. P. Hertel, and W. Thirring, *Ann. Phys. (N.Y.)* **63**, 520 (1971).
31. J. Barré, D. Mukamel, and S. Ruffo, *Phys. Rev. Lett.* **87**, 030601 (2001); F. Leyvraz, and S.Ruffo, *Ensemble Inequivalence in Systems with Long-Range Interactions*, cond-mat/0112124, (2001); R. S. Ellis, K. Haven, and B. Turkington, *J. Stat. Phys.* **101**, 999 (2000); and *Nonlinearity* **15**, 239 (2002); and references quoted in these papers.
32. M. W. Zemansky, *Heat and Thermodynamics* (McGraw-Hill, New York, 1968).
33. K. Huang, *Statistical Mechanics* (J.Wiley & Sons, New York, 1987).
34. L. D. Landau, and E. M. Lifschitz, *Statistical Physics* (Pergamon, Oxford, 1980).
35. E. G. D. Cohen, *Am. J. Phys.* **58**, 619 (1990).
36. C. N. Yang and T. D. Lee, *Phys. Rev.* **87**, 404 (1952).
37. G. Gallavotti, *Statistical Mechanics. A Short Treatise* (Springer, New York, 1999), Section 6.8.
38. T. D. Lee, and C. N. Yang, *Phys. Rev.* **87**, 410 (1952).
39. H. Poincaré, *Les Méthodes Nouvelles de la Mécanique Celeste* (Gauthier-Villars, Paris, 1892) and (Blanchard, Paris, 1987).
40. E. Fermi, *Nuovo Cimento* **25**, 267 (1923); **26**, 105 (1923).
41. A. Giorgilli, and L. Galgani, *Celest. Mech.* **17**, 267 (1978).
42. A. N. Kolmogorov, *Dokl. Akad. Nauk. SSSR* **98**, 527 (1954); V. I. Arnold, *Russ. Math. Surv.* **18**, 9 (1963); J. Moser, *Nachr. Akad. Wiss. Goettingen Math. Phys. Kl. 2* **1**, 1 (1962).
43. H. Rüssmann, in *Dynamical Systems, Theory and Applications, Lectures Notes in Physics* **38**, ed. J. Moser (Springer, Berlin, 1975); L. Chierchia and G. Gallavotti, *Nuovo Cimento B* **67**, 277 (1982).
44. E. Wayne, *Comm. Math. Phys.* **96**, 311 (1984); **96**, 331 (1984).
45. N. N. Nekhoroshev, *Funct. Anal. Appl.* **5**, 338 (1971); *Russ. Math. Surv.* **32**, 1 (1977); G. Benettin, L. Galgani, and A. Giorgilli, *Celest. Mech.* **37**, 1 (1985); A. Giorgilli, and L. Galgani, *Celest. Mech.* **37**, 95 (1985); P. Lochak, *Phys. Lett. A* **143**, 39 (1990).
46. G. Benettin, and G. Gallavotti, *J. Stat. Phys.* **44**, 293 (1986); G. Gallavotti, *Quasi-Integrable Mechanical Systems*, in K. Osterwalder, and R. Stora (eds.), *Les Houches, Session* **43** (1984).
47. G. Benettin, L. Galgani, and A. Giorgilli, *Comm. Math. Phys.* **113**, 87 (1987); *Comm. Math. Phys.* **121**, 557 (1989).
48. M. Hénon and C. Heiles, *Astron. J.* **69**, 73 (1964).
49. A. J. Lichtenberg and M. A. Lieberman, *Regular and Chaotic Dynamics*, Second Edition (Springer, New York, 1992).
50. V. K. Melnikov, *Trans. Moscow Math. Soc.* **12**, 1 (1963).
51. S. Smale, *Bull. Am. Math. Soc.* **73**, 747 (1967).

52. G. D. Birkhoff, *Mem. Pont. Acad. Sci. Novi Lyncaei* **1**, 85 (1935). S. Smale, *Diffeomorphism with Many Periodic Points*, in *Differential and Combinatorial Topology*, ed. S. S. Cairns (Princeton University Press, Princeton, 1963).

53. J. Guckenheimer, and P. Holmes, *Nonlinear Oscillations, Dynamical Systems and Bifurcation of Vector Fields* (Springer, New York, 1983).

54. S. Wiggins, *Chaotic Transport in Dynamical Systems* (Springer, New York, 1992).

55. V. I. Arnold, *Dokl. Akad. Nauk. SSSR* **156**, 9 (1964).

56. M. Tabor, *Chaos and Integrability in Nonlinear Dynamics* (Wiley, New York, 1989).

57. V. I. Oseledeč, *Trans. Moscow Math. Soc.* **19**, 197 (1969).

58. C. P. Dettmann, and G. P. Morriss, *Phys. Rev. E* **55**, 3693 (1997); and references therein.

59. H. Fürstenberg, and H. Kesten, *Ann. Math. Stat.* **31**, 457 (1960).

60. J.-P. Eckmann and D. Ruelle, *Rev. Mod. Phys.* **57**, 617 (1985).

61. G. Benettin, L. Galgani, and J.-M. Strelcyn, *Phys. Rev. A* **14** 2338 (1976);

62. G. Benettin, L. Galgani, A. Giorgilli, and J.-M. Strelcyn, *Meccanica* **15**, 1 (1980).

63. R. Livi, A. Politi, and S. Ruffo, *J. Stat. Phys.* **19**, 2083 (1986).

64. Ya. G. Sinai, Princeton preprint (1995).

65. S. Ruffo, *Lyapunov Spectra and Characterization of Chaotic Dynamics*, in *Complex Dynamics*, edited by R. Livi and J.-P. Nadal (Nova Publishing, 1994).

66. E. Fermi, J. Pasta, and S. Ulam (with M. Tsingou), Los Alamos report LA-1940, reprinted in *The Many-Body Problem*, edited by D. C. Mattis (World Scientific, Singapore, 1992), and also in *Collected Papers of Enrico Fermi*, edited by E. Segré (University of Chicago, Chicago, 1965), Vol. 2, p. 978.

67. S. Lepri, R. Livi, and A. Politi, *Phys. Rep.* **377**, 1 (2003).

68. G. Gallavotti, and E. G. D. Cohen, *Phys. Rev. Lett.* **74**, 2694 (1995); and *J. Stat. Phys.* **80**, 931 (1995).

69. See the Focus Issue of *Chaos* **15**, (2005), commemorating the 50th anniversary of the publication of the Los Alamos report LA-1940 by E. Fermi, J. Pasta, and S. Ulam. This Focus Issue contains a number of contributions beginning with page 015101. See also: R. Livi, A. J. Lichtenberg, M. Pettini, and S. Ruffo *Dynamics of Oscillator Chains*, in *The Fermi-Pasta-Ulam Experiment 50 Years Later*, edited by G. Gallavotti, (Springer, New York, 2006), pages 144–246.

70. N. J. Zabusky, and M. D. Kruskal, *Phys. Rev. Lett.* **15**, 240 (1965); N. J. Zabusky, and G. S. Deem, *J. Comp. Phys.* **2**, 126 (1967); N. J. Zabusky, *J. Phys. Soc. Jpn.* **26**, 196 (1969).

71. C. F. Driscoll, and T. M. O'Neil, *Phys. Rev. Lett.* **37**, 69 (1976).

72. J. Ford, and J. Waters, *J. Math. Phys.* **4**, 1293 (1963).

73. F. M. Izrailev, and B. V. Chirikov, *Dokl. Akad. Nauk SSSR* **166**, 57 (1966); [*Sov. Phys. Dokl.* **11**, 30 (1966).]

74. L. Casetti, M. Cerruti-Sola, M. Pettini, and E. G. D. Cohen, *Phys. Rev.* **55**, 6566 (1997).

75. D. L. Shepelyansky, *Nonlinearity* **10**, 1331 (1997).

76. J. De Luca, A. J. Lichtenberg, and M. A. Lieberman, *Chaos* **5**, 283 (1995).

77. R. Livi, M. Pettini, S. Ruffo, M. Sparpaglione, and A. Vulpiani, *Phys. Rev. A* **31**, 1039 (1985).

78. M. Pettini, and M. Landolfi, *Phys. Rev. A* **41**, 768 (1990).

79. M. Pettini, and M. Cerruti-Sola, *Phys. Rev. A* **44**, 975 (1991).

80. G. Tsaur, and J. Wang, *Phys. Rev. E* **54**, 4657 (1996).

81. L. Casetti, R. Livi, A. Macchi, and M. Pettini, *Europhys. Lett.* **32**, 549 (1995).

82. M. Cerruti-Sola, M. Pettini, and E. G. D. Cohen, *Phys. Rev.* **62**, 6078 (2000).

83. K. Yoshimura, *Physica D* **104**, 148 (1997).

84. V. Constantoudis, and N. Theodorakopoulos, *Phys. Rev.* **55**, 7612 (1997).

85. D. Escande, H. Kantz, R. Livi, and S. Ruffo, *J. Stat. Phys.* **76**, 539 (1987).

86. L. Casetti, C. Clementi, and M. Pettini, *Phys. Rev. E* **54**, 5969 (1996).

87. M.-C. Firpo, *Phys. Rev. E* **57**, 6599 (1998).

88. A. Mossa, M. Pettini, and C. Clementi, *Phys. Rev. E* **74**, 041805 (2006).

89. L. Caiani, L. Casetti, and M. Pettini, *J. Phys. A: Math. Gen.* **31**, 3357 (1998).

90. L. Caiani, L. Casetti, C. Clementi, G. Pettini, M. Pettini, and R. Gatto, *Phys. Rev. E* **57**, 3886 (1998).

91. P. Butera, and G. Caravati, *Phys. Rev. A* **36**, 962 (1987).

92. L. Caiani, L. Casetti, C. Clementi, and M. Pettini, *Phys. Rev. Lett.* **79**, 4361 (1997).

93. S. Wiggins, *Global Bifurcations and Chaos – Analytical Methods* (Springer, New York, 1988).

94. D. Lynden-Bell, and R. Wood, *Monthly Not. Roy. Astron. Soc.* **138**, 495 (1968); D. Lynden-Bell, and R. M. Lynden-Bell, *Monthly Not. Roy. Astron. Soc.* **181**, 405 (1977); D. Lynden-Bell, *Physica A* **263**, 293 (1999).

95. M. Cerruti-Sola, P. Cipriani, and M. Pettini, *Monthly Not. Roy. Astron. Soc.* **328**, 339 (2001).

96. D. H. E. Gross, *Microcanonical Thermodynamics: Phase Transitions in "Small" Systems*, volume 66 of Lecture Notes in Physics (World Scientific, Singapore, 2001); D. H. E. Gross, A. Ecker, and X. Z. Zhang, *Ann. Physik* **5**, 446 (1996).

97. A. Hüller, *Z. Phys. B* **95**, 63 (1994).

98. D. Ruelle, *Statistical Mechanics. Rigorous Results*, (Benjamin, Reading, 1969).

99. P. J. Channel, and C. Scovel, *Nonlinearity* **3**, 231 (1990).

100. P. MacLachlan, and P. Atela, *Nonlinearity* **5**, 541 (1992).

101. L. Casetti, *Physica Scripta* **51** , 29 (1995).

102. J. Moser, *Mem. Am. Math. Soc.* **81**, 1 (1968).

103. G. Benettin, and A. Giorgilli, *J. Stat. Phys.* **73**, 1117 (1994).

104. E. M. Pearson, T. Halicioglu, and W. A. Tiller, *Phys. Rev. A* **32**, 3030 (1985).

105. M. Cerruti-Sola, C. Clementi, and M. Pettini, *Phys. Rev. E* **61**, 5171 (2000).

106. K. Binder, *Z. Phys. B* **43**, 119 (1981).

107. B. Dünweg, in *Monte Carlo and Molecular Dynamics of Condensed Matter Systems, Conference Proceedings S.I.F.*, Vol. **79**, edited by K. Binder and G. Ciccotti (Editrice Compositori, Bologna, 1996).

108. J. L. Lebowitz, J. K. Percus, and L. Verlet, *Phys. Rev.* **153**, 250 (1967).

109. N. Goldenfeld, *Lectures on Phase Transitions and the Renormalization Group* (Addison-Wesley, New York, 1992).

110. M. Antoni, and S. Ruffo, *Phys. Rev. E* **52**, 2361 (1995).

111. Ch. Dellago, H. A. Posch, and W. G Hoover, *Phys. Rev. E* **53**, 1485 (1996).

112. Ch. Dellago, and H. A. Posch, *Physica A* **230**, 364 (1996).

113. Ch. Dellago, and H. A. Posch, *Physica A* **237**, 95 (1997).

114. Ch. Dellago, and H. A. Posch, *Physica A* **240**, 68 (1997).

115. X. Leoncini, A. D. Verga, and S. Ruffo, *Phys. Rev. E* **57**, 6377 (1998).

116. M. Antoni, S. Ruffo, and A. Torcini, *Dynamics and Statistics of Simple Models with Infinite-Range Attractive Interaction*, Proceedings of the Conference "The Chaotic Universe", Rome-Pescara in February 1999; archived in:*cond-mat/9908336*.

117. L. Caiani, Laurea Thesis in Physics (University of Florence, 1995).

118. H. Rund, *The Differential Geometry of Finsler Spaces* (Springer, Berlin, 1959); M. Matsumoto, *Foundations of Finsler Geometry and Special Finsler Spaces* (Kaiseisha Press, Japan, 1986).

119. L. P. Eisenhart, *Ann. of Math. (Princeton)* **30**, 591 (1929).

120. A. Lichnerowicz, *Théories Rélativistes de la Gravitation et de l'Éléctromagnétisme* (Masson, Paris, 1955).

121. M. Abate, and G. Patrizio, *Finsler Metrics – A Global Approach* (Springer, Berlin, 1994).

122. G. S. Asanov, *Finsler Geometry, Relativity, and Gauge Theories* (Reidel, Dordrecht, 1985).

123. P. Cipriani, and M. Di Bari, *Planetary & Space Sci.* **46**, 1499 (1998); ibidem, 1543 (1998);

124. P. Cipriani, and M. Di Bari, *Phys. Rev. Lett.* **81**, 5532 (1998).

125. E. Musso, and F. Tricerri, *Ann. Math. Pur. Appl.* **150**, 1 (1988).

126. M. P. do Carmo, *Riemannian Geometry* (Birkhäuser, Boston - Basel, 1993).

127. W. Klingenberg, *Riemannian Geometry* (W. de Gruyter, Berlin, 1982).

128. K. Yano, and W. Ishihara, *Tangent and Cotangent Bundles* (Dekker, New York, 1973).

129. M. Pettini, *Phys. Rev. E* **47**, 828 (1993).

130. M. Cerruti-Sola, R. Franzosi, and M. Pettini, *Phys. Rev. E* **56**, 4872 (1997).

131. L. Casetti, and M. Pettini, *Phys. Rev. E* **48**, 4320 (1993).

132. M. Di Bari, D. Boccaletti, P. Cipriani, and G. Pucacco, *Phys. Rev. E* **55**, 6448 (1997).

133. V. I. Arnold, *Mathematical Methods of Classical Mechanics* (Springer, Berlin, 1978).

134. L. P. Eisenhart, *Riemannian Geometry* (Princeton University Press, Princeton, 1964).

135. G. Caviglia, *J. Math. Phys.* **24**, 2065 (1983).

136. M. Crampin, *Rep. Math. Phys.* **20**, 31 (1984).

137. R. Abraham, and J. E. Marsden, *Foundations of Mechanics* (Addison-Wesley, Redwood City, 1987).

138. M. A. Olshanetsky, and A. M. Perelomov, *Phys. Rep.* **71**, 313 (1981).

139. H. W. Guggenheimer, *Differential Geometry* (Dover, New York, 1977).

140. D. V. Choodnovsky, and G. V. Choodnovsky, *Nuovo Cimento Lett.* **19**, 291 (1977); *Nuovo Cimento Lett.* **22**, 31 (1978); ibidem, **22**, 47.

141. M. Hénon, 1974, *Phys. Rev. B* **9**, 1921 (1974).

142. L. Bates, *Rep. Math. Phys.* **26**, 413 (1988).

143. D. Boccaletti, and G. Pucacco, *Nuovo Cimento B* **112**, 181 (1997).

144. A. R. Forsyth, *Theory of Differential Equations*, Vol. 5 and Vol. 6 *Partial Differential Equations* (Dover Publications, New York, 1959).

145. C. Clementi, Laurea Thesis in Physics (University of Florence, 1995).

146. C. Clementi, and M. Pettini, *Cel. Mech. & Dyn. Astron.* **84**, 263 (2002).

147. Y. F. Chang, M. Tabor, and J. Weiss, *J. Math. Phys.* **23**, 531 (1982).

148. D. Baleanu, *Gen. Relativ. Gravit.* **30**, 195 (1998).

149. G. W. Gibbons, R. H. Rietdijk, and J. W. van Holten, *Nucl. Phys. B* **404**, 42 (1993).
150. P. Sommers, *J. Math. Phys.* **14**, 787 (1973).
151. K. Rosquist, and G. Pucacco, *J. Phys. A: Math. Gen.* **28**, 3235 (1995).
152. K. Yano, and S. Bochner, *Curvature and Betti Numbers* (Princeton University Press, Princeton, 1953).
153. K. Rosquist, *J. Math. Phys.* **30**, 2319 (1989).
154. E. Gozzi, and M. Reuter, *Chaos, Solitons and Fractals* **4**, 1117 (1994).
155. V. I. Arnold (ed.), *Dynamical Systems III*, Encyclopædia of Mathematical Sciences **3** (Springer, Berlin, 1988).
156. G. Knieper, and H. Weiss, *J. Diff. Geom.* **39**, 229 (1994).
157. L. Casetti, R. Livi, and M. Pettini, *Phys. Rev. Lett.* **74**, 375 (1995).
158. A. H. Nayfeh, and D. T. Mook, *Nonlinear Oscillations* (Wiley, New York, 1979).
159. M. C. Gutzwiller, *Chaos in Classical and Quantum Mechanics* (Springer, New York, 1990).
160. M. Abramowitz, and I. A. Stegun, *Handbook of Mathematical Functions* (Dover, New York, 1965).
161. M. Cerruti-Sola, and M. Pettini, *Phys. Rev. E* **53**, 179 (1996).
162. M. Pettini and R. Valdettaro, *Chaos* **5**, 646 (1995).
163. S. I. Goldberg, *Curvature and Homology* (Dover, New York, 1965).
164. G. Ciraolo, and M. Pettini, *Cel. Mech. & Dyn. Astron.* **83**, 171 (2002).
165. N. G. van Kampen, *Phys. Rep.* **24**, 71 (1976).
166. L. Casetti, and A. Macchi, *Phys. Rev. E* **55**, 2539 (1997).
167. L. Casetti, R. Gatto, and M. Pettini, *J. Phys. A: Math. Gen.* **32**, 3055 (1999).
168. R. Livi, M. Pettini, S. Ruffo, and A. Vulpiani, *J. Stat. Phys.* **48**, 539 (1987).
169. L. Casetti, *Aspects of Dynamics, Geometry, and Statistical Mechanics in Hamiltonian Systems* (PhD Thesis, Scuola Normale Superiore, Pisa, 1997).
170. P. Poggi, and S. Ruffo, *Physica D* **103**, 251 (1997).
171. D. F. Lawden, *Elliptic Functions and Applications* (Springer, New York, 1989).
172. R. Franzosi, P. Poggi, and M. Cerruti-Sola, *Phys. Rev. E* **71**, 036218 (2005).
173. D. Escande, H. Kantz, R. Livi, and S. Ruffo, *J. Stat. Phys.* **76**, 605 (1994).
174. L. Bunimovič, G. Casati, and I. Guarneri, *Phys. Rev. Lett.* **77**, 2941 (1996).
175. V. Latora, A. Rapisarda, and S. Ruffo, *Phys. Rev. Lett.* **80**, 692 (1998).
176. Y. Y. Yamaguchi, *Progr. Theor. Phys.* **95**, 717 (1996).
177. J. Barré, and T. Dauxois, *Europhys. Lett.* **55**, 164 (2001).
178. P. Grinza, and A. Mossa, *Phys. Rev. Lett.* **92**, 158102 (2004).
179. K. G. Wilson, and J. Kogut, *Phys. Rep.* **12**, 75 (1974).
180. S.-K. Ma, *Modern Theory of Critical Phenomena* (Benjamin, New York, 1976).
181. G. Parisi, *Statistical Field Theory* (Addison-Wesley, New York, 1988).
182. J. Zinn-Justin, *Quantum Field Theory and Critical Phenomena* (Oxford University press, Oxford, 1989).
183. M. Le Bellac, *Quantum and Statistical Field Theory* (Oxford University press, Oxford, 1991).
184. A. Bonasera, V. Latora, and A. Rapisarda, *Phys. Rev. Lett.* **75**, 3434 (1995).
185. C. S. O'Hern, D. A. Egolf, and H. S. Greenside, *Phys. Rev. E* **53**, 3374 (1996).
186. V. Mehra, and R. Ramaswamy, *Phys. Rev. E* **56**, 2508 (1997).
187. V. Latora, A. Rapisarda, and S. Ruffo, *Physica D* **131**, 38 (1999).
188. R. Livi, and A. Maritan, *Phys. Rev. D* **23**, 2252 (1981).

189. C. Clementi, *Dynamics of Homopolymer Chain Models* (MSc Thesis, SISSA/ISAS, Trieste, 1996).

190. A. Baumgärtner, *J. Chem. Phys.* **72**, 871 (1980).

191. G. Iori, E. Marinari, and G. Parisi, *J. Phys: Math. Gen. A* **24**, 5349 (1992).

192. G. Iori, E. Marinari, G. Parisi, and M. V. Struglia, *Physica A* **185**, 98 (1992).

193. C. Anteneodo, and C. Tsallis, *Phys. Rev. Lett.* **80**, 5313 (1998).

194. R. Franzosi, L. Casetti, L. Spinelli, and M. Pettini, *Phys. Rev. E* **60**, R5009 (1999).

195. R. Franzosi, *Geometrical and Topological Aspects in the Study of Phase Transitions* (PhD Thesis, University of Florence, 1998).

196. M. Spivak, *A Comprehensive Introduction to Differential Geometry* (Publish or Perish, Berkeley, 1979).

197. J. A. Thorpe, *Elementary Topics in Differential Geometry* (Springer, New York, 1979).

198. R. Franzosi, M. Pettini, and L. Spinelli, *Phys. Rev. Lett.* **84**, 2774 (2000).

199. L. Casetti, E. G. D. Cohen, and M. Pettini, *Phys. Rev. Lett.* **82**, 4160 (1999).

200. This standard co-area formula can be found in [201].

201. H. Federer, *Geometric Measure Theory*, (Springer, New York, 1969).

202. To solve equations (7.20) for a general (i.e., not mean-field-like) potential energy function one could borrow methods from: J. Vollmer, W. Breymann, and R. Schilling, *Phys. Rev. B* **47**, 11767 (1993).

203. C. Nash, *Differential Topology and Quantum Field Theory* (Academic, London, 1991).

204. A. Dhar, *Physica A* **259**, 119 (1989).

205. K. Binder, *MonteCarlo Methods in Statistical Physics* (Springer, Berlin, 1979).

206. L. Casetti, M. Pettini, and E. G. D. Cohen, *J. Stat. Phys.* **111**, 1091 (2003).

207. S. S. Chern, and R. K. Lashof, *Am. J. Math.* **79**, 306 (1957); *Michigan Math. J.* **5**, 5 (1958).

208. U. Pinkall, *Math. Zeit.* **193**, 241 (1986).

209. P. Laurence, *Zeitschrifts für Angewändte Mathematik und Physik* **40**, 258 (1989).

210. R. S. Palais, and C. Terng, *Critical Point Theory and Submanifold Geometry*, *Lecture Notes in Mathematics* **1353** (Springer, Berlin, 1988).

211. L. Onsager, *Phys. Rev.* **65**, 117 (1944).

212. D. Ruelle, *Thermodynamic Formalism*, *Encyclopaedia of Mathematics and its Applications* (Addison-Wesley, New York, 1978).

213. H. O. Georgii, *Gibbs Measures and Phase Transitions*, (Walter de Gruyter, Berlin, 1988).

214. L. Spinelli, *A Topological Approach to Phase Transitions* (PhD Thesis, University of Provence, Marseille, 1999).

215. R. Franzosi, and M. Pettini, *Phys. Rev. Lett.* **92**, 060601 (2004).

216. R. Franzosi, M. Pettini, and L. Spinelli, *Topology and Phase Transitions I. Preliminary results*, *Nucl. Phys. B*, (2007) in press; archived in math-ph/0505057.

217. R. Franzosi and M. Pettini, *Topology and Phase Transitions II. Theorem on a necessary relation*, *Nucl. Phys. B*, (2007) in press; archived in math-ph/0505058.

218. L. Schwartz, *Analyse. Topologie Générale et Analyse Fonctionelle*, (Hermann, Paris, 1970), Deuxième Partie.

219. R. Bott, and J. Mather, *Topics in Topology and Differential Geometry*, in *Battelle Rencontres*, eds. C. M. De Witt, and J. A. Wheeler, p.460.

220. P. Brémaud, *Markov Chains*, (Springer, New York 2001), Chapter 7.
221. A. I. Khinchin, *Mathematical Foundations of Statistical Mechanics*, (Dover Publications, Inc., New York 1949).
222. L. Angelani, L. Casetti, M. Pettini, G. Ruocco, and F. Zamponi, *Europhys. Lett.* **62**, 775 (2003).
223. L. Angelani, L. Casetti, M. Pettini, G. Ruocco, and F. Zamponi, *Phys. Rev. E* **71**, 036152 (2005).
224. T. Dauxois, V. Latora, A. Rapisarda, S. Ruffo, and A. Torcini, *The Hamiltonian Mean-Field Model: From Dynamics to Statistical Mechanics and Back*, in [225].
225. T. Dauxois et al. (eds.), *Dynamics and Thermodynamics of Systems with Long-Range Interactions, Lecture Notes in Physics* **602** (Springer, Berlin, 2002).
226. J. Barré, F. Bouchet, T. Dauxois, ans S. Ruffo, *J. Stat. Phys.* **119**, 677 (2005).
227. J. Milnor, *Morse Theory, Ann. Math. Studies* **51** (Princeton University Press, Princeton, 1963).
228. B. A. Dubrovin, A. T. Fomenko, and S. P. Novikov, *Modern Geometry-Methods and Applications* (Volume III, Springer, New York, 1985).
229. J. H. Wilkinson, *The Algebraic Eigenvalue Problem* (Clarendon Press, Oxford, 1965).
230. G. H. Golub, and C. F. van Loan, *Matrix computations* (Johns Hopkins University Press, Baltimore, 1996).
231. B. Madan, and T. Keyes, *J. Chem. Phys.* **98**, 3342 (1993).
232. L. Angelani, R. Di Leonardo, G. Ruocco, A. Scala, and F. Sciortino, *Phys. Rev. Lett.* **85**, 5356 (2000); *J. Chem. Phys.* **116**, 10297 (2002); L. Angelani, G. Ruocco, M. Sampoli, and F. Sciortino, *J. Chem. Phys.* **119**, 2120 (2003).
233. L. Angelani, G. Ruocco, and F. Zamponi, *J. Chem. Phys.* **118**, 8301 (2003); F. Zamponi, L. Angelani, L. F. Cugliandolo, J. Kurchan, and G. Ruocco, *J. Phys. A: Math. Gen.* **36** 8565 (2003).
234. M. Morse and S. S. Cairns, *Critical Point Theory in Global Analysis and Differential Topology*, (Academic Press, New York, 1969).
235. L. Angelani, G. Ruocco, and F. Zamponi, *Phys. Rev. E* **72**, 016122 (2005).
236. M. Kastner, *Phys. Rev. Lett.* **93**, 150601 (2004).
237. A. Andronico, L. Angelani, G. Ruocco, and F. Zamponi, *Phys. Rev. E* **70**, 041101 (2004).
238. D. A. Garanin, R. Schilling, and A. Scala, *Phys. Rev. E* **70**, 036125 (2004).
239. F. Baroni, *Phase Transitions and Topological Changes in Configuration Space for Mean-Field Models*, Laurea Thesis in Physics (University of Florence, 2002).
240. I. Hahn, and M. Kastner, *Phys. Rev. E* **72**, 056134 (2005).
241. M. Kastner, *Physica A* **365**, 128 (2006).
242. A. C. Ribeiro Teixeira, and D. A. Stariolo, *Phys. Rev. E* **70**, 016113 (2004).
243. M. Kastner, and O. Schnetz, *J. Stat. Phys.* **122**, 1195 (2006).
244. M. Kastner, *Physica A* **359**, 447 (2006).
245. S. Risau-Gusman, A. C. Ribeiro Teixeira, and D. A. Stariolo, *Phys. Rev. Lett.* **95**, 145702 (2005).
246. D. Greb, *Localisation in Equivariant Cohomology and the Duistermaat-Heckman Theorem*, lecture notes of the SFB-TR 12 *"Symmetries and Universality in Mesoscopic Systems"*, April 2006; T. Kärki, and A. J. Niemi, *On the Duistermaat-Heckman Formula and Integrable Models*, archived in hep-th/9402041.

247. V. Guillemin, and S. Sternberg, *Geometric Asymptotics*, (American Mathematical Society, Providence, 1977).
248. A. Moro, private communication.
249. P. G. Debenedetti, and F. H. Stillinger, *Nature* **410**, 259 (2001); C. L. Brooks III, J. N. Onuchic, and D. J. Wales, *Science* **293**, 612 (2001); D. J. Wales, *Science* **293**, 2067 (2001).
250. L. Angelani, R. Di Leonardo, G. Parisi, and G. Ruocco, *Phys. Rev. Lett.* **87**, 055502 (2001).
251. L. Mazzoni, and L. Casetti, *Phys. Rev. Lett.* **97**, 218104 (2006).
252. T. Veitshans, D. Klimov, and D. Thirumalai, *Folding Des.* **2**, 1 (1997).
253. E. Witten, *J. Geom. Phys.* **9**, 303 (1992).
254. E. Keski-Vakkuri, A. J. Niemi, G. Semenoff, and O. Tirkkonen, *Phys. Rev. D* **44**, 3899 (1991) A. J. Niemi, and O.Tirkkonen, *Ann. Phys. (NY)* **235**, 318 (1994).
255. H. Umezawa, *Advanced Field Theory: Micro, Macro, and Thermal Physics* (American Institute of Physics, 1993).
256. R. Giachetti, and V. Tognetti, *Phys. Rev. Lett.* **55**, 912 (1985); R. P. Feynman, and H. Kleinert, *Phys. Rev. A* **34**, 5080 (1986); A. Cuccoli, R. Giachetti, V. Tognetti, R Vaia, and P. Verrucchi, *J. Phys.: Cond. Mat.* **7**, 7891-7938 (1995).
257. D. Bohm, and B. J. Hiley, *The Undivided Universe*, (Routledge and Kegan Paul, London, 1993).
258. P. R. Holland, *The Quantum Theory of Motion*, (Cambridge University Press, Cambridge, 1993).
259. G. Iacomelli, and M. Pettini, *Phys. Lett. A* **212**, 29 (1996).
260. D. C. Brody, D. W. Hook, and L. P. Hughston, *Quantum Phase Transitions without Thermodynamic Limits*, archived in: quant-ph/0511162.
261. M. Nakahara, *Geometry, Topology and Physics* (Adam Hilger, Bristol, 1991).
262. M. Morse, *Calculus of Variations in the Large* (American Mathematical Society, Providence, 1934).

Author Index

Subject Index

443

Interdisciplinary Applied Mathematics